Birkhäuser

Progress in Probability

Volume 80

Progress in Probability is designed for the publication of workshops, seminars and conference proceedings on all aspects of probability theory and stochastic processes, as well as their connections with and applications to other areas such as mathematical statistics and statistical physics.

All manuscripts are peer-reviewed to meet the highest standards of scientific literature:

- Series editors provide their final decision for inclusion in the series after manuscripts have been fully peer-reviewed;
- For contributed volumes, volume editors conduct the anonymous peer-review process before final submission;
- The peer-review process is anonymous and confidential in order to ensure reviewers provide impartial opinions

Interested authors may submit proposals by email to the series editors or to the relevant Birkhäuser editor listed under "Contacts."

Radosław Adamczak • Nathael Gozlan •
Karim Lounici • Mokshay Madiman
Editors

High Dimensional Probability IX

The Ethereal Volume

Birkhäuser

Editors
Radosław Adamczak
Institute of Mathematics
University of Warsaw
Warszawa, Poland

Nathael Gozlan
MAP5
Université Paris Cité
Paris, France

Karim Lounici
Centre de Mathématiques Appliquées
École Polytechnique
Palaiseau, France

Mokshay Madiman
Department of Mathematical Sciences
University of Delaware
Newark, DE, USA

ISSN 1050-6977 ISSN 2297-0428 (electronic)
Progress in Probability
ISBN 978-3-031-26978-3 ISBN 978-3-031-26979-0 (eBook)
https://doi.org/10.1007/978-3-031-26979-0

Mathematics Subject Classification: 60-06, 60B20, 60E15, 60Fxx, 60G15, 62E20, 62Gxx

This book is published under the imprint Birkhäuser, www.birkhauser-science.com by the registered company Springer Nature Switzerland AG
The registered company address is: Gewerbestrasse 11, 6330 Cham, Switzerland

Dedicated to the memory of
Richard M. Dudley,
Elizabeth Meckes,
and Joel Zinn,
who greatly enriched both our mathematics
and the HDP community.

Preface

The tradition of the High Dimensional Probability (HDP) conferences can be traced back to the International Conferences on Probability in Banach Spaces. The first of these took place in Oberwolfach, Germany, in 1975, and the last in Bowdoin College, United States, in 1991. In 1994, after eight Probability in Banach Spaces meetings, in order to reflect the growing community and scope of the conference, a decision was taken to give the series its current name: the International Conference on High Dimensional Probability. The first High Dimensional Probability conference was also held in Oberwolfach, in 1996, and the proceedings were published in 1998.

The present volume is an outgrowth of the Ninth High Dimensional Probability Conference (HDP IX), which due to the COVID-19 epidemic had to be held online from June 15th to June 19th, 2020. The thematic range and quality of the talks and contributed papers demonstrate that high-dimensional probability remains a very active field of research, with connections to diverse fields of pure and applied mathematics.

The origins of high-dimensional probability are related to the investigation of limit theorems for random vectors and regularity of stochastic processes. The first investigations were motivated by the study of necessary and sufficient conditions for the boundedness and continuity of trajectories of Gaussian processes and the extension of classical limit theorems, such as laws of large numbers, laws of the iterated logarithm, and central limit theorems, to Hilbert and Banach space-valued random variables as well as empirical processes.

The techniques developed to this end turned out to be extremely fruitful beyond the original area. The methods of high-dimensional probability and especially its offshoots, the concentration of measure phenomenon and generic chaining, soon found numerous applications in seemingly distant areas of mathematics, as well as statistics and computer science, in particular in random matrix theory, convex geometry, asymptotic geometric analysis, nonparametric statistics, empirical process theory, statistical learning theory, compressed sensing, strong and weak approximations, distribution function estimation in high dimensions, combinatorial optimization, random graph theory, stochastic analysis in infinite dimensions, and information and coding theory. At the same time, these areas gave new momentum to the high-dimensional probability community, suggesting new problems and directions.

The fruitful interactions with various fields last to this day, and have given rise to many substantial developments in the area in recent years. In particular, numerous important results have been obtained concerning the connections between various functional inequalities related to the concentration of measure phenomenon, the Kannan-Lovász-Simonovits conjecture for the spectral gap of log-concave probability measures, the application of generic chaining methods to study the suprema of stochastic processes and norms of random matrices, the Malliavin–Stein theory of Gaussian approximation, functional inequalities for high-dimensional models of statistical physics, connections between convex geometry and information theory, and various stochastic inequalities and their applications in high-dimensional statistics and computer science. This breadth is duly reflected in the diverse contributions contained in the present volume.

Before presenting the content of this volume, let us mention that the HDP community sadly lost in the last few years three of its most esteemed and prominent members: Richard Dudley, Elizabeth Meckes, and Joel Zinn. This book is dedicated to their memory.

The contributed papers of this volume are divided into four general areas: inequalities and convexity, limit theorems, stochastic processes, and high-dimensional statistics. To give readers an idea of their scope, in the following we briefly describe them by subject area and in the order they appear in this volume.

Inequalities and Convexity

- *Covariance representations, L^p-Poincaré inequalities, Stein's kernels and high-dimensional CLTs*, by B. Arras and C. Houdré. This paper explores connections between covariance representations, Bismut-type formulas, and Stein's method. The authors establish covariance representations for several well-known probability measures on $\mathbb{R}^d$ and apply them to the study of L^p-L^q covariance estimates. They also revisit the L^p-Poincaré inequality for the standard Gaussian measure from the point of view of covariance representations. Further, using Bismut-type formulas they obtain L^p-Poincaré and pseudo-Poincaré inequalities for α-stable measures. Finally using the construction of Stein's kernels by closed forms techniques, they obtain quantitative high-dimensional CLTs in 1-Wasserstein distance with sharp convergence rates and explicit dependence on parameters when the limiting Gaussian probability measure is anisotropic.
- *Volume properties of high-dimensional Orlicz balls*, by F. Barthe and P. Wolff. The authors obtain asymptotic estimates for the volume of families of Orlicz balls in high dimensions. As an application, they describe a large family of Orlicz balls which verify the famous conjecture of Kannan, Lovász, and Simonovits about spectral gaps. The paper also studies the asymptotic independence of coordinates of uniform random vectors on Orlicz balls, as well as integrability properties of their linear functionals.

- *Entropic isoperimetric inequalities*, by S. G. Bobkov and C. Roberto. The paper discusses a family of inequalities, known in Information Theory as entropic isoperimetric inequalities, that compare Rényi entropies to the Fisher information. Connecting these inequalities to the Gagliardo–Nirenberg inequality enables the authors to obtain optimal bounds for several ranges of parameters.
- *Transport proofs of some functional inverse Santaló inequalities*, by M. Fradelizi, N. Gozlan and S. Zugmeyer. This paper presents a simple proof of a recent result of the second author which establishes that functional inverse Santaló inequalities follow from Entropy-Transport inequalities. Then, using transport arguments together with elementary correlation inequalities, the authors prove these sharp Entropy-Transport inequalities in dimension 1, which therefore gives an alternative transport proof of the sharp functional Mahler conjecture in dimension 1, for both the symmetric and the general case. The proof of the functional inverse Santaló inequalities in the n-dimensional unconditional case is also revisited using these transport ideas.
- *Tail bounds for sums of independent two-sided exponential random variables*, by J. Li and T. Tkocz. The authors establish upper and lower bounds with matching leading terms for tails of weighted sums of two-sided exponential random variables. This extends Janson's recent results for one-sided exponentials.
- *Boolean functions with small second order influences on the discrete cube*, by K. Oleszkiewicz. The author studies the concept of second-order influences, introduced by Tanguy, in a specific context of Boolean functions on the discrete cube. Some bounds which Tanguy obtained as applications of his more general approach are extended and complemented. The main result asserts that a Boolean function with uniformly small second-order influences is close to a constant function or a dictatorship/antidictatorship function, which may be regarded a modification of the classical theorems of Kahn–Kalai–Linial and Friedgut–Kalai–Naor.
- *Some notes on concentration for α-subexponential random variables*, by H. Sambale. This paper provides extensions of classical concentration inequalities for random variables which have α-subexponential tail decay for any $\alpha \in (0, 2]$. This includes Hanson–Wright type and convex concentration inequalities in various situations. In particular, uniform Hanson–Wright inequalities and convex concentration results for simple random tensors are obtained in the spirit of recent works by Klochkov–Zhivotovskiy and Vershynin.

Limit Theorems

- *Limit theorems for random sums of random summands*, by D. Grzybowski. This paper establishes limit theorems for sums of randomly chosen random variables conditioned on the summands. Several versions of the corner growth setting are considered, including specific cases of dependence amongst the summands and summands with heavy tails. A version of Hoeffding's combinatorial central limit theorem and results for summands taken uniformly from a random sample are also obtained.

- *A note on central limit theorems for trimmed subordinated subordinators*, by D. M. Mason. This article establishes self-standardized central limit theorems (CLTs) for trimmed subordinated subordinators. Two ways are considered to trim a subordinated subordinator. One way leads to CLTs for the usual trimmed subordinator and a second way to a closely related subordinated trimmed subordinator and CLTs for it.
- *Functional central limit theorem via nonstationary projective conditions*, by F. Merlevède and M. Peligrad. This paper surveys some recent progress on the Gaussian approximation for nonstationary dependent structures via martingale methods. It presents general theorems involving projective conditions for triangular arrays of random variables as well as various applications for rho-mixing and alpha-dependent triangular arrays, stationary sequences in a random time scenery, application to the quenched FCLT, application to linear statistics with alpha-dependent innovations, and application to functions of a triangular stationary Markov chain.

Stochastic Processes

- *Sudakov minoration*, by W. Bednorz. The author considers the problem of bounding suprema of canonical processes based on a class of log-concave random vectors satisfying some additional structural conditions. The main result is a version of Sudakov minoration, providing lower bounds for expected suprema over well separated sets. Under additional assumptions on the growth of moments, this leads to a Fernique–Talagrand type theorem giving two-sided bounds for expected suprema over general sets expressed in terms of geometric functionals based on appropriate partition schemes.
- *Lévy measures of infinitely divisible positive processes: examples and distributional identities*, by N. Eisenbaum and J. Rosiński. The authors present a variety of distributional identities for Lévy processes based on the knowledge of their Lévy measures, treating in particular the case of Poisson processes, Sato processes, stochastic convolutions, and tempered stable subordinators. Connections with infinitely divisible random measures and transference principles for path properties of general nonnegative infinitely divisible processes are also discussed
- *Bounding suprema of canonical processes via convex hull*, by R. Latała. This paper discusses the method of bounding suprema of canonical processes based on the inclusion of their index set into a convex hull of a well-controlled set of points. While the upper bound is immediate, the reverse estimate was established to date only for a narrow class of regular stochastic processes. It is shown that for specific index sets, including arbitrary ellipsoids, regularity assumptions may be substantially weakened.

High-Dimensional Statistics

- *Random geometric graph: some recent developments and perspectives*, by Q. Duchemin and Y. De Castro. The Random Geometric Graph (RGG) is a random graph model for network data with an underlying spatial representation. This paper surveys various models of RGG and presents the recent developments from the lens of high-dimensional settings and nonparametric inference. The authors discuss tools used in the analysis of these models, involving results from probability, statistics, combinatorics, or information theory. They also present a list of challenging open problems motivated by applications and of purely theoretical interest.

- *Functional estimation in log-concave location families*, by V. Koltchinskii and M. Wahl. The authors investigate estimation of smooth functionals of the location parameter in location families given as shifts of log-concave densities on $\mathbb{R}^d$, generalizing Gaussian shift models. Under appropriate regularity assumptions, they construct estimators based on i.i.d. samples with minimax optimal error rates measured by the L_2-distance as well as by more general Orlicz norm distances.

Warszawa, Poland — Radosław Adamczak
Paris, France — Nathael Gozlan
Palaiseau, France — Karim Lounici
Newark, DE, USA — Mokshay Madiman

Contents

Part I Inequalities and Convexity

Covariance Representations, L^p-Poincaré Inequalities, Stein's Kernels, and High-Dimensional CLTs 3
Benjamin Arras and Christian Houdré

Volume Properties of High-Dimensional Orlicz Balls 75
Franck Barthe and Paweł Wolff

Entropic Isoperimetric Inequalities 97
Sergey G. Bobkov and Cyril Roberto

Transport Proofs of Some Functional Inverse Santaló Inequalities 123
Matthieu Fradelizi, Nathael Gozlan, and Simon Zugmeyer

Tail Bounds for Sums of Independent Two-Sided Exponential Random Variables 143
Jiawei Li and Tomasz Tkocz

Boolean Functions with Small Second-Order Influences on the Discrete Cube 155
Krzysztof Oleszkiewicz

Some Notes on Concentration for α-Subexponential Random Variables .. 167
Holger Sambale

Part II Limit Theorems

Limit Theorems for Random Sums of Random Summands 195
David Grzybowski

A Note on Central Limit Theorems for Trimmed Subordinated Subordinators 211
David M. Mason

Functional Central Limit Theorem via Nonstationary Projective Conditions 229
Florence Merlevède and Magda Peligrad

Part III Stochastic Processes

Sudakov Minoration for Products of Radial-Type Log-Concave Measures 257
Witold Bednorz

Lévy Measures of Infinitely Divisible Positive Processes: Examples and Distributional Identities 297
Nathalie Eisenbaum and Jan Rosiński

Bounding Suprema of Canonical Processes via Convex Hull 325
Rafał Latała

Part IV High-Dimensional Statistics

Random Geometric Graph: Some Recent Developments and Perspectives 347
Quentin Duchemin and Yohann De Castro

Functional Estimation in Log-Concave Location Families 393
Vladimir Koltchinskii and Martin Wahl

Part I
Inequalities and Convexity

Covariance Representations, L^p-Poincaré Inequalities, Stein's Kernels, and High-Dimensional CLTs

Benjamin Arras and Christian Houdré

2020 Mathematics Subject Classification: 26D10; 35R11; 47D07; 60E07; 60F05

1 Introduction

Covariance representations and Bismut-type formulas play a major role in modern probability theory. The most striking (and simple) instances are without a doubt the ones regarding the standard Gaussian probability measure on $\mathbb{R}^d$. These identities have many applications, ranging from functional inequalities to concentration phenomena, regularization along semigroups, continuity of certain singular integral operators, and Stein's method. The main objective of the present manuscript is to illustrate this circle of ideas. While some of the results presented here might be well known to specialists, others seem to be new. Let us further describe the main content of these notes.

In the first section, we revisit covariance identities based on closed form techniques and semigroup arguments. In particular, when strong gradient bounds are available, L^p-L^q asymmetric covariance estimates ($p \in [1, +\infty)$ and $q = p/(p-1)$) are put forward.

In the second section, based on various representation formulas, we discuss L^p-Poincaré inequalities ($p \geq 2$) and pseudo-Poincaré inequality for the standard Gaussian measure and the nondegenerate symmetric α-stable probability measures

Research supported in part by the grant # **524678** from the Simons Foundation.

B. Arras (✉)
Univ. Lille, CNRS, UMR 8524 – Laboratoire Paul Painlevé, Lille, France
e-mail: benjamin.arras@univ-lille.fr

C. Houdré
School of Mathematics, Georgia Institute of Technology, Atlanta, GA, USA
e-mail: houdre@math.gatech.edu

R. Adamczak et al. (eds.), *High Dimensional Probability IX*, Progress in Probability 80, https://doi.org/10.1007/978-3-031-26979-0_1

on $\mathbb{R}^d$ with $\alpha \in (1, 2)$. In particular, in Proposition 3.2, we give proof of the Gaussian L^p-Poincaré inequalities, $p \geq 2$, with sharp constants based on the covariance representation (2.26). Propositions 3.1 and 3.4 are concerned with L^p-Poincaré-type and pseudo-Poincaré inequalities for the nondegenerate symmetric α-stable, $\alpha \in (1, 2)$, probability measures on $\mathbb{R}^d$. These two results are based on the various Bismut-type formulas obtained in [9, Proposition 2.1] and in Proposition 3.3.

Finally, in the third section, as an application of our methodology, we build Stein's kernels in order to obtain, in the 1-Wasserstein distance, stability results and rates of convergence for high-dimensional central limit theorems (CLTs) when the limiting probability measure is a centered Gaussian measure with a nondegenerate covariance matrix. Theorems 4.2 and 4.3 are the main results of this section. The methodology is based on Stein's method for multivariate Gaussian probability measures and on closed form techniques under a finite Poincaré-type constant assumption.

2 Notations and Preliminaries

In the sequel, let us adopt and recall the notations of [9, Section 1]. Throughout, the Euclidean norm on $\mathbb{R}^d$ is denoted by $\|\cdot\|$ and the Euclidean inner product by $\langle ; \rangle$. Then $X \sim ID(b, \Sigma, \nu)$ indicates that the d-dimensional random vector X is infinitely divisible with a characteristic triplet (b, Σ, ν). In other words, its characteristic function φ_X is given, for all $\xi \in \mathbb{R}^d$, by

$$\varphi_X(\xi) = \exp\left(i\langle b; \xi\rangle - \frac{1}{2}\langle \Sigma\xi; \xi\rangle + \int_{\mathbb{R}^d} (e^{i\langle \xi; u\rangle} - 1 - i\langle \xi; u\rangle \mathbb{1}_{\|u\|\leq 1})\nu(du)\right),$$

where $b \in \mathbb{R}^d$, where Σ is a symmetric positive semi-definite $d \times d$ matrix and where ν, the Lévy measure, is a positive Borel measure on $\mathbb{R}^d$ such that $\nu(\{0\}) = 0$ and $\int_{\mathbb{R}^d}(1 \wedge \|u\|^2)\nu(du) < +\infty$. In particular, if $b = 0$, $\Sigma = I_d$, (the $d \times d$ identity matrix), and $\nu = 0$, then X is a standard Gaussian random vector with law γ, and its characteristic function is given, for all $\xi \in \mathbb{R}^d$, by

$$\hat{\gamma}(\xi) := \int_{\mathbb{R}^d} e^{i\langle y; \xi\rangle}\gamma(dy) = \exp\left(-\frac{\|\xi\|^2}{2}\right). \tag{2.1}$$

For $\alpha \in (0, 2)$, let ν_α be a Lévy measure such that, for all $c > 0$,

$$c^{-\alpha}T_c(\nu_\alpha)(du) = \nu_\alpha(du), \tag{2.2}$$

where $T_c(\nu_\alpha)(B) := \nu_\alpha(B/c)$, for all B Borel set of $\mathbb{R}^d \setminus \{0\}$. Recall that such a Lévy measure admits the polar decomposition

$$\nu_\alpha(du) = \mathbb{1}_{(0,+\infty)}(r)\mathbb{1}_{\mathbb{S}^{d-1}}(y)\frac{dr}{r^{\alpha+1}}\sigma(dy), \tag{2.3}$$

where σ is a positive finite measure on the Euclidean unit sphere of $\mathbb{R}^d$, denoted by $\mathbb{S}^{d-1}$. In the sequel, it is assumed that the measure σ is symmetric and that ν_α is nondegenerate in that

$$\inf_{y\in\mathbb{S}^{d-1}} \int_{\mathbb{S}^{d-1}} |\langle y; x\rangle|^\alpha \lambda_1(dx) \neq 0, \tag{2.4}$$

where λ_1, the spectral measure, is a symmetric finite positive measure on $\mathbb{S}^{d-1}$ proportional to σ (namely, $\lambda_1(dx) = -\cos(\alpha\pi/2)\Gamma(2-\alpha)/(\alpha(\alpha-1))\sigma(dx)$, $\alpha \in (1,2)$, where Γ is the Euler gamma function). Let μ_α be the α-stable probability measure on $\mathbb{R}^d$ defined through the corresponding characteristic function, for all $\xi \in \mathbb{R}^d$, by

$$\varphi_\alpha(\xi) := \int_{\mathbb{R}^d} e^{i\langle y;\xi\rangle}\mu_\alpha(dy) = \begin{cases} \exp\left(\int_{\mathbb{R}^d}(e^{i\langle u;\xi\rangle} - 1 - i\langle\xi; u\rangle)\nu_\alpha(du)\right), & \alpha \in (1,2),\\ \exp\left(\int_{\mathbb{R}^d}\left(e^{i\langle u;\xi\rangle} - 1 - i\langle\xi; u\rangle\mathbb{1}_{\|u\|\le 1}\right)\nu_1(du)\right), & \alpha = 1,\\ \exp\left(\int_{\mathbb{R}^d}(e^{i\langle u;\xi\rangle} - 1)\nu_\alpha(du)\right), & \alpha \in (0,1). \end{cases} \tag{2.5}$$

For σ symmetric, [66, Theorem 14.13.] provides a useful alternative representation for the characteristic function φ_α given, for all $\xi \in \mathbb{R}^d$, by

$$\varphi_\alpha(\xi) = \exp\left(-\int_{\mathbb{S}^{d-1}} |\langle y;\xi\rangle|^\alpha \lambda_1(dy)\right). \tag{2.6}$$

Let λ denote a uniform measure on the Euclidean unit sphere of $\mathbb{R}^d$ (i.e., λ is a positive finite measure on $\mathbb{S}^{d-1}$ proportional to the spherical part of the d-dimensional Lebesgue measure). For $\alpha \in (1,2)$, let $\nu_\alpha^{\mathrm{rot}}$ be the Lévy measure on $\mathbb{R}^d$ with polar decomposition

$$\nu_\alpha^{\mathrm{rot}}(du) = c_{\alpha,d}\mathbb{1}_{(0,+\infty)}(r)\mathbb{1}_{\mathbb{S}^{d-1}}(y)\frac{dr}{r^{\alpha+1}}\lambda(dy) \tag{2.7}$$

and with

$$c_{\alpha,d} = \frac{-\alpha(\alpha-1)\Gamma\left(\frac{\alpha+d}{2}\right)}{4\cos\left(\frac{\alpha\pi}{2}\right)\Gamma\left(\frac{\alpha+1}{2}\right)\pi^{\frac{d-1}{2}}\Gamma(2-\alpha)}. \tag{2.8}$$

Finally, denote by $\mu_\alpha^{\mathrm{rot}}$ the rotationally invariant α-stable probability measure on $\mathbb{R}^d$ with the Lévy measure given by (2.7) and with the choice of λ ensuring that, for all $\xi \in \mathbb{R}^d$,

$$\varphi_\alpha^{\mathrm{rot}}(\xi) = \hat{\mu}_\alpha^{\mathrm{rot}}(\xi) = \exp\left(-\frac{\|\xi\|^\alpha}{2}\right). \tag{2.9}$$

As well known, the probability measure $\mu_\alpha^{\mathrm{rot}}$ is absolutely continuous with respect to the d-dimensional Lebesgue measure, and its Lebesgue density, denoted by p_α^{rot}, is infinitely differentiable and is such that, for all $x \in \mathbb{R}^d$,

$$\frac{C_2}{(1+\|x\|)^{\alpha+d}} \leq p_\alpha^{\mathrm{rot}}(x) \leq \frac{C_1}{(1+\|x\|)^{\alpha+d}},$$

for some constants $C_1, C_2 > 0$ depending only on α and d. For $\alpha \in (1,2)$, let $\nu_{\alpha,1}$ be the Lévy measure on $\mathbb{R}$ given by

$$\nu_{\alpha,1}(du) = c_\alpha \frac{du}{|u|^{\alpha+1}}, \tag{2.10}$$

with

$$c_\alpha = \left(\frac{-\alpha(\alpha-1)}{4\Gamma(2-\alpha)\cos\left(\frac{\alpha\pi}{2}\right)}\right). \tag{2.11}$$

Next, let $\mu_{\alpha,1}$ be the α-stable probability measure on $\mathbb{R}$ with Lévy measure $\nu_{\alpha,1}$ and with the corresponding characteristic function defined, for all $\xi \in \mathbb{R}$, by

$$\hat{\mu}_{\alpha,1}(\xi) = \exp\left(\int_{\mathbb{R}} \left(e^{i\langle u;\xi\rangle} - 1 - i\langle u;\xi\rangle\right)\nu_{\alpha,1}(du)\right) = \exp\left(-\frac{|\xi|^\alpha}{2}\right). \tag{2.12}$$

Finally, throughout, let $\mu_{\alpha,d} = \mu_{\alpha,1} \otimes \cdots \otimes \mu_{\alpha,1}$ be the product probability measure on $\mathbb{R}^d$ with the corresponding characteristic function given, for all $\xi \in \mathbb{R}^d$, by

$$\hat{\mu}_{\alpha,d}(\xi) = \prod_{k=1}^d \hat{\mu}_{\alpha,1}(\xi_k) = \exp\left(\int_{\mathbb{R}^d} \left(e^{i\langle \xi;u\rangle} - 1 - i\langle \xi;u\rangle\right)\nu_{\alpha,d}(du)\right) \tag{2.13}$$

and with

$$\nu_{\alpha,d}(du) = \sum_{k=1}^d \delta_0(du_1) \otimes \cdots \otimes \delta_0(du_{k-1}) \otimes \nu_{\alpha,1}(du_k) \otimes \delta_0(du_{k+1}) \otimes \cdots \otimes \delta_0(du_d), \tag{2.14}$$

where δ_0 is the Dirac measure at 0.

In the sequel, $\mathcal{S}(\mathbb{R}^d)$ is the Schwartz space of infinitely differentiable functions, which, with their derivatives of any order, are rapidly decreasing, and $\mathcal{F}$ is the Fourier transform operator given, for all $f \in \mathcal{S}(\mathbb{R}^d)$ and all $\xi \in \mathbb{R}^d$, by

$$\mathcal{F}(f)(\xi) = \int_{\mathbb{R}^d} f(x)e^{-i\langle x;\xi\rangle}dx.$$

On $\mathcal{S}(\mathbb{R}^d)$, the Fourier transform is an isomorphism and the following well-known inversion formula holds:

$$f(x) = \frac{1}{(2\pi)^d}\int_{\mathbb{R}^d} \mathcal{F}(f)(\xi)e^{i\langle \xi;x\rangle}d\xi, \quad x \in \mathbb{R}^d.$$

$\mathcal{C}_c^{\infty}(\mathbb{R}^d)$ is the space of infinitely differentiable functions on $\mathbb{R}^d$ with compact support, and $\|\cdot\|_{\infty,\mathbb{R}}$ denotes the supremum norm on $\mathbb{R}$. Let $\mathcal{C}_b(\mathbb{R}^d)$ be the space of bounded continuous functions on $\mathbb{R}^d$ and let $\mathcal{C}_b^1(\mathbb{R}^d)$ be the space of continuously differentiable functions which are bounded on $\mathbb{R}^d$ together with their first derivatives. For $p \in (1, +\infty)$, $L^p(\mu_\alpha)$ denotes the space of equivalence classes (with respect to μ_α-almost everywhere equality) of functions which are Borel measurable and which are p-summable with respect to the probability measure μ_α. This space is endowed with the usual norm $\|\cdot\|_{L^p(\mu_\alpha)}$ defined, for all suitable f, by

$$\|f\|_{L^p(\mu_\alpha)} := \left(\int_{\mathbb{R}^d} |f(x)|^p \mu_\alpha(dx)\right)^{\frac{1}{p}}.$$

Similarly, for $p \in (1, +\infty)$, $L^p(\mathbb{R}^d, dx)$ denotes the classical Lebesgue space where the reference measure is the Lebesgue measure. It is endowed with the norm $\|\cdot\|_{L^p(\mathbb{R}^d,dx)}$ defined, for all suitable f, by

$$\|f\|_{L^p(\mathbb{R}^d,dx)} := \left(\int_{\mathbb{R}^d} |f(x)|^p dx\right)^{\frac{1}{p}}.$$

Next, let us introduce two semigroups of operators acting on $\mathcal{S}(\mathbb{R}^d)$ naturally associated with γ and μ_α. Let $(P_t^{\gamma})_{t\geq 0}$ and $(P_t^{\nu_\alpha})_{t\geq 0}$ be defined, for all $f \in \mathcal{S}(\mathbb{R}^d)$, all $x \in \mathbb{R}^d$, and all $t \geq 0$, by

$$P_t^{\gamma}(f)(x) = \int_{\mathbb{R}^d} f(xe^{-t} + \sqrt{1-e^{-2t}}y)\gamma(dy), \tag{2.15}$$

$$P_t^{\nu_\alpha}(f)(x) = \int_{\mathbb{R}^d} f(xe^{-t} + (1-e^{-\alpha t})^{\frac{1}{\alpha}}y)\mu_\alpha(dy). \tag{2.16}$$

The semigroup (2.15) is the classical Gaussian Ornstein-Uhlenbeck semigroup, and the semigroup (2.16) is the Ornstein-Uhlenbeck semigroup associated with the α-stable probability measure μ_α and recently put forward in the context of Stein's method for self-decomposable distributions (see [6–8]). Finally, denoting by $((P_t^{\nu_\alpha})^*)_{t\geq 0}$ the adjoint of the semigroup $(P_t^{\nu_\alpha})_{t\geq 0}$, the "carré de Mehler" semigroup is defined, for all $t \geq 0$, by

$$\mathcal{P}_t = (P^{\nu_\alpha}_{\frac{t}{\alpha}})^* \circ P^{\nu_\alpha}_{\frac{t}{\alpha}} = P^{\nu_\alpha}_{\frac{t}{\alpha}} \circ (P^{\nu_\alpha}_{\frac{t}{\alpha}})^*. \tag{2.17}$$

In the sequel, ∂_k denotes the partial derivative of order 1 in the variable x_k, ∇ the gradient operator, Δ the Laplacian operator, and $D^{\alpha-1}$, $(D^{\alpha-1})^*$, and $\mathbf{D}^{\alpha-1}$ the fractional operators defined, for all $f \in \mathcal{S}(\mathbb{R}^d)$ and all $x \in \mathbb{R}^d$, by

$$D^{\alpha-1}(f)(x) := \int_{\mathbb{R}^d} (f(x+u) - f(x)) u \nu_\alpha(du), \tag{2.18}$$

$$(D^{\alpha-1})^*(f)(x) := \int_{\mathbb{R}^d} (f(x-u) - f(x)) u \nu_\alpha(du), \tag{2.19}$$

$$\mathbf{D}^{\alpha-1}(f)(x) := \frac{1}{2}\left(D^{\alpha-1}(f)(x) - (D^{\alpha-1})^*(f)(x)\right). \tag{2.20}$$

Let us introduce also a gradient length, ∇_ν, which is linked to the energy form appearing in the Poincaré-type inequality for the infinitely divisible probability measures (see [47, Corollary 2]). For any Lévy measure ν on $\mathbb{R}^d$, all $f \in \mathcal{S}(\mathbb{R}^d)$, and all $x \in \mathbb{R}^d$,

$$\nabla_\nu(f)(x) = \left(\int_{\mathbb{R}^d} |f(x+u) - f(x)|^2 \nu(du)\right)^{\frac{1}{2}}. \tag{2.21}$$

Also, let us recall the definition of the gamma transform of order $r > 0$. For $(P_t)_{t\geq 0}$ a C_0-semigroup of contractions on a Banach space, with generator $\mathcal{A}$, the gamma transform of order $r > 0$ is defined, for all suitable f, by

$$(E - \mathcal{A})^{-\frac{r}{2}} f = \frac{1}{\Gamma(\frac{r}{2})} \int_0^{+\infty} \frac{e^{-t}}{t^{1-\frac{r}{2}}} P_t(f) dt, \tag{2.22}$$

where E is the identity operator and where the integral on the right-hand side has to be understood in the Bochner sense. Moreover, for all $\lambda > 0$, all $r > 0$, and all f suitable,

$$(\lambda E - \mathcal{A})^{-\frac{r}{2}} f = \frac{1}{\Gamma(\frac{r}{2})} \int_0^{+\infty} \frac{e^{-\lambda t}}{t^{1-\frac{r}{2}}} P_t(f) dt,$$

In particular, when this makes sense, as λ tends to 0^+,

$$(-\mathcal{A})^{-\frac{r}{2}} f = \frac{1}{\Gamma(\frac{r}{2})} \int_0^{+\infty} t^{\frac{r}{2}-1} P_t(f) dt.$$

Finally, the generators of the two semigroups can be obtained through Fourier representation formulas, and it is straightforward to check that the respective generators are given, for $\alpha \in (1, 2)$, for all $f \in \mathcal{S}(\mathbb{R}^d)$, and for all $x \in \mathbb{R}^d$, by

$$\mathcal{L}^{\gamma}(f)(x) = -\langle x; \nabla(f)(x)\rangle + \Delta(f)(x), \tag{2.23}$$

$$\mathcal{L}^{\alpha}(f)(x) = -\langle x; \nabla(f)(x)\rangle + \int_{\mathbb{R}^d} \langle \nabla(f)(x+u) - \nabla(f)(x); u\rangle \nu_{\alpha}(du). \tag{2.24}$$

Recall also one of the main results of [9], giving a Bismut-type formula for the nondegenerate symmetric α-stable probability measures on $\mathbb{R}^d$ with $\alpha \in (1, 2)$: for all $f \in \mathcal{S}(\mathbb{R}^d)$, all $x \in \mathbb{R}^d$, and all $t > 0$,

$$D^{\alpha-1} P_t^{\nu_{\alpha}}(f)(x) = \frac{e^{-(\alpha-1)t}}{\left(1 - e^{-\alpha t}\right)^{1-\frac{1}{\alpha}}} \int_{\mathbb{R}^d} y f\left(xe^{-t} + (1 - e^{-\alpha t})^{\frac{1}{\alpha}} y\right) \mu_{\alpha}(dy). \tag{2.25}$$

Finally, recall the covariance representation obtained in [47, Proposition 2] for $X \sim ID(b, \Sigma, \nu)$: for all $f, g \in \mathcal{S}(\mathbb{R}^d)$,

$$\begin{aligned} \mathrm{Cov}(f(X), g(X)) = \int_0^1 & \mathbb{E}\Big[\langle \Sigma \nabla(f)(X_z); \nabla(g)(Y_z)\rangle \\ & + \int_{\mathbb{R}^d} \Delta_u(f)(X_z)\Delta_u(g)(Y_z)\nu(du)\Big] dz, \end{aligned} \tag{2.26}$$

where $\Delta_u(f)(x) = f(x+u) - f(x)$ and where, for all $z \in [0, 1]$, (X_z, Y_z) has a characteristic function given, for all $\xi_1, \xi_2 \in \mathbb{R}^d$, by

$$\varphi(\xi_1, \xi_2) = (\varphi_X(\xi_1)\varphi_X(\xi_2))^{1-z}\varphi_X(\xi_1 + \xi_2)^z.$$

Next, let us investigate new covariance identities based on semigroup techniques (inspired from [6, Section 5]) and the corresponding asymmetric covariance estimates (see, e.g., [28] for log-concave measures and [44] for convex measures, which include Cauchy-type probability measures on $\mathbb{R}^d$). But, first, as a simple consequence of the covariance identity (2.26) combined with Hölder's inequality, one has the following proposition for the Gaussian and the general infinitely divisible cases (we refer to [47] and to [66] for any unexplained definitions regarding infinitely divisible distributions).

Proposition 2.1

(i) Let $X \sim \gamma$. Then, for all $f, g \in \mathcal{S}(\mathbb{R}^d)$, all $p \in [1, +\infty)$, and $q = p/(p-1)$ (with $q = +\infty$ when $p = 1$),

$$|\operatorname{Cov}(f(X), g(X))| \le \|\nabla(f)\|_{L^p(\gamma)}\|\nabla(g)\|_{L^q(\gamma)}. \tag{2.27}$$

(ii) Let $X \sim ID(b, 0, \nu)$. Then, for all $f, g \in \mathcal{S}(\mathbb{R}^d)$, all $p \in [1, +\infty)$, and $q = p/(p-1)$ (with $q = +\infty$ when $p = 1$),

$$|\operatorname{Cov}(f(X), g(X))| \le \|\nabla_\nu(f)\|_{L^p(\mu)}\|\nabla_\nu(g)\|_{L^q(\mu)}, \tag{2.28}$$

where $X \sim \mu$.

In the context of Stein's method for self-decomposable laws, a general covariance identity has been obtained in [8, Theorem 5.10] in the framework of closed symmetric nonnegative definite bilinear forms with dense domain under some coercive assumption. Indeed, the identity (5.15) there can be understood as a generalization of (2.5) in [28] and of (3.2) in [44] from which asymmetric covariance estimates can be obtained. Let us generalize Proposition 2.1 beyond the scope of infinitely divisible distributions. Key properties in order to establish these Brascamp-Lieb-type inequalities are subcommutation and/or some form of regularization (see, e.g., [5, 28]). Actually, let us provide, first, a slight extension of [8, Theorem 5.10].

Theorem 2.1 *Let H be a real Hilbert space with inner product $\langle \cdot; \cdot \rangle_H$ and induced norm $\|\cdot\|_H$. Let $\mathcal{E}$ be a closed symmetric nonnegative definite bilinear form with dense linear domain $\mathcal{D}(\mathcal{E})$. Let $\{G_\alpha : \alpha > 0\}$ and $\{P_t : t > 0\}$ be, respectively, the strongly continuous resolvent and the strongly continuous semigroup on H associated with $\mathcal{E}$. Moreover, let there exist a closed linear subspace $H_0 \subset H$ and a function ψ continuous on $(0, +\infty)$ with values in $(0, 1]$ such that $\lim_{t \to +\infty} \psi(t) = 0$,*

$$\int_0^{+\infty} \psi(t)dt < +\infty,$$

and such that, for all $t > 0$ and all $u \in H_0$,

$$\|P_t(u)\|_H \le \psi(t)\|u\|_H. \tag{2.29}$$

Let G_{0^+} be the operator defined by

$$G_{0^+}(u) := \int_0^{+\infty} P_t(u)dt, \quad u \in H_0, \tag{2.30}$$

where the above integral is understood to be in the Bochner sense. Then, for all $u \in H_0$, $G_{0^+}(u)$ belongs to $\mathcal{D}(\mathcal{E})$ and, for all $v \in \mathcal{D}(\mathcal{E})$,

$$\mathcal{E}\left(G_{0^+}(u), v\right) = \langle u; v\rangle_H. \tag{2.31}$$

Moreover, for all $u \in H_0$,

$$\mathcal{E}\left(G_{0^+}(u), G_{0^+}(u)\right) \leq \left(\int_0^{+\infty} \psi(t)dt\right) \|u\|_H^2. \tag{2.32}$$

Proof The proof follows closely the lines of the one of [8, Theorem 5.10] and so is omitted. □

Remark 2.1

(i) Note that from the semigroup property, for all $n \geq 1$, all $t > 0$, and all $f \in H_0$,

$$\|P_t(f)\|_H \leq \left(\psi\left(\frac{t}{n}\right)\right)^n \|f\|_H.$$

Then the behavior at 0^+ of the function ψ can lead to an exponential convergence result. Indeed, assuming that the function ψ is regular near 0 with $\psi(0^+) = 1$ and with $\psi'(0^+) < 0$, one gets, for all $t > 0$,

$$\lim_{n\to+\infty} \left(\psi\left(\frac{t}{n}\right)\right)^n = \exp\left(\psi'(0)t\right).$$

(ii) Let us next consider a rather long list of examples where our results provide covariance identities and L^2-estimates. In some situations, where strong gradient bounds are known, it is possible to obtain $L^p - L^q$ asymmetric covariance estimates. First, very classically, the Dirichlet form associated with the standard Gaussian probability measure on $\mathbb{R}^d$ is given, for all (real-valued) $f, g \in \mathcal{S}(\mathbb{R}^d)$, by

$$\mathcal{E}_\gamma(f, g) = \int_{\mathbb{R}^d} \langle\nabla(f)(x); \nabla(g)(x)\rangle\gamma(dx), \tag{2.33}$$

and an integration by parts formula ensures that $\mathcal{E}_\gamma$ is closable. The associated semigroup is the Ornstein-Uhlenbeck semigroup $(P_t^\gamma)_{t\geq 0}$ given in (2.15) with the generator given by (2.23). It is well known, thanks to the Gaussian Poincaré inequality, that, for all $f \in \mathcal{S}(\mathbb{R}^d)$ with $\gamma(f) := \int_{\mathbb{R}^d} f(x)\gamma(dx) = 0$ and all $t > 0$,

$$\|P_t^\gamma(f)\|_{L^2(\gamma)} \leq e^{-t}\|f\|_{L^2(\gamma)}.$$

Thus, Theorem 2.1 provides the following covariance representation: for all $f, g \in \mathcal{S}(\mathbb{R}^d)$ with $\int_{\mathbb{R}^d} f(x)\gamma(dx) = 0$,

$$\int_{\mathbb{R}^d} f(x)g(x)\gamma(dx) = \int_{\mathbb{R}^d} \langle \nabla(\tilde{f}_\gamma)(x); \nabla(g)(x)\rangle \gamma(dx) = \mathcal{E}_\gamma(g, \tilde{f}_\gamma), \tag{2.34}$$

where

$$\tilde{f}_\gamma = \int_0^{+\infty} P_t^\gamma(f)dt = (-\mathcal{L}^\gamma)^{-1}(f).$$

A straightforward application of Theorem 2.1 ensures, for all $f \in \mathcal{S}(\mathbb{R}^d)$ with $\int_{\mathbb{R}^d} f(x)\gamma(dx) = 0$, that

$$\mathcal{E}_\gamma(\tilde{f}_\gamma, \tilde{f}_\gamma) \leq \|f\|^2_{L^2(\gamma)} \leq \mathcal{E}_\gamma(f, f),$$

so that, by the Cauchy-Schwarz inequality,

$$\left|\int_{\mathbb{R}^d} f(x)g(x)\gamma(dx)\right| \leq \mathcal{E}_\gamma(g, g)^{\frac{1}{2}} \mathcal{E}_\gamma(f, f)^{\frac{1}{2}},$$

which is a particular instance of Proposition 2.1 (i), with $p = q = 2$. To obtain the general $L^p - L^q$ covariance estimates based on the covariance representation (2.34), one can use the commutation formula $\nabla(P_t^\gamma(f)) = e^{-t} P_t^\gamma(\nabla(f))$ so that the following inequality holds true: for all $p \in (1, +\infty)$ and all $f \in \mathcal{S}(\mathbb{R}^d)$ with $\int_{\mathbb{R}^d} f(x)\gamma(dx) = 0$,

$$\|\nabla(\tilde{f}_\gamma)\|_{L^p(\gamma)} \leq \|\nabla(f)\|_{L^p(\gamma)}. \tag{2.35}$$

A direct application of Hölder's inequality combined with (2.34) and with (2.35) provides the general case of Proposition 2.1 (i). Note that the previous lines of reasoning do not depend on the dimension of the ambient space so that the covariance estimate (2.27) extends to the infinite-dimensional setting and the Malliavin calculus framework (see, e.g., [55, Section 2.9]). The details are left to the interested reader (see, also [46, 48]).

(iii) Next, let us consider another hypercontractive semigroup related to a classical probability measure on $\mathbb{R}$ with finite exponential moments. (The corresponding multidimensional version follows by tensorization.) Let $\alpha \geq 1/2$ and let $\gamma_{\alpha,1}$ be the gamma probability measure on $(0, +\infty)$ with shape parameter α and scale parameter 1. Let $\mathcal{E}_\alpha$ denote the Dirichlet form associated with the Laguerre dynamics and given, for all $f, g \in \mathcal{C}_c^\infty((0, +\infty))$, by

$$\mathcal{E}_\alpha(f, g) = \int_0^{+\infty} x f'(x) g'(x) \gamma_{\alpha,1}(dx).$$

This closable symmetric bilinear form generates the well-known Laguerre semigroup $(P_t^{\alpha,1})_{t\geq 0}$ (see, e.g., [10, 13]) with the generator given, for all $f \in \mathcal{C}_c^\infty((0, +\infty))$ and all $x \in (0, +\infty)$, by

$$\mathcal{L}^{\alpha,1}(f)(x) = xf''(x) + (\alpha - x)f'(x).$$

Recall that the Poincaré inequality for this dynamics follows, e.g., from the spectral expansion of a test function belonging to $\mathcal{D}(\mathcal{E}_\alpha)$ along the Laguerre orthonormal polynomials which are the eigenfunctions of $\mathcal{L}^{\alpha,1}$. Moreover, letting ∂_σ be the differential operator defined, for all $f \in \mathcal{C}_c^\infty((0,+\infty))$ and all $x \in (0,+\infty)$, by

$$\partial_\sigma(f)(x) := \sqrt{x} f'(x),$$

the following intertwining formula has been proved in [10, Lemma 11]: for all $f \in \mathcal{C}_c^\infty((0,+\infty))$, all $x \in (0,+\infty)$, and all $t > 0$,

$$\partial_\sigma(P_t^{\alpha,1}(f))(x) = e^{-\frac{t}{2}} \mathbb{E}\left(\frac{\left(e^{-\frac{t}{2}}\sqrt{x} + \sqrt{\frac{1-e^{-t}}{2}} Z\right)}{(X_t^x)^{\frac{1}{2}}} \partial_\sigma(f)(X_t^x) \right), \tag{2.36}$$

where Z is a standard normal random variable and where X_t^x is given by

$$X_t^x = (1-e^{-t}) X_{\alpha-\frac{1}{2},1} + \left(e^{-\frac{t}{2}}\sqrt{x} + \sqrt{\frac{1-e^{-t}}{2}} Z\right)^2,$$

with $X_{\alpha-1/2,1} \sim \gamma_{\alpha-1/2,1}$ independent of Z. Using (2.36), the following subcommutation inequality holds true: for all $f \in \mathcal{C}_c^\infty((0,+\infty))$, all $x > 0$, and all $t > 0$,

$$|\partial_\sigma(P_t^{\alpha,1}(f))(x)| \leq e^{-\frac{t}{2}} P_t^{\alpha,1}(|\partial_\sigma(f)|)(x). \tag{2.37}$$

Performing a reasoning similar to the one in the Gaussian case, one gets the following asymmetric covariance estimate: for all $f, g \in \mathcal{C}_c^\infty((0,+\infty))$, all $p \in (1,+\infty)$, and $q = p/(p-1)$,

$$\left|\mathrm{Cov}(f(X_{\alpha,1}), g(X_{\alpha,1}))\right| \leq 2\|\partial_\sigma(g)\|_{L^q(\gamma_{\alpha,1})} \|\partial_\sigma(f)\|_{L^p(\gamma_{\alpha,1})}.$$

The previous subcommutation inequality can be seen as a direct consequence of the Bakry-Emery criterion since for this Markov diffusion semigroup, $\Gamma_2(f) \geq \Gamma(f)/2$, for all $f \in \mathcal{C}_c^\infty((0,+\infty))$.

(iv) Next, let us consider the Jacobi semigroup $(Q_t^{\alpha,\beta})_{t\geq 0}$, related to the beta probability measures on $[-1,1]$ of the form

$$\mu_{\alpha,\beta}(dx) = C_{\alpha,\beta}(1-x)^{\alpha-1}(1+x)^{\beta-1}\mathbb{1}_{[-1,1]}(x)dx,$$

with $\alpha > 0$, $\beta > 0$ such that $\min(\alpha, \beta) > 3/2$ and where $C_{\alpha,\beta} > 0$ is a normalization constant. The generator of the Jacobi semigroup is given, for all $f \in \mathcal{C}_c^\infty([-1, 1])$ and all $x \in [-1, 1]$, by

$$\mathcal{L}_{\alpha,\beta}(f)(x) = (1 - x^2) f''(x) + ((\beta - \alpha) - (\alpha + \beta)x) f'(x).$$

Moreover, the corresponding "carré du champs" operator is $\Gamma_{\alpha,\beta}(f)(x) = (1 - x^2)(f'(x))^2$, for all $f \in \mathcal{C}_c^\infty([-1, 1])$ and all $x \in [-1, 1]$, so that the natural gradient associated with the Jacobi operator is given, for all $f \in \mathcal{C}_c^\infty([-1, 1])$ and all $x \in [-1, 1]$, by

$$\partial_{\alpha,\beta}(f)(x) := \sqrt{1 - x^2} f'(x).$$

According to [13, Section 2.1.7], this Markov diffusion operator satisfies a curvature-dimension condition of type $CD(\kappa_{\alpha,\beta}, n_{\alpha,\beta})$ for some $\kappa_{\alpha,\beta}, n_{\alpha,\beta} > 0$ depending only on α and β. In particular, it satisfies a curvature dimension condition $CD(\kappa_{\alpha,\beta}, \infty)$, and so one has the following subcommutation formula: for all $f \in \mathcal{C}_c^\infty([-1, 1])$, all $t > 0$, and all $x \in [-1, 1]$,

$$\left|\partial_{\alpha,\beta}\left(Q_t^{\alpha,\beta}(f)\right)(x)\right| \leq e^{-\kappa_{\alpha,\beta} t} Q_t^{\alpha,\beta}\left(\left|\partial_{\alpha,\beta}(f)\right|\right)(x).$$

The covariance representation then reads as follows: for all $f, g \in \mathcal{C}_c^\infty([-1, 1])$ with $\mu_{\alpha,\beta}(f) = 0$,

$$\mathrm{Cov}(f(X_{\alpha,\beta}), g(X_{\alpha,\beta})) = \int_{[-1,1]} (1 - x^2) g'(x) \tilde{f}'(x) \mu_{\alpha,\beta}(dx),$$
$$\tilde{f} = \int_0^{+\infty} Q_t^{\alpha,\beta}(f) dt,$$

where $X_{\alpha,\beta} \sim \mu_{\alpha,\beta}$. Applying the same strategy as before gives the following asymmetric covariance estimate: for all $f, g \in \mathcal{C}_c^\infty([-1, 1])$, all $p \in (1, +\infty)$, and $q = p/(p - 1)$,

$$\left|\mathrm{Cov}(f(X_{\alpha,\beta}), g(X_{\alpha,\beta}))\right| \leq \frac{1}{\kappa_{\alpha,\beta}} \|\partial_{\alpha,\beta}(f)\|_{L^p(\mu_{\alpha,\beta})} \|\partial_{\alpha,\beta}(g)\|_{L^q(\mu_{\alpha,\beta})}.$$

(v) Let μ be a centered probability measure on $\mathbb{R}^d$ given by

$$\mu(dx) = \frac{1}{Z(\mu)} \exp(-V(x))\, dx,$$

where $Z(\mu) > 0$ is a normalization constant and where V is a nonnegative smooth function on $\mathbb{R}^d$ such that

$$\mathrm{Hess}(V)(x) \geq \kappa I_d, \quad x \in \mathbb{R}^d,$$

for some $\kappa > 0$, and where I_d is the identity matrix. It is well known that such a probability measure satisfies a Poincaré inequality with respect to the classical "carré du champs" (see, e.g., [13, Theorem 4.6.3]), namely, for all f smooth enough on $\mathbb{R}^d$,

$$\mathrm{Var}_\mu(f) \leq C_\kappa \int_{\mathbb{R}^d} \|\nabla(f)(x)\|^2 \mu(dx),$$

for some $C_\kappa > 0$ depending on κ and on $d \geq 1$ (here and in the sequel, $\mathrm{Var}_\mu(f)$ denotes the variance of f under μ). In particular, from the Brascamp and Lieb inequality, $C_\kappa \leq 1/\kappa$. For all $f, g \in \mathcal{C}_c^\infty(\mathbb{R}^d)$, let

$$\mathcal{E}_\mu(f, g) = \int_{\mathbb{R}^d} \langle \nabla(f)(x); \nabla(g)(x) \rangle \mu(dx).$$

The bilinear form $\mathcal{E}_\mu$ is clearly symmetric on $L^2(\mu)$, and let us discuss briefly its closability following [20, Section 2.6]. The Lebesgue density of μ, is such that, for all $p \in (1, +\infty)$ and for any compact subset, K, of $\mathbb{R}^d$,

$$\int_K \rho_\mu(x)^{-\frac{1}{p-1}} dx < +\infty. \tag{2.38}$$

Based on (2.38), it is not difficult to see that the weighted Sobolev norms $\|\cdot\|_{1,p,\mu}$, defined, for all $f \in \mathcal{C}_c^\infty(\mathbb{R}^d)$, by

$$\|f\|_{1,p,\mu} = \|f\|_{L^p(\mu)} + \sum_{k=1}^{d} \|\partial_k(f)\|_{L^p(\mu)},$$

are closable. In particular, for $p = 2$, this provides the closability of the form $\mathcal{E}_\mu$. Then, thanks to Theorem 2.1, for all $f, g \in \mathcal{S}(\mathbb{R}^d)$ with $\mu(f) = 0$,

$$\mathcal{E}_\mu(g, \tilde{f}_\mu) = \langle f; g \rangle_{L^2(\mu)}, \quad \tilde{f}_\mu = \int_0^{+\infty} P_t^\mu(f) dt, \tag{2.39}$$

with $(P_t^\mu)_{t \geq 0}$ being the symmetric Markovian semigroup generated by the smallest closed extension of the symmetric bilinear form $\mathcal{E}_\mu$. Finally, for all $f \in \mathcal{S}(\mathbb{R}^d)$,

$$\Gamma_2(f)(x) = \sum_{j,k} \left(\partial^2_{j,k}(f)(x)\right)^2 + \langle \nabla(f)(x); \mathrm{Hess}(V)(x) \nabla(f)(x) \rangle \geq \kappa \Gamma(f)(x).$$

In other words, the curvature-dimension condition $CD(\kappa, \infty)$ is satisfied so that the following strong gradient bound holds true: for all $t > 0$ and all $f \in \mathcal{S}(\mathbb{R}^d)$,

$$\sqrt{\Gamma(P_t^\mu)(f)(x)} \leq e^{-\kappa t} P_t^\mu \left(\sqrt{\Gamma(f)}\right)(x).$$

Then the following asymmetric covariance estimate holds true: for all $f, g \in \mathcal{S}(\mathbb{R}^d)$, all $p \in (1, +\infty)$, and $q = p/(p-1)$,

$$|\mathrm{Cov}(f(X); g(X))| \leq \frac{1}{\kappa} \|\nabla(g)\|_{L^q(\mu)} \|\nabla(f)\|_{L^p(\mu)},$$

with $X \sim \mu$. Combining (2.39) with estimates from [28], one retrieves the Brascamp and Lieb inequality for strictly log-concave measures (i.e., such that $\mathrm{Hess}(V)(x) > 0$, for all $x \in \mathbb{R}^d$), as well as its asymmetric versions (see, e.g., [28, Theorem 1.1]).

(vi) Again, let us discuss a class of probability measures that lies at the interface of the nonlocal and local frameworks. Let $m > 0$ and let μ_m be the probability measure on $\mathbb{R}^d$ given by

$$\mu_m(dx) = c_{m,d} \left(1 + \|x\|^2\right)^{-m-d/2} dx,$$

for some normalization constant $c_{m,d} > 0$ depending only on m and d. First, the characteristic function of a random vector with law μ_m is given (see [70, Theorem II]), for all $\xi \in \mathbb{R}^d$, by

$$\varphi_m(\xi) = \exp\left(\int_{\mathbb{R}^d} \left(e^{i\langle \xi; u\rangle} - 1 - \frac{i\langle u; \xi\rangle}{1 + \|u\|^2}\right) \nu_m(du)\right). \tag{2.40}$$

Above, ν_m is the Lévy measure on $\mathbb{R}^d$ given by

$$\nu_m(du) = \frac{2}{\|u\|^d} \left(\int_0^{+\infty} g_m(2w) L_{\frac{d}{2}}\left(\sqrt{2w}\|u\|\right) dw\right) du, \tag{2.41}$$

with, for all $w > 0$,

$$g_m(w) = \frac{2}{\pi^2 w} \frac{1}{J_m^2(\sqrt{w}) + Y_m^2(\sqrt{w})}, \quad L_{\frac{d}{2}}(w) = \frac{1}{(2\pi)^{\frac{d}{2}}} w^{\frac{d}{2}} K_{\frac{d}{2}}(w),$$

with J_m, Y_m, and $K_{d/2}$ denoting respectively the Bessel functions of the first and second kind and the modified Bessel function of order $d/2$. Based on the representations (2.40) and (2.41), it is clear that μ_m is self-decomposable so that it is naturally associated with the nonlocal Dirichlet form given, for all $f, g \in C_c^\infty(\mathbb{R}^d)$, by

$$\mathcal{E}_m(f, g) = \int_{\mathbb{R}^d}\int_{\mathbb{R}^d} \Delta_u(f)(x)\Delta_u(g)(x)\nu_m(du)\mu_m(dx).$$

Moreover, since μ_m is infinitely divisible, [47, Corollary 2] ensures that μ_m satisfies the following Poincaré-type inequality, for all f smooth enough on $\mathbb{R}^d$:

$$\mathrm{Var}_m(f) \leq \int_{\mathbb{R}^d}\int_{\mathbb{R}^d} |f(x+u) - f(x)|^2 \nu_m(du)\mu_m(dx)$$

(see [8, Proposition 5.1] for proof based on a semigroup argument when $m > 1/2$). Since $\mu_m * \nu_m \ll \mu_m$, the symmetric bilinear form $\mathcal{E}_m$ is closable so that $(\mathcal{P}_t^m)_{t\geq 0}$, the symmetric semigroup generated by the smallest closed extension $(\mathcal{E}_m, \mathcal{D}(\mathcal{E}_m))$, verifies the following ergodic property: for all $f \in L^2(\mu_m)$ with $\int_{\mathbb{R}^d} f(x)\mu_m(dx) = 0$,

$$\|\mathcal{P}_t^m(f)\|_{L^2(\mu_m)} \leq e^{-t}\|f\|_{L^2(\mu_m)}.$$

Then one can apply Theorem 2.1 to obtain the following covariance representation: for all $f \in \mathcal{S}(\mathbb{R}^d)$ with $\mu_m(f) = 0$ and all $g \in \mathcal{S}(\mathbb{R}^d)$,

$$\mathcal{E}_m(g, \tilde{f}_m) = \langle f; g\rangle_{L^2(\mu_m)}, \quad \tilde{f}_m = \int_0^{+\infty} \mathcal{P}_t^m(f)dt. \tag{2.42}$$

Now, this class of probability measures has been investigated in the context of weighted Poincaré-type inequality (see, e.g., [19, 22]). Indeed, for all f smooth enough on $\mathbb{R}^d$ and all $m > 0$,

$$\mathrm{Var}_m(f) \leq C_{m,d} \int_{\mathbb{R}^d} f(x)(-\mathcal{L}_m^\sigma)(f)(x)\mu_m(dx),$$

where the operator $\mathcal{L}_m^\sigma$ is given, on smooth functions, by

$$\mathcal{L}_m^\sigma(f)(x) = (1 + \|x\|^2)\Delta(f)(x) + 2\left(1 - m - \frac{d}{2}\right)\langle x; \nabla(f)(x)\rangle$$

and where $C_{m,d} > 0$ is a constant depending on m and d, which can be explicitly computed or bounded depending on the relationships between m and d (see [22, Corollaries 5.2 and 5.3]). The corresponding Dirichlet form is given, for all $f, g \in \mathcal{C}_c^\infty(\mathbb{R}^d)$, by

$$\mathcal{E}_m^\sigma(f, g) = \int_{\mathbb{R}^d} \left(1 + \|x\|^2\right)\langle \nabla(f)(x); \nabla(g)(x)\rangle \mu_m(dx).$$

Once again, using the exponential $L^2(\mu_m)$-convergence to the equilibrium of the semigroup induced by the form $\mathcal{E}_m^\sigma$ (denoted by $(\mathcal{P}_t^{m,\sigma})_{t\geq 0}$), one obtains the following covariance representation formula: for all $f, g \in \mathcal{C}_c^\infty(\mathbb{R}^d)$ with $\mu_m(f) = 0$,

$$\mathcal{E}_m^\sigma(\tilde{f}_m^\sigma, g) = \langle f; g \rangle_{L^2(\mu_m)}, \quad \tilde{f}_m^\sigma = \int_0^{+\infty} \mathcal{P}_t^{m,\sigma}(f) dt. \tag{2.43}$$

Now, using either (2.42) or (2.43), one gets

$$|\mathrm{Cov}(f(X_m), g(X_m))| \le \|\nabla_{\nu_m}(f)\|_{L^2(\mu_m)} \|\nabla_{\nu_m}(g)\|_{L^2(\mu_m)},$$

$$|\mathrm{Cov}(f(X_m), g(X_m))| \le C_{m,d} \|\sigma \nabla(f)\|_{L^2(\mu_m)} \|\sigma \nabla(g)\|_{L^2(\mu_m)},$$

with $X_m \sim \mu_m$ and $\sigma(x)^2 = (1 + \|x\|^2)$, for all $x \in \mathbb{R}^d$.

(vii) Let us return to the α-stable probability measures on $\mathbb{R}^d$, $\alpha \in (0, 2)$. Let $\mathcal{E}$ be the nonnegative definite symmetric bilinear form given, for all $f, g \in \mathcal{S}(\mathbb{R}^d)$, by

$$\mathcal{E}(f, g) = \int_{\mathbb{R}^d} \int_{\mathbb{R}^d} \Delta_u(f)(x) \Delta_u(g)(x) \nu_\alpha(du) \mu_\alpha(dx).$$

Recall that since $\mu_\alpha * \nu_\alpha \ll \mu_\alpha$, the bilinear form $\mathcal{E}$ is closable. The associated semigroup, $(\mathcal{P}_t)_{t \ge 0}$, is the "carré de Mehler" semigroup defined in (2.17), whose $L^2(\mu_\alpha)$-generator is given, for all $f \in \mathcal{S}(\mathbb{R}^d)$, by

$$\mathcal{L}(f) = \frac{1}{\alpha} \left(\mathcal{L}^\alpha + (\mathcal{L}^\alpha)^* \right)(f)$$

and already put forward in [8, 9]. Moreover, the Poincaré-type inequality for the α-stable probability measure implies that, for all $f \in L^2(\mu_\alpha)$ with $\mu_\alpha(f) = 0$ and all $t > 0$,

$$\|\mathcal{P}_t(f)\|_{L^2(\mu_\alpha)} \le e^{-t} \|f\|_{L^2(\mu_\alpha)}.$$

Thus, Theorem 2.1 provides the following covariance representation: for all $f, g \in \mathcal{S}(\mathbb{R}^d)$ with $\int_{\mathbb{R}^d} f(x) \mu_\alpha(dx) = 0$,

$$\int_{\mathbb{R}^d} f(x) g(x) \mu_\alpha(dx) = \int_{\mathbb{R}^d} \int_{\mathbb{R}^d} \Delta_u(g)(x) \Delta_u(\tilde{f})(x) \nu_\alpha(du) \mu_\alpha(dx) = \mathcal{E}(g, \tilde{f}),$$

where

$$\tilde{f} = \int_0^{+\infty} \mathcal{P}_t(f) dt = (-\mathcal{L})^{-1}(f).$$

Moreover, still from Theorem 2.1, for all $f \in \mathcal{S}(\mathbb{R}^d)$ with $\int_{\mathbb{R}^d} f(x) \mu_\alpha(dx) = 0$,

$$\mathcal{E}(\tilde{f}, \tilde{f}) \leq \|f\|^2_{L^2(\mu_\alpha)} \leq \mathcal{E}(f, f),$$

so that, by the Cauchy-Schwarz inequality,

$$\left| \int_{\mathbb{R}^d} f(x)g(x)\mu_\alpha(dx) \right|^2 \leq \mathcal{E}(g, g)\mathcal{E}(f, f),$$

which is a particular instance of Proposition 2.1 (ii), with $p = q = 2$ and with $\nu = \nu_\alpha$.

(viii) Finally, let us discuss the case of the infinitely divisible probability measures on $\mathbb{R}^d$, $d \geq 1$, in full generality. Let ν be a Lévy measure on $\mathbb{R}^d$, as defined at the beginning of this section. Let μ be the infinitely divisible probability measure on $\mathbb{R}^d$, defined through its Fourier transform, for all $\xi \in \mathbb{R}^d$, by

$$\hat{\mu}(\xi) = \exp\left(\int_{\mathbb{R}^d} \left(e^{i\langle \xi; u \rangle} - 1 - i\langle u; \xi \rangle \mathbb{1}_{\|u\| \leq 1} \right) \nu(du) \right). \tag{2.44}$$

Now, let $\mathcal{E}_\nu$ be the associated bilinear symmetric nonnegative definite form defined, for all $f, g \in \mathcal{S}(\mathbb{R}^d)$, by

$$\mathcal{E}_\nu(f, g) = \int_{\mathbb{R}^d} \int_{\mathbb{R}^d} (f(x+u) - f(x))(g(x+u) - g(x))\mu(dx)\nu(du). \tag{2.45}$$

Thanks to [30, Lemma 4.1], $\mu * \nu \ll \mu$ so that the form $\left(\mathcal{E}_\nu, \mathcal{S}(\mathbb{R}^d)\right)$ is closable. Then let us denote by $(\mathcal{E}_\nu, \mathcal{D}(\mathcal{E}_\nu))$ its smallest closed extension with dense linear domain $\mathcal{D}(\mathcal{E}_\nu)$ in $L^2(\mu)$. Then in the sequel, let us denote by $(\mathcal{P}^\nu_t)_{t \geq 0}$ and by $\mathcal{L}^\nu$ the corresponding symmetric contraction semigroup on $L^2(\mu)$ and its self-adjoint $L^2(\mu)$-generator. In particular, recall that, for all $f \in \mathcal{D}(\mathcal{L}^\nu)$ and all $g \in \mathcal{D}(\mathcal{E}_\nu)$,

$$\langle (-\mathcal{L}^\nu)(f); g \rangle_{L^2(\mu)} = \mathcal{E}_\nu(f, g). \tag{2.46}$$

Moreover, the following Poincaré-type inequality holds true: for all $f \in \mathcal{D}(\mathcal{E}_\nu)$ such that $\mathbb{E} f(X) = 0$,

$$\mathbb{E} f(X)^2 \leq \int_{\mathbb{R}^d} \int_{\mathbb{R}^d} |f(x+u) - f(x)|^2 \nu(du)\mu(dx), \tag{2.47}$$

where $X \sim \mu$. Then by standard arguments, for all $f \in L^2(\mu)$ with $\mathbb{E} f(X) = 0$ and all $t \geq 0$,

$$\|\mathcal{P}^\nu_t(f)\|_{L^2(\mu)} \leq e^{-t} \|f\|_{L^2(\mu)}.$$

Reasoning as in the previous cases implies the following covariance identity: for all $f, g \in \mathcal{S}(\mathbb{R}^d)$ with $\mathbb{E}f(X) = 0$,

$$\mathbb{E}f(X)g(X) = \mathcal{E}_\nu(\tilde{f}, g),$$

where $\tilde{f}$ is defined by

$$\tilde{f} = \int_0^{+\infty} \mathcal{P}_t^\nu(f)dt = (-\mathcal{L}^\nu)^{-1}(f).$$

In particular, one retrieves, as previously, Proposition 2.1 (ii), with $p = q = 2$ and with a general ν.

As detailed next, it is possible to refine the existence result given by Theorem 2.1 by using a celebrated characterization of surjective, closed and densely defined linear operators on a Hilbert space by a priori estimates on their adjoints. This abstract existence result is well known in the theory of partial differential equations (see, e.g., [26, Theorem 2.20]) and seems to go back to [45, Lemma 1.1]. Combined with integration by parts and the Poincaré inequality, it allows the retrieval of the covariance representations of Remark 2.1. In particular, this characterization result allows going beyond the assumption of Poincaré inequality for the underlying probability measure in order to prove the existence of Stein's kernels.

Theorem 2.2 *Let H be a separable real Hilbert space with inner product $\langle \cdot; \cdot \rangle_H$ and induced norm $\| \cdot \|_H$. Let $\mathcal{A}$ be a closed and densely defined linear operator on H with domain $\mathcal{D}(\mathcal{A})$ and such that, for all $u \in \mathcal{D}(\mathcal{A}^*)$,*

$$\|u\|_H \leq C\|\mathcal{A}^*(u)\|_H, \tag{2.48}$$

for some $C > 0$ not depending on u and where $(\mathcal{A}^*, \mathcal{D}(\mathcal{A}^*))$ is the adjoint of $(\mathcal{A}, \mathcal{D}(\mathcal{A}))$. Then, for all $u \in H$, there exists $G(u) \in \mathcal{D}(\mathcal{A})$ such that, for all $v \in H$,

$$\langle \mathcal{A}(G(u)); v\rangle_H = \langle u; v\rangle_H.$$

Moreover, if $\mathcal{A}$ is self-adjoint, then, for all $u \in H$,

$$|\langle \mathcal{A}(G(u)); G(u)\rangle| \leq C\|u\|_H^2.$$

Let us further provide the Banach-space version of the previous result (see, e.g., [26, Theorem 2.20] for proof).

Theorem 2.3 *Let E and F be two Banach spaces with respective norms $\| \cdot \|_E$ and $\| \cdot \|_F$. Let $\mathcal{A}$ be a closed and densely defined linear operator on E with domain $\mathcal{D}(\mathcal{A})$ and such that, for all $u^* \in \mathcal{D}(\mathcal{A}^*)$,*

$$\|u^*\|_{F^*} \leq C\|\mathcal{A}^*(u^*)\|_{E^*}, \tag{2.49}$$

for some $C > 0$ not depending on u^* and where $(\mathcal{A}^*, \mathcal{D}(\mathcal{A}^*))$ is the adjoint of $(\mathcal{A}, \mathcal{D}(\mathcal{A}))$. Then, for all $u \in F$, there exists $G(u) \in \mathcal{D}(\mathcal{A})$ such that

$$\mathcal{A}(G(u)) = u.$$

Conversely, if $\mathcal{A}$ is surjective, then, for all $u^* \in \mathcal{D}(\mathcal{A}^*)$,

$$\|u^*\|_{F^*} \leq C\|\mathcal{A}^*(u^*)\|_{E^*}, \tag{2.50}$$

for some $C > 0$ not depending on u^*.

Remark 2.2 Let us briefly explain how one can apply the previous existence theorem in the classical Gaussian setting. Let $H = L^2(\gamma)$, let $H_0 = \{f \in H, \int_{\mathbb{R}^d} f(x)\gamma(dx) = 0\}$, and let $\mathcal{A} = -\mathcal{L}^\gamma$. Note that if $f \in \mathcal{D}(\mathcal{A})$ then, for all $c \in \mathbb{R}$, $f + c \in \mathcal{D}(\mathcal{A})$ so that $f_0 = f - \int_{\mathbb{R}^d} f(x)\gamma(dx) \in \mathcal{D}(\mathcal{A})$. Moreover, $\mathcal{L}^\gamma$ is a linear densely defined self-adjoint operator on H. Finally, from the Gaussian Poincaré inequality, for all $f \in \mathcal{D}(\mathcal{A})$ such that $\int_{\mathbb{R}^d} f(x)\gamma(dx) = 0$,

$$\|f\|_H^2 \leq \int_{\mathbb{R}^d} f(x)\left(-\mathcal{L}^\gamma\right)(f)(x)\gamma(dx),$$

and so the Cauchy-Schwarz inequality gives, for all such f,

$$\|f\|_H \leq \|(-\mathcal{L}^\gamma)^*(f)\|_H.$$

Thus, from Theorem 2.2, for all $f \in H_0$, there exists $\tilde{f} \in \mathcal{D}_0(\mathcal{A})$ such that, for all $v \in H_0$,

$$\langle \mathcal{A}(\tilde{f}); v\rangle_H = \langle f; v\rangle_H, \tag{2.51}$$

where $\mathcal{D}_0(\mathcal{A}) = \mathcal{D}(\mathcal{A}) \cap H_0$. Equation (2.51) extends to all $v \in H$ by translation.

To conclude, this section discusses an example where the underlying probability measure does not satisfy an L^2-L^2 Poincaré inequality with respect to the classical energy form but for which it is possible to obtain a covariance identity with the standard "carré du champs operator" and the corresponding estimates.

Proposition 2.2 *Let $\delta \in (0, 1)$ and let μ_δ be the probability measure on $\mathbb{R}$ given by*

$$\mu_\delta(dx) = C_\delta \exp(-|x|^\delta)dx = p_\delta(x)dx,$$

for some normalizing constant $C_\delta > 0$. Let $p \in [2, +\infty)$ and let $g \in L^p(\mu_\delta)$ be such that $\int_{\mathbb{R}} g(x)\mu_\delta(dx) = 0$. Let f_δ be defined, for all $x \in \mathbb{R}$, by

$$f_\delta(x) = \frac{1}{p_\delta(x)}\int_x^{+\infty} g(y)p_\delta(y)dy = -\frac{1}{p_\delta(x)}\int_{-\infty}^x g(y)p_\delta(y)dy. \tag{2.52}$$

Then, for all $r \in [1, +\infty)$ *such that* $r/(r-1) > q = p/(p-1)$,

$$\|f_\delta\|_{L^r(\mu_\delta)} \leq C(p, \delta, r)\|g\|_{L^p(\mu_\delta)},$$

for some $C(p, \delta, r) > 0$ *depending only on* p, δ, *and* r. *Moreover,* f_δ *is a weak solution to*

$$\mathcal{A}_\delta(f_\delta) = g,$$

where, for all $f \in \mathcal{C}_c^\infty(\mathbb{R})$ *and all* $x \in \mathbb{R} \setminus \{0\}$,

$$\mathcal{A}_\delta(f)(x) = -f'(x) + \delta|x|^{\delta-1}\operatorname{sign}(x) f(x). \tag{2.53}$$

Proof First, applying Hölder's inequality, for all $x \in \mathbb{R}$,

$$|f_\delta(x)| \leq \frac{1}{p_\delta(x)}\left(\int_x^{+\infty} p_\delta(y)dy\right)^{\frac{1}{q}} \|g\|_{L^p(\mu_\delta)} = G_\delta(x)\|g\|_{L^p(\mu_\delta)}, \tag{2.54}$$

with $G_\delta(x) = \frac{1}{p_\delta(x)}\left(\int_x^{+\infty} p_\delta(y)dy\right)^{\frac{1}{q}}$, for all $x \in \mathbb{R}$. Moreover, by a change of variables, for all $x > 0$,

$$\int_x^{+\infty} p_\delta(y)dy = \frac{1}{\delta}\int_{x^\delta}^{+\infty} z^{\frac{1}{\delta}-1}e^{-z}dz = \frac{1}{\delta}\Gamma\left(\frac{1}{\delta}, x^\delta\right),$$

where $\Gamma\left(\frac{1}{\delta}, x\right)$ is the incomplete gamma function at $x > 0$. Now,

$$\Gamma\left(\frac{1}{\delta}, x\right) \underset{x\to+\infty}{\sim} x^{\frac{1}{\delta}-1}e^{-x} \Rightarrow \int_x^{+\infty} p_\delta(y)dy \underset{x\to+\infty}{\sim} \frac{1}{\delta}x^{1-\delta}e^{-x^\delta},$$

and similarly as $x \to -\infty$ for the integral $\int_{-\infty}^x p_\delta(y)dy$ (recall that μ_δ is symmetric). Now,

$$\begin{aligned}\int_0^{+\infty} |f_\delta(x)|^r \mu_\delta(dx) &\leq \|g\|_{L^p(\mu_\delta)}^r \left(\int_0^{+\infty} p_\delta(x)(G_\delta(x))^r dx\right), \\ &\leq C_{1,p,\delta,r}\|g\|_{L^p(\mu_\delta)}^r,\end{aligned}$$

with

$$C_{1,p,\delta,r} := \int_0^{+\infty} p_\delta(x)(G_\delta(x))^r dx < +\infty$$

since $r/(r-1) > q$. A similar estimate holds true for the integral of $|f_\delta|^r$ on $(-\infty, 0)$, thanks to the second integral representation of f_δ. Finally, for all $\psi \in \mathcal{C}_c^\infty(\mathbb{R})$,

$$\int_{\mathbb{R}} \mathcal{A}_\delta(f_\delta)(x)\psi(x)p_\delta(x)dx = \int_{\mathbb{R}} f_\delta(x)\psi'(x)p_\delta(x)dx = \int_{\mathbb{R}} g(x)\psi(x)p_\delta(x)dx.$$

The conclusion follows. □

The next proposition investigates the properties of the unique primitive function F_δ of f_δ such that $\int_{\mathbb{R}} F_\delta(x)p_\delta(x)dx = 0$.

Proposition 2.3 *Let $\delta \in (0, 1)$, let $p \in [2, +\infty)$, let $g \in L^p(\mu_\delta)$ with $\mu_\delta(g) = 0$, and let f_δ be given by* (2.52). *Let F_δ be defined, for all $x \in \mathbb{R}$, by*

$$F_\delta(x) = F_\delta(0) + \mathbb{1}_{(0,+\infty)}(x)\int_0^x f_\delta(y)dy - \mathbb{1}_{(-\infty,0)}(x)\int_x^0 f_\delta(y)dy, \tag{2.55}$$

with

$$F_\delta(0) = \int_{-\infty}^0 \left(\int_x^0 f_\delta(y)dy\right)p_\delta(x)dx - \int_0^{+\infty}\left(\int_0^x f_\delta(y)dy\right)p_\delta(x)dx.$$

Then, for all $r \in [1, +\infty)$ such that $r/(r-1) > q = p/(p-1)$,

$$\|F_\delta\|_{L^r(\mu_\delta)} \le C_2(p, \delta, r)\|g\|_{L^p(\mu_\delta)}, \tag{2.56}$$

for some $C_2(p, \delta, r) > 0$ depending only on p, δ, and r. Moreover, F_δ is a weak solution to

$$(-\mathcal{L}_\delta)(F_\delta) = g,$$

where, for all $f \in \mathcal{C}_c^\infty(\mathbb{R})$ and all $x \in \mathbb{R}\setminus\{0\}$,

$$\mathcal{L}_\delta(f)(x) = f''(x) - \delta|x|^{\delta-1}\operatorname{sgn}(x)f'(x). \tag{2.57}$$

Proof Let δ, p, and r be as in the statement of the lemma. Then by convexity and Fubini's theorem,

$$\begin{aligned}\int_0^{+\infty}\left|\int_0^x f_\delta(y)dy\right|^r p_\delta(x)dx &\le \int_0^{+\infty} x^{r-1}\left(\int_0^x |f_\delta(y)|^r dy\right)p_\delta(x)dx,\\ &\le \int_0^{+\infty}|f_\delta(y)|^r\left(\int_y^{+\infty} x^{r-1}p_\delta(x)dx\right)dy.\end{aligned}$$

Using (2.54),

$$\int_0^{+\infty}\left|\int_0^x f_\delta(y)dy\right|^r p_\delta(x)dx \leq \|g\|^r_{L^p(\mu_\delta)}\int_0^{+\infty}(G_\delta(y))^r\left(\int_y^{+\infty}x^{r-1}p_\delta(x)dx\right)dy.$$

Now, since $r/(r-1) > q = p/(p-1)$,

$$\int_0^{+\infty}(G_\delta(y))^r\left(\int_y^{+\infty}x^{r-1}p_\delta(x)dx\right)dy < +\infty.$$

A similar analysis can be performed for the integral of $|\int_x^0 f_\delta(y)dy|^r$ over $(-\infty, 0)$. This proves the $L^p - L^r$ estimate (2.56). Noticing that $(-\mathcal{L}_\delta)(F_\delta) = \mathcal{A}_\delta(f_\delta)$, the end of the proof follows. □

To finish this section, let us discuss the higher dimensional situations, namely $d \geq 2$. As above, the second order differential operator under consideration is given, for all $f \in \mathcal{C}_c^\infty(\mathbb{R}^d)$ and all $x \in \mathbb{R}^d \setminus \{0\}$, by

$$\mathcal{L}_{d,\delta}(f)(x) = \Delta(f)(x) - \delta\|x\|^{\delta-2}\langle x; \nabla(f)(x)\rangle.$$

Once again, integration by parts ensures that the operator is symmetric on $\mathcal{C}_c^\infty(\mathbb{R}^d)$ with

$$\int_{\mathbb{R}^d}(-\mathcal{L}_{d,\delta})(f)(x)g(x)\mu_{d,\delta}(dx) = \int_{\mathbb{R}^d}\langle\nabla(f)(x); \nabla(g)(x)\rangle\mu_{d,\delta}(dx),$$

and thus closable. This operator is essentially self-adjoint as soon as the logarithmic derivative of the Lesbegue density of the probability measure $\mu_{d,\delta}$ belongs to the Lesbegue space $L^4(\mu_{d,\delta})$, which is the case when $\delta \in (1-d/4, 1)$ (see [53]). Note that if $d \geq 4$, this is true for all $\delta \in (0, 1)$. Next, let φ_δ be the function defined, for all $t \in (0, +\infty)$, by

$$\varphi_\delta(t) = \exp\left(-t^{\frac{\delta}{2}}\right)$$

so that $p_{d,\delta}(x) = C_{d,\delta}\varphi_\delta(\|x\|^2)$, for all $x \in \mathbb{R}^d$, and let us consider the $\mathbb{R}^d$-valued function $\tau_\delta = (\tau_{\delta,1}, \ldots, \tau_{\delta,d})$ defined, for all $k \in \{1, \ldots, d\}$ and all $x \in \mathbb{R}^d$, by

$$\tau_{\delta,k}(x) = \frac{1}{2\varphi_\delta(\|x\|^2)}\int_{\|x\|^2}^{+\infty}\varphi_\delta(t)dt.$$

Then by a straightforward integration by parts, for all $\psi \in \mathcal{C}_c^\infty(\mathbb{R}^d)$ and all $k \in \{1, \ldots, d\}$,

$$\int_{\mathbb{R}^d}\tau_{k,\delta}(x)\partial_k(\psi)(x)\mu_{d,\delta}(dx) = \int_{\mathbb{R}^d}x_k\psi(x)\mu_{d,\delta}(dx).$$

Moreover, it is clear that $\|\tau_{\delta,k}\|_{L^2(\mu_{d,\delta})} < +\infty$, for all $k \in \{1, \ldots, d\}$. More generally, let $g = (g_1, \ldots, g_d) \in L^2(\mu_{d,\delta})$ with $\int_{\mathbb{R}^d} g_k(x)\mu_{d,\delta}(dx) = 0$, for all $k \in \{1, \ldots, d\}$, and let us study the following weak formulation problem:

$$\mathcal{L}_{d,\delta}(f) = g, \quad \int_{\mathbb{R}^d} f(x)\mu_{d,\delta}(dx) = 0.$$

A first partial answer to the previous weak formulation problem is through the use of semigroup techniques combined with weak Poincaré-type inequality. From [64, Example 1.4, c)], the semigroup, $(P_t^\delta)_{t\geq 0}$, generated by the self-adjoint extension of $\mathcal{L}_{d,\delta}$ satisfies the following estimate: for all $g \in L^\infty(\mu_{d,\delta})$ with $\int_{\mathbb{R}^d} g(x)\mu_{d,\delta}(dx) = 0$ and all $t \geq 0$,

$$\|P_t^\delta(g)\|_{L^2(\mu_{d,\delta})} \leq c_1 \|g\|_{L^\infty(\mu_{d,\delta})} \exp\left(-c_2 t^{\frac{\delta}{4-3\delta}}\right),$$

for some $c_1, c_2 > 0$ depending only on d, δ. Thus, setting

$$\tilde{f}_\delta = \int_0^{+\infty} P_t^\delta(g)dt$$

and reasoning as in Theorem 2.1, for all $\psi \in \mathcal{C}_c^\infty(\mathbb{R}^d)$,

$$\int_{\mathbb{R}^d} \langle \nabla(\tilde{f}_\delta)(x); \nabla(\psi)(x)\rangle \mu_{d,\delta}(dx) = \langle g; \psi\rangle_{L^2(\mu_{d,\delta})},$$

namely, $\tilde{f}_\delta$ is a solution to the weak formulation problem with $g \in L^\infty(\mu_{d,\delta})$ such that $\mu_{d,\delta}(g) = 0$ and with

$$\|\tilde{f}_\delta\|_{L^2(\mu_{d,\delta})} \leq C_{d,\delta}\|g\|_{L^\infty(\mu_{d,\delta})}.$$

3 Representation Formulas and L^p-Poincaré Inequalities

Let us start this section with a new result valid for the nondegenerate symmetric α-stable probability measures on $\mathbb{R}^d$, $\alpha \in (1, 2)$.

Proposition 3.1 *Let $d \geq 1$, let $\alpha \in (1, 2)$, and let μ_α be a nondegenerate symmetric α-stable probability measure on $\mathbb{R}^d$. Let $p \in (1, +\infty)$ and let p_1, p_2 be such that $1/p = 1/p_1 + 1/p_2$ with $1 < p_1 < \alpha$. Then, for all f smooth enough on $\mathbb{R}^d$,*

$$\|f - \mu_\alpha(f)\|_{L^p(\mu_\alpha)} \leq \left(\int_0^{+\infty} q_\alpha(t)dt\right) \left(\mathbb{E}\|X\|^{p_1}\right)^{\frac{1}{p_1}} \|\nabla(f)\|_{L^{p_2}(\mu_\alpha)},$$

where $X \sim \mu_\alpha$ and q_α is defined, for all $t > 0$, by

$$q_\alpha(t) = \frac{e^{-t}}{(1-e^{-\alpha t})^{1-\frac{1}{\alpha}}}\left((1-e^{-\alpha t})^{\alpha-1} + e^{-\alpha^2 t+\alpha t}\right)^{\frac{1}{\alpha}}.$$

Proof A straightforward application of the Bismut formula (2.25) (see also [9, Proposition 2.1]) for the action of the operator $D^{\alpha-1}$ on $P_t^{\nu_\alpha}(f)$, $t > 0$ and $f \in \mathcal{S}(\mathbb{R}^d)$, together with the decomposition of the nonlocal part of the generator of the α-stable Ornstein-Uhlenbeck semigroup, implies, for all $f \in \mathcal{S}(\mathbb{R}^d)$ and all $x \in \mathbb{R}^d$, that

$$f(x) - \mathbb{E}f(X) = \int_0^{+\infty}\int_{\mathbb{R}^d}\langle xe^{-t} - \frac{e^{-\alpha t}y}{(1-e^{-\alpha t})^{1-\frac{1}{\alpha}}}; \nabla(f)(xe^{-t} + (1-e^{-\alpha t})^{\frac{1}{\alpha}}y)\rangle \mu_\alpha(dy)dt,$$

where $X \sim \mu_\alpha$. Therefore,

$$|f(x) - \mathbb{E}f(X)| \leq \int_0^{+\infty}\mathbb{E}_Y\left|\langle xe^{-t} - \frac{e^{-\alpha t}Y}{(1-e^{-\alpha t})^{1-\frac{1}{\alpha}}}; \nabla(f)(xe^{-t} + (1-e^{-\alpha t})^{\frac{1}{\alpha}}Y)\rangle\right| dt,$$

with $Y \sim \mu_\alpha$. Thus, by Minkowski's integral inequality and Jensen's inequality,

$$(\mathbb{E}_X|f(X) - \mathbb{E}f(X)|^p)^{\frac{1}{p}} \leq \int_0^{+\infty}\left(\mathbb{E}_{X,Y}\left|\langle Xe^{-t} - \frac{e^{-\alpha t}}{(1-e^{-\alpha t})^{1-\frac{1}{\alpha}}}Y; \nabla(f)\left(Xe^{-t} + (1-e^{-\alpha t})^{\frac{1}{\alpha}}Y\right)\rangle\right|^p\right)^{\frac{1}{p}} dt.$$

Now, thanks to the stability property, under the product measure $\mu_\alpha \otimes \mu_\alpha$, $Xe^{-t} + (1-e^{-\alpha t})^{\frac{1}{\alpha}}Y$ is distributed according to μ_α. Moreover,

$$Xe^{-t} - \frac{e^{-\alpha t}Y}{(1-e^{-\alpha t})^{1-\frac{1}{\alpha}}} =_{\mathcal{L}} \left(e^{-\alpha t} + \frac{e^{-\alpha^2 t}}{(1-e^{-\alpha t})^{\alpha-1}}\right)^{\frac{1}{\alpha}} X.$$

Finally, observe that, for all $t > 0$,

$$\begin{aligned} q_\alpha(t) &= \left(e^{-\alpha t}\frac{(1-e^{-\alpha t})^{\alpha-1} + e^{-\alpha^2 t+\alpha t}}{(1-e^{-\alpha t})^{\alpha-1}}\right)^{\frac{1}{\alpha}} \\ &= \frac{e^{-t}}{(1-e^{-\alpha t})^{1-\frac{1}{\alpha}}}\left((1-e^{-\alpha t})^{\alpha-1} + e^{-\alpha^2 t+\alpha t}\right)^{\frac{1}{\alpha}}. \end{aligned}$$

Let F_1 and F_2 be the two functions defined, for all $x, y \in \mathbb{R}^d$ and all $t > 0$, by

$$F_1(x, y, t) = xe^{-t} - \frac{e^{-\alpha t} y}{(1 - e^{-\alpha t})^{1-\frac{1}{\alpha}}},$$

$$F_2(x, y, t) = \nabla(f)\left(xe^{-t} + (1 - e^{-\alpha t})^{\frac{1}{\alpha}} y\right).$$

Then from the generalized Hölder's inequality, for all $p \in (1, +\infty)$,

$$\|\langle F_1(, t); F_2(, t)\rangle\|_{L^p(\mu_\alpha \otimes \mu_\alpha)} \leq \|F_1(, t)\|_{L^{p_1}(\mu_\alpha \otimes \mu_\alpha)} \|F_2(, t)\|_{L^{p_2}(\mu_\alpha \otimes \mu_\alpha)},$$

where $1/p_1 + 1/p_2 = 1/p$. Take $1 < p_1 < \alpha$. From the previous identities in law, one gets

$$\|f - \mu_\alpha(f)\|_{L^p(\mu_\alpha)} \leq \|X\|_{L^{p_1}(\mu_\alpha)} \|\nabla(f)\|_{L^{p_2}(\mu_\alpha)} \left(\int_0^{+\infty} q_\alpha(t) dt\right).$$

This concludes the proof of the proposition. □

Before moving on, let us briefly comment on the Gaussian situation. Let γ be the standard Gaussian probability measure on $\mathbb{R}^d$. Following lines of reasoning as above, it is not difficult to obtain the corresponding inequality for the standard Gaussian probability measure on $\mathbb{R}^d$. However, a crucial difference with the general symmetric α-stable situation is that, under the product probability measure $\gamma \otimes \gamma$, the Gaussian random vectors given, for all $t > 0$, by

$$Xe^{-t} + \sqrt{1 - e^{-2t}} Y, \quad Xe^{-t} - \frac{e^{-2t} Y}{\sqrt{1 - e^{-2t}}},$$

where $(X, Y) \sim \gamma \otimes \gamma$, are independent of each other and are equal in law to a standard $\mathbb{R}^d$-valued Gaussian random vector (up to some constant depending on t for the second one). Finally, conditioning, one gets the following classical dimension-free inequality, which is a particular case of a result of Pisier (see, e.g., [59, Theorem 2.2]): for all f smooth enough on $\mathbb{R}^d$ and all $p \in (1, +\infty)$,

$$\|f - \gamma(f)\|_{L^p(\gamma)} \leq \frac{\pi}{2} \left(\mathbb{E}|Z|^p\right)^{\frac{1}{p}} \|\nabla(f)\|_{L^p(\gamma)}, \tag{3.1}$$

where $Z \sim \mathcal{N}(0, 1)$. Note that the constant in (3.1) is not optimal since for $p = 2$, the best constant is known to be equal to 1 and for large p, it is of the order $\sqrt{p}$. Note also that the previous lines of reasoning continue to hold in the vector-valued setting (namely, when f and g are vector valued in a general Banach space). Finally, a different estimate has been obtained in [18, Theorem 7.1 and Remark 7.2], which is linked to the isoperimetric constant and to the product structure of the standard Gaussian probability measure on $\mathbb{R}^d$. Note that the dependency on p in [18, inequality 7.5] is of the order p. Moreover, in [56, Proposition 3.1], the following fine version of the L^p-Poincaré inequality on the Wiener space is proved: for all *even integers* $p \geq 2$ and all $F \in \mathbb{D}^{1,p}$ such that $\mathbb{E}F = 0$,

$$\left(\mathbb{E}|F|^p\right)^{\frac{1}{p}} \leq (p-1)^{\frac{1}{2}} \left(\mathbb{E}\|DF\|^p_{\mathcal{H}}\right)^{\frac{1}{p}},$$

where DF is the Malliavin derivative of F, $\mathcal{H}$ is a real separable Hilbert space on which the isonormal Gaussian process is defined, and $\mathbb{D}^{1,p}$ is the L^p-Sobolev-Watanabe-Kree space of order 1. Finally, recently, it has been proved in [2, Theorem 2.6] that, for all $p \geq 2$ and all $F \in \mathbb{D}^{1,p}$ such that $\mathbb{E}F = 0$,

$$\left(\mathbb{E}|F|^p\right)^{\frac{1}{p}} \leq (p-1)^{\frac{1}{2}} \left(\mathbb{E}\|DF\|^p_{\mathcal{H}}\right)^{\frac{1}{p}}.$$

As shown next, with an argument based on the covariance identity (2.26), it is possible to easily retrieve, for $p \geq 2$, such estimates (see also the discussion in [67, pages 1806–1808]).

Proposition 3.2 *Let $d \geq 1$, let γ be the standard Gaussian probability measure on $\mathbb{R}^d$ and let $p \in [2, +\infty)$. Then, for all $f \in \mathcal{S}(\mathbb{R}^d)$ such that $\int_{\mathbb{R}^d} f(x)\gamma(dx) = 0$,*

$$\|f\|_{L^p(\gamma)} \leq \sqrt{p-1}\|\nabla(f)\|_{L^p(\gamma)}. \tag{3.2}$$

Proof From (2.26), for all f, g smooth enough and real valued with $\int_{\mathbb{R}^d} f(x)$ $\gamma(dx) = 0$,

$$\mathbb{E}f(X)g(X) = \int_0^1 \mathbb{E}\langle\nabla(f)(X_z); \nabla(g)(Y_z)\rangle dz,$$

where $X_z =_{\mathcal{L}} Y_z =_{\mathcal{L}} X \sim \gamma$, for all $z \in [0, 1]$. Next, let $p \geq 2$ and take $g = \Phi'_p(f)/p$ where $\Phi_p(x) = |x|^p$, for $x \in \mathbb{R}$. Then since Φ_p is twice continuously differentiable on $\mathbb{R}$,

$$\nabla(g)(x) = \frac{1}{p}\nabla(f)(x)\Phi''_p(f(x)) = (p-1)\nabla(f)(x)|f(x)|^{p-2}, \quad x \in \mathbb{R}^d.$$

Thus, for all f smooth enough with mean 0 with respect to the Gaussian measure γ,

$$\mathbb{E}|f(X)|^p = (p-1)\int_0^1 \mathbb{E}|f(Y_z)|^{p-2}\langle\nabla(f)(X_z); \nabla(f)(Y_z)\rangle dz.$$

Using Hölder's inequality with $r = p/(p-2)$ and $r^* = p/2$ as well as the Cauchy-Schwarz inequality,

$$|\mathbb{E}|f(Y_z)|^{p-2}\langle\nabla(f)(X_z);\nabla(f)(Y_z)\rangle| \leq \left(\mathbb{E}\,|f(X)|^p\right)^{1-\frac{2}{p}}\left(\mathbb{E}\|\nabla(f)(X_z)\|^{\frac{p}{2}}\|\nabla(f)(Y_z)\|^{\frac{p}{2}}\right)^{\frac{2}{p}},$$
$$\leq \left(\mathbb{E}\,|f(X)|^p\right)^{1-\frac{2}{p}}\left(\mathbb{E}\|\nabla(f)(X)\|^p\right)^{\frac{2}{p}}.$$

Assuming that $f \neq 0$, the rest of the proof easily follows. □

Remark 3.1 From the covariance representation (2.26) in the general case, it is possible to obtain a version of the L^p-Poincaré inequality for the centered Gaussian probability measure with covariance matrix Σ. Namely, for all $p \in [2, +\infty)$ and all $f \in \mathcal{S}(\mathbb{R}^d)$ such that $\int_{\mathbb{R}^d} f(x)\gamma_\Sigma(dx) = 0$,

$$\|f\|_{L^p(\gamma_\Sigma)} \leq \sqrt{p-1}\|\Sigma^{\frac{1}{2}}\nabla(f)\|_{L^p(\gamma_\Sigma)}. \tag{3.3}$$

As a corollary of the previous L^p-Poincaré inequality, let us prove a Sobolev-type inequality with respect to the standard Gaussian measure on $\mathbb{R}^d$.

Corollary 3.1 *Let $d \geq 1$, let γ be the standard Gaussian probability measure on $\mathbb{R}^d$, and let $p \in [2, +\infty)$. Then, for all $f \in \mathcal{S}(\mathbb{R}^d)$ such that $\int_{\mathbb{R}^d} f(x)\gamma(dx) = 0$ and all $\lambda > 0$,*

$$\|f\|_{L^p(\gamma)} \leq \sqrt{p-1}C(\lambda, p)\left(\lambda\|f\|_{L^p(\gamma)} + \|(-\mathcal{L}^\gamma)(f)\|_{L^p(\gamma)}\right),$$

where $C(\lambda, p)$ is given by

$$C(\lambda, p) := \gamma_2(q)\left(\int_0^{+\infty}\frac{e^{-(\lambda+1)t}}{\sqrt{1-e^{-2t}}}dt\right), \qquad \gamma_2(q) = (\mathbb{E}|X|^q)^{\frac{1}{q}},$$

where $X \sim \gamma$ and where $q = p/(p-1)$. In particular,

$$\|f\|_{L^p(\gamma)} \leq \frac{\pi}{2}\sqrt{p-1}\gamma_2(q)\|(-\mathcal{L}^\gamma)(f)\|_{L^p(\gamma)}.$$

Proof The proof is rather straightforward and is a consequence of the Bismut formula for the standard Gaussian measure on $\mathbb{R}^d$. For all $t > 0$, all $f \in \mathcal{S}(\mathbb{R}^d)$ with $\int_{\mathbb{R}^d} f(x)\gamma(dx) = 0$, and all $x \in \mathbb{R}^d$,

$$\nabla P_t^\gamma(f)(x) = \frac{e^{-t}}{\sqrt{1-e^{-2t}}}\int_{\mathbb{R}^d} yf(xe^{-t} + y\sqrt{1-e^{-2t}})\gamma(dy).$$

Now, based on the previous formula, it is clear that

$$\nabla \circ \left(\lambda E - \mathcal{L}^\gamma\right)^{-1}(f)(x) = \int_0^{+\infty} e^{-\lambda t}\nabla P_t^\gamma(f)(x)dt,$$
$$= \int_{\mathbb{R}^d} yI_{2,\lambda}(f)(x, y)\gamma(dy),$$

with

$$I_{2,\lambda}(f)(x,y) = \int_0^{+\infty} \frac{e^{-(1+\lambda)t}}{\sqrt{1-e^{-2t}}} f\left(xe^{-t} + y\sqrt{1-e^{-2t}}\right) dt.$$

Next, by duality and Hölder's inequality,

$$\begin{aligned}\|\nabla \circ \left(\lambda E - \mathcal{L}^\gamma\right)^{-1}(f)(x)\| &= \sup_{z\in\mathbb{R}^d,\ \|z\|=1} \left| \int_{\mathbb{R}^d} \langle z; y\rangle I_{2,\lambda}(f)(x,y)\gamma(dy)\right|, \\ &\leq \gamma_2(q)\left(\int_{\mathbb{R}^d} |I_{2,\lambda}(f)(x,y)|^p \gamma(dy)\right)^{\frac{1}{p}}.\end{aligned}$$

Taking the $L^p(\gamma)$-norm and applying Minkowski's integral inequality give the following:

$$\begin{aligned}\|\nabla \circ \left(\lambda E - \mathcal{L}^\gamma\right)^{-1}(f)\|_{L^p(\gamma)} &\leq \gamma_2(q)\|I_{2,\lambda}(f)\|_{L^p(\gamma\otimes\gamma)}, \\ &\leq \gamma_2(q)\left(\int_0^{+\infty} \frac{e^{-(\lambda+1)t}}{\sqrt{1-e^{-2t}}}dt\right)\|f\|_{L^p(\gamma)}.\end{aligned}$$

Thus, for all $f \in \mathcal{S}(\mathbb{R}^d)$ such that $\int_{\mathbb{R}^d} f(x)\gamma(dx) = 0$,

$$\begin{aligned}\|f\|_{L^p(\gamma)} \leq \sqrt{p-1}\|\nabla(f)\|_{L^p(\gamma)} &\leq \sqrt{p-1}\gamma_2(q)\left(\int_0^{+\infty} \frac{e^{-(1+\lambda)t}}{\sqrt{1-e^{-2t}}}dt\right)\|(\lambda E - \mathcal{L}^\gamma)(f)\|_{L^p(\gamma)}, \\ &\leq \sqrt{p-1}\gamma_2(q)\left(\int_0^{+\infty} \frac{e^{-(1+\lambda)t}}{\sqrt{1-e^{-2t}}}dt\right)\left(\lambda\|f\|_{L^p(\gamma)} + \|(-\mathcal{L}^\gamma)(f)\|_{L^p(\gamma)}\right).\end{aligned}$$

The conclusion easily follows since

$$\int_0^{+\infty} \frac{e^{-t}}{\sqrt{1-e^{-2t}}}dt = \frac{\pi}{2}.$$

□

Let us return to the nondegenerate symmetric α-stable case with $\alpha \in (1,2)$. Based on the following decomposition of the nonlocal part of the generator of the stable Ornstein-Uhlenbeck semigroup (and on Bismut-type formulas),

$$\mathcal{L}^\alpha(f)(x) = -\langle x; \nabla(f)(x)\rangle + \sum_{j=1}^d \partial_j D_j^{\alpha-1}(f)(x), \tag{3.4}$$

L^p-Poincaré-type inequalities for the symmetric nondegenerate α-stable probability measures on $\mathbb{R}^d$ with $\alpha \in (1,2)$ and with $p \in [2,+\infty)$ are discussed.

First, let us provide an analytic formula for the dual semigroup $((P_t^{\nu_\alpha})^*)_{t\geq 0}$ of the α-stable Ornstein-Uhlenbeck semigroup. This representation follows from (2.16). Recall that p_α, the Lebesgue density of a nondegenerate α-stable probability measure with $\alpha \in (1,2)$, is positive on $\mathbb{R}^d$ (see, e.g., [72, Lemma 2.1]). This result appears to be new.

Lemma 3.1 *Let $d \geq 1$, let $\alpha \in (1,2)$, let μ_α be a nondegenerate symmetric α-stable probability measure on $\mathbb{R}^d$, and let p_α be its Lebesgue density. Then, for all $g \in \mathcal{S}(\mathbb{R}^d)$, all $t > 0$, and all $x \in \mathbb{R}^d$,*

$$\begin{aligned}(P_t^{\nu_\alpha})^*(g)(x) &= \frac{1}{(1-e^{-\alpha t})^{\frac{d}{\alpha}}} \int_{\mathbb{R}^d} g(u) p_\alpha(u) p_\alpha\left(\frac{x - ue^{-t}}{(1-e^{-\alpha t})^{\frac{1}{\alpha}}}\right) \frac{du}{p_\alpha(x)}, \\ &= \frac{e^{td}}{(1-e^{-\alpha t})^{\frac{d}{\alpha}}} \int_{\mathbb{R}^d} g(e^t x + e^t z) \frac{p_\alpha(xe^t + ze^t)}{p_\alpha(x)} p_\alpha\left(\frac{z}{(1-e^{-\alpha t})^{\frac{1}{\alpha}}}\right) dz.\end{aligned} \tag{3.5}$$

Proof Let $f, g \in \mathcal{S}(\mathbb{R}^d)$ and let $t > 0$. Then

$$\int_{\mathbb{R}^d} P_t^{\nu_\alpha}(f)(x) g(x) p_\alpha(x) dx = \int_{\mathbb{R}^{2d}} f\left(xe^{-t} + (1-e^{-\alpha t})^{\frac{1}{\alpha}} y\right) p_\alpha(y) g(x) p_\alpha(x) dx dy.$$

Now, let us perform several changes of variables: first, change y into $z/(1-e^{-\alpha t})^{\frac{1}{\alpha}}$, then x into $e^t u$, and, finally, $(u+z, u)$ into (x, y). Then

$$\int_{\mathbb{R}^d} P_t^{\nu_\alpha}(f)(x) g(x) p_\alpha(x) dx = \int_{\mathbb{R}^{2d}} f(x) g(e^t y) p_\alpha(ye^t) p_\alpha\left(\frac{x-y}{(1-e^{-\alpha t})^{\frac{1}{\alpha}}}\right) \frac{e^{td} dx dy}{(1-e^{-\alpha t})^{\frac{d}{\alpha}}}.$$

This concludes the proof of the lemma. □

Note that this representation generalizes completely the case $\alpha = 2$ for which $(P_t^\gamma)^* = P_t^\gamma$, for all $t \geq 0$. Also, note that the previous representation ensures that, for all $g \in \mathcal{S}(\mathbb{R}^d)$ and all $t > 0$,

$$\int_{\mathbb{R}^d} (P_t^{\nu_\alpha})^*(g)(x) \mu_\alpha(dx) = \int_{\mathbb{R}^d} g(x) \mu_\alpha(dx),$$

which can be seen using a duality argument and the fact that $P_t^{\nu_\alpha}$, $t > 0$, is mass conservative. Based on (3.5), let us give a specific representation of the dual semigroup as the composition of three elementary operators. For this purpose, denote by M_α the multiplication operator by the stable density p_α. Namely, for all $g \in \mathcal{S}(\mathbb{R}^d)$ and all $x \in \mathbb{R}^d$,

$$M_\alpha(g)(x) = g(x)p_\alpha(x).$$

The inverse of M_α corresponds to multiplication by $1/p_\alpha$. Now, denote by $(T_t^\alpha)_{t\geq 0}$ the continuous family of operators, defined, for all $g \in \mathcal{S}(\mathbb{R}^d)$, all $x \in \mathbb{R}^d$, and all $t > 0$, by

$$T_t^\alpha(g)(x) = \int_{\mathbb{R}^d} g(u)p_\alpha\left(\frac{x - ue^{-t}}{(1-e^{-\alpha t})^{\frac{1}{\alpha}}}\right)\frac{du}{(1-e^{-\alpha t})^{\frac{d}{\alpha}}},$$

with the convention that $T_0^\alpha(g) = g$. For fixed $t > 0$, the previous operator admits a representation that is close in spirit to the Mehler representation of the semigroup $(P_t^{\nu_\alpha})_{t\geq 0}$: for all $g \in \mathcal{S}(\mathbb{R}^d)$, all $x \in \mathbb{R}^d$, and all $t \geq 0$,

$$T_t^\alpha(g)(x) = e^{td}\int_{\mathbb{R}^d} g\left(e^t x + (1-e^{-\alpha t})^{\frac{1}{\alpha}}e^t z\right)\mu_\alpha(dz).$$

Moreover, from Fourier inversion, for all $g \in \mathcal{S}(\mathbb{R}^d)$, all $x \in \mathbb{R}^d$, and all $t \geq 0$,

$$T_t^\alpha(g)(x) = e^{td}\int_{\mathbb{R}^d} \mathcal{F}(g)(\xi)e^{i\langle \xi; xe^t\rangle}\frac{\varphi_\alpha(e^t\xi)}{\varphi_\alpha(\xi)}\frac{d\xi}{(2\pi)^d}. \tag{3.6}$$

In particular, the Fourier transform of $T_t^\alpha(g)$ is given, for all $\xi \in \mathbb{R}^d$ and all $t \geq 0$, by

$$\mathcal{F}(T_t^\alpha(g))(\xi) = \mathcal{F}(g)(e^{-t}\xi)\frac{\varphi_\alpha(\xi)}{\varphi_\alpha(e^{-t}\xi)}.$$

Then, thanks to Lemma 3.1, for all $g \in \mathcal{S}(\mathbb{R}^d)$ and all $t > 0$,

$$(P_t^{\nu_\alpha})^*(g) = ((M_\alpha)^{-1} \circ T_t^\alpha \circ M_\alpha)(g). \tag{3.7}$$

The semigroup of operators $((P_t^{\nu_\alpha})^*)_{t\geq 0}$ is the h-transform of the semigroup $(T_t^\alpha)_{t\geq 0}$ by the positive function p_α (see, e.g., [13, Section 1.15.8]), which is harmonic for the generator of $(T_t^\alpha)_{t\geq 0}$. The next technical lemma gathers standard properties of the continuous family of operators $(T_t^\alpha)_{t\geq 0}$. First, define the following bilinear form, which appears as a remainder in the product rule for the nonlocal operator $D^{\alpha-1}$: for all $f, g \in \mathcal{S}(\mathbb{R}^d)$ and all $x \in \mathbb{R}^d$,

$$R^\alpha(f,g)(x) = \int_{\mathbb{R}^d} (f(x+u) - f(x))(g(x+u) - g(x))u\nu_\alpha(du). \tag{3.8}$$

In particular, this remainder term is null when $\alpha = 2$ since the classical product rule holds in this diffusive situation. Finally, for all $x \in \mathbb{R}^d$,

$$D^{\alpha-1}(p_\alpha)(x) = -xp_\alpha(x), \quad (D^{\alpha-1})^*(p_\alpha)(x) = xp_\alpha(x). \tag{3.9}$$

The next lemma states and proves many rather elementary properties of the family of operators $(T_t^\alpha)_{t\geq 0}$.

Lemma 3.2 *For all $f \in \mathcal{C}_b(\mathbb{R}^d)$ and all $s, t \geq 0$,*

$$T_{s+t}^\alpha(f) = (T_t^\alpha \circ T_s^\alpha)(f) = (T_s^\alpha \circ T_t^\alpha)(f).$$

For all $f \in \mathcal{S}(\mathbb{R}^d)$,

$$\lim_{t\to+\infty} T_t^\alpha(f)(x) = M_\alpha\left(\int_{\mathbb{R}^d} f(x)dx\right), \quad \lim_{t\to 0^+} T_t^\alpha(f)(x) = f(x).$$

For all $f \in \mathcal{C}_b(\mathbb{R}^d)$ and all $t \geq 0$,

$$\int_{\mathbb{R}^d} T_t^\alpha(p_\alpha f)(x)dx = \int_{\mathbb{R}^d} p_\alpha(x)f(x)dx.$$

For all $f \in \mathcal{C}_b(\mathbb{R}^d)$ with $f \geq 0$ and all $t \geq 0$,

$$T_t^\alpha(f) \geq 0, \quad T_t^\alpha(1) = e^{td}, \quad T_t^\alpha(p_\alpha) = p_\alpha.$$

For all $f, g \in \mathcal{S}(\mathbb{R}^d)$ and all $t \geq 0$,

$$\int_{\mathbb{R}^d} T_t^\alpha(f)(x)g(x)dx = \int_{\mathbb{R}^d} f(x)P_t^{\nu_\alpha}(g)(x)dx.$$

Namely, for all $t > 0$, the dual operator of T_t^α in standard Lebesgue spaces is given, for all $f \in \mathcal{S}(\mathbb{R}^d)$, by

$$(T_t^\alpha)^*(f) = P_t^{\nu_\alpha}(f).$$

The generator A_α of $(T_t^\alpha)_{t\geq 0}$ is given, for all $f \in \mathcal{S}(\mathbb{R}^d)$ and all $x \in \mathbb{R}^d$, by

$$A_\alpha(f)(x) = df(x) + \langle x; \nabla(f)(x)\rangle + \int_{\mathbb{R}^d} \langle \nabla(f)(x+u) - \nabla(f)(x); u\rangle \nu_\alpha(du).$$

For all $x \in \mathbb{R}^d$,

$$A_\alpha(p_\alpha)(x) = 0.$$

The "carré du champs operator" associated with A_α is given, for all $f \in \mathcal{S}(\mathbb{R}^d)$ and all $x \in \mathbb{R}^d$, by

$$\Gamma_\alpha(f,g)(x) = -\frac{df(x)g(x)}{2} + \frac{\alpha}{2}\int_{\mathbb{R}^d}(f(x+u)-f(x))(g(x+u)-g(x))\nu_\alpha(du).$$

For all $f \in \mathcal{S}(\mathbb{R}^d)$ and all $x \in \mathbb{R}^d$,

$$A_\alpha(p_\alpha f)(x) = \sum_{k=1}^{d}\left(R_k^\alpha(\partial_k(p_\alpha), f)(x) + R_k^\alpha(\partial_k(f), p_\alpha)(x)\right)$$
$$+\sum_{k=1}^{d}\left(\partial_k(p_\alpha)(x)D_k^{\alpha-1}(f)(x) + p_\alpha(x)\partial_k D_k^{\alpha-1}(f)(x)\right). \tag{3.10}$$

For all $g \in \mathcal{C}_b^1(\mathbb{R}^d)$, all $u \in \mathbb{R}^d$, all $x \in \mathbb{R}^d$ and all $t \geq 0$,

$$\Delta_u\left(T_t^\alpha(g)\right)(x) = e^{td}\int_{\mathbb{R}^d}\Delta_{ue^t}(g)(xe^t + (1-e^{-\alpha t})^{\frac{1}{\alpha}}e^t z)\mu_\alpha(dz),$$

and,

$$\nabla_{\nu_\alpha}(T_t^\alpha(g))(x) \leq e^{\frac{\alpha t}{2}}T_t^\alpha\left(\nabla_{\nu_\alpha}(g)\right)(x).$$

In particular, for all $f \in \mathcal{S}(\mathbb{R}^d)$ and all $x \in \mathbb{R}^d$,

$$(\mathcal{L}^\alpha)^*(f)(x) = ((M_\alpha)^{-1}\circ A_\alpha \circ M_\alpha)(f)(x),$$
$$= \frac{1}{p_\alpha(x)}\sum_{k=1}^{d}\left(R_k^\alpha(\partial_k(p_\alpha), f)(x) + R_k^\alpha(\partial_k(f), p_\alpha)(x)\right)$$
$$+\sum_{k=1}^{d}\left(\frac{\partial_k(p_\alpha)(x)}{p_\alpha(x)}D_k^{\alpha-1}(f)(x) + \partial_k D_k^{\alpha-1}(f)(x)\right). \tag{3.11}$$

Finally, the "carré du champs" operator associated with the generator $(\mathcal{L}^\alpha)^*$ is given, for all $f, g \in \mathcal{S}(\mathbb{R}^d)$ and all $x \in \mathbb{R}^d$, by

$$\Gamma^*(f,g)(x) = \frac{\alpha}{2}\int_{\mathbb{R}^d}(f(x+u)-f(x))(g(x+u)-g(x))\nu_\alpha(du)$$
$$+\frac{1}{2}\sum_{k=1}^{d}\frac{\partial_k(p_\alpha)(x)}{p_\alpha(x)}R_k^\alpha(f,g)(x)$$
$$+\frac{1}{2p_\alpha(x)}\sum_{k=1}^{d}\left(\partial_k R_k^\alpha(p_\alpha, fg)(x) - g(x)\partial_k R_k^\alpha(p_\alpha, f)(x)\right.$$
$$\left.-f(x)\partial_k R_k^\alpha(p_\alpha, g)(x)\right).$$

Proof The proof is very classical and based on a characteristic function methodology and on the Fourier representation (3.6). The only non-trivial identity is given by (3.10). So, for all $f \in \mathcal{S}(\mathbb{R}^d)$ and all $x \in \mathbb{R}^d$,

$$A_\alpha(p_\alpha f)(x) = dp_\alpha(x)f(x) + \langle x; p_\alpha(x)\nabla(f)(x)\rangle + \langle x; f(x)\nabla(p_\alpha)(x)\rangle + \sum_{k=1}^{d} \partial_k D_k^{\alpha-1}(p_\alpha f)(x).$$

Moreover,

$$D_k^{\alpha-1}(p_\alpha f)(x) = p_\alpha(x)D_k^{\alpha-1}(f)(x) + f(x)D_k^{\alpha-1}(p_\alpha)(x) + R_k^\alpha(p_\alpha, f)(x).$$

Thus,

$$\partial_k D_k^{\alpha-1}(p_\alpha f)(x) = A + B + C,$$

where,

$$A = \partial_k R_k^\alpha(p_\alpha, f)(x), \quad B = \partial_k\left(p_\alpha(x)D_k^{\alpha-1}(f)\right)(x), \quad C = \partial_k\left(f(x)D_k^{\alpha-1}(p_\alpha)\right)(x).$$

Now, using the classical product rule,

$$A = R_k^\alpha\left(\partial_k(p_\alpha), f\right)(x) + R_k^\alpha\left(\partial_k(f), p_\alpha\right)(x),$$

and,

$$B = \partial_k(p_\alpha)(x)D_k^{\alpha-1}(f)(x) + p_\alpha(x)\partial_k D_k^{\alpha-1}(f)(x).$$

Finally, using (3.9),

$$C = -x_k p_\alpha(x)\partial_k(f)(x) + f(x)\left(-p_\alpha(x) - x_k\partial_k(p_\alpha)(x)\right),$$

and putting everything together concludes the proof of (3.10). □

From the previous lemma and the decomposition (3.7), it is clear, by duality, that the linear operator $(P_t^{\nu_\alpha})^*$ is continuous on every $L^p(\mu_\alpha)$, for $p \in (1, +\infty)$. Indeed, for all $f, g \in \mathcal{S}(\mathbb{R}^d)$ and all $t \geq 0$,

$$\begin{aligned}\langle (P_t^{\nu_\alpha})^*(g); f\rangle_{L^2(\mu_\alpha)} &= \langle T_t^\alpha(M_\alpha(g)); f\rangle_{L^2(\mathbb{R}^d, dx)} \\ &= \langle M_\alpha(g); P_t^{\nu_\alpha}(f)\rangle_{L^2(\mathbb{R}^d, dx)} = \langle g; P_t^{\nu_\alpha}(f)\rangle_{L^2(\mu_\alpha)}.\end{aligned}$$

Moreover, based on the last statements of Lemma 3.2, one can infer the corresponding formulas for the generator of the "carré de Mehler" semigroup on $\mathcal{S}(\mathbb{R}^d)$ and for its corresponding square field operator: for all $f, g \in \mathcal{S}(\mathbb{R}^d)$ and all $x \in \mathbb{R}^d$,

$$\mathcal{L}(f)(x) = \frac{1}{\alpha}\Bigg(- \langle x; \nabla(f)(x)\rangle + \sum_{k=1}^{d} \partial_k D_k^{\alpha-1}(f)(x) + \frac{1}{p_\alpha(x)} \sum_{k=1}^{d} \partial_k R_k^{\alpha}(p_\alpha, f)(x)$$
$$+ \sum_{k=1}^{d} \left(\frac{\partial_k(p_\alpha)(x)}{p_\alpha(x)} D_k^{\alpha-1}(f)(x) + \partial_k D_k^{\alpha-1}(f)(x) \right) \Bigg),$$

and,

$$\tilde{\Gamma}(f,g)(x) = \frac{1}{\alpha}\left(\Gamma(f,g)(x) + \Gamma^*(f,g)(x)\right).$$

Let us now prove two Bismut-type formulas associated with $(P_t^{\nu_\alpha})_{t\geq 0}$ and $((P_t^{\nu_\alpha})^*)_{t\geq 0}$ for integro-differential operators appearing in the generators of the respective semigroups.

Proposition 3.3 *Let $d \geq 1$, let $\alpha \in (1,2)$, let μ_α be a nondegenerate symmetric α-stable probability measure on $\mathbb{R}^d$, and let p_α be its positive Lebesgue density. Then, for all $f \in \mathcal{S}(\mathbb{R}^d)$, all $x \in \mathbb{R}^d$, and all $t > 0$,*

$$\nabla P_t^{\nu_\alpha}(f)(x) = -\frac{e^{-t}}{(1-e^{-\alpha t})^{\frac{1}{\alpha}}} \int_{\mathbb{R}^d} \frac{\nabla(p_\alpha)(y)}{p_\alpha(y)} f(xe^{-t} + (1-e^{-\alpha t})^{\frac{1}{\alpha}} y)\mu_\alpha(dy) \tag{3.12}$$

and

$$D^{\alpha-1}\left((P_t^{\nu_\alpha})^*(f)\right)(x) + \frac{1}{p_\alpha(x)} R^{\alpha}\left(p_\alpha, (P_t^{\nu_\alpha})^*(f)\right)(x) = \frac{-xe^{-\alpha t}}{(1-e^{-\alpha t})}(P_t^{\nu_\alpha})^*(f)(x)$$
$$+ \frac{e^{-t}}{\left(1-e^{-\alpha t}\right)}(P_t^{\nu_\alpha})^*(xf)(x), \tag{3.13}$$

for all $x \in \mathbb{R}^d$.

Proof The identity (3.12) is a direct consequence of the commutation relation and standard integration by parts. Let us prove (3.13). For this purpose, for all $x \in \mathbb{R}^d$ and all $t > 0$ fixed, denote by $F_{\alpha,x,t}$ the function defined, for all $u \in \mathbb{R}^d$, by

$$F_{\alpha,x,t}(u) = \frac{1}{\left(1-e^{-\alpha t}\right)^{\frac{d}{\alpha}}} \frac{p_\alpha(u)}{p_\alpha(x)} p_\alpha\left(\frac{x - ue^{-t}}{(1-e^{-\alpha t})^{\frac{1}{\alpha}}}\right).$$

Note that $F_{\alpha,x,t}$ is a probability density on $\mathbb{R}^d$. Moreover, for all $t > 0$ and all $u \in \mathbb{R}^d$,

$$\int_{\mathbb{R}^d} F_{\alpha,x,t}(u) p_\alpha(x) dx = p_\alpha(u).$$

First, for all $x \in \mathbb{R}^d$, all $u \in \mathbb{R}^d$, and all $t > 0$,

$$\begin{aligned}\Delta_u\left((P_t^{\nu_\alpha})^*(f)\right)(x) &= \int_{\mathbb{R}^d} f(v)(F_{\alpha,x+u,t}(v) - F_{\alpha,x,t}(v))dv,\\ &= \int_{\mathbb{R}^d} f(v)\Delta_u(F_{\alpha,.,t}(v))(x)dv.\end{aligned}$$

Thus,

$$D^{\alpha-1}\left((P_t^{\nu_\alpha})^*(f)\right)(x) = \int_{\mathbb{R}^d} f(v)\left(\int_{\mathbb{R}^d} \Delta_u(F_{\alpha,.,t}(v))(x)u\nu_\alpha(du)\right)dv.$$

Similarly, by linearity,

$$\begin{aligned}\frac{1}{p_\alpha(x)}R^\alpha\left(p_\alpha, (P_t^{\nu_\alpha})^*(f)\right)(x) &= \int_{\mathbb{R}^d} u\nu_\alpha(du)\left(\frac{p_\alpha(x+u)}{p_\alpha(x)} - 1\right)\Delta_u\left((P_t^{\nu_\alpha})^*(f)\right)(x),\\ &= \int_{\mathbb{R}^d} f(v)\left(\int_{\mathbb{R}^d} u\nu_\alpha(du)\left(\frac{p_\alpha(x+u)}{p_\alpha(x)} - 1\right)\Delta_u(F_{\alpha,.,t}(v))(x)\right)dv.\end{aligned}$$

Then, for all $t > 0$ and all $x \in \mathbb{R}^d$,

$$D^{\alpha-1}\left((P_t^{\nu_\alpha})^*(f)\right)(x) + \frac{1}{p_\alpha(x)}R^\alpha\left(p_\alpha, (P_t^{\nu_\alpha})^*(f)\right)(x) = \int_{\mathbb{R}^d} f(v)\left(\int_{\mathbb{R}^d} u\nu_\alpha(du)\frac{p_\alpha(x+u)}{p_\alpha(x)}\Delta_u(F_{\alpha,.,t}(v))(x)\right)dv. \tag{3.14}$$

Let us fix $x, v \in \mathbb{R}^d$ and $t > 0$. Then

$$\begin{aligned}\int_{\mathbb{R}^d} u\nu_\alpha(du)\frac{p_\alpha(x+u)}{p_\alpha(x)}\Delta_u(F_{\alpha,.,t}(v))(x) &= \frac{1}{(1-e^{-\alpha t})^{\frac{d}{\alpha}}}\int_{\mathbb{R}^d} u\nu_\alpha(du)\frac{p_\alpha(x+u)}{p_\alpha(x)}\left(\frac{p_\alpha(v)}{p_\alpha(x+u)}p_\alpha\left(\frac{x+u-ve^{-t}}{(1-e^{-\alpha t})^{\frac{1}{\alpha}}}\right)\right.\\ &\quad\left. - \frac{p_\alpha(v)}{p_\alpha(x)}p_\alpha\left(\frac{x-ve^{-t}}{(1-e^{-\alpha t})^{\frac{1}{\alpha}}}\right)\right),\\ &= \frac{p_\alpha(v)}{p_\alpha(x)^2}\frac{1}{(1-e^{-\alpha t})^{\frac{d}{\alpha}}}\int_{\mathbb{R}^d} u\nu_\alpha(du)\left(p_\alpha(x)p_\alpha\left(\frac{x+u-ve^{-t}}{(1-e^{-\alpha t})^{\frac{1}{\alpha}}}\right)\right.\\ &\quad\left. - p_\alpha(x+u)p_\alpha\left(\frac{x-ve^{-t}}{(1-e^{-\alpha t})^{\frac{1}{\alpha}}}\right)\right),\\ &= \frac{p_\alpha(v)}{p_\alpha(x)^2}\frac{1}{(1-e^{-\alpha t})^{\frac{d}{\alpha}}}\int_{\mathbb{R}^d} u\nu_\alpha(du)p_\alpha(x)\left(p_\alpha\left(\frac{x+u-ve^{-t}}{(1-e^{-\alpha t})^{\frac{1}{\alpha}}}\right)\right.\\ &\quad\left. - p_\alpha\left(\frac{x-ve^{-t}}{(1-e^{-\alpha t})^{\frac{1}{\alpha}}}\right)\right) - \frac{p_\alpha(v)}{p_\alpha(x)^2}\frac{1}{(1-e^{-\alpha t})^{\frac{d}{\alpha}}}\\ &\quad\times\int_{\mathbb{R}^d} u\nu_\alpha(du)(p_\alpha(x+u) - p_\alpha(x))p_\alpha\left(\frac{x-ve^{-t}}{(1-e^{-\alpha t})^{\frac{1}{\alpha}}}\right).\end{aligned}$$

Recalling that, for all $x \in \mathbb{R}^d$,

$$D^{\alpha-1}(p_\alpha)(x) = -xp_\alpha(x).$$

Thus,

$$\frac{p_\alpha(v)}{p_\alpha(x)^2}\frac{1}{(1-e^{-\alpha t})^{\frac{d}{\alpha}}}D^{\alpha-1}(p_\alpha)(x)p_\alpha\left(\frac{x-ve^{-t}}{(1-e^{-\alpha t})^{\frac{1}{\alpha}}}\right) = \frac{p_\alpha(v)}{p_\alpha(x)}\frac{(-x)}{(1-e^{-\alpha t})^{\frac{d}{\alpha}}}p_\alpha\left(\frac{x-ve^{-t}}{(1-e^{-\alpha t})^{\frac{1}{\alpha}}}\right),$$

and from scale invariance,

$$\begin{aligned}\frac{p_\alpha(v)}{p_\alpha(x)}\frac{1}{(1-e^{-\alpha t})^{\frac{d}{\alpha}}}\int_{\mathbb{R}^d}u\nu_\alpha(du)\Delta_{\frac{u}{(1-e^{-\alpha t})^{\frac{1}{\alpha}}}}(p_\alpha)\left(\frac{x-ve^{-t}}{(1-e^{-\alpha t})^{\frac{1}{\alpha}}}\right) &= \frac{p_\alpha(v)}{p_\alpha(x)}\frac{(1-e^{-\alpha t})^{\frac{1}{\alpha}-1}}{(1-e^{-\alpha t})^{\frac{d}{\alpha}}}\\ &\quad\times D^{\alpha-1}(p_\alpha)\left(\frac{x-ve^{-t}}{(1-e^{-\alpha t})^{\frac{1}{\alpha}}}\right),\\ &= -\frac{p_\alpha(v)}{p_\alpha(x)}\frac{(1-e^{-\alpha t})^{-1}}{(1-e^{-\alpha t})^{\frac{d}{\alpha}}}\left(x-ve^{-t}\right)\\ &\quad\times p_\alpha\left(\frac{x-ve^{-t}}{(1-e^{-\alpha t})^{\frac{1}{\alpha}}}\right).\end{aligned}$$

Then using (3.14),

$$\begin{aligned}D^{\alpha-1}\left((P_t^{\nu_\alpha})^*(f)\right)(x) + \frac{1}{p_\alpha(x)}R^\alpha\left(p_\alpha, (P_t^{\nu_\alpha})^*(f)\right)(x) &= \frac{-xe^{-\alpha t}}{(1-e^{-\alpha t})}(P_t^{\nu_\alpha})^*(f)(x)\\ &\quad+\frac{e^{-t}}{\left(1-e^{-\alpha t}\right)}(P_t^{\nu_\alpha})^*(hf)(x),\end{aligned}$$

where $h(v) = v$, for all $v \in \mathbb{R}^d$. This concludes the proof of the proposition. □

Before moving on, let us prove a technical lemma providing a sharp upper bound for the asymptotic behavior of

$$\frac{1}{p_\alpha(x)}R^\alpha(p_\alpha, f)(x),$$

as $\|x\| \to +\infty$, with $f \in C_c^\infty(\mathbb{R}^d)$, and when the associated Lévy measure on $\mathbb{R}^d$ is given by $\nu_\alpha(du) = du/\|u\|^{\alpha+d}$.

Lemma 3.3 *Let $d \geq 1$, let $\alpha \in (1, 2)$, and let $\nu_\alpha(du) = du/\|u\|^{\alpha+d}$. Let R^α be given by* (3.8) *and let p_α be the positive Lebesgue density of the nondegenerate symmetric α-stable probability measure μ_α with Lévy measure ν_α. Then, for all $f \in C_c^\infty(\mathbb{R}^d)$ and all x large enough,*

$$\left\| \frac{1}{p_\alpha(x)} R^\alpha(p_\alpha, f)(x) \right\| \le C\left(1 + \|x\|\right),$$

for some positive constant C depending on α, d, and f.

Proof Without loss of generality, assume that $f \in C_c^\infty(\mathbb{R}^d)$ is a bump function: i.e., $\mathrm{Supp}(f) \subset \mathcal{B}(0,1)$, the Euclidean unit ball of $\mathbb{R}^d$, and $f(x) \in [0,1]$, for all $x \in \mathbb{R}^d$. Then, for all $x \in \mathbb{R}^d$ such that $\|x\| \ge 3$,

$$\begin{aligned}\frac{1}{p_\alpha(x)} R^\alpha(p_\alpha, f)(x) &= \frac{1}{p_\alpha(x)} \int_{\mathbb{R}^d} u\nu_\alpha(du)\left(p_\alpha(x+u) - p_\alpha(x)\right) f(x+u), \\ &= \frac{1}{p_\alpha(x)} \int_{\mathcal{B}(0,1)} (u-x) \frac{du}{\|u-x\|^{\alpha+d}} \left(p_\alpha(u) - p_\alpha(x)\right) f(u).\end{aligned}$$

Thus, since $\|x\| \ge 3$,

$$\left\| \frac{1}{p_\alpha(x)} R^\alpha(p_\alpha, f)(x) \right\| \le \frac{C}{\|x\|^{\alpha+d}} \frac{1}{p_\alpha(x)} \int_{\mathcal{B}(0,1)} \|u-x\| du \, |p_\alpha(u) - p_\alpha(x)| \, |f(u)|.$$

Moreover, for all $x \in \mathbb{R}^d$,

$$\frac{C_1}{(1+\|x\|)^{\alpha+d}} \le p_\alpha(x) \le \frac{C_2}{(1+\|x\|)^{\alpha+d}}, \tag{3.15}$$

for some $C_1, C_2 > 0$ two positive constants depending on α and d. Thus, for all $\|x\| \ge 3$,

$$\begin{aligned}\left\| \frac{1}{p_\alpha(x)} R^\alpha(p_\alpha, f)(x) \right\| &\le C_{\alpha,d,f} \frac{(1+\|x\|)^{\alpha+d}}{\|x\|^{\alpha+d}} \left(1+\|x\|\right), \\ &\le C_{\alpha,d,f}\left(1+\|x\|\right).\end{aligned}$$

This concludes the proof of the lemma. □

Next, let us investigate pseudo-Poincaré inequality (see, e.g., [51] and the references therein) for the dual semigroup $((P_t^{\nu_\alpha})^*_{t\ge 0})$ in $L^p(\mu_\alpha)$, for all $p \in (1,\alpha)$. To start, let $(R^\alpha)^*$ be defined, for all $f, g \in \mathcal{S}(\mathbb{R}^d)$ and all $x \in \mathbb{R}^d$, by

$$(R^\alpha)^*(g, f)(x) = \int_{\mathbb{R}^d} (g(x-u) - g(x))(f(x-u) - f(x)) u\nu_\alpha(du).$$

Proposition 3.4 *Let $d \ge 1$, let $\alpha \in (1,2)$, let μ_α be a nondegenerate symmetric α-stable probability measure on $\mathbb{R}^d$, and let p_α be its positive Lebesgue density. Further, assume that*

$$\left\| \frac{\nabla(p_\alpha)}{p_\alpha} \right\|_{L^p(\mu_\alpha)} < +\infty, \quad p \in (1, +\infty) \tag{3.16}$$

and that, for all $p \in (1, \alpha)$ and all $f \in \mathcal{C}_c^\infty(\mathbb{R}^d)$,

$$\left\| \frac{1}{p_\alpha} (R^\alpha)^*(p_\alpha, f) \right\|_{L^p(\mu_\alpha)} < +\infty. \tag{3.17}$$

Then, for all $p \in (1, \alpha)$, all $f \in \mathcal{C}_c^\infty(\mathbb{R}^d)$, and all $t > 0$,

$$\|(P_t^{\nu_\alpha})^*(f) - f\|_{L^p(\mu_\alpha)} \leq C_\alpha t^{1-\frac{1}{\alpha}} \left\| (D^{\alpha-1})^*(f) + \frac{1}{p_\alpha}(R^\alpha)^*(p_\alpha, f) \right\|_{L^p(\mu_\alpha)} \left\| \frac{\nabla(p_\alpha)}{p_\alpha} \right\|_{L^p(\mu_\alpha)}, \tag{3.18}$$

for some $C_\alpha > 0$ depending only on α.

Proof The argument is based on duality and on (3.12). Let $f, g \in \mathcal{C}_c^\infty(\mathbb{R}^d)$, let $p \in (1, \alpha)$, and let $p^* = p/(p-1)$. Then by standard semigroup arguments,

$$\begin{aligned}
\langle (P_t^{\nu_\alpha})^*(f) - f; g\rangle_{L^2(\mu_\alpha)} &= \int_0^t \langle f; \mathcal{L}^\alpha P_s^{\nu_\alpha}(g)\rangle_{L^2(\mu_\alpha)} ds, \\
&= -\int_0^t \langle xf; \nabla P_s^{\nu_\alpha}(g)\rangle_{L^2(\mu_\alpha)} ds + \int_0^t \langle f; \nabla . D^{\alpha-1} P_s^{\nu_\alpha}(g)\rangle_{L^2(\mu_\alpha)} ds, \\
&= \int_0^t \left\langle -xf + \frac{(D^{\alpha-1})^*(p_\alpha f)}{p_\alpha}; \nabla P_s^{\nu_\alpha}(g) \right\rangle_{L^2(\mu_\alpha)} ds.
\end{aligned}$$

First, thanks to (3.12), for all $s \in (0, t]$ and all $x \in \mathbb{R}^d$,

$$\nabla P_s^{\nu_\alpha}(g)(x) = -\frac{e^{-s}}{\left(1 - e^{-\alpha s}\right)^{\frac{1}{\alpha}}} \int_{\mathbb{R}^d} \frac{\nabla(p_\alpha)(y)}{p_\alpha(y)} g\left(xe^{-s} + (1 - e^{-s\alpha})^{\frac{1}{\alpha}} y\right) \mu_\alpha(dy).$$

Moreover, for all $x \in \mathbb{R}^d$,

$$\frac{(D^{\alpha-1})^*(p_\alpha f)(x)}{p_\alpha(x)} = (D^{\alpha-1})^*(f)(x) + xf(x) + \frac{1}{p_\alpha(x)}(R^\alpha)^*(p_\alpha, f)(x),$$

where

$$(R^\alpha)^*(p_\alpha, f)(x) = \int_{\mathbb{R}^d} (p_\alpha(x-u) - p_\alpha(x))(f(x-u) - f(x)) u \nu_\alpha(du).$$

Thus,

$$\langle (P_t^{\nu_\alpha})^*(f) - f; g\rangle_{L^2(\mu_\alpha)} = \int_0^t \left\langle (D^{\alpha-1})^*(f) + \frac{1}{p_\alpha}(R^\alpha)^*(p_\alpha, f); \nabla P_s^{\nu_\alpha}(g) \right\rangle_{L^2(\mu_\alpha)} ds.$$

Now, for all $s \in (0, t]$,

$$\left\langle (D^{\alpha-1})^*(f) + \frac{1}{p_\alpha}(R^\alpha)^*(p_\alpha, f); \nabla P_s^{\nu_\alpha}(g) \right\rangle_{L^2(\mu_\alpha)} = -\frac{e^{-s}}{\left(1-e^{-\alpha s}\right)^{\frac{1}{\alpha}}} \int_{\mathbb{R}^d} \int_{\mathbb{R}^d} \langle (D^{\alpha-1})^*(f)(x)$$
$$+ \frac{1}{p_\alpha(x)}(R^\alpha)^*(p_\alpha, f)(x); \frac{\nabla(p_\alpha)(y)}{p_\alpha(y)} \rangle$$
$$\times g\left(xe^{-s} + (1-e^{-\alpha s})^{\frac{1}{\alpha}} y\right) \mu_\alpha(dx)\mu_\alpha(dy).$$

By Hölder's inequality,

$$\left| \left\langle (D^{\alpha-1})^*(f) + \frac{1}{p_\alpha}(R^\alpha)^*(p_\alpha, f); \nabla P_s^{\nu_\alpha}(g) \right\rangle_{L^2(\mu_\alpha)} \right| \leq \frac{e^{-s}}{(1-e^{-\alpha s})^{\frac{1}{\alpha}}} \left\| \frac{\nabla(p_\alpha)}{p_\alpha} \right\|_{L^p(\mu_\alpha)} \|g\|_{L^{p^*}(\mu_\alpha)}$$
$$\times \left\| (D^{\alpha-1})^*(f) + \frac{1}{p_\alpha}(R^\alpha)^*(p_\alpha, f) \right\|_{L^p(\mu_\alpha)}.$$

Standard arguments allow to conclude the proof of the proposition. □

The inequality (3.18) is a straightforward generalization of the Gaussian pseudo-Poincaré inequality. Before moving on, let us discuss the condition (3.16). In the rotationally invariant case, recall the following classical pointwise bounds: for all $x \in \mathbb{R}^d$,

$$\frac{C_2}{(1+\|x\|)^{\alpha+d}} \leq p_\alpha^{\mathrm{rot}}(x) \leq \frac{C_1}{(1+\|x\|)^{\alpha+d}}, \tag{3.19}$$

for some C_1, C_2 positive constants. Moreover (see, e.g., [33]), for all $x \in \mathbb{R}^d$,

$$\|\nabla(p_\alpha^{\mathrm{rot}})(x)\| \leq \frac{C_3}{(1+\|x\|)^{\alpha+d+1}},$$

for some positive constant C_3, so that the logarithmic derivative of p_α^{rot} is uniformly bounded on $\mathbb{R}^d$ and so belongs to $L^p(\mu_\alpha)$, for all $p \geq 1$. Another interesting case is when the coordinates are independent and distributed according to the same symmetric α-stable law on $\mathbb{R}$ with $\alpha \in (1, 2)$. It is straightforward to check that, in this case, the logarithmic derivative is uniformly bounded on $\mathbb{R}^d$.

Remark 3.2 Let us end the α-stable case, $\alpha \in (1, 2)$, with a discussion on L^p-Poincaré inequalities, for $p \geq 2$. Classically, by formal semigroup arguments,

$$\|f\|_{L^p(\mu_\alpha)}^p = \mathbb{E}f(X_\alpha)g(X_\alpha) = -\int_0^{+\infty} \mathbb{E}(\mathcal{L}_\alpha)^*(P_t^{\nu_\alpha})^*(f)(X_\alpha)g(X_\alpha)dt,$$

with $f \in C_c^\infty(\mathbb{R}^d)$ such that $\mu_\alpha(f) = 0$, with $p \geq 2$ and with $g(x) = \mathrm{sign}(f(x))|f(x)|^{p-1}$. Moreover, using standard integration by parts and (3.13), for all $t > 0$,

$$\mathbb{E}(\mathcal{L}_\alpha)^*(P_t^{\nu_\alpha})^*(f)(X_\alpha)g(X_\alpha) = -\int_{\mathbb{R}^d}\left\langle \frac{-xe^{-\alpha t}}{(1-e^{-\alpha t})}(P_t^{\nu_\alpha})^*(f)(x) + \frac{e^{-t}}{(1-e^{-\alpha t})}(P_t^{\nu_\alpha})^*(hf)(x); \nabla(g)(x)\right\rangle \mu_\alpha(dx).$$

Now, based on Proposition 3.4 and on the fact that $p \geq 2$, it does not seem possible to reproduce the semigroup proof of the L^p-Poincaré inequality presented in the Gaussian case. Indeed, the bad concentration properties of the α-stable probability measures, with $\alpha \in (1,2)$, as well as the occurrence of the remainder terms R^α and $(R^\alpha)^*$ prohibit the use of Hölder's inequality, followed by the Cauchy-Schwarz inequality.

Very recently, moment estimates for heavy-tailed probability measures on $\mathbb{R}^d$ of Cauchy-type have been obtained in [1, Corollary 4.3.] (see, also (4.2) and (4.3) and the discussion above these) based on weighted Beckner-type inequalities. Note that the right-hand sides of these inequalities put into play weighted norms of the classical gradient operator. Let us observe that it is possible to obtain these weighted Poincaré inequalities from the non-local ones in some cases, as shown in the next proposition.

Proposition 3.5 *Let μ be the standard exponential probability measure on $(0,+\infty)$ and let ν be the associated Lévy measure on $(0,+\infty)$. Then, for all $f \in \mathcal{S}(\mathbb{R})$,*

$$\int_{(0,+\infty)}\int_{(0,+\infty)} |f(x+u)-f(x)|^2\nu(du)\mu(dx) \leq \int_{(0,+\infty)} w|f'(w)|^2\mu(dw).$$

Proof First, by Jensen's inequality,

$$|f(x+u)-f(x)|^2 \leq u^2\int_0^1 |f'(x+tu)|^2dt.$$

Thus, since $\nu(du)/du = e^{-u}/u$, $u>0$,

$$\int_{(0,+\infty)^2} |f(x+u)-f(x)|^2\nu(du)\mu(dx) \leq \int_{(0,+\infty)^2} u\left(\int_0^1 |f'(x+tu)|^2dt\right)e^{-u}e^{-x}dxdu.$$

Now, for all $t \in (0,1)$, let $\mathcal{D}_t = \{(w,z) \in (0,+\infty)^2 : w > tz\}$ and let Φ_t be the $\mathcal{C}^1$-diffeomorphism from $(0,+\infty)^2$ to $\mathcal{D}_t$ defined, for all $(x,u) \in (0,+\infty)^2$, by

$$\Phi_t(x,u) = (x+tu, u).$$

Thus, by the change of variables with Φ_t,

$$\int_{(0,+\infty)^2} |f(x+u)-f(x)|^2 \nu(du)\mu(dx) \leq \int_{(0,+\infty)^2} z\left(\int_0^1 |f'(w)|^2 dt\right) \mathbb{1}_{\mathcal{D}_t}(w,z) e^{-z} e^{-(w-tz)} dw dz,$$

$$\leq \int_0^{+\infty}\int_0^1 |f'(w)|^2 e^{-w}\left(\int_0^{+\infty} z e^{-z} e^{tz} \mathbb{1}_{\mathcal{D}_t}(w,z) dz\right) dt dw.$$

Now,

$$\int_0^1\left(\int_0^{\frac{w}{t}} z e^{-z} e^{zt} dz\right) dt = \int_0^1\left(\int_0^w \frac{y}{t} e^{-\frac{y}{t}} e^y \frac{dy}{t}\right) dt,$$

$$= \int_0^w y e^y \left(\int_0^1 e^{-\frac{y}{t}} \frac{dt}{t^2}\right) dy = \int_0^w y e^y \frac{e^{-y}}{y} dy = w.$$

This concludes the proof of the lemma. □

4 Stein's Kernels and High-Dimensional CLTs

This section shows how to apply [8, Theorem 5.10.] or Theorem 2.1 to build Stein's kernels to provide stability results for Poincaré-type inequality and rates of convergence, in the 1-Wasserstein distance, in a high-dimensional central limit theorem. Let $d \geq 1$ and let Σ be a covariance matrix, which is not identically null, and let γ_Σ be the centered Gaussian probability measure on $\mathbb{R}^d$ with covariance matrix Σ; i.e., the characteristic function of the corresponding Gaussian random vector is given, for all $\xi \in \mathbb{R}^d$, by

$$\hat{\gamma}_\Sigma(\xi) = \exp\left(-\frac{\langle \xi; \Sigma(\xi)\rangle}{2}\right).$$

Next, let U_Σ be the Poincaré functional formally defined, for all suitable $\mu \in \mathcal{M}_1(\mathbb{R}^d)$ ($\mathcal{M}_1(\mathbb{R}^d)$ is the set of probability measures on $\mathbb{R}^d$), by

$$U_\Sigma(\mu) := \sup_{f \in \mathcal{H}_\Sigma(\mu)} \frac{\mathrm{Var}_\mu(f)}{\int_{\mathbb{R}^d} \langle \nabla(f)(x); \Sigma\left(\nabla(f)(x)\right)\rangle \mu(dx)},$$

where $\mathcal{H}_\Sigma(\mu)$ is the set of Borel measurable real-valued functions f defined on $\mathbb{R}^d$ such that

$$\int_{\mathbb{R}^d} |f(x)|^2 \mu(dx) < +\infty, \quad 0 < \int_{\mathbb{R}^d} \langle \nabla(f)(x); \Sigma\left(\nabla(f)(x)\right)\rangle \mu(dx) < +\infty$$

and such that $\mathrm{Var}_\mu(f) > 0$. It is well known since the works [23, Theorem 3] and [31, Theorem 2.1] that the functional U_Σ is rigid. Let us adopt the methodology developed in [7, 8, 34, 41] using Stein's method to obtain a stability result that generalizes the one for the isotropic case. For the sake of completeness, the rigidity result is re-proved next via semigroup methods, although the result is rather immediate from (2.26). Note that in [31, Theorem 2.1], the given covariance matrix $\Sigma = \left(\sigma_{i,j}\right)_{1\leq i,j\leq d}$ is such that $\sigma_{i,i} > 0$, for all $i \in \{1, \ldots, d\}$.

Lemma 4.1 *Let $d \geq 1$ and let Σ be a not identically null $d \times d$ covariance matrix. Then*

$$U_\Sigma(\gamma_\Sigma) = 1.$$

Proof The proof is very classical and relies on a semigroup argument to prove the Poincaré inequality for the Gaussian probability measure γ_Σ and on the fact that the functions $x \mapsto x_j$, for all $j \in \{1, \ldots, d\}$, are eigenfunctions of the Ornstein-Uhlenbeck operator associated with γ_Σ. Let $(P_t^\Sigma)_{t\geq 0}$ be the Ornstein-Uhlenbeck semigroup given, for all $f \in \mathcal{C}_b(\mathbb{R}^d)$, all $t \geq 0$, and all $x \in \mathbb{R}^d$, by

$$P_t^\Sigma(f)(x) = \int_{\mathbb{R}^d} f\left(xe^{-t} + \sqrt{1-e^{-2t}}y\right)\gamma_\Sigma(dy).$$

From the above Mehler formula, it is clear that the probability measure γ_Σ is an invariant measure for the semigroup $(P_t^\Sigma)_{t\geq 0}$; that $\mathcal{S}(\mathbb{R}^d)$ is a core for the generator, denoted by $\mathcal{L}^\Sigma$, of $(P_t^\Sigma)_{t\geq 0}$; and that, for all $f \in \mathcal{S}(\mathbb{R}^d)$ and all $x \in \mathbb{R}^d$,

$$\mathcal{L}^\Sigma(f)(x) = -\langle x; \nabla(f)(x)\rangle + \Delta^\Sigma(f)(x),$$

with

$$\Delta^\Sigma(f)(x) = \frac{1}{(2\pi)^d}\int_{\mathbb{R}^d} \mathcal{F}(f)(\xi)e^{i\langle x;\xi\rangle}\langle i\xi; \Sigma(i\xi)\rangle d\xi = \langle \Sigma; \mathrm{Hess}(f)(x)\rangle_{HS},$$

where $\langle A; B\rangle_{HS} = \mathrm{Tr}(A^t B)$. Next, let $f \in \mathcal{S}(\mathbb{R}^d)$ be such that $\int_{\mathbb{R}^d} f(x)\gamma_\Sigma(dx) = 0$. Differentiating the variance of $P_t^\Sigma(f)$ with respect to the time parameter gives

$$\frac{d}{dt}\left(\mathbb{E}P_t^\Sigma(f)(X)^2\right) = 2\mathbb{E}P_t^\Sigma(f)(X)\mathcal{L}^\Sigma P_t^\Sigma(f)(X),$$

where $X \sim \gamma_\Sigma$. Hence, for all $t \geq 0$,

$$\frac{d}{dt}\left(\mathbb{E}P_t^\Sigma(f)(X)^2\right) = 2\mathbb{E}P_t^\Sigma(f)(X)\left(-\langle X; \nabla(P_t^\Sigma(f))(X)\rangle + \langle \Sigma; \mathrm{Hess}\left(P_t^\Sigma(f)\right)(X)\rangle_{HS}\right).$$

Now, since $\mathcal{S}(\mathbb{R}^d)$ is a core for $\mathcal{L}^\Sigma$; invariant with respect to P_t^Σ, for all $t \geq 0$; and stable for the pointwise multiplication of functions,

$$\mathbb{E}\langle X; \nabla\left(P_t^\Sigma(f)^2\right)(X)\rangle = \mathbb{E}\langle \Sigma; \mathrm{Hess}\left(P_t^\Sigma(f)^2\right)(X)\rangle_{HS}.$$

Thus, using the Leibniz formula, for all $t \geq 0$,

$$\frac{d}{dt}\left(\mathbb{E}P_t^{\Sigma}(f)(X)^2\right) = -\left(\mathbb{E}\langle\Sigma; \text{Hess}\left(P_t^{\Sigma}(f)^2\right)(X)\rangle_{HS} - 2\mathbb{E}P_t^{\Sigma}(f)(X)\langle\Sigma; \text{Hess}\left(P_t^{\Sigma}(f)\right)(X)\rangle_{HS}\right),$$
$$= -\mathbb{E}\langle\Sigma; \text{Hess}\left(P_t^{\Sigma}(f)^2\right)(X) - 2P_t^{\Sigma}(f)(X)\,\text{Hess}\left(P_t^{\Sigma}(f)\right)(X)\rangle_{HS},$$
$$= -2\mathbb{E}\langle\nabla\left(P_t^{\Sigma}(f)\right)(X); \Sigma\left(\nabla\left(P_t^{\Sigma}(f)\right)(X)\right)\rangle.$$

Now, the commutation formula, $\nabla\left(P_t^{\Sigma}(f)\right) = e^{-t}P_t^{\Sigma}(\nabla(f))$, ensures that

$$\mathbb{E}\langle\nabla\left(P_t^{\Sigma}(f)\right)(X); \Sigma\left(\nabla\left(P_t^{\Sigma}(f)\right)(X)\right)\rangle = e^{-2t}\mathbb{E}\langle P_t^{\Sigma}(\nabla(f))(X); \Sigma P_t^{\Sigma}(\nabla(f))(X)\rangle,$$
$$= e^{-2t}\mathbb{E}\langle\sqrt{\Sigma}P_t^{\Sigma}(\nabla(f))(X); \sqrt{\Sigma}P_t^{\Sigma}(\nabla(f))(X)\rangle,$$
$$= e^{-2t}\mathbb{E}\langle P_t^{\Sigma}\left(\sqrt{\Sigma}(\nabla(f))\right)(X); P_t^{\Sigma}\left(\sqrt{\Sigma}(\nabla(f))\right)(X)\rangle,$$
$$= e^{-2t}\mathbb{E}\left\|P_t^{\Sigma}\left(\sqrt{\Sigma}(\nabla(f))\right)(X)\right\|^2,$$
$$\leq e^{-2t}\mathbb{E}P_t^{\Sigma}\left(\left\|\sqrt{\Sigma}(\nabla(f))\right\|^2\right)(X),$$
$$\leq e^{-2t}\mathbb{E}\left\|\sqrt{\Sigma}(\nabla(f))(X)\right\|^2.$$

Thus, for all $t \geq 0$,

$$\frac{d}{dt}\left(\mathbb{E}P_t^{\Sigma}(f)(X)^2\right) \geq -2e^{-2t}\mathbb{E}\left\|\sqrt{\Sigma}(\nabla(f))(X)\right\|^2.$$

Integrating with respect to t between 0 and $+\infty$ ensures that

$$\mathbb{E}f(X)^2 \leq \mathbb{E}\left\|\sqrt{\Sigma}(\nabla(f))(X)\right\|^2. \tag{4.1}$$

This last inequality implies that $U_{\Sigma}(\gamma_{\Sigma}) \leq 1$. Next, for all $j \in \{1, \ldots, d\}$, let g_j be the function defined, for all $x \in \mathbb{R}^d$, by $g_j(x) = x_j$. Now, for all $j \in \{1, \ldots, d\}$,

$$\text{Var}_{\gamma_{\Sigma}}(g_j) = \int_{\mathbb{R}^d} x_j^2\gamma_{\Sigma}(dx) = \sigma_{j,j} = \int_{\mathbb{R}^d}\langle\nabla(g_j)(x); \Sigma(\nabla(g_j))(x)\rangle\gamma_{\Sigma}(dx),$$

where $\Sigma = (\sigma_{i,j})_{1\leq i,j\leq d}$. Thus, $U_{\Sigma}(\gamma_{\Sigma}) \geq 1$. This concludes the proof of the lemma. □

Remark 4.1 The proof of the Poincaré-type inequality (4.1) for the probability measure γ_{Σ} could have been performed without using the semigroup $(P_t^{\Sigma})_{t\geq 0}$. Instead, one could use the covariance representation (2.26). Indeed, taking $g = f$ and using the Cauchy-Schwarz inequality, one retrieves the inequality (4.1). Following the end of the proof of Lemma 4.1, one can conclude that $U_{\Sigma}(\gamma_{\Sigma}) = 1$ also when Σ is generic but different from 0.

The next lemma provides the rigidity result for U_Σ.

Lemma 4.2 *Let $d \geq 1$ and let $\Sigma = (\sigma_{i,j})_{1\leq i,j\leq d}$ be a not identically null $d \times d$ covariance matrix. Let μ be a probability measure on $\mathbb{R}^d$ with a finite second moment such that, for all $i \in \{1, \dots, d\}$,*

$$\int_{\mathbb{R}^d} x\mu(dx) = 0, \quad \int_{\mathbb{R}^d} x_i^2 \mu(dx) = \sigma_{i,i}.$$

Then $U_\Sigma(\mu) = 1$ if and only if $\mu = \gamma_\Sigma$.

Proof The sufficiency is a direct consequence of Lemma 4.1 or Remark 4.1. Thus, let us prove the direct implication. Assume that $U_\Sigma(\mu) = 1$. Then, for all $f \in \mathcal{H}_\Sigma(\mu)$,

$$\mathrm{Var}_\mu(f) \leq \int_{\mathbb{R}^d} \Gamma_\Sigma(f, f)(x)\mu(dx) =: \mathcal{E}_\Sigma(f, f), \tag{4.2}$$

with, for all $x \in \mathbb{R}^d$,

$$\Gamma_\Sigma(f, f)(x) = \langle \nabla(f)(x); \Sigma(\nabla(f)(x))\rangle = \left\| \sqrt{\Sigma}(\nabla(f)(x)) \right\|^2.$$

Now, for all $j \in \{1, \dots, d\}$ and all $\varepsilon \in \mathbb{R}$ with $\varepsilon \neq 0$, let f_j be defined, for all $x \in \mathbb{R}^d$, by

$$f_j(x) = g_j(x) + \varepsilon f(x),$$

for some $f \in \mathcal{S}(\mathbb{R}^d)$ and with $g_j(x) = x_j$. Then, for all $j \in \{1, \dots, d\}$,

$$\mathrm{Var}_\mu(f_j) = \mathrm{Cov}_\mu(g_j + \varepsilon f, g_j + \varepsilon f) = \mathrm{Var}_\mu(g_j) + 2\varepsilon \,\mathrm{Cov}_\mu(g_j, f) + \varepsilon^2 \,\mathrm{Var}_\mu(f)$$

and

$$\mathcal{E}_\Sigma(f_j, f_j) = \mathcal{E}_\Sigma(g_j, g_j) + 2\varepsilon\mathcal{E}_\Sigma(g_j, f) + \varepsilon^2\mathcal{E}_\Sigma(f, f).$$

(Here and in the sequel, $\mathrm{Cov}_\mu(f, g)$ indicates the covariance of f and g under μ). Thus, thanks to (4.2), for all $j \in \{1, \dots, d\}$,

$$\mathrm{Cov}_\mu(g_j, f) = \mathcal{E}_\Sigma(g_j, f).$$

Namely, for all $j \in \{1, \dots, d\}$ and all $f \in \mathcal{S}(\mathbb{R}^d)$,

$$\mathbb{E}X_j f(X) = \mathbb{E}\langle \Sigma(e_j); \nabla(f)(X)\rangle, \quad X \sim \mu,$$

with $e_j = (0, \ldots, 0, 1, 0, \ldots, 0)^T$. The end of the proof follows easily by a standard argument involving the characteristic function. Indeed, via Fourier inversion and duality, for all $j \in \{1, \ldots, d\}$ and all $\xi \in \mathbb{R}^d$,

$$\partial_{\xi_j}\left(\varphi_\mu\right)(\xi) = -\langle \Sigma(e_j); \xi\rangle \varphi_\mu(\xi),$$

where φ_μ is the characteristic function of μ, which is $\mathcal{C}^1$ on $\mathbb{R}^d$, since μ has a finite second moment. Then, for all $\xi \in \mathbb{R}^d$,

$$\langle \xi; \nabla(\varphi_\mu)(\xi)\rangle = -\langle \xi; \Sigma(\xi)\rangle \varphi_\mu(\xi).$$

Passing to spherical coordinates, for all $(r, \theta) \in (0, +\infty) \times \mathbb{S}^{d-1}$,

$$\partial_r\left(\varphi_\mu\right)(r\theta) = -r\langle \theta; \Sigma(\theta)\rangle \varphi_\mu(r\theta).$$

Fixing $\theta \in \mathbb{S}^{d-1}$, integrating with respect to r, and using $\varphi_\mu(0) = 1$, for all $(r, \theta) \in (0, +\infty) \times \mathbb{S}^{d-1}$,

$$\varphi_\mu(r\theta) = \exp\left(-\frac{r^2}{2}\langle \theta; \Sigma(\theta)\rangle\right).$$

This concludes the proof of the lemma. □

Before moving to the proof of the stability result, let us recall some well-known facts about Stein's method for the multivariate Gaussian probability measure γ_Σ on $\mathbb{R}^d$. The standard references are [14, 24, 29, 42, 43, 54, 55, 57, 58, 60, 62, 63, 65, 68, 69]. In the sequel, let $h \in \mathcal{C}_c^\infty(\mathbb{R}^d)$ be such that

$$\|h\|_{\mathrm{Lip}} := \sup_{x, y \in \mathbb{R}^d,\, x \neq y} \frac{|h(x) - h(y)|}{\|x - y\|} = \sup_{x \in \mathbb{R}^d} \|\nabla(h)(x)\| \leq 1$$

and let f_h be defined, for all $x \in \mathbb{R}^d$, by

$$f_h(x) = -\int_0^{+\infty} \left(P_t^\Sigma(h)(x) - \mathbb{E}h(X)\right) dt, \quad X \sim \gamma_\Sigma. \tag{4.3}$$

The next lemma recalls regularity results for f_h as well as a representation formula for its Hessian matrix, which allows to obtain dimension-free bounds for the supremum norms involving the operator or the Hilbert-Schmidt norms of $\mathrm{Hess}(f_h)$.

Lemma 4.3 *Let $d \geq 1$ and let Σ be a nondegenerate $d \times d$ covariance matrix. Let $h \in \mathcal{C}_c^\infty(\mathbb{R}^d)$ be such that $\|h\|_{\mathrm{Lip}} \leq 1$ and let f_h be given by (4.3). Then, f_h is well defined, twice continuously differentiable on $\mathbb{R}^d$, and*

$$\sup_{x\in\mathbb{R}^d}\|\nabla(f_h)(x)\|\leq 1,\quad \sup_{x\in\mathbb{R}^d}\|\operatorname{Hess}(f_h)(x)\|_{op}\leq\sqrt{\frac{2}{\pi}}\|\Sigma^{-\frac{1}{2}}\|_{op},\quad \sup_{x\in\mathbb{R}^d}\|\operatorname{Hess}(f_h)(x)\|_{HS}\leq\|\Sigma^{-\frac{1}{2}}\|_{op}.$$

Moreover, if $h\in\mathcal{C}_c^\infty(\mathbb{R}^d)$ is such that

$$\sup_{x\in\mathbb{R}^d}\|\operatorname{Hess}(h)(x)\|_{op}\leq 1,$$

then

$$\sup_{x\in\mathbb{R}^d}\|\operatorname{Hess}(f_h)(x)\|_{op}\leq\frac{1}{2}.$$

Finally, if $h\in\mathcal{C}_c^\infty(\mathbb{R}^d)$ is such that

$$\tilde{M}_2(h):=\sup_{x\in\mathbb{R}^d}\|\operatorname{Hess}(h)(x)\|_{HS}\leq 1,$$

then

$$\sup_{x\in\mathbb{R}^d}\|\operatorname{Hess}(f_h)(x)\|_{HS}\leq\frac{1}{2}.$$

Proof Thanks to the Mehler formula, for all $t\geq 0$ and all $x\in\mathbb{R}^d$,

$$\begin{aligned}\left|P_t^\Sigma(h)(x)-\mathbb{E}h(X)\right|&\leq\int_{\mathbb{R}^d}\left|h\left(xe^{-t}+\sqrt{1-e^{-2t}}y\right)-h(y)\right|\gamma_\Sigma(dy),\\&\leq\|h\|_{\mathrm{Lip}}\left(e^{-t}\|x\|+\left|1-\sqrt{1-e^{-2t}}\right|\int_{\mathbb{R}^d}\|y\|\gamma_\Sigma(dy)\right).\end{aligned}$$

The right-hand side of the previous inequality is clearly integrable, with respect to t, on $(0,+\infty)$. Thus, f_h is well defined on $\mathbb{R}^d$. The fact that f_h is twice continuously differentiable on $\mathbb{R}^d$ follows from the commutation formula $\nabla P_t^\Sigma(h)=e^{-t}P_t^\Sigma(\nabla(h))$. Now, for all $x\in\mathbb{R}^d$,

$$\nabla(f_h)(x)=-\int_0^{+\infty}e^{-t}P_t^\Sigma(\nabla(h))(x)dt.$$

Thus, for all $u\in\mathbb{R}^d$ such that $\|u\|=1$,

$$\langle\nabla(f_h)(x);u\rangle=-\int_0^{+\infty}e^{-t}P_t^\Sigma(\langle u;\nabla(h)\rangle)(x)dt.$$

Then, for all $x\in\mathbb{R}^d$ and all $u\in\mathbb{R}^d$ such that $\|u\|=1$,

$$|\langle \nabla(f_h)(x); u\rangle| \leq \left(\int_0^{+\infty} e^{-t}dt\right) \|h\|_{\mathrm{Lip}} \leq 1.$$

Next, let us deal with the Hessian matrix of f_h. For all $k, \ell \in \{1, \ldots, d\}$ and all $x \in \mathbb{R}^d$,

$$\partial_k \partial_\ell (f_h)(x) = -\int_0^{+\infty} e^{-2t} P_t^{\Sigma}(\partial_k \partial_\ell(h))(x)dt.$$

Moreover, thanks to Bismut's formula, for all $k, \ell \in \{1, \ldots, d\}$, all $x \in \mathbb{R}^d$, and all $t > 0$,

$$\begin{aligned} P_t^{\Sigma}(\partial_k \partial_\ell(h))(x) &= \int_{\mathbb{R}^d} \partial_k \partial_\ell(h)\left(xe^{-t} + \sqrt{1-e^{-2t}}y\right)\gamma_\Sigma(dy) \\ &= \frac{1}{\sqrt{1-e^{-2t}}} \int_{\mathbb{R}^d} \langle \Sigma^{-1}(e_\ell); y\rangle \partial_k(h)\left(xe^{-t} + \sqrt{1-e^{-2t}}y\right)\gamma_\Sigma(dy), \end{aligned}$$

and so, for all $\ell, k \in \{1, \ldots, d\}$ and all $x \in \mathbb{R}^d$,

$$\partial_k \partial_\ell (f_h)(x) = -\int_0^{+\infty} \frac{e^{-2t}}{\sqrt{1-e^{-2t}}} \left(\int_{\mathbb{R}^d} \langle \Sigma^{-1}(e_\ell); y\rangle \partial_k(h)\left(xe^{-t} + \sqrt{1-e^{-2t}}y\right)\gamma_\Sigma(dy)\right)dt.$$

Thus, for all $x \in \mathbb{R}^d$,

$$\mathrm{Hess}(f_h)(x) = -\int_0^{+\infty} \frac{e^{-2t}}{\sqrt{1-e^{-2t}}} \int_{\mathbb{R}^d} \Sigma^{-1}(y)\left(\nabla(h)\left(xe^{-t} + \sqrt{1-e^{-2t}}y\right)\right)^T \gamma_\Sigma(dy)dt. \tag{4.4}$$

Now, let $u, v \in \mathbb{R}^d$ be such that $\|u\| = \|v\| = 1$. Then, for all $x \in \mathbb{R}^d$,

$$\begin{aligned} |\langle \mathrm{Hess}(f_h)(x)u; v\rangle| &= \left|\int_0^{+\infty} \frac{e^{-2t}}{\sqrt{1-e^{-2t}}} \left(\int_{\mathbb{R}^d} \langle \Sigma^{-1}(y); v\rangle \langle \nabla(h)\left(xe^{-t} + \sqrt{1-e^{-2t}}y\right); u\rangle \gamma_\Sigma(dy)\right)dt\right|, \\ &\leq \left(\int_0^{+\infty} \frac{e^{-2t}}{\sqrt{1-e^{-2t}}}dt\right) \int_{\mathbb{R}^d} \left|\langle \Sigma^{-1}(y); v\rangle\right| \gamma_\Sigma(dy), \\ &\leq \int_{\mathbb{R}^d} \left|\langle \Sigma^{-\frac{1}{2}}(y); v\rangle\right| \gamma(dy), \\ &\leq \|\Sigma^{-\frac{1}{2}}\|_{op} \left(\int_{\mathbb{R}} |x| e^{-\frac{x^2}{2}} \frac{dx}{\sqrt{2\pi}}\right), \\ &\leq \sqrt{\frac{2}{\pi}} \|\Sigma^{-\frac{1}{2}}\|_{op}, \end{aligned}$$

where γ is the standard Gaussian measure on $\mathbb{R}^d$ (i.e., with the covariance matrix given by I_d) and since, under γ, for all $u \in \mathbb{S}^{d-1}$, $\langle \Sigma^{-\frac{1}{2}}(y); u\rangle$ is a centered normal random variable with variance $\|\Sigma^{-\frac{1}{2}}(u)\|^2$. It remains to estimate the Hilbert-Schmidt norm of $\mathrm{Hess}(f_h)(x)$ based on (4.4). Through a similar argument and using Hölder's inequality for Schatten norms, for all $x \in \mathbb{R}^d$,

$$\|\,\mathrm{Hess}(f_h)(x)\|_{HS} \leq \sup_{A \in \mathcal{M}_{d\times d}(\mathbb{R}),\ \|A\|_{HS}=1} \|A\Sigma^{-\frac{1}{2}}\|_{HS} \leq \|\Sigma^{-\frac{1}{2}}\|_{op},$$

where $\mathcal{M}_{d\times d}(\mathbb{R})$ denotes the set of $d \times d$ matrices with real coefficients. Finally, let us prove that when $h \in \mathcal{C}_c^{\infty}(\mathbb{R}^d)$ with

$$\sup_{x\in\mathbb{R}^d} \|\,\mathrm{Hess}(h)(x)\|_{op} \leq 1,$$

then

$$\sup_{x\in\mathbb{R}^d} \|\,\mathrm{Hess}(f_h)(x)\|_{op} \leq \frac{1}{2}$$

(and similarly with the operator norm replaced by the Hilbert-Schmidt norm). From the previous computations, for all $\ell, k \in \{1, \ldots, d\}$ and all $x \in \mathbb{R}^d$,

$$\partial_\ell \partial_k \left(f_h\right)(x) = -\int_0^{+\infty} e^{-2t} P_t^{\Sigma}(\partial_k \partial_\ell(h))(x)dt.$$

Then, for all $u, v \in \mathbb{S}^{d-1}$ and all $x \in \mathbb{R}^d$,

$$\langle \mathrm{Hess}(f_h)(x)u; v\rangle = -\int_0^{+\infty} e^{-2t} P_t^{\Sigma}\left(\langle \mathrm{Hess}(h)u; v\rangle\right)(x)dt.$$

The conclusion easily follows. □

Remark 4.2

(i) To the best of our knowledge, the bound,

$$\sup_{x\in\mathbb{R}^d} \|\,\mathrm{Hess}(f_h)(x)\|_{HS} \leq \|\Sigma^{-\frac{1}{2}}\|_{op}, \tag{4.5}$$

for $h \in \mathcal{C}^1(\mathbb{R}^d)$ with $\|h\|_{\mathrm{Lip}} \leq 1$, is the best available in the literature. It generalizes the bound obtained in [29, Lemma 2.2] for the isotropic case, amends [54, Proof of Lemma 2], and improves the bound obtained in [58, Inequality (13)] (see also [57, Lemma 3.3]).

(ii) Assuming that $d = 2$ and that $\Sigma = I_2$, let us compute the quantity

$$J(A) := \int_{\mathbb{R}^2} \|A(y)\|\gamma(dy),$$

for some specific values of A, a 2×2 matrix with $\|A\|_{HS} = 1$. Take, for instance, $A = I_2/\sqrt{2}$. Then, by standard computations using polar coordinates,

$$J\left(\frac{1}{\sqrt{2}} I_2\right) = \frac{1}{\sqrt{2}} \int_{\mathbb{R}^2} \|y\| \exp\left(-\frac{\|y\|^2}{2}\right) \frac{dy}{2\pi} = \frac{\sqrt{\pi}}{2},$$

which seems to question the bound obtained in [54, Proof of Lemma 2, page 161].

(iii) In [38, proof of Theorem 1.2], the bound (4.5) appears as the corrected inequality (3.13) in there.

Next, let us discuss the notion and the existence of Stein's kernels with respect to the Gaussian probability measure γ_Σ. The idea is the following: let μ be a probability measure on $\mathbb{R}^d$ with a finite second moment such that $\int_{\mathbb{R}^d} x\mu(dx) = 0$ and $\mathrm{Cov}(X_\mu, X_\mu) = \Sigma$, where $X_\mu \sim \mu$. Moreover, assume that there exists τ_μ, a function defined on $\mathbb{R}^d$ with values in $\mathcal{M}_{d\times d}(\mathbb{R})$, such that, for all appropriate vector-valued functions f defined on $\mathbb{R}^d$,

$$\int_{\mathbb{R}^d} \langle \tau_\mu(x); \nabla(f)(x)\rangle_{HS}\mu(dx) = \int_{\mathbb{R}^d} \langle x; f(x)\rangle\mu(dx). \tag{4.6}$$

In the vector-valued case, $\nabla(f)$ denotes the Jacobian matrix of f. Then the classical argument for bounding distances goes as follows: let $h \in \mathcal{C}^1(\mathbb{R}^d)$ be such that $\|h\|_{\mathrm{Lip}} \le 1$, and let f_h be given by (4.3). (Actually, in finite dimension, one can take, without loss of generality, $h \in \mathcal{C}_c^\infty(\mathbb{R}^d)$, the main point being that $\|h\|_{\mathrm{Lip}} \le 1$.) Then f_h is a strong solution to the following partial differential equation: for all $x \in \mathbb{R}^d$,

$$-\langle x; \nabla(f_h)(x)\rangle + \langle \Sigma; \mathrm{Hess}(f_h)(x)\rangle_{HS} = h(x) - \mathbb{E}h(X), \quad X \sim \gamma_\Sigma.$$

Integrating with respect to μ and using the formal definition of τ_μ give

$$\begin{aligned}\left|\mathbb{E}h(X_\mu) - \mathbb{E}h(X)\right| &= \left|\mathbb{E}\left(-\langle X_\mu; \nabla(f_h)(X_\mu)\rangle + \langle \Sigma; \mathrm{Hess}(f_h)(X_\mu)\rangle_{HS}\right)\right|, \\ &= \left|\mathbb{E}\left(\langle \Sigma - \tau_\mu(X_\mu); \mathrm{Hess}(f_h)(X_\mu)\rangle_{HS}\right)\right|,\end{aligned}$$

with $X_\mu \sim \mu$. Then by the Cauchy-Schwarz inequality and the bound obtained in Lemma 4.3,

$$\left|\mathbb{E}h(X_\mu) - \mathbb{E}h(X)\right| \le \|\Sigma^{-\frac{1}{2}}\|_{op}\left(\mathbb{E}\left(\|\tau_\mu(X_\mu) - \Sigma\|_{HS}^2\right)\right)^{\frac{1}{2}}. \tag{4.7}$$

Observe that the right-hand side of the previous inequality does not depend on h anymore. In the sequel, let us explain how to prove the existence of τ_μ and how to bound the Stein discrepancy based on closed form techniques. For this purpose, let us consider the following bilinear symmetric nonnegative definite form defined, for all $f, g \in \mathcal{C}_c^\infty(\mathbb{R}^d, \mathbb{R}^d)$, by

$$\mathcal{E}_{\Sigma,\mu}(f,g)=\int_{\mathbb{R}^d}\langle\Sigma\left(\nabla(g)(x)\right);\nabla(f)(x)\rangle_{HS}\mu(dx),$$

where Σ is a nondegenerate covariance matrix and where μ is a probability measure on $\mathbb{R}^d$ with a finite second moment such that

$$\int_{\mathbb{R}^d}x\mu(dx)=0,\quad\int_{\mathbb{R}^d}xx^T\mu(dx)=\Sigma.$$

Proposition 4.1 *Let $d\geq 1$ and let Σ be a nondegenerate $d\times d$ covariance matrix. Let μ be a probability measure on $\mathbb{R}^d$ with a finite second moment such that*

$$\int_{\mathbb{R}^d}x\mu(dx)=0,\quad\int_{\mathbb{R}^d}xx^T\mu(dx)=\Sigma.$$

Let the form $\left(\mathcal{E}_{\Sigma,\mu},\mathcal{C}_c^\infty(\mathbb{R}^d,\mathbb{R}^d)\right)$ be closable. Finally, let there exists $U_{\Sigma,\mu}>0$ such that, for all $f\in\mathcal{C}_c^\infty(\mathbb{R}^d,\mathbb{R}^d)$ with $\int_{\mathbb{R}^d}f(x)\mu(dx)=0$,

$$\int_{\mathbb{R}^d}\|f(x)\|^2\mu(dx)\leq U_{\Sigma,\mu}\int_{\mathbb{R}^d}\langle\Sigma\left(\nabla(f)(x)\right);\nabla(f)(x)\rangle_{HS}\mu(dx).\tag{4.8}$$

Then there exists τ_μ such that, for all $f\in\mathcal{D}(\mathcal{E}_{\Sigma,\mu})$,

$$\int_{\mathbb{R}^d}\langle x;f(x)\rangle\mu(dx)=\int_{\mathbb{R}^d}\langle\nabla(f)(x);\tau_\mu(x)\rangle_{HS}\mu(dx).$$

Moreover,

$$\int_{\mathbb{R}^d}\|\tau_\mu(x)\|_{HS}^2\mu(dx)\leq U_{\Sigma,\mu}\|\Sigma\|_{HS}^2.$$

Proof First, let us build Stein's kernel τ_μ. Since the form $\left(\mathcal{E}_{\Sigma,\mu},\mathcal{C}_c^\infty(\mathbb{R}^d,\mathbb{R}^d)\right)$ is closable, consider its smallest closed extension denoted by $\left(\mathcal{E}_{\Sigma,\mu},\mathcal{D}(\mathcal{E}_{\Sigma,\mu})\right)$, where $\mathcal{D}(\mathcal{E}_{\Sigma,\mu})$ is its dense linear domain. Moreover, let $\mathcal{L}^\mu$, $(G_\delta^\mu)_{\delta>0}$, and $(P_t^\mu)_{t\geq 0}$ be the corresponding generator, the strongly continuous resolvent, and the strongly continuous semigroup. In particular, recall that, for all $f\in L^2(\mathbb{R}^d,\mathbb{R}^d,\mu)$ and all $\delta>0$,

$$G_\delta^\mu(f)=\int_0^{+\infty}e^{-\delta t}P_t^\mu(f)dt.$$

Next, let $f\in\mathcal{C}_c^\infty(\mathbb{R}^d,\mathbb{R}^d)$ be such that $\int_{\mathbb{R}^d}f(x)\mu(dx)=0$. Then by the very definition of the generator $\mathcal{L}^\mu$ and integration by parts,

$$\frac{d}{dt}\left(\mathbb{E}\|P_t^{\mu}(f)(X_{\mu})\|^2\right) = 2\mathbb{E}\langle P_t^{\mu}(f)(X_{\mu}); \mathcal{L}^{\mu} P_t^{\mu}(f)(X_{\mu})\rangle = -2\mathcal{E}_{\Sigma,\mu}\left(P_t^{\mu}(f), P_t^{\mu}(f)\right)$$
$$\leq -\frac{2}{U_{\Sigma,\mu}}\mathbb{E}\|P_t^{\mu}(f)(X_{\mu})\|^2,$$

with $X_{\mu} \sim \mu$. Then, for all $f \in \mathcal{C}_c^{\infty}(\mathbb{R}^d, \mathbb{R}^d)$ such that $\int_{\mathbb{R}^d} f(x)\mu(dx) = 0$,

$$\left\| P_t^{\mu}(f)\right\|^2_{L^2(\mathbb{R}^d,\mathbb{R}^d,\mu)} \leq \exp\left(-\frac{2t}{U_{\Sigma,\mu}}\right) \|f\|^2_{L^2(\mathbb{R}^d,\mathbb{R}^d,\mu)},$$

which, via a density argument, clearly extends to all $f \in L^2(\mathbb{R}^d, \mathbb{R}^d, \mu)$ with $\int_{\mathbb{R}^d} f(x)\mu(dx) = 0$. Then from Theorem 2.1, for all $g \in L^2(\mathbb{R}^d, \mathbb{R}^d, \mu)$ such that $\int_{\mathbb{R}^d} g(x)\mu(dx) = 0$ and all $f \in \mathcal{D}(\mathcal{E}_{\Sigma,\mu})$,

$$\mathcal{E}_{\Sigma,\mu}(G^{\mu}_{0+}(g), f) = \int_{\mathbb{R}^d} \langle g(x); f(x)\rangle \mu(dx). \tag{4.9}$$

Now, since μ has a finite second moment and since $\int_{\mathbb{R}^d} x\mu(dx) = 0$, then set $g(x) = x$, for all $x \in \mathbb{R}^d$, and so for μ-a.e. $x \in \mathbb{R}^d$,

$$\tau_{\mu}(x) = \Sigma\nabla\left(G^{\mu}_{0+}(g)\right)(x).$$

Finally, taking $f(x) = \Sigma G^{\mu}_{0+}(g)(x)$, μ-a.e. $x \in \mathbb{R}^d$, in (4.9) gives

$$\int_{\mathbb{R}^d} \|\tau_{\mu}(x)\|^2_{HS}\mu(dx) = \int_{\mathbb{R}^d} \langle \Sigma x; G^{\mu}_{0+}(g)(x)\rangle\mu(dx)$$
$$= \int_{\mathbb{R}^d} \langle \Sigma^{\frac{1}{2}}x; \Sigma^{\frac{1}{2}} G^{\mu}_{0+}(g)(x)\rangle\mu(dx).$$

Then by the Cauchy-Schwarz inequality,

$$\int_{\mathbb{R}^d} \|\tau_{\mu}(x)\|^2_{HS}\mu(dx) \leq U_{\Sigma,\mu}\,\|\Sigma\|^2_{HS}.$$

This concludes the proof of the proposition. □

Remark 4.3

(i) Let us analyze the closability assumption on the bilinear form $\left(\mathcal{E}_{\Sigma,\mu}, \mathcal{C}_c^{\infty}(\mathbb{R}^d, \mathbb{R}^d)\right)$. If $\mu = \gamma$, then by the Gaussian integration by parts, for all $f, g \in \mathcal{C}_c^{\infty}(\mathbb{R}^d, \mathbb{R}^d)$,

$$\int_{\mathbb{R}^d} \langle \nabla(f)(x); \nabla(g)(x)\rangle_{HS}\gamma(dx) = \int_{\mathbb{R}^d} \langle f(x); (-\mathcal{L})(g)(x)\rangle\gamma(dx),$$

where, for all $x \in \mathbb{R}^d$ and all $j \in \{1, \ldots, d\}$,

$$\mathcal{L}(g_j)(x) = -\langle x; \nabla(g_j)(x)\rangle + \Delta(g_j)(x).$$

Now, let $(f_n)_{n\geq 1}$ be a sequence of functions such that, for all $n \geq 1$, $f_n \in \mathcal{C}_c^\infty(\mathbb{R}^d, \mathbb{R}^d)$, $\|f_n\|_{L^2(\mathbb{R}^d,\mathbb{R}^d,\gamma)} \to 0$, as n tends to $+\infty$, and $(\nabla(f_n))_{n\geq 1}$ is a Cauchy sequence in $L^2(\mathbb{R}^d, \mathcal{H}, \gamma)$, where $(\mathcal{H}, \langle\cdot;\cdot\rangle_{\mathcal{H}}) = (\mathcal{M}_{d\times d}(\mathbb{R}), \langle\cdot;\cdot\rangle_{HS})$. Since $L^2(\mathbb{R}^d, \mathcal{H}, \gamma)$ is complete, there exists $F \in L^2(\mathbb{R}^d, \mathcal{H}, \gamma)$ such that $\nabla(f_n) \to F$, as n tends to $+\infty$. Moreover, for all $\psi \in \mathcal{C}_c^\infty(\mathbb{R}^d, \mathbb{R}^d)$,

$$\begin{aligned}\int_{\mathbb{R}^d} \langle F; \nabla(\psi)(x)\rangle_{HS}\gamma(dx) &= \lim_{n\to+\infty} \int_{\mathbb{R}^d} \langle \nabla(f_n)(x); \nabla(\psi)(x)\rangle_{HS}\gamma(dx),\\ &= \lim_{n\to+\infty} \int_{\mathbb{R}^d} \langle f_n(x); (-\mathcal{L})(\psi)(x)\rangle\gamma(dx),\\ &= \lim_{n\to+\infty} \langle f_n; (-\mathcal{L})(\psi)\rangle_{L^2(\mathbb{R}^d,\mathbb{R}^d,\gamma)} = 0.\end{aligned}$$

Since this is true for all $\psi \in \mathcal{C}_c^\infty(\mathbb{R}^d, \mathbb{R}^d)$, $F = 0$ in $L^2(\mathbb{R}^d, \mathcal{H}, \gamma)$, and therefore, the form is closable.

(ii) Let us assume that $\mu(dx) = \psi(x)dx$, where ψ is the positive Radon-Nikodym derivative of μ with respect to the Lebesgue measure. Moreover, let us assume that $\Sigma = I_d$, the $d \times d$ identity matrix, and that $\psi \in \mathcal{C}^1(\mathbb{R}^d)$ with

$$\int_{\mathbb{R}^d} \left|\frac{\partial_j(\psi)(x)}{\psi(x)}\right|^2 \mu(dx) < +\infty, \quad j \in \{1, \ldots, d\}. \tag{4.10}$$

Then through standard integration by parts, for all $f, g \in \mathcal{C}_c^\infty(\mathbb{R}^d, \mathbb{R}^d)$,

$$\int_{\mathbb{R}^d} \langle \nabla(f)(x); \nabla(g)(x)\rangle_{HS}\mu(dx) = \int_{\mathbb{R}^d} \langle f(x); (-\mathcal{L}^\psi)(g)(x)\rangle\mu(dx),$$

with, for all $i \in \{1, \ldots, d\}$ and all $x \in \mathbb{R}^d$,

$$\mathcal{L}^\psi(g_i)(x) = \Delta(g_i)(x) + \left\langle \frac{\nabla(\psi)(x)}{\psi(x)}; \nabla(g_i)(x)\right\rangle.$$

Finally, reasoning as in (i), one can prove that the form $\left(\mathcal{E}_\mu, \mathcal{C}_c^\infty(\mathbb{R}^d, \mathbb{R}^d)\right)$ is closable since $\psi \in \mathcal{C}^1(\mathbb{R}^d)$ and since the condition (4.10) holds.

(iii) In [27] (see also [3]), sharper sufficient conditions are put forward which ensure that the form $\left(\mathcal{E}_\mu, \mathcal{C}_c^\infty(\mathbb{R}^d, \mathbb{R}^d)\right)$ is closable. Indeed, assume that $\mu(dx) = \psi(x)^2 dx$ with $\psi \in H^1_{loc}(\mathbb{R}^d, dx)$, where $H^1_{loc}(\mathbb{R}^d, dx)$ is the set of functions in $L^2_{loc}(\mathbb{R}^d, dx)$ such that their weak gradient belongs to

$L^2_{loc}(\mathbb{R}^d, dx)$ (here, $L^2_{loc}(\mathbb{R}^d, dx)$ is the space of locally square-integrable functions on $\mathbb{R}^d$). Then reasoning as in (i) and (ii), one can prove that the induced form is closable since, for any K compact subset of $\mathbb{R}^d$,

$$\int_K \|\nabla(\psi)(x)\|^2 \, dx < +\infty.$$

Let us pursue the discussion with a first stability result.

Theorem 4.1 *Let $d \geq 1$ and let Σ be a nondegenerate $d \times d$ covariance matrix. Let γ_Σ be the centered Gaussian probability measure with covariance matrix Σ and let μ be a probability measure on $\mathbb{R}^d$ with a finite second moment such that*

$$\int_{\mathbb{R}^d} x\mu(dx) = 0, \quad \int_{\mathbb{R}^d} xx^T \mu(dx) = \Sigma.$$

Let the form $\left(\mathcal{E}_{\Sigma,\mu}, \mathcal{C}_c^\infty(\mathbb{R}^d, \mathbb{R}^d)\right)$ be closable. Finally, let there exists $U_{\Sigma,\mu} > 0$ such that, for all $f \in \mathcal{C}_c^\infty(\mathbb{R}^d, \mathbb{R}^d)$ with $\int_{\mathbb{R}^d} f(x)\mu(dx) = 0$,

$$\int_{\mathbb{R}^d} \|f(x)\|^2 \mu(dx) \leq U_{\Sigma,\mu} \int_{\mathbb{R}^d} \langle \Sigma\left(\nabla(f)(x)\right); \nabla(f)(x)\rangle_{HS} \mu(dx).$$

Then

$$W_1(\mu, \gamma_\Sigma) \leq \|\Sigma^{-\frac{1}{2}}\|_{op} \|\Sigma\|_{HS} \sqrt{U_{\Sigma,\mu} - 1}. \tag{4.11}$$

Proof Recall that

$$W_1(\mu, \gamma_\Sigma) = \sup_{\|h\|_{\text{Lip}} \leq 1} \left|\mathbb{E}h(X_\mu) - \mathbb{E}h(X)\right|,$$

with $X_\mu \sim \mu$ and $X \sim \gamma_\Sigma$. Moreover, by standard approximation arguments (see Lemmas 5.1 and 5.2 of the Appendix),

$$W_1(\mu, \gamma_\Sigma) = \sup_{h \in \mathcal{C}_c^\infty(\mathbb{R}^d),\, \|h\|_{\text{Lip}} \leq 1} \left|\mathbb{E}h(X_\mu) - \mathbb{E}h(X)\right|.$$

Now, let $h \in \mathcal{C}_c^\infty(\mathbb{R}^d)$ be such that $\|h\|_{\text{Lip}} \leq 1$. Thus, thanks to Lemma 4.3 and to Stein's method applied to the multivariate Gaussian probability measure γ_Σ,

$$\left|\mathbb{E}h(X_\mu) - \mathbb{E}h(X)\right| \leq \|\Sigma^{-\frac{1}{2}}\|_{op} \left(\mathbb{E}\left(\|\tau_\mu(X_\mu) - \Sigma\|^2_{HS}\right)\right)^{\frac{1}{2}}.$$

Next, from (4.6) and Proposition 4.1,

$$
\begin{aligned}
\mathbb{E}\left\|\tau_\mu\left(X_\mu\right)-\Sigma\right\|_{HS}^2 &= \mathbb{E}\|\tau_\mu(X_\mu)\|_{HS}^2+\|\Sigma\|_{HS}^2-2\mathbb{E}\langle\tau_\mu(X_\mu);\Sigma\rangle_{HS},\\
&= \mathbb{E}\|\tau_\mu(X_\mu)\|_{HS}^2+\|\Sigma\|_{HS}^2-2\mathbb{E}\langle X_\mu;\Sigma X_\mu\rangle,\\
&= \mathbb{E}\|\tau_\mu(X_\mu)\|_{HS}^2-\|\Sigma\|_{HS}^2,\\
&\le \|\Sigma\|_{HS}^2\left(U_{\Sigma,\mu}-1\right).
\end{aligned}
$$

□

Remark 4.4

(i) When $\Sigma = I_d$, the inequality (4.11) boils down to

$$W_1(\mu,\gamma) \le \sqrt{d}\sqrt{U_{I_d,\mu}-1}, \tag{4.12}$$

which matches the upper bound obtained in [34, Theorem 4.1] for the 2-Wasserstein distance based on [52, Proposition 3.1].

(ii) Note that the previous reasoning ensures as well the following bound (which is relevant in an infinite dimensional setting): for all μ as in Theorem 4.1,

$$\tilde{d}_{W_2}(\mu,\gamma_\Sigma) \le \frac{1}{2}\|\Sigma\|_{HS}\sqrt{U_{\Sigma,\mu}-1}, \tag{4.13}$$

with

$$\tilde{d}_{W_2}(\mu,\gamma_\Sigma) := \sup_{h\in\mathcal{C}^2(\mathbb{R}^d),\,\|h\|_{\mathrm{Lip}}\le 1,\,\tilde{M}_2(h)\le 1}\left|\int_{\mathbb{R}^d} h(x)\mu(dx) - \int_{\mathbb{R}^d} h(x)\gamma_\Sigma(dx)\right|.$$

In the forthcoming result, a regularization argument shows how to remove the closability assumption.

Theorem 4.2 *Let $d \ge 1$ and let Σ be a nondegenerate $d\times d$ covariance matrix. Let γ_Σ be the centered Gaussian probability measure with covariance matrix Σ and let μ be a probability measure on $\mathbb{R}^d$ with a finite second moment such that*

$$\int_{\mathbb{R}^d} x\mu(dx) = 0, \quad \int_{\mathbb{R}^d} xx^T\mu(dx) = \Sigma.$$

Finally, let there exists $U_{\Sigma,\mu} > 0$ such that for all $f \in \mathcal{C}_c^\infty(\mathbb{R}^d,\mathbb{R}^d)$ with $\int_{\mathbb{R}^d} f(x)\mu(dx) = 0$,

$$\int_{\mathbb{R}^d}\|f(x)\|^2\mu(dx) \le U_{\Sigma,\mu}\int_{\mathbb{R}^d}\langle\Sigma\left(\nabla(f)(x)\right);\nabla(f)(x)\rangle_{HS}\mu(dx).$$

Then

$$W_1(\mu, \gamma_\Sigma) \leq \|\Sigma^{-\frac{1}{2}}\|_{op}\|\Sigma\|_{HS}\sqrt{U_{\Sigma,\mu} - 1}. \tag{4.14}$$

Proof Let $d \geq 1$ and let Σ be a nondegenerate covariance matrix. Let $\varepsilon > 0$ and let γ_ε be the centered Gaussian probability measure on $\mathbb{R}^d$ with the covariance matrix given by $\varepsilon^2\Sigma$. Let μ be a centered probability measure on $\mathbb{R}^d$ with a finite second moment such that

$$\int_{\mathbb{R}^d} xx^T \mu(dx) = \Sigma,$$

and satisfying the Poincaré-type inequality (4.8) with constant $U_{\Sigma,\mu}$. Next, let μ_ε be the probability measure on $\mathbb{R}^d$ defined through the following characteristic function: for all $\xi \in \mathbb{R}^d$,

$$\hat{\mu}_\varepsilon(\xi) := \hat{\mu}(\xi) \exp\left(-\frac{\varepsilon^2\langle\xi; \Sigma(\xi)\rangle}{2}\right),$$

and let $X_\varepsilon \sim \mu_\varepsilon$. Then

$$X_\varepsilon =_{\mathcal{L}} X_\mu + Z_\varepsilon,$$

where (X_μ, Z_ε) are independent with $X_\mu \sim \mu$ and $Z_\varepsilon \sim \gamma_\varepsilon$ and where $=_{\mathcal{L}}$ stands for equality in distribution. Next, let $\mathcal{E}_\varepsilon$ be the bilinear symmetric form defined, for all $f, g \in \mathcal{C}_c^\infty(\mathbb{R}^d, \mathbb{R}^d)$, by

$$\mathcal{E}_\varepsilon(f, g) = \int_{\mathbb{R}^d} \langle\Sigma_\varepsilon(\nabla(f)(x)); \nabla(g)(x)\rangle_{HS}\mu_\varepsilon(dx),$$

where $\Sigma_\varepsilon := (1 + \varepsilon^2)\Sigma$. In particular, the probability measure μ_ε is absolutely continuous with respect to the Lebesgue measure with density ψ_ε given, for all $x \in \mathbb{R}^d$, by

$$\psi_\varepsilon(x) = \int_{\mathbb{R}^d} p_\varepsilon(x - y)\mu(dy),$$

where p_ε is the density of the nondegenerate Gaussian probability measure γ_ε. Let us prove that the form $\left(\mathcal{E}_\varepsilon, \mathcal{C}_c^\infty(\mathbb{R}^d, \mathbb{R}^d)\right)$ is closable. Let $(f_n)_{n\geq1}$ be a sequence of functions in $\mathcal{C}_c^\infty(\mathbb{R}^d, \mathbb{R}^d)$ such that $\|f_n\|_{L^2(\mathbb{R}^d,\mathbb{R}^d,\mu_\varepsilon)}$ tends to 0 as n tends to $+\infty$ and such that $(\nabla(f_n))_{n\geq1}$ is a Cauchy sequence in $L^2(\mathbb{R}^d, \mathcal{H}, \mu_\varepsilon)$, where $(\mathcal{H}, \langle\cdot; \cdot\rangle_\mathcal{H})$ is given by $(\mathcal{M}_{d\times d}(\mathbb{R}), \langle\Sigma_\varepsilon\cdot; \cdot\rangle_{HS})$. Since Σ_ε is nondegenerate, $L^2(\mathbb{R}^d, \mathcal{H}, \mu_\varepsilon)$ is complete. Thus, there exists $F \in L^2(\mathbb{R}^d, \mathcal{H}, \mu_\varepsilon)$ such that

$$\nabla(f_n) \underset{n\to+\infty}{\longrightarrow} F, \quad L^2(\mathbb{R}^d, \mathcal{H}, \mu_\varepsilon).$$

Next, let $\Psi \in \mathcal{C}_c^\infty\left(\mathbb{R}^d, \mathcal{M}_{d\times d}(\mathbb{R})\right)$. Then integrating by parts,

$$\begin{aligned}\int_{\mathbb{R}^d}(1+\varepsilon^2)\langle \Sigma F(x); \Psi(x)\rangle_{HS}\psi_\varepsilon(x)dx &= (1+\varepsilon^2)\lim_{n\longrightarrow+\infty}\int_{\mathbb{R}^d}\langle \nabla(f_n)(x); \Sigma\Psi(x)\rangle_{HS}\psi_\varepsilon(x)dx,\\ &= (1+\varepsilon^2)\lim_{n\longrightarrow+\infty}\sum_{i,j=1}^d\int_{\mathbb{R}^d}\partial_i(f_{n,j})(x)(\Sigma\Psi)_{i,j}(x)\psi_\varepsilon(x)dx,\\ &= (1+\varepsilon^2)\lim_{n\longrightarrow+\infty}\sum_{i,j=1}^d-\int_{\mathbb{R}^d}f_{n,j}(x)\partial_i\left((\Sigma\Psi)_{i,j}(x)\psi_\varepsilon(x)\right)dx,\\ &= (1+\varepsilon^2)\lim_{n\longrightarrow+\infty}\sum_{i,j=1}^d-\int_{\mathbb{R}^d}f_{n,j}(x)\Big(\partial_i\left((\Sigma\Psi)_{i,j}\right)(x)\psi_\varepsilon(x)\\ &\quad+(\Sigma\Psi)_{i,j}(x)\partial_i\left(\psi_\varepsilon\right)(x)\Big)dx.\end{aligned}$$

Now, by the Cauchy-Schwarz inequality, for all $n \geq 1$, all $i, j \in \{1, \ldots, d\}$, and all $\varepsilon > 0$,

$$\left|\int_{\mathbb{R}^d}f_{n,j}(x)\partial_i\left((\Sigma\Psi)_{i,j}\right)(x)\psi_\varepsilon(x)dx\right| \leq C_{i,j}(\Psi,\Sigma)\|f_{n,j}\|_{L^2(\mathbb{R}^d,\mathbb{R},\mu_\varepsilon)},$$

where $C_{i,j}(\Psi, \Sigma) > 0$ depends on i, j, Ψ, and Σ only. Moreover, by the Cauchy-Schwarz inequality again, for all $n \geq 1$, all $i, j \in \{1, \ldots, d\}$, and all $\varepsilon > 0$,

$$\left|\int_{\mathbb{R}^d}f_{n,j}(x)(\Sigma\Psi)_{i,j}(x)\frac{\partial_i\left(\psi_\varepsilon\right)(x)}{\psi_\varepsilon(x)}\psi_\varepsilon(x)dx\right| \leq \tilde{C}_{i,j}\left(\Psi,\Sigma\right)\|f_{n,j}\|_{L^2(\mathbb{R}^d,\mathbb{R},\mu_\varepsilon)}\left\|\frac{\partial_i(\psi_\varepsilon)}{\psi_\varepsilon}\right\|_{L^2(K_\Psi,\mathbb{R},\mu_\varepsilon)},$$

for some $\tilde{C}_{i,j}(\Psi, \Sigma) > 0$ only depending on i, j, Ψ, and Σ and for some compact subset K_Ψ of $\mathbb{R}^d$ depending only on Ψ. In particular, note that, for all $\varepsilon > 0$ and all compact subsets K of $\mathbb{R}^d$,

$$\left\|\frac{\partial_i(\psi_\varepsilon)}{\psi_\varepsilon}\right\|^2_{L^2(K,\mathbb{R},\mu_\varepsilon)} = \int_K\left|\frac{\partial_i(\psi_\varepsilon)(x)}{\psi_\varepsilon(x)}\right|^2\psi_\varepsilon(x)dx < +\infty$$

since $\psi_\varepsilon \in \mathcal{C}^1(\mathbb{R}^d)$ and $\psi_\varepsilon > 0$. Thus, for all $\Psi \in \mathcal{C}_c^\infty(\mathbb{R}^d, \mathcal{M}_{d\times d}(\mathbb{R}))$,

$$\int_{\mathbb{R}^d}(1+\varepsilon^2)\langle \Sigma F(x); \Psi(x)\rangle_{HS}\psi_\varepsilon(x)dx = 0,$$

which ensures that the form is closable. Moreover, for all $f \in \mathcal{C}_c^\infty(\mathbb{R}^d, \mathbb{R}^d)$ with $\int_{\mathbb{R}^d} f(x)\mu_\varepsilon(dx) = 0$,

$$\int_{\mathbb{R}^d}\|f(x)\|^2\mu_\varepsilon(dx) \leq U_{\Sigma,\mu,\varepsilon}\int_{\mathbb{R}^d}\langle \Sigma(\nabla(f))(x); \nabla(f)(x)\rangle_{HS}\mu_\varepsilon(dx).$$

Finally, note that the nondegenerate Gaussian probability measure γ_ε verifies the following Poincaré-type inequality: for all $f \in \mathcal{C}_c^\infty(\mathbb{R}^d, \mathbb{R}^d)$ such that $\int_{\mathbb{R}^d} f(x)\gamma_\varepsilon(dx) = 0$,

$$\int_{\mathbb{R}^d} \|f(x)\|^2 \gamma_\varepsilon(dx) \leq \varepsilon^2 \int_{\mathbb{R}^d} \langle \Sigma \nabla(f)(x); \nabla(f)(x) \rangle_{HS} \gamma_\varepsilon(dx),$$

so that, with obvious notation, $U_\Sigma(\gamma_\varepsilon) = \varepsilon^2 U_{\Sigma,\varepsilon}(\gamma_\varepsilon) = \varepsilon^2$. So, to conclude, let us find an upper bound for the Poincaré constant $U_{\Sigma,\mu,\varepsilon}$ based on the fact that the probability measure μ_ε is the convolution of μ and γ_ε, both satisfying a Poincaré-type inequality with the energy form given, for all $f \in \mathcal{C}_c^\infty(\mathbb{R}^d, \mathbb{R}^d)$, by

$$\mathcal{E}_\Sigma(f, f) = \int_{\mathbb{R}^d} \langle \Sigma \nabla(f)(x); \nabla(f)(x) \rangle_{HS} \mu(dx)$$

and with respective constants $U_{\Sigma,\mu}$ and ε^2. The proof follows closely the one of [23, Theorem 2, (vii)]. Let $f \in \mathcal{C}_c^\infty(\mathbb{R}^d, \mathbb{R}^d)$ be such that $\int_{\mathbb{R}^d} f(x)\mu_\varepsilon(dx) = 0$ and let $x \in \mathbb{R}^d$ be fixed. Then

$$\int_{\mathbb{R}^d} \|\tau_x(f)(y) - \int_{\mathbb{R}^d} \tau_x(f)(y)\mu(dy)\|^2 \mu(dy) \leq U_{\Sigma,\mu} \int_{\mathbb{R}^d} \langle \Sigma \nabla(\tau_x(f))(y); \nabla(\tau_x(f))(y) \rangle_{HS} \mu(dy), \tag{4.15}$$

where τ_x is the translation operator defined, for all f smooth enough and all $y \in \mathbb{R}^d$, by $\tau_x(f)(y) = f(x + y)$. Developing the square gives

$$\int_{\mathbb{R}^d} \|\tau_x(f)(y)\|^2 \mu(dy) \leq U_{\Sigma,\mu} \int_{\mathbb{R}^d} \langle \Sigma \nabla(\tau_x(f))(y); \nabla(\tau_x(f))(y) \rangle_{HS} \mu(dy) + \left\| \int_{\mathbb{R}^d} \tau_x(f)(y)\mu(dy) \right\|^2 .$$

Integrating the previous inequality in the x variable with respect to the probability measure γ_ε gives

$$\int_{\mathbb{R}^d} \|f(z)\|^2 \mu_\varepsilon(dz) \leq U_{\Sigma,\mu} \int_{\mathbb{R}^d} \langle \Sigma \nabla(f)(z); \nabla(f)(z) \rangle_{HS} \mu_\varepsilon(dz) + \int_{\mathbb{R}^d} \left\| \int_{\mathbb{R}^d} \tau_x(f)(y)\mu(dy) \right\|^2 \gamma_\varepsilon(dx).$$

Now, since $f \in \mathcal{C}_c^\infty(\mathbb{R}^d, \mathbb{R}^d)$, the function G defined, for all $x \in \mathbb{R}^d$, by

$$G(x) := \int_{\mathbb{R}^d} \tau_x(f)(y)\mu(dy)$$

is in $\mathcal{C}^1(\mathbb{R}^d)$ with the Jacobian matrix given, for all $x \in \mathbb{R}^d$, by

$$\nabla(G)(x) = \int_{\mathbb{R}^d} \nabla(f)(x+y)\mu(dy).$$

Thus,

$$\int_{\mathbb{R}^d} \|G(x)\|^2 \gamma_\varepsilon(dx) \le \varepsilon^2 \int_{\mathbb{R}^d} \langle \Sigma \nabla(G)(x); \nabla(G)(x)\rangle_{HS} \gamma_\varepsilon(dx) + \left\| \int_{\mathbb{R}^d} G(x)\gamma_\varepsilon(dx) \right\|^2.$$

But $\int_{\mathbb{R}^d} G(x)\gamma_\varepsilon(dx) = 0$. Finally, by Jensen's inequality, for all $x \in \mathbb{R}^d$,

$$\begin{aligned} \langle \Sigma \nabla(G)(x); \nabla(G)(x)\rangle_{HS} &= \left\| \Sigma^{\frac{1}{2}} \nabla(G)(x) \right\|_{HS}^2, \\ &= \left\| \Sigma^{\frac{1}{2}} \int_{\mathbb{R}^d} \nabla(f)(x+y)\mu(dy) \right\|_{HS}^2, \\ &\le \int_{\mathbb{R}^d} \left\| \Sigma^{\frac{1}{2}} \nabla(f)(x+y) \right\|_{HS}^2 \mu(dy). \end{aligned}$$

Then

$$\int_{\mathbb{R}^d} \|G(x)\|^2 \gamma_\varepsilon(dx) \le \varepsilon^2 \int_{\mathbb{R}^d \times \mathbb{R}^d} \left\| \Sigma^{\frac{1}{2}} \nabla(f)(x+y) \right\|_{HS}^2 \mu(dy)\gamma_\varepsilon(dx),$$

which implies that $U_{\Sigma,\mu,\varepsilon} \le U_{\Sigma,\mu} + \varepsilon^2$. So from Theorem 4.1,

$$W_1(\mu_\varepsilon, \tilde{\gamma}_\varepsilon) \le \|\Sigma_\varepsilon^{-\frac{1}{2}}\|_{op} \|\Sigma_\varepsilon\|_{HS} \sqrt{U_{\Sigma,\mu,\varepsilon} - 1}, \tag{4.16}$$

where $\tilde{\gamma}_\varepsilon$ is a nondegenerate centered Gaussian probability measure with the covariance matrix given by $\Sigma_\varepsilon = (1+\varepsilon^2)\Sigma$. Now, by Lévy's continuity theorem, it is clear that μ_ε and $\tilde{\gamma}_\varepsilon$ converge weakly, respectively, to μ and γ_Σ, as $\varepsilon \to 0^+$. Thus, let $h \in \mathcal{C}_c^\infty(\mathbb{R}^d)$ be such that $\|h\|_{\text{Lip}} \le 1$. From the proof of Theorem 4.1,

$$\left| \int_{\mathbb{R}^d} h(x)\mu_\varepsilon(dx) - \int_{\mathbb{R}^d} h(x)\tilde{\gamma}_\varepsilon(dx) \right| \le \|\Sigma_\varepsilon^{-\frac{1}{2}}\|_{op} \|\Sigma_\varepsilon\|_{HS} \sqrt{U_{\Sigma,\mu} + \varepsilon^2 - 1}.$$

Letting $\varepsilon \to 0^+$ and then taking the supremum over all $h \in \mathcal{C}_c^\infty(\mathbb{R}^d)$ with $\|h\|_{\text{Lip}} \le 1$ leads to

$$W_1(\mu, \gamma_\Sigma) \le \|\Sigma^{-\frac{1}{2}}\|_{op} \|\Sigma\|_{HS} \sqrt{U_{\Sigma,\mu} - 1},$$

which concludes the proof of the theorem. □

As an application of the previous techniques, let us provide a rate of convergence in the 1-Wasserstein distance in the multivariate central limit theorem where the limiting centered Gaussian probability measure on $\mathbb{R}^d$ has a covariance matrix given by Σ. The argument is based on a specific representation of the Stein kernel of the standardized sum given, for all $n \geq 1$, by

$$S_n = \frac{1}{\sqrt{n}} \sum_{k=1}^{n} X_k, \tag{4.17}$$

where $(X_k)_{k\geq 1}$ is a sequence of independent and identically distributed (iid) centered random vectors of $\mathbb{R}^d$ with a finite second moment such that

$$\mathbb{E} X_1 X_1^T = \Sigma$$

and whose law satisfies the Poincaré-type inequality (4.8). This result is new.

Theorem 4.3 *Let $d \geq 1$ and let Σ be a nondegenerate $d \times d$ covariance matrix. Let γ_Σ be the nondegenerate centered Gaussian probability measure on $\mathbb{R}^d$ with the covariance matrix given by Σ. Let μ be a centered probability measure on $\mathbb{R}^d$ with finite second moments such that*

$$\int_{\mathbb{R}^d} x x^T \mu(dx) = \Sigma,$$

and which satisfies the Poincaré-type inequality (4.8) *for some $U_{\Sigma,\mu} > 0$. Let $(X_k)_{k\geq 1}$ be a sequence of independent and identically distributed random vectors of $\mathbb{R}^d$ with law μ and let $(S_n)_{n\geq 1}$ be the sequence of normalized sums defined by* (4.17) *and with respective laws $(\mu_n)_{n\geq 1}$. Then, for all $n \geq 1$,*

$$W_1(\mu_n, \gamma_\Sigma) \leq \frac{\|\Sigma^{-\frac{1}{2}}\|_{op} \|\Sigma\|_{HS}}{\sqrt{n}} \sqrt{U_{\Sigma,\mu} - 1}. \tag{4.18}$$

Proof First, let us assume that μ is such that the form $\left(\mathcal{E}_\mu, C_c^\infty(\mathbb{R}^d, \mathbb{R}^d)\right)$ is closable. Then by Proposition 4.1, there exists τ_μ such that, for all $f \in \mathcal{D}(\mathcal{E}_{\Sigma,\mu})$,

$$\int_{\mathbb{R}^d} \langle x; f(x) \rangle \mu(dx) = \int_{\mathbb{R}^d} \langle \nabla(f)(x); \tau_\mu(x) \rangle_{HS} \mu(dx).$$

Next, let τ_n be defined, for all $n \geq 1$, by

$$\tau_n(x) = \mathbb{E}\left[\frac{1}{n} \sum_{k=1}^{n} \tau_\mu(X_k) | S_n = x \right].$$

Observe that, for all $n \geq 1$ and all f smooth enough,

$$\int_{\mathbb{R}^d} \langle \tau_n(x); \nabla(f)(x)\rangle_{HS}\mu_n(dx) = \frac{1}{n}\sum_{k=1}^{n} \mathbb{E}\langle \tau_\mu(X_k); \nabla(f)(S_n)\rangle_{HS},$$
$$= \mathbb{E}\langle S_n; f(S_n)\rangle,$$

as the sequence $(X_k)_{k\geq 1}$ is a sequence of iid random vectors of $\mathbb{R}^d$ and that τ_μ is a Stein kernel for the law of X_1. Next, let $h \in \mathcal{C}_c^\infty(\mathbb{R}^d)$ be such that $\|h\|_{\text{Lip}} \leq 1$. Then from the proof of Theorem 4.1, for all $n \geq 1$

$$|\mathbb{E}h(S_n) - \mathbb{E}h(X)| \leq \|\Sigma^{-\frac{1}{2}}\|_{op} \left(\mathbb{E}\left(\|\tau_n(S_n) - \Sigma\|_{HS}^2\right)\right)^{\frac{1}{2}},$$

where $X \sim \gamma_\Sigma$. So let us estimate the Stein discrepancy, i.e., the last term on the right-hand side of the above inequality. By Jensen's inequality, independence, since $\mathbb{E}\tau_\mu(X_1) = \Sigma$ and from the proof of Theorem 4.1,

$$\mathbb{E}\|\tau_n(S_n) - \Sigma\|_{HS}^2 \leq \frac{1}{n^2}\sum_{k=1}^{n} \mathbb{E}\|\tau_\mu(X_1) - \Sigma\|_{HS}^2,$$
$$\leq \frac{1}{n}\mathbb{E}\|\tau_\mu(X_1) - \Sigma\|_{HS}^2$$
$$\leq \frac{\|\Sigma\|_{HS}^2}{n}\left(U_{\Sigma,\mu} - 1\right).$$

Thus, for all $n \geq 1$,

$$|\mathbb{E}h(S_n) - \mathbb{E}h(X)| \leq \frac{\|\Sigma^{-\frac{1}{2}}\|_{op}\|\Sigma\|_{HS}}{\sqrt{n}}\sqrt{U_{\Sigma,\mu} - 1},$$

and so the bound (4.18) is proved when the form $\left(\mathcal{E}_\mu, \mathcal{C}_c^\infty(\mathbb{R}^d, \mathbb{R}^d)\right)$ is closable. A regularization argument as in the proof of Theorem 4.2 allows to get the bound (4.18) for the general case, concluding the proof of the theorem. □

Remark 4.5

(i) In Theorem 4.3, one could have assumed that $(X_k)_{k\geq 1}$ is a sequence of independent random vectors of $\mathbb{R}^d$ with laws $(\tilde{\mu}_k)_{k\geq 1}$ such that, for all $k \geq 1$,

$$\int_{\mathbb{R}^d} x\tilde{\mu}_k(dx) = 0, \quad \int_{\mathbb{R}^d} xx^T \tilde{\mu}_k(dx) = \Sigma,$$

and with Poincaré constants $(U_{\Sigma,\tilde{\mu}_k})_{k\geq 1}$. Then by a completely similar argument, for all $n \geq 1$,

$$W_1(\mu_n, \gamma_\Sigma) \leq \frac{\|\Sigma^{-\frac{1}{2}}\|_{op}\|\Sigma\|_{HS}}{n}\left(\sum_{k=1}^{n}(U_{\Sigma,\tilde{\mu}_k} - 1)\right)^{\frac{1}{2}}, \tag{4.19}$$

with $S_n \sim \mu_n$.

(ii) When $\Sigma = I_d$, the bound (4.18) boils down to

$$W_1(\mu_n, \gamma) \leq \sqrt{\frac{d}{n}}\sqrt{U_{d,\mu} - 1},$$

which matches exactly the bound obtained in [34, Theorem 4.1] for the 2-Wasserstein distance (recall that by Hölder's inequality, $W_1(\mu, \nu) \leq W_2(\mu, \nu)$, with μ, ν two probability measures on $\mathbb{R}^d$ with a finite second moment). Recently, a large amount of work has been dedicated to rates of convergence in transportation distances for high-dimensional central limit theorems. Let us briefly recall some of these most recent results. In [73], under the condition that $\|X_1\| \leq \beta$ a.s., for some $\beta > 0$, the bound,

$$W_2(\mu_n, \gamma_\Sigma) \leq \frac{5\sqrt{d}\beta\,(1 + \log n)}{\sqrt{n}}, \tag{4.20}$$

is proved, where μ_n is the law of the normalized sum S_n defined by (4.17). In particular, note that [73, Theorem 1.1] is established for all covariance matrices Σ and not only for $\Sigma = I_d$. Thus, our bound gets rid of the term $\beta\sqrt{d}$ in the general nondegenerate case under a finite Poincaré constant assumption, which is not directly comparable to the condition $\|X_1\| \leq \beta$ a.s. (see the discussion after [34, Theorem 4.1] and [15]). Note that, in the isotropic case, the bound (4.20) scales linearly with the dimension since $\beta \geq \sqrt{d}$. An improvement of the bound (4.20) has been obtained in [36, Theorem 1] with $\log n$ replaced by $\sqrt{\log n}$. Similarly, in [39, Theorem B.1], using Stein's method and the Bismut formula, the following bound is obtained at the level of the 1-Wasserstein distance:

$$W_1(\mu_n, \gamma) \leq \frac{Cd\beta\,(1 + \log n)}{\sqrt{n}},$$

for some $C > 0$ and under the assumption that $\|X_i\| \leq \beta$ a.s. for all $i \geq 1$. The anisotropic case is not covered by this last result, but the nonidentically distributed case is. In [21, Theorem 1], under $\mathbb{E}\|X_1\|^4 < +\infty$, the following holds true:

$$W_2(\mu_n, \gamma) \leq \frac{Cd^{\frac{1}{4}}\|\mathbb{E}X_1X_1^T\|X_1\|^2\|_{HS}^{\frac{1}{2}}}{\sqrt{n}},$$

for some $C \in (0, 14)$. The previous bound scales at least linearly with the dimension as well. Finally, let us mention [40], where sharp rates of convergence in the p-Wasserstein distances, for $p \geq 2$, are obtained under various (strong) convexity assumptions.

(iii) Thanks to an inequality of Talagrand (see [71]), quantitative rates of convergence, in relative entropy, toward the Gaussian probability measure γ on $\mathbb{R}^d$ imply quantitative rates of convergence in the 2-Wasserstein distance. In [11, Theorem 1] and [49, Theorem 1.3], under a spectral gap assumption, a rate of convergence of order $1/n$ is obtained in relative entropy in dimension 1, while [12, Theorem 1.1] provides a quantitative entropy jump result under log-concavity in any dimension. Finally, [34] and [36] contain quantitative high-dimensional entropic CLTs under various assumptions.

(iv) There is vast literature on quantitative multivariate central limit theorems for different probability metrics. For example, in [16], a rate of convergence is obtained for the convex distance. Namely, for all $n \geq 1$,

$$\sup_{A \in \mathcal{C}} |\mathbb{P}(S_n \in A) - \mathbb{P}(Z \in A)| \leq \frac{cd^{\frac{1}{4}}}{\sqrt{n}} \mathbb{E}\|\Sigma^{-\frac{1}{2}} X_1\|^3,$$

where $\mathcal{C}$ is the set of all measurable convex subsets of $\mathbb{R}^d$, Z is a centered Gaussian random vector of $\mathbb{R}^d$ with nondegenerate covariance matrix Σ, c is a positive constant which can be made explicit (see [61]), and $(X_i)_{i\geq 1}$ is a sequence of iid random vectors of $\mathbb{R}^d$ such that $\mathbb{E}X_1 = 0$, $\mathbb{E}X_1 X_1^T = \Sigma$, and $\mathbb{E}\|X_1\|^3 < +\infty$. In [37], quantitative high-dimensional CLTs are investigated by means of Stein's method but where the set $\mathcal{C}$ is replaced by the set of hyperrectangles of $\mathbb{R}^d$. In particular, [37, Theorem 1.1] provides an error bound using Stein's kernels, which holds in the nondegenerate anisotropic case. Note that [37, Corollary 1.1] uses the results contained in [40] in order to build Stein's kernels when the sampling distribution is a centered probability measure on $\mathbb{R}^d$ with a log-concave density and a nondegenerate covariance matrix Σ with diagonal entries equal to 1. Finally, in [35], the optimal growth rate of the dimension with the sample size for the probability metric over all hyperrectangles is completely identified under general moment conditions.

Remark 4.6

(i) Let $d \geq 1$, let Σ be a $d \times d$ nondegenerate covariance matrix, and let V be a nonnegative function defined on $\mathbb{R}^d$, which is twice continuously differentiable everywhere on $\mathbb{R}^d$. Let

$$C_V \int_{\mathbb{R}^d} e^{-V(x)} dx = 1,$$

for some constant $C_V > 0$ which depends on V and d. Let μ_V denote the induced probability measure on $\mathbb{R}^d$ and let us assume that

$$\int_{\mathbb{R}^d} x\mu_V(dx) = 0, \quad \int_{\mathbb{R}^d} xx^T \mu_V(dx) = \Sigma.$$

Assume further that, for all $x \in \mathbb{R}^d$,

$$\text{Hess}(V)(x) \geq \kappa \Sigma^{-1}, \tag{4.21}$$

for some $\kappa \in (0, 1]$ (where the order is in the sense of positive semi-definite matrices). Since Σ is nondegenerate, (4.21) ensures that the probability measure μ_V is strongly log-concave. Thus, by the Brascamp and Lieb inequality (see, e.g., [25, 28]), for all $f \in \mathcal{C}_c^\infty(\mathbb{R}^d)$ with $\int_{\mathbb{R}^d} f(x)\mu_V(dx) = 0$,

$$\int_{\mathbb{R}^d} |f(x)|^2 \mu_V(dx) \leq \int_{\mathbb{R}^d} \langle \nabla(f)(x); \text{Hess}(V)(x)^{-1}\nabla(f)(x)\rangle \mu_V(dx).$$

The previous inequality readily implies, for all $f \in \mathcal{C}_c^\infty(\mathbb{R}^d, \mathbb{R}^d)$ with $\int_{\mathbb{R}^d} f(x)\mu_V(dx) = 0$, that

$$\int_{\mathbb{R}^d} \|f(x)\|^2 \mu_V(dx) \leq \int_{\mathbb{R}^d} \langle \nabla(f)(x); \text{Hess}(V)(x)^{-1}\nabla(f)(x)\rangle_{HS} \mu_V(dx),$$

which gives, thanks to (4.21),

$$\int_{\mathbb{R}^d} \|f(x)\|^2 \mu_V(dx) \leq \frac{1}{\kappa} \int_{\mathbb{R}^d} \langle \nabla(f)(x); \Sigma\nabla(f)(x)\rangle_{HS} \mu_V(dx).$$

In particular, if $\kappa = 1$, then $\mu_V = \gamma_\Sigma$. Moreover, Theorem 4.3 provides the following bound for $X_1 \sim \mu_V$: for all $n \geq 1$,

$$W_1(\mu_n, \gamma_\Sigma) \leq \frac{\|\Sigma^{-\frac{1}{2}}\|_{op}\|\Sigma\|_{HS}}{\sqrt{n}} \sqrt{\frac{1}{\kappa} - 1}.$$

(ii) Since [17], it is well known that log-concave probability measures (i.e., probability measures with log-concave densities with respect to the Lebesgue measure) on $\mathbb{R}^d$ satisfy a Poincaré-type inequality. Let $d \geq 1$, let $\Sigma = I_d$, and let μ be a log-concave probability measure on $\mathbb{R}^d$ in an isotropic position, i.e., such that

$$\int_{\mathbb{R}^d} x\mu(dx) = 0, \quad \int_{\mathbb{R}^d} xx^T \mu(dx) = I_d.$$

Then, for all $f \in \mathcal{C}_c^\infty(\mathbb{R}^d)$ such that $\int_{\mathbb{R}^d} f(x)\mu(dx) = 0$,

$$\int_{\mathbb{R}^d} |f(x)|^2 \mu(dx) \leq C_p(\mu) \int_{\mathbb{R}^d} \langle \nabla(f)(x); \nabla(f)(x)\rangle \mu(dx),$$

where $C_p(\mu) > 0$ is the best constant for which the previous inequality holds. According to the well-known Kannan-Lovász-Simonovits (KLS) conjecture (see, e.g., [4]), the constant $C_p(\mu)$ should be uniformly upper bounded by some universal constant $C \geq 1$ (independent of the dimension) for all log-concave probability measures on $\mathbb{R}^d$ in an isotropic position. In Theorem 4.3, this conjecture would imply that, for all $n \geq 1$,

$$W_1(\mu_n, \gamma) \leq \sqrt{\frac{d}{n}}\sqrt{C-1},$$

if $X_1 \sim \mu$. To date, the best-known bound on the constant $C_P(\mu)$ is provided by the very recent result in [32], which ensures a lower bound on the isoperimetric constant of an isotropic log-concave probability measure μ on $\mathbb{R}^d$:

$$I_P(\mu) \geq d^{-c'\left(\frac{\log\log d}{\log d}\right)^{\frac{1}{2}}},$$

for some $c' > 0$ (see also [50] for an even more recent improvement). According to Cheeger's inequality (see, e.g., [17]),

$$C_P(\mu) \leq c_1 d^{c_2\left(\frac{\log\log d}{\log d}\right)^{\frac{1}{2}}},$$

for some positive numerical constants c_1 and c_2.

In analogy with the Kannan, Lovász, and Simonovits (KLS) conjecture and with regard to the general anisotropic case with a nondegenerate covariance matrix Σ, it seems natural to wonder whether the functional $U_{\Sigma,\mu}$ is uniformly bounded over the class of centered log-concave probability measures on $\mathbb{R}^d$ with covariance structure Σ and whether the corresponding upper bound is dimension free.

5 Appendix

Lemma 5.1 *Let $d \geq 1$ and let μ, ν be two probability measures on $\mathbb{R}^d$ with a finite first moment. Then*

$$W_1(\mu, \nu) = \sup_{h\in\mathcal{C}^\infty(\mathbb{R}^d),\ \|h\|_{\mathrm{Lip}\leq 1}} \left|\int_{\mathbb{R}^d} h(x)\mu(dx) - \int_{\mathbb{R}^d} h(x)\nu(dx)\right|. \tag{5.1}$$

Proof Recall that by duality,

$$W_1(\mu, \nu) = \sup_{h \in \mathrm{Lip}, \|h\|_{\mathrm{Lip}} \leq 1} \left| \int_{\mathbb{R}^d} h(x)\mu(dx) - \int_{\mathbb{R}^d} h(x)\nu(dx) \right|, \tag{5.2}$$

where Lip is the space of Lipschitz functions on $\mathbb{R}^d$ with the Lipschitz semi-norm

$$\|h\|_{\mathrm{Lip}} = \sup_{x, y \in \mathbb{R}^d, x \neq y} \frac{|h(x) - h(y)|}{\|x - y\|}.$$

So, at first, it is clear that

$$W_1(\mu, \nu) \geq \sup_{h \in \mathcal{C}^\infty(\mathbb{R}^d), \|h\|_{\mathrm{Lip}} \leq 1} \left| \int_{\mathbb{R}^d} h(x)\mu(dx) - \int_{\mathbb{R}^d} h(x)\nu(dx) \right|.$$

Next, let $h \in \mathcal{C}^1(\mathbb{R}^d)$ be such that $\|h\|_{\mathrm{Lip}} \leq 1$. Let $\varepsilon > 0$ and let p_ε be the centered multivariate Gaussian density with covariance matrix εI_d; i.e., for all $y \in \mathbb{R}^d$,

$$p_\varepsilon(y) = \frac{1}{(2\pi\varepsilon)^{\frac{d}{2}}} \exp\left(-\frac{\|y\|^2}{2\varepsilon}\right).$$

Moreover, let

$$h_\varepsilon(x) = \int_{\mathbb{R}^d} h(x - y) p_\varepsilon(y) dy, \quad x \in \mathbb{R}^d.$$

It is clear that $h_\varepsilon \in \mathcal{C}^\infty(\mathbb{R}^d)$ and that

$$\|h_\varepsilon\|_{\mathrm{Lip}} \leq 1, \quad \varepsilon > 0,$$

since $\|h\|_{\mathrm{Lip}} \leq 1$. Moreover, for all $\varepsilon > 0$ and all $x \in \mathbb{R}^d$,

$$\begin{aligned} |h(x) - h_\varepsilon(x)| &\leq \|h\|_{\mathrm{Lip}} \int_{\mathbb{R}^d} \|z\| p_\varepsilon(z) dz, \\ &\leq C_d \sqrt{\varepsilon}, \end{aligned}$$

for some constant $C_d > 0$ depending only on $d \geq 1$. Thus,

$$W_1(\mu, \nu) \leq \sup_{h \in \mathcal{C}^\infty(\mathbb{R}^d), \|h\|_{\mathrm{Lip}} \leq 1} \left| \int_{\mathbb{R}^d} h(x)\mu(dx) - \int_{\mathbb{R}^d} h(x)\nu(dx) \right| + 2C_d\sqrt{\varepsilon}.$$

Letting $\varepsilon \to 0^+$ concludes the proof of this reduction principle. □

Lemma 5.2 *Let $d \geq 1$ and let μ, ν be two probability measures on $\mathbb{R}^d$ with a finite first moment. Then*

$$W_1(\mu,\nu) = \sup_{h\in\mathcal{C}_c^\infty(\mathbb{R}^d),\ \|h\|_{\mathrm{Lip}\le 1}} \left| \int_{\mathbb{R}^d} h(x)\mu(dx) - \int_{\mathbb{R}^d} h(x)\nu(dx) \right|. \tag{5.3}$$

Proof Let h be a Lipschitz function on $\mathbb{R}^d$ such that

$$\|h\|_{\mathrm{Lip}} = \sup_{x,y\in\mathbb{R}^d, x\neq y} \frac{|h(x)-h(y)|}{\|x-y\|} \le 1.$$

Recall that by Rademacher's theorem, such a function h is differentiable almost everywhere on $\mathbb{R}^d$. Actually, as shown next, it is possible to restrict the supremum appearing in (5.2) to bounded Lipschitz functions h defined on $\mathbb{R}^d$ and such that $\|h\|_{\mathrm{Lip}} \le 1$. Indeed, let $R > 0$ and let G_R be the function defined, for all $y \in \mathbb{R}$, by

$$G_R(y) = (-R) \vee (y \wedge R).$$

G_R is clearly bounded on $\mathbb{R}$ by R, and, for all $y \in \mathbb{R}$ fixed,

$$\lim_{R\to+\infty} G_R(y) = y.$$

Moreover, $\|G_R\|_{\mathrm{Lip}} \le 1$ by construction. Thus, for all $R > 0$, let h_R be defined, for all $x \in \mathbb{R}^d$, by $h_R(x) = G_R(h(x))$. The function h_R is bounded on $\mathbb{R}^d$ and 1-Lipschitz by composition; moreover, $\lim_{R\to+\infty} G_R(h(x)) = h(x)$, for all $x \in \mathbb{R}^d$. Then

$$\left|\int_{\mathbb{R}^d} h(x)\mu(dx) - \int_{\mathbb{R}^d} h(x)\nu(dx)\right| \le \left|\int_{\mathbb{R}^d} h_R(x)\mu(dx) - \int_{\mathbb{R}^d} h_R(x)\nu(dx)\right| \\ + \left|\int_{\mathbb{R}^d} (h_R(x)-h(x))\mu(dx)\right| + \left|\int_{\mathbb{R}^d} (h_R(x)-h(x))\nu(dx)\right|.$$

Thus,

$$\left|\int_{\mathbb{R}^d} h(x)\mu(dx) - \int_{\mathbb{R}^d} h(x)\nu(dx)\right| \le \sup_{h\in\mathrm{Lip}_b,\ \|h\|_{\mathrm{Lip}}\le 1} \left|\int_{\mathbb{R}^d} h(x)\mu(dx) - \int_{\mathbb{R}^d} h(x)\nu(dx)\right| \\ + \left|\int_{\mathbb{R}^d} (h_R(x)-h(x))\mu(dx)\right| + \left|\int_{\mathbb{R}^d} (h_R(x)-h(x))\nu(dx)\right|,$$

where Lip_b is the set of bounded Lipschitz functions on $\mathbb{R}^d$. Next, without loss of generality, let us assume that $h(0) = 0$. Now, since, for all $x \in \mathbb{R}^d$ and all $R > 0$,

$$|h_R(x) - h(x)| \le 2|h(x)| \le 2\|x\|,$$

the dominated convergence theorem ensures that

$$\lim_{R\to+\infty}\int_{\mathbb{R}^d}(h_R(x)-h(x))\mu(dx)=0,$$

and similarly for ν. Thus,

$$\left|\int_{\mathbb{R}^d}h(x)\mu(dx)-\int_{\mathbb{R}^d}h(x)\nu(dx)\right|\leq\sup_{h\in\mathrm{Lip}_b,\,\|h\|_{\mathrm{Lip}}\leq1}\left|\int_{\mathbb{R}^d}h(x)\mu(dx)-\int_{\mathbb{R}^d}h(x)\nu(dx)\right|.$$

Next, applying the regularization procedure of Lemma 5.1, one has

$$W_1(\mu,\nu)=\sup_{h\in C_b^\infty(\mathbb{R}^d),\,\|h\|_{\mathrm{Lip}}\leq1}\left|\int_{\mathbb{R}^d}h(x)\mu(dx)-\int_{\mathbb{R}^d}h(x)\nu(dx)\right|,$$

where $C_b^\infty(\mathbb{R}^d)$ is the space of infinitely differentiable bounded functions on $\mathbb{R}^d$. Finally, let $h\in C_b^\infty(\mathbb{R}^d)$ be such that $\|h\|_{\mathrm{Lip}}\leq1$ and let ψ be a smooth compactly supported function with values in $[0,1]$ and with support included in the Euclidean ball centered at the origin and of radius 2 such that, for all $x\in\mathbb{R}^d$ with $\|x\|\leq1$, $\psi(x)=1$. Then let

$$\tilde{h}_R(x)=\psi\left(\frac{x}{R}\right)h(x),\quad R\geq1,\quad x\in\mathbb{R}^d.$$

Clearly, $\tilde{h}_R\in C_c^\infty(\mathbb{R}^d)$; moreover, for all $x\in\mathbb{R}^d$ and all $R\geq1$,

$$\nabla(\tilde{h}_R)(x)=\frac{1}{R}\nabla(\psi)(\frac{x}{R})h(x)+\psi\left(\frac{x}{R}\right)\nabla(h)(x).$$

Then, for all $x\in\mathbb{R}^d$,

$$\|\nabla(\tilde{h}_R)(x)\|\leq1+\frac{1}{R}\|h\|_\infty\|\nabla(\psi)\|_\infty.$$

Thus,

$$\begin{aligned}\left|\int_{\mathbb{R}^d}h(x)\mu(dx)-\int_{\mathbb{R}^d}h(x)\nu(dx)\right|&\leq\left|\int_{\mathbb{R}^d}\tilde{h}_R(x)\mu(dx)-\int_{\mathbb{R}^d}\tilde{h}_R(x)\nu(dx)\right|+\left|\int_{\mathbb{R}^d}(h(x)-\tilde{h}_R(x))\mu(dx)\right|\\&\quad+\left|\int_{\mathbb{R}^d}(h(x)-\tilde{h}_R(x))\nu(dx)\right|,\\&\leq\left(1+\frac{1}{R}\|h\|_\infty\|\nabla(\psi)\|_\infty\right)\sup_{h\in C_c^\infty(\mathbb{R}^d),\,\|h\|_{\mathrm{Lip}}\leq1}\left|\int_{\mathbb{R}^d}h(x)\mu(dx)-\int_{\mathbb{R}^d}h(x)\nu(dx)\right|\\&\quad+\left|\int_{\mathbb{R}^d}h(x)\left(1-\psi\left(\frac{x}{R}\right)\right)\mu(dx)\right|+\left|\int_{\mathbb{R}^d}h(x)\left(1-\psi\left(\frac{x}{R}\right)\right)\nu(dx)\right|.\end{aligned}$$

But

$$\left| \int_{\mathbb{R}^d} h(x) \left(1 - \psi\left(\frac{x}{R}\right)\right) \mu(dx) \right| \leq \|h\|_\infty \int_{\|x\| \geq R} \mu(dx),$$

and similarly for ν. Thus, letting $R \to +\infty$,

$$\left| \int_{\mathbb{R}^d} h(x)\mu(dx) - \int_{\mathbb{R}^d} h(x)\nu(dx) \right| \leq \sup_{h \in \mathcal{C}_c^\infty(\mathbb{R}^d),\ \|h\|_{\mathrm{Lip}} \leq 1} \left| \int_{\mathbb{R}^d} h(x)\mu(dx) - \int_{\mathbb{R}^d} h(x)\nu(dx) \right|,$$

which concludes the proof of the lemma. □

Acknowledgments CH thanks G. Reinert for pointing out to him, at the Stein's Golden Jubilee Conference in June 2022, that the bound (4.5) has independently been obtained in the joint work [38]. We are grateful to an anonymous referee for the insightful remarks, which led to an improvement in the presentation of this manuscript.

References

1. R. Adamczak, B. Polaczyk, M. Strzelecki, Modified log-Sobolev inequalities, Beckner inequalities and moment estimates. J. Funct. Anal. **282**(7), 109349 (2022)
2. D. Addona, M. Muratori, M. Rossi, On the equivalence of Sobolev norms in Malliavin spaces. J. Funct. Anal. **283**(7), 109600 (2022)
3. S. Albeverio, M. Röckner, Classical dirichlet forms on topological vector spaces – closability and a Cameron-Martin formula. J. Funct. Anal. **88**(2), 395–436 (1990)
4. D. Alonso-Gutiérrez, J. Bastero, *Approaching the Kannan-Lovász-Simonovits and Variance Conjectures*. Lecture Notes in Mathematics (Springer, Cham, 2015)
5. M. Arnaudon, M. Bonnefont, A. Joulin, Intertwinings and generalized Brascamp-Lieb inequalities. Rev. Mat. Iberoam. **34**(3), 1021–1054 (2018)
6. B. Arras, C. Houdré, *On Stein's Method for Infinitely Divisible Laws with Finite First Moment*. Springer Briefs in Probability and Mathematical Statistics (Springer, Cham, 2019)
7. B. Arras, C. Houdré, On Stein's method for multivariate self-decomposable laws with finite first moment. Electron. J. Probab. **24**(29), 1–33 (2019)
8. B. Arras, C. Houdré, On Stein's method for multivariate self-decomposable laws. Electron. J. Probab. **24**(128), 63 (2019)
9. B. Arras, C. Houdré, On some operators associated with non-degenerate symmetric α-stable probability measures. Potential Anal. (2022). https://doi.org/10.1007/s11118-022-10026-9
10. B. Arras, Y. Swan, A stroll along the gamma. Stoch. Process. Appl. **127**, 3661–3688 (2017)
11. S. Artstein, K.M. Ball, F. Barthe, A. Naor, On the rate of convergence in the entropic central limit theorem. Probab. Theory Relat. Fields **129**, 381–390 (2004)
12. K. Ball, V.H. Nguyen, Entropy jumps for isotropic log-concave random vectors and spectral gap. Studia Math. **213**(1), 81–96 (2012)
13. D. Bakry, I. Gentil, M. Ledoux, *Analysis and Geometry of Markov Diffusion Operators* (Springer, Cham, 2014)
14. A.D. Barbour, Stein's method for diffusion approximations. Probab. Theory Relat. Fields. **84**(3), 297–322 (1990)
15. J.-B. Bardet, N. Gozlan, F. Malrieu, P.-A. Zitt, Functional inequalities for Gaussian convolutions of compactly supported measures: explicit bounds and dimension dependence. Bernoulli **24**(1), 333–353 (2018)
16. V. Bentkus, A Lyapunov-type bound in $\mathbb{R}^d$. Theory Probab. Appl. **49**(2), 311–323 (2005)

17. S.G. Bobkov, Isoperimetric and analytic inequalities for log-concave probability measures. Ann. Probab. **27**(4), 1903–1921 (1999)
18. S.G. Bobkov, C. Houdré, Isoperimetric constants for product probability measures. Ann. Probab. **25**(1), 184–205 (1997)
19. S.G. Bobkov, M. Ledoux, Weighted Poincaré-type inequalities for cauchy and other convex measures. Ann. Probab. **37**(2), 403–427 (2009)
20. V.I. Bogachev, *Differentiable Measures and the Malliavin Calculus* (American Mathematical Society, Providence, 2010)
21. T. Bonis, Stein's method for normal approximation in Wasserstein distances with application to the multivariate central limit theorem. Probab. Theory Relat. Fields **178**, 827–860 (2020)
22. M. Bonnefont, A. Joulin, Y. Ma, Spectral gap for spherically symmetric log-concave probability measures, and beyond. J. Funct. Anal. **270**, 2456–2482 (2016)
23. A.A. Borovkov, S.A. Utev, On an inequality and a related characterization of the normal distribution. Theory Probab. Appl. **28**(2), 219–228 (1984)
24. S. Bourguin, S. Campese, Approximation of Hilbert-Valued Gaussians on Dirichlet structures. Electron. J. Probab. **25**, 1–30 (2020)
25. H.J. Brascamp, E.H. Lieb, On extensions of the Brunn-Minkowski and Prékopa-Leindler theorems, including inequalities for log-concave functions, and with an application to the diffusion equation. J. Funct. Anal. **22**, 366–389 (1976)
26. H. Brézis, *Functional Analysis, Sobolev Spaces and Partial Differential Equations* (Springer, New York, 2011)
27. P. Cattiaux, M. Fradon, Entropy, reversible diffusion processes, and Markov uniqueness. J. Funct. Anal. **138**(1), 243–272 (1996)
28. E.A. Carlen, D. Cordero-Erausquin, E.H. Lieb, Asymmetric covariance estimates of Brascamp-Lieb type and related inequalities for log-concave measures. Ann. Inst. Henri Poincaré Probab. Stat. **49**(1), 1–12 (2013)
29. S. Chatterjee, E. Meckes, Multivariate normal approximation using exchangeable pairs. ALEA Lat. Am. J. Probab. Math. Stat. **4**, 257–283 (2008)
30. L.H.Y. Chen, Poincaré-type inequalities via Stochastic integrals. Z. Wahrscheinlichkeitstheorie verw. Gebiete **69**, 251–277 (1985)
31. L.H.Y. Chen, J.H. Lou, Characterization of probability distributions by Poincaré-type inequalities. Ann. Inst. Henri Poincaré Probab. Stat. **23**(1), 91–110 (1987)
32. Y. Chen, An almost constant lower bound of the isoperimetric coefficient in the KLS conjecture. Geom. Funct. Anal. **31**, 34–61 (2021)
33. Z.-Q. Chen, X. Zhang, Heat kernels and analyticity of non-symmetric jump diffusions semigroups. Probab. Theory Relat. Fields **165**, 267–312 (2016)
34. T.A. Courtade, M. Fathi, A. Pananjady, Existence of Stein's kernels under a spectral gap, and discrepancy bounds. Ann. Inst. Henri Poincaré Probab. Stat. **55**(2), 777–790 (2019)
35. D. Das, S. Lahiri, Central limit theorem in high dimensions: the optimal bound on dimension growth rate. Trans. Am. Math. Soc. **374**(10), 6991–7009 (2021)
36. R. Eldan, D. Mikulincer, A. Zhai, The CLT in high dimensions: quantitative bounds via Martingale embedding. Ann. Probab. **48**(5), 2494–2524 (2020)
37. X. Fang, Y. Koike, High-dimensional central limit theorems by Stein's method. Ann. Appl. Probab. **31**(4), 1660–1686 (2021)
38. X. Fang, Y. Koike, New error bounds in multivariate normal approximations via exchangeable pairs with applications to Wishart matrices and fourth moment theorems. Ann. Appl. Probab. **32**(1), 602–631 (2022)
39. X. Fang, Q.-M. Shao, L. Xu, Multivariate approximations in Wasserstein distance by Stein's method and Bismut's formula. Probab. Theory Relat. Fields **174**, 945–979 (2019)
40. M. Fathi, Stein's kernels and moment maps. Ann. Probab. **47**(4), 2172–2185 (2019)
41. M. Fathi, Higher-order Stein's kernels for Gaussian approximation. Studia Math. **256**, 241–258 (2021)
42. L. Goldstein, Y. Rinott, Multivariate normal approximations by Stein's method and size bias couplings. J. Appl. Probab. **33**(1), 1–17 (1996)

43. F. Götze, On the rate of convergence in the multivariate CLT. Ann. Probab. **19**(2), 724–739 (1991)
44. V. Hoang Nguyen, Φ-entropy inequalities and asymmetric covariance estimates for convex measures. Bernoulli **25**(4A), 3090–3108 (2019)
45. L. Hörmander, On the theory of general partial differential operators. Acta Math. **94**, 161–248 (1955)
46. C. Houdré, V. Pérez-Abreu, Covariance identities and inequalities for functionals on Wiener and Poisson spaces. Ann. Probab. **23**(1), 400–419 (1995)
47. C. Houdré, V. Pérez-Abreu, D. Surgailis, Interpolation, correlation identities and inequalities for infinitely divisible variables. J. Fourier Anal. Appl. **4**(6), 651–668 (1998)
48. C. Houdré, N. Privault, Concentration and deviation inequalities in infinite dimensions via covariance representations. Bernoulli **8**(6), 697–720 (2002)
49. O. Johnson, A.R. Barron, Fisher information inequalities and the central limit theorem. Probab. Theory Relat. Fields **129**, 391–409 (2004)
50. B. Klartag, J. Lehec, Bourgain's slicing problem and KLS isoperimetry up to polylog. Geom. Funct. Anal. **32**, 1134–1159 (2022)
51. M. Ledoux, On improved Sobolev embedding theorems. Math. Res. Lett. **10**, 659–669 (2003)
52. M. Ledoux, I. Nourdin, G. Peccati, Stein's method, logarithmic Sobolev and transport inequalities. Geom. Funct. Anal. **25**, 256–306 (2015)
53. V.A. Liskevich, Y.A. Semenov, Dirichlet operators: a priori estimates and the uniqueness problem. J. Funct. Anal. **109**, 199–213 (1992)
54. E. Meckes, On Stein's method for multivariate normal approximation, in *High Dimensional Probability V: The Luminy Volume*. Institute of Mathematical Statistics (IMS) Collections, vol. 5 (Institute of Mathematical Statistics, Beachwood, 2009), pp. 153–178
55. I. Nourdin, G. Peccati, *Normal Approximations with Malliavin Calculus: From Stein's Method to Universality* (Cambridge University Press, Cambridge, 2012)
56. I. Nourdin, G. Peccati, G. Reinert, Second order Poincaré inequalities and CLTs on Wiener space. J. Funct. Anal. **257**, 593–609 (2009)
57. I. Nourdin, G. Peccati, A. Réveillac, Multivariate normal approximation using Stein's method and Malliavin calculus. Ann. Inst. H. Poincaré Probab. Statist. **46**(1), 45–58 (2010)
58. I. Nourdin, G. Peccati, X. Yang, Multivariate normal approximation on the Wiener space: new bounds in the convex distance. J. Theor. Probab. **35**, 2020–2037 (2021)
59. G. Pisier, Probabilistic methods in the geometry of Banach spaces, in *Probability and Analysis* (Springer, Berlin/Heidelberg, 1986), pp. 167–241
60. M. Raic, A multivariate CLT for decomposable random vectors with finite second moments. J. Theor. Probab. **17**(3), 573–603 (2004)
61. M. Raic, A multivariate Berry-Esseen theorem with explicit constants. Bernoulli **25**(4A), 2824–2853 (2019)
62. G. Reinert, A. Röllin, Multivariate normal approximation with Stein's method of exchangeable pairs under a general linearity condition. Ann. Probab. **37**(6), 2150–2173 (2009)
63. Y. Rinott, V. Rotar, A multivariate CLT for local dependence with $n^{-1/2}\log(n)$ rate and applications to multivariate graph related statistics. J. Multivariate Anal. **56**, 333–350 (1996)
64. M. Röckner, F.-Y. Wang, Weak Poincaré inequalities and L^2-convergence rates of Markov semigroups. J. Funct. Anal. **185**, 564–603 (2001)
65. A. Röllin, Stein's method in high dimensions with applications. Ann. Inst. Henri Poincaré Probab. Stat. **49**(2), 529–549 (2013)
66. K.-I. Sato, *Lévy Processes and Infinitely Divisible Distributions* (Cambridge Studies in Advanced Mathematics, Cambridge University Press, Cambridge 2013), **68**, xiv+521 pp
67. A. Saumard, J.A. Wellner, On the isoperimetric constant, covariance inequalities and Lp-Poincaré inequalities in dimension one. Bernoulli **25**(3), 1794–1815 (2019)
68. H.H. Shih, On Stein's method for infinite-dimensional Gaussian approximation in abstract Wiener spaces. J. Funct. Anal. **261**(5), 1236–1283 (2011)

69. C. Stein, *Approximate Computation of Expectations*. Institute of Mathematical Statistics Lecture Notes. Monograph Series, vol. 7 (Institute of Mathematical Statistics, Hayward, 1986)
70. K. Takano, The Lévy representation of the characteristic function of the probability density $\Gamma(m + \frac{d}{2})(\pi^{d/2}\Gamma(m))^{-1}(1 + \|x\|^2)^{-m-d/2}$. Bull. Fac. Sci. Ibaraki Univ. Ser. A **21**, 21–27 (1989)
71. M. Talagrand, Transportation cost for Gaussian and other product measures. Geom. Funct. Anal. **6**(3), 587–600 (1996)
72. T. Watanabe, Asymptotic estimates of multi-dimensional stable densities and their applications. Trans. Am. Math. Soc. **359**(6), 2851–2879 (2007)
73. A. Zhai, A high-dimensional CLT in W2 distance with near optimal convergence rate. Probab. Theory Relat. Fields **170**, 821–845 (2018)

Volume Properties of High-Dimensional Orlicz Balls

Franck Barthe and Paweł Wolff

2020 Mathematics Subject Classification: 52A23; 60E15; 60F05

Lebesgue spaces play a central role in functional analysis and enjoy remarkable structural properties. A natural extension of this family is given by the class of Orlicz spaces, which also enjoy a wealth of remarkable properties, see, e.g., [21]. Similarly, for $p \geq 1$, the unit balls of $\mathbb{R}^n$ equipped with the ℓ_p-norm, often denoted by $B_p^n = \{x \in \mathbb{R}^n;\ \sum_i |x_i|^p \leq 1\}$, are well studied convex bodies, and usually the first family of test cases for new conjectures. Their simple analytic description allows for many explicit calculations, for instance, of their volume. A simple probabilistic representation of uniform random vectors on B_p^n, given in terms of i.i.d. random variables of law $\exp(-|t|^p)\,dt/K_p$, is available, see [4]. It allows to investigate various fine properties of the volume distribution on B_p^n. The study of general Orlicz balls is more difficult, due to the lack of explicit formulas, in particular for the volume of the set itself.

In this note, we show that probabilistic methods allow to derive precise asymptotic estimates of the volume of Orlicz balls when the dimension tends to infinity, and rough estimates which are valid in every dimension. This allows us to complement a result of Kolesnikov and Milman [13] on the spectral gap of uniform measures on Orlicz balls, by giving an explicit description of the range of parameters where their result applies, see Sect. 5. In Sect. 6, we show, among other results, the asymptotic independence of a fixed set of coordinates of uniform random vectors on some families of Orlicz balls of increasing dimensions. This is a natural extension of a classical observation (going back to Maxwell) about uniform vectors on Euclidean spheres and balls. The last section deals with properties of linear functionals of random vectors on Orlicz balls.

F. Barthe (✉) · P. Wolff
Institut de Mathématiques de Toulouse, Université de Toulouse & CNRS, Toulouse, France
e-mail: barthe@math.univ-toulouse.fr; pwolff@mimuw.edu.pl

R. Adamczak et al. (eds.), *High Dimensional Probability IX*, Progress in Probability 80, https://doi.org/10.1007/978-3-031-26979-0_2

After this research work was completed, we learned by J. Prochno of his independent work [10] with Z. Kabluchko about similar volume asymptotics for Orlicz balls. Their paper uses sophisticated methods from the theory of large deviations, which have the potential to give more precise results for a given sequence of balls in increasing dimensions. Our approach is more elementary and focuses on uniform convergence over some wide range of parameters, as required by our applications to the spectral gap conjecture.

1 Notation and Statement

Throughout this paper, a Young function is a non-negative convex function on $\mathbb{R}$ which vanishes only at 0. Note that we do not assume symmetry at this stage. For a given Young function $\Psi : \mathbb{R} \to \mathbb{R}^+$, denote

$$B^n_\Psi = \Big\{x \in \mathbb{R}^n : \sum_{i=1}^n \Psi(x_i) \le 1\Big\}$$

the corresponding n-dimensional Orlicz ball. Our aim is to estimate the asymptotic volume of $B^n_{\Psi/E_n} = \{x \in \mathbb{R}^n : \sum_{i=1}^n \Psi(x_i) \le E_n\}$ for relevant sequences E_n of linear order in the dimension.

Let $\lambda > 0$. Consider the following probability measure on $\mathbb{R}$,

$$\mu_\lambda(dt) = e^{-\lambda\Psi(t)} \frac{dt}{Z_\lambda},$$

with Z_λ being a normalization constant. Let X be a random variable with the distribution μ_λ. Set

$$m = m_\lambda = \mathbb{E}\Psi(X), \qquad \sigma^2 = \sigma_\lambda^2 = \mathrm{Var}\big(\Psi(X)\big).$$

Our aim is to prove

Theorem 1.1 *Consider a Young function Ψ and $\lambda > 0$. Let $n \ge 1$ be an integer and $\alpha \in \mathbb{R}$. Set*

$$E := m_\lambda n + \alpha\sigma_\lambda\sqrt{n},$$

then

$$\begin{aligned}\mathrm{Vol}\big(B^n_{\Psi/E}\big) &= (Z_\lambda e^{\lambda m_\lambda})^n \frac{1}{\lambda\sigma_\lambda\sqrt{2\pi n}} e^{-\alpha^2/2} e^{\lambda\sigma_\lambda\sqrt{n}\alpha}(1 + O(n^{-1/2}))\\ &= \frac{Z_\lambda^n e^{\lambda E}}{\lambda\sigma_\lambda\sqrt{2\pi n}} e^{-\alpha^2/2}(1 + O(n^{-1/2})),\end{aligned}$$

where the term $O(n^{-1/2})$ depends on λ, Ψ, and non-decreasingly in $|\alpha|$.

Corollary 1.2 *Consider a Young function Ψ and $\lambda > 0$. Let $(a_n)_{n\geq 1}$ be a bounded sequence, and $E_n := m_\lambda n + a_n\sqrt{n}$. Then when the dimension n tends to ∞,*

$$\mathrm{Vol}\big(B^n_{\Psi/E_n}\big) \sim \frac{Z^n_\lambda e^{\lambda E_n}}{\lambda\sigma_\lambda\sqrt{2\pi n}}\, e^{-a_n^2/(2\sigma_\lambda^2)}.$$

Let us mention that the above results can be applied to B^n_{Ψ/E_n} when $E_n = mn + a_n\sqrt{n}$ where $m > 0$ is fixed and $(a_n)_n$ is a bounded sequence. Indeed the next lemma ensures the existence of a $\lambda > 0$ such that $m = m_\lambda$.

Lemma 1.3 *Let Ψ be as above. Then the map defined $(0, +\infty)$ to $(0, +\infty)$ by*

$$\lambda \mapsto R(\lambda) := \frac{\int \Psi(t)e^{-\lambda\Psi(t)}dt}{\int e^{-\lambda\Psi(t)}dt}$$

is onto.

Proof By hypothesis, $\int \exp(-\lambda\Psi) < \infty$ for all $\lambda > 0$. This fact allows us to apply the dominated convergence theorem, and to show that the ratio $R(\lambda)$ is a continuous function of $\lambda > 0$. Let us show that $\lim_{\lambda\to 0^+} R(\lambda) = \infty$ and $\lim_{\lambda\to\infty} R(\lambda) = 0$. The claim will then follow by continuity.

Consider an arbitrary $K > 0$. Since $\Psi \geq 0$,

$$\frac{\int \Psi e^{-\lambda\Psi}}{\int e^{-\lambda\Psi}} \geq \frac{K\int_{\Psi\geq K} e^{-\lambda\Psi}}{\int e^{-\lambda\Psi}} = K\left(1 - \frac{\int_{\Psi<K} e^{-\lambda\Psi}}{\int e^{-\lambda\Psi}}\right)$$
$$\geq K\left(1 - \frac{\mathrm{Vol}(\{x;\, \Psi(x) < K\})}{\int e^{-\lambda\Psi}}\right).$$

By monotone convergence, $\lim_{\lambda\to 0^+}\int e^{-\lambda\Psi} = \infty$. Hence, $\liminf_{\lambda\to 0^+} R(\lambda) \geq K$. Since this holds for every $K > 0$, we conclude that $\lim_{\lambda\to 0^+} R(\lambda) = \infty$.

Let $\varepsilon > 0$. As above, since $\Psi \geq 0$,

$$\frac{\int \Psi e^{-\lambda\Psi}}{\int e^{-\lambda\Psi}} \leq \varepsilon + \frac{\int_{\Psi>\varepsilon} \Psi e^{-\lambda\Psi}}{\int e^{-\lambda\Psi}}.$$

Next, using $x \leq e^x$ and for $\lambda > 2$,

$$\int_{\Psi>\varepsilon} \Psi e^{-\lambda\Psi} \leq \int_{\Psi>\varepsilon} e^{-(\lambda-1)\Psi} = \int_{\Psi>\varepsilon} e^{-\Psi} e^{-(\lambda-2)\Psi} \leq e^{-(\lambda-2)\varepsilon}\int e^{-\Psi},$$

and

$$\int e^{-\lambda\Psi} \geq \int_{\Psi\leq\varepsilon/2} e^{-\lambda\Psi} \geq e^{-\lambda\varepsilon/2}\mathrm{Vol}(\{x;\,\Psi(x)\leq\varepsilon/2\}).$$

Since $\Psi(0)=0$ and Ψ is continuous, the latter quantity is positive. Combining the above three estimates, we get

$$\frac{\int \Psi e^{-\lambda\Psi}}{\int e^{-\lambda\Psi}} \leq \varepsilon + e^{-(\frac{\lambda}{2}-2)\varepsilon}\frac{\int e^{-\Psi}}{\mathrm{Vol}(\{x;\,\Psi(x)\leq\varepsilon/2\})}.$$

Letting $\lambda\to\infty$ yields $\limsup_{\lambda\to\infty} R(\lambda)\leq\varepsilon$, for all $\varepsilon>0$. □

2 Probabilistic Formulation

We start with a formula relating the volume with an expectation expressed in terms of independent random variables. Let $\lambda>0$. Let $(X_i)_{i\in\mathbb{N}^*}$ be i.i.d. r.v.'s with the distribution $\mu_\lambda(dt) = e^{-\lambda\Psi(t)}\,dt/Z_\lambda$. Recall that $m_\lambda = \mathbb{E}\Psi(X_i)$ and $\sigma_\lambda^2 = \mathrm{Var}\big(\Psi(X_i)\big)$. We denote by S_n the normalized central limit sums:

$$S_n = \frac{1}{\sigma_\lambda\sqrt{n}}\sum_{i=1}^{n}(\Psi(X_i)-m_\lambda).$$

With this notation, we get the following representation for any $\lambda>0$

$$\begin{aligned}\mathrm{Vol}\big(B^n_{\Psi/E}\big) &= \int \mathbf{1}_{\{\sum_{i=1}^n \Psi(x_i)\leq E\}}dx\\ &= \int \mathbf{1}_{\{\sum_{i=1}^n \Psi(x_i)\leq E\}} Z_\lambda^n e^{\lambda\sum_{i=1}^n \Psi(x_i)}\prod_{i=1}^n \mu_\lambda(dx_i)\\ &= Z_\lambda^n\mathbb{E}\left(e^{\lambda\sum_{i=1}^n \Psi(X_i)}\mathbf{1}_{\{\sum_{i=1}^n \Psi(X_i)\leq E\}}\right)\\ &= (Z_\lambda e^{\lambda m_\lambda})^n\mathbb{E}\left(e^{\lambda\sigma_\lambda\sqrt{n}S_n}\mathbf{1}_{\{S_n\leq\frac{E-m_\lambda n}{\sigma_\lambda\sqrt{n}}\}}\right). \qquad (2.1)\end{aligned}$$

By the Central Limit Theorem, S_n converges in distribution to a standard Gaussian random variable when n tends to infinity. Such Gaussian approximation results allow to estimate the asymptotic behavior of the above expectations. Nevertheless, a direct application of the CLT or the Berry–Esseen bounds does not seem to be sufficient for our purposes. A more refined analysis is required, built on classical results and techniques on the distribution of sums of independent random variables which go back to Cramér [7] (see also [2]).

Theorem 1.1 is a direct consequence of the following one, applied to $Y_i = (\Psi(X_i) - m_\lambda)/\sigma_\lambda$. For a random variable V, let $\mathbb{P}_V$ and φ_V denote the distribution and the characteristic function.

Theorem 2.1 *Let $(Y_i)_{i\geq 1}$ be a sequence of i.i.d. real random variables such that $\mathbb{E}|Y_i|^3 < \infty$, $\mathbb{E}Y_i = 0$ and $\mathrm{Var}(Y_i) = 1$. Suppose $\varepsilon, \delta > 0$ are such that so-called Cramér's condition is satisfied for Y_i:*

$$|\varphi_{Y_i}(t)| \leq 1 - \varepsilon \quad \textit{for } |t| > \delta. \tag{2.2}$$

For $n \geq 1$, let $S_n = (Y_1 + \cdots + Y_n)/\sqrt{n}$. Then for $\ell > 0$ and $\alpha \in \mathbb{R}$,

$$\mathbb{E}\left(e^{\ell\sqrt{n}\,S_n}\mathbf{1}_{S_n\leq\alpha}\right) = \frac{1}{\ell\sqrt{2\pi n}}e^{\ell\sqrt{n}\alpha-\alpha^2/2}\big(1 + O(n^{-1/2})\big).$$

Remark 2.2 The term $O(n^{-1/2})$ involves an implicit dependence in ℓ, α and the law of Y_1. For $n \geq 16\ell^2 + (2|\alpha| + 1)^2\ell^{-2}$, our argument provides a term $O(n^{-1/2})$ which depends only on $(\ell, \delta, 1/\varepsilon, \nu_3 := \mathbb{E}|Y_i|^3, |\alpha|)$. Moreover the dependence is continuous in the parameters, and non-decreasing in all the parameters but ℓ. This allows for uniform bounds when the parameters are in compact subsets of their domain.

Remark 2.3 Note that non-trivial ε and δ exist by the Riemann–Lebesgue lemma as soon has the law of Y_i is absolutely continuous.

3 Probabilistic Preliminaries

We start with some useful lemmas. The first one is a key estimate for quantitative central limit theorems, quoted from Petrov's book [18].

Lemma 3.1 ([18], Lemma V.2.1, p. 109) *Let $X_1, \ldots, X_n$ be independent random variables, $\mathbb{E}X_j = 0$, $\mathbb{E}|X_j|^3 < \infty$ ($j = 1, \ldots, n$). Denote $B_n = \sum_{j=1}^n \mathbb{E}X_j^2$, $L_n = B_n^{-3/2}\sum_{j=1}^n \mathbb{E}|X_j|^3$ and $S_n = B_n^{-1/2}\sum_{j=1}^n X_j$. Then*

$$|\varphi_{S_n}(t) - e^{-t^2/2}| \leq 16L_n|t|^3e^{-t^2/3}$$

for $|t| \leq \frac{1}{4L_n}$.

Lemma 3.2 ([18], Lemma I.2.1, p. 10) *For any characteristic function φ,*

$$1 - |\varphi(2t)|^2 \leq 4(1 - |\varphi(t)|^2)$$

holds for all $t \in \mathbb{R}$.

Lemma 3.3 *Let (S_n) be as in Theorem 2.1. Let T be independent of (S_n) and assume that its characteristic function φ_T is Lebesgue integrable. Then for all $n \geq 1$, the density of $S_n + \frac{T}{n}$ is bounded by a number $C = C(1/\varepsilon, \delta, \nu_3, \|\varphi_T\|_1)$, which is non-decreasing in each of its parameters.*

***Proof of Lemma* 3.3** Since $\varphi_{S_n+T/n} = \varphi_{S_n}\varphi_T(\cdot/n)$ is Lebesgue integrable, the inversion formula ensures that the density of $S_n + \frac{1}{n}T$ at x equals

$$\begin{aligned}
g_{S_n+\frac{1}{n}T}(x) &= \frac{1}{2\pi}\int_{-\infty}^{\infty} e^{-itx}\varphi_{S_n}(t)\varphi_T\big(t/n\big)\,dt \\
&= \frac{1}{2\pi}\int_{-\infty}^{\infty} e^{-itx}e^{-t^2/2}\varphi_T\big(t/n\big)\,dt \\
&+ \frac{1}{2\pi}\int_{-\infty}^{\infty} e^{-itx}(\varphi_{S_n}(t) - e^{-t^2/2})\varphi_T\big(t/n\big)\,dt \\
&\leq \frac{1}{\sqrt{2\pi}} + \frac{1}{2\pi}\int_{-\infty}^{\infty} |\varphi_{S_n}(t) - e^{-t^2/2}|\,|\varphi_T\big(t/n\big)|\,dt.
\end{aligned}$$

To bound the last integral, we apply Lemma 3.1 with $B_n = n$ and $L_n = \nu_3 n^{-1/2}$. We get

$$\begin{aligned}
g_{S_n+T/n}(x) &\leq \frac{1}{\sqrt{2\pi}} + \frac{1}{2\pi}\int_{|t|\leq\frac{\sqrt{n}}{4\nu_3}} \frac{16\nu_3}{\sqrt{n}}|t|^3e^{-t^2/3}\,dt \\
&+ \frac{1}{2\pi}\int_{|t|>\frac{\sqrt{n}}{4\nu_3}} (|\varphi_{S_n}(t)| + e^{-t^2/2})|\varphi_T\big(t/n\big)|\,dt \\
&\leq \frac{1}{\sqrt{2\pi}} + \frac{72\nu_3}{\pi\sqrt{n}} \\
&+ \underbrace{\frac{1}{2\pi}\int_{|t|>\frac{\sqrt{n}}{4\nu_3}} |\varphi_{S_n}(t)|\,|\varphi_T\big(t/n\big)|\,dt}_{I} + \frac{1}{2\pi}\int_{|t|>\frac{\sqrt{n}}{4\nu_3}} e^{-t^2/2}\,dt.
\end{aligned}$$

For integral (I) from the last line we use (2.2) which implies

$$|\varphi_{S_n}(t)| = \left|\varphi_{Y_1}\big(t/\sqrt{n}\big)^n\right| \leq (1-\varepsilon)^n \quad \text{for } |t| \geq \delta\sqrt{n}.$$

However, δ might be larger than $\frac{1}{4\nu_3}$, i.e., $4\nu_3\delta \geq 1$. If this is so, we use Lemma 3.2 on characteristic functions: since (2.2) implies

$$1 - |\varphi_{Y_i}(t)|^2 \geq \varepsilon \quad \text{for } |t| \geq \delta,$$

Lemma 3.2 implies that for any non-negative integer k,

$$1 - |\varphi_{Y_i}(t)|^2 \geq 4^{-k}\varepsilon \quad \text{for } |t| \geq 2^{-k}\delta.$$

Taking $k = \lceil \log_2(4\nu_3\delta) \rceil)$ implies $2^{-k}\delta \leq \frac{1}{4\nu_3}$ and $4^{-k} \geq \frac{1}{(8\nu_3\delta)^2}$ and hence

$$|\varphi_{Y_i}(t)|^2 \leq 1 - \frac{\varepsilon}{(8\nu_3\delta)^2} \quad \text{for } |t| \geq \frac{1}{4\nu_3}.$$

In any case, we obtain that

$$|\varphi_{S_n}(t)| \leq \left(1 - \frac{\varepsilon}{\max(1, (8\nu_3\delta)^2)}\right)^{n/2} \quad \text{for } |t| \geq \frac{\sqrt{n}}{4\nu_3}. \tag{3.1}$$

Using the above we estimate the integral (I) as follows. Using the rough estimate

$$(1-x)^m = e^{m\log(1-x)} \leq e^{-mx} = \frac{1}{e^{mx}} \leq \frac{1}{mx},$$

valid for any $m > 0$ and $x \in (0, 1)$, we get

$$\begin{aligned} I &\leq \left(1 - \frac{\varepsilon}{\max(1, (8\nu_3\delta)^2)}\right)^{n/2} n \int_0^\infty |\varphi_T(u)|\, du \\ &\leq n \frac{2\max(1, (8\nu_3\delta)^2)}{n\varepsilon} \|\varphi_T\|_1 \leq 2\|\varphi_T\|_1 \frac{1 + (8\nu_3\delta)^2}{\varepsilon}. \end{aligned}$$

Finally we obtain that the density of $S_n + \frac{1}{n}T$ is bounded by $C_1 + C_2\nu_3 + C_3\|\varphi_T\|_1 \frac{1+(\nu_3\delta)^2}{\varepsilon}$ for some constants $C_1, C_2, C_3 > 0$. □

Denote by ϕ the density of the standard normal distribution on $\mathbb{R}$ and let Φ be its cumulative distribution function. Our last two preliminary statements are easy consequences of the equality $e^{\gamma t}\phi(t) = e^{\gamma^2/2}\phi(t - \gamma)$ satisfied by the Gaussian density

Lemma 3.4 *Let Z be a standard normal random variable. For any Borel set $A \subset \mathbb{R}$,*

$$\mathbb{E}e^{\gamma Z}\mathbf{1}_{Z\in A} = e^{\gamma^2/2}\mathbb{P}(Z \in A - \gamma). \tag{3.2}$$

Lemma 3.5 *For any $s > 0$ and $\alpha \in \mathbb{R}$, and λ such $\lambda s - \frac{\alpha}{s} > 1$, it holds*

$$\int_0^\infty \lambda e^{-\lambda x} \frac{1}{\sqrt{2\pi}s} e^{-\frac{(x-\alpha)^2}{2s^2}}\, dx = \frac{1}{\sqrt{2\pi}s} e^{-\frac{\alpha^2}{2s^2}} \left(1 + O\left(\frac{1 + \frac{|\alpha|}{s}}{\lambda s - \frac{\alpha}{s}}\right)\right).$$

In particular if s and α stay bounded in the sense that $s \in [1/S, S]$, $|\alpha| \le A$ holds for some $A, S > 0$, then for $\lambda > 2AS^{-2} + S^{-1}$, the last factor simplifies to $1 + O_{A,S}\left(\frac{1}{\lambda}\right)$.

Proof Using a standard Gaussian random variable Z, we rewrite the left-hand side as

$$\begin{aligned}\mathcal{T} := \int_0^\infty \lambda e^{-\lambda x} \frac{1}{\sqrt{2\pi}\, s} e^{-\frac{(x-\alpha)^2}{2s^2}}\, dx &= \lambda \mathbb{E} e^{-\lambda(sZ+\alpha)} \mathbf{1}_{Z > -\frac{\alpha}{s}} \\ &= \lambda e^{-\lambda\alpha} e^{\lambda^2 s^2/2} \mathbb{P}\Big(Z > \lambda s - \frac{\alpha}{s}\Big) \\ &= \lambda e^{-\frac{\alpha^2}{2s^2}} e^{(\lambda s - \frac{\alpha}{s})^2/2}\Big(1 - \Phi\big(\lambda s - \frac{\alpha}{s}\big)\Big),\end{aligned}$$

where the second equality follows from (3.2). Next we use the classical bound, for $t > 0$,

$$\frac{1}{t} \ge \sqrt{2\pi}\, e^{t^2/2}\big(1 - \Phi(t)\big) \ge \frac{1}{\sqrt{t^2+2}},$$

which implies that for $t > 1$, $\sqrt{2\pi}\, t e^{t^2/2}\big(1 - \Phi(t)\big) = 1 + O(1/t^2)$. When $\lambda s - \frac{\alpha}{s} > 1$ we obtain that

$$\begin{aligned}\sqrt{2\pi}\, s\, e^{\frac{\alpha^2}{2s^2}} \mathcal{T} &= \frac{\lambda s}{\lambda s - \frac{\alpha}{s}} \left(1 + O\Big(\frac{1}{(\lambda s - \frac{\alpha}{s})^2}\Big)\right) \\ &= \left(1 + \frac{\frac{\alpha}{s}}{\lambda s - \frac{\alpha}{s}}\right) \cdot \left(1 + O\Big(\frac{1}{(\lambda s - \frac{\alpha}{s})^2}\Big)\right) = 1 + O\left(\frac{1 + \frac{|\alpha|}{s}}{\lambda s - \frac{\alpha}{s}}\right).\end{aligned}$$

The case when α and s are bounded readily follows. □

4 Proof of Theorem 2.1

Our aim is to show that for any $\alpha \in \mathbb{R}$, $\mathcal{I} = \mathcal{J} \times (1 + O(n^{-1/2}))$ where

$$\mathcal{I} = \mathbb{E} e^{\ell\sqrt{n} S_n} \mathbf{1}_{\{S_n \le \alpha\}} \quad \text{and} \quad \mathcal{J} = \frac{1}{\ell\sqrt{2\pi n}} e^{\ell\sqrt{n}\alpha} e^{-\alpha^2/2}.$$

Let Z be a standard Gaussian random variable, independent of the Y_i's. The first step is to introduce the modified quantity

$$\mathcal{I}_2 = \mathbb{E} e^{\ell\sqrt{n}(S_n + n^{-1}Z)} \mathbf{1}_{\{S_n + n^{-1}Z \le \alpha\}},$$

and to check that it is enough for our purpose to establish $\mathcal{I}_2 = \mathcal{J} \times (1 + O(n^{-1/2}))$. In order to do so we estimate the difference between $\mathcal{I}$ and $\mathcal{I}_2$.

By the triangle inequality:

$$\begin{aligned}|\mathcal{I}_2 - \mathcal{I}| &\le \mathbb{E}e^{\ell\sqrt{n}S_n}|e^{\ell n^{-1/2}Z} - 1|\mathbf{1}_{\{S_n\le\alpha\}} \\ &\quad + \mathbb{E}e^{\ell\sqrt{n}(S_n+n^{-1}Z)}\left|\mathbf{1}_{\{S_n+n^{-1}Z\le\alpha\}} - \mathbf{1}_{\{S_n\le\alpha\}}\right| \\ &= \mathcal{I}_3 + \mathcal{I}_4 + \mathcal{I}_5,\end{aligned}$$

where

$$\begin{aligned}\mathcal{I}_3 &= \mathbb{E}e^{\ell\sqrt{n}S_n}|e^{\ell n^{-1/2}Z} - 1|\mathbf{1}_{\{S_n\le\alpha\}} \\ \mathcal{I}_4 &= \mathbb{E}e^{\ell\sqrt{n}(S_n+n^{-1}Z)}\mathbf{1}_{\{\alpha<S_n\le\alpha-n^{-1}Z\}} \\ \mathcal{I}_5 &= \mathbb{E}e^{\ell\sqrt{n}(S_n+n^{-1}Z)}\mathbf{1}_{\{\alpha-n^{-1}Z<S_n\le\alpha\}}.\end{aligned}$$

By independence $\mathcal{I}_3 = \mathcal{I} \cdot \mathbb{E}|e^{\ell n^{-1/2}Z} - 1|$. Next, we use that for $t \in [0, 1]$,

$$\mathbb{E}|e^{tZ} - 1| \le \sqrt{\mathbb{E}\big(e^{2tZ} - 2e^{tZ} + 1\big)} = \sqrt{e^{2t^2} - 2e^{t^2/2} + 1} \le 3t.$$

Thus, under the hypothesis $n \ge 16\ell^2$ we obtain that $\mathcal{I}_3 \le \frac{3\ell}{\sqrt{n}}\mathcal{I} \le 3\mathcal{I}/4$.

For the term $\mathcal{I}_4$, we introduce $T = U + U'$ where U and U' are independent random variables uniformly distributed in $(-1, 1)$ and note that $\varphi_T(u) = (\sin(u)/u)^2$ is Lebesgue integrable. Since $|T| \le 2$ a.s.,

$$\begin{aligned}\mathcal{I}_4 &\le e^{\ell\sqrt{n}\alpha}\int_0^\infty \mathbb{P}(\alpha < S_n \le \alpha + n^{-1}x)\phi(x)\,dx \\ &\le e^{\ell\sqrt{n}\alpha}\int_0^\infty \mathbb{P}(\alpha - 2/n < S_n + T/n \le \alpha + (x+2)/n)\phi(x)\,dx.\end{aligned}$$

By Lemma 3.3, $S_n + T/n$ has a density which is bounded by a constant, say $C > 0$. Then

$$\begin{aligned}\mathcal{I}_4 &\le e^{\ell\sqrt{n}\alpha}\int_0^\infty C\frac{x+4}{n}\phi(x)\,dx = \frac{C}{n}e^{\ell\sqrt{n}\alpha}(\pi^{-1/2} + 2) \\ &= \frac{C}{\sqrt{n}} \cdot \ell\sqrt{2\pi}\,e^{\alpha^2/2}\mathcal{J} \cdot (\pi^{-1/2} + 2) = \mathcal{J} \cdot O(n^{-1/2}).\end{aligned}$$

The term $\mathcal{I}_5$ is estimated in a similar way:

$$\begin{aligned}\mathcal{I}_5 &\le e^{\ell\sqrt{n}\alpha}\int_0^\infty e^{\ell x n^{-1/2}}\mathbb{P}(\alpha - x/n < S_n \le \alpha)\phi(x)\,dx\\ &\le e^{\ell\sqrt{n}\alpha}\int_0^\infty e^{\ell x n^{-1/2}}\mathbb{P}(\alpha-(x+2)/n < S_n + T/n \le \alpha + 2/n)\phi(x)\,dx\\ &\le e^{\ell\sqrt{n}\alpha}\int_0^\infty e^{\ell x}C\frac{x+4}{n}\phi(x)\,dx = \mathcal{J}\cdot O(n^{-1/2}).\end{aligned}$$

This concludes the first step of the proof, which guarantees that for $n \ge 16\ell^2$

$$|\mathcal{I}_2 - \mathcal{I}| \le \frac{3\ell}{\sqrt{n}}\mathcal{I} + O\Big(\frac{1}{\sqrt{n}}\Big)\mathcal{J}. \tag{4.1}$$

Our next task is to prove that $\mathcal{I}_2 = \mathcal{J}\times(1+O(n^{-1/2}))$. We use the Fourier transform approach. It relies on the Parseval formula, which ensures that whenever random variables V and W have square integrable densities g_V and g_W, their characteristic functions are also square integrable and the following relation holds:

$$\int_{-\infty}^{\infty} g_V(x)g_W(x)\,dx = \frac{1}{2\pi}\int_{-\infty}^{\infty}\varphi_V(t)\overline{\varphi_W(t)}\,dt. \tag{4.2}$$

Given n, set $W = \alpha - (S_n + \frac{1}{n}Z)$. Then

$$\begin{aligned}\mathcal{I}_2 &= \mathbb{E}e^{\ell\sqrt{n}(S_n+n^{-1}Z)}\mathbf{1}_{\{S_n+n^{-1}Z\le\alpha\}} = e^{\ell\sqrt{n}\alpha}\mathbb{E}e^{-\ell\sqrt{n}W}\mathbf{1}_{W\ge 0}\\ &= \frac{e^{\ell\sqrt{n}\alpha}}{\ell\sqrt{n}}\int_0^\infty \ell\sqrt{n}e^{-\ell\sqrt{n}x}\,d\mathbb{P}_W(x).\end{aligned}$$

Let V be a random variable having exponential distribution with parameter $\ell\sqrt{n}$. We have proved that

$$\widetilde{\mathcal{I}_2} := \ell\sqrt{n}e^{-\ell\sqrt{n}\alpha}\mathcal{I}_2 = \int g_V(x)\,d\mathbb{P}_W(x).$$

Observe that our goal is to establish that $\widetilde{\mathcal{I}_2} = \frac{1}{\sqrt{2\pi}}e^{-\alpha^2/2}(1+O(n^{-1/2}))$.

Since $\mathbb{P}_W$ is given by the convolution of a probability measure and of the bounded density of Z/n, it is absolutely continuous with bounded (and thus square integrable) density. Hence, we may apply the Parseval formula (4.2) to V and W. Since $\varphi_W(t) = e^{i\alpha t}\overline{\varphi_{S_n}(t)}e^{-t^2/(2n^2)}$, we obtain

$$\widetilde{\mathcal{I}_2} = \frac{1}{2\pi}\int_{-\infty}^{\infty}\frac{1}{1-\frac{it}{\ell\sqrt{n}}}e^{-i\alpha t}\varphi_{S_n}(t)e^{-t^2/(2n^2)}\,dt = \frac{\mathcal{M}+\mathcal{E}}{2\pi},$$

where

$$\mathcal{M} = \int_{-\infty}^{\infty} \frac{e^{-i\alpha t}}{1 - \frac{it}{\ell\sqrt{n}}} e^{-t^2/2} e^{-t^2/(2n^2)}\, dt$$

$$\mathcal{E} = \int_{-\infty}^{\infty} \frac{e^{-i\alpha t}}{1 - \frac{it}{\ell\sqrt{n}}} (\varphi_{S_n}(t) - e^{-t^2/2}) e^{-t^2/(2n^2)}\, dt.$$

Applying Parseval's formula as before, but replacing S_n with and independent standard Gaussian variable G yields $\mathcal{M}/(2\pi) = \int g_V d\mathbb{P}_{\widetilde{W}}$ where $\widetilde{W} = \alpha - (G + Z/n)$ has $\mathcal{N}(\alpha, 1 + n^{-2})$ distribution. Therefore

$$\frac{\mathcal{M}}{2\pi} = \int_0^{\infty} \ell\sqrt{n} e^{-\ell\sqrt{n}x} \frac{e^{-\frac{(x-\alpha)^2}{2(1+n^{-2})}}}{\sqrt{2\pi(1+n^{-2})}} dx.$$

Lemma 3.5 with $\lambda := \ell\sqrt{n}$ and $s^2 := 1 + n^{-2}$ yields, provided $\ell\sqrt{n} \geq 2|\alpha| + 1$,

$$\begin{aligned}\frac{\mathcal{M}}{2\pi} &= \frac{1}{\sqrt{2\pi(1+n^{-2})}} e^{-\frac{\alpha^2}{2(1+n^{-2})}} (1 + O(n^{-1/2})) \\ &= \frac{1}{\sqrt{2\pi}} e^{-\frac{\alpha^2}{2}} (1 + O(n^{-1/2})).\end{aligned}$$

It remains to bound the error term:

$$\begin{aligned}|\mathcal{E}| &= \left| \int_{-\infty}^{\infty} \frac{e^{-i\alpha t}}{1 - \frac{it}{\ell\sqrt{n}}} (\varphi_{S_n}(t) - e^{-t^2/2}) e^{-t^2/(2n^2)}\, dt \right| \\ &\leq \int_{-\infty}^{\infty} \left|\varphi_{S_n}(t) - e^{-t^2/2}\right| e^{-t^2/(2n^2)}\, dt \\ &\leq \int_{|t| \leq \sqrt{n}/(4\nu_3)} 16\nu_3 n^{-1/2} |t|^3 e^{-t^2/3}\, dt \\ &\quad + \int_{|t| > \sqrt{n}/(4\nu_3)} \left|\varphi_{S_n}(t)\right| e^{-t^2/(2n^2)}\, dt + \int_{|t| > \sqrt{n}/(4\nu_3)} e^{-t^2/2}\, dt \\ &\leq C\nu_3 n^{-1/2} + I + II,\end{aligned}$$

where the second inequality follows from Lemma 3.1. The estimate of the term II is immediate:

$$II \leq 2e^{-n/(32\nu_3^2)}.$$

In order to estimate I, we use (3.1) and a variant of its previous application using the bound $(1-x)^m \le 1/e^{mx} \le 2/(mx)^2$ for $x \in (0,1)$:

$$I \le \left(1 - \frac{\varepsilon}{\max(1, (8\nu_3\delta)^2)}\right)^{n/2} n\sqrt{2\pi} = O_{\nu_3, \frac{1}{\varepsilon}, \delta}(n^{-1/2}).$$

Hence $\mathcal{E} = O(n^{-1/2}) = e^{-\alpha^2/2} O(e^{\alpha^2/2} n^{-1/2}) = e^{-\alpha^2/2} O_{|\alpha|}(n^{-1/2})$. This ends the proof of the second step, asserting $\mathcal{I}_2 = \mathcal{J} \times (1 + O(n^{-1/2}))$. Combining the latter with (4.1) yields the claim of the theorem.

5 Application to Spectral Gaps

Our volume asymptotics for Orlicz balls allow to complement a result of Kolesnikov and Milman [13] about a famous conjecture by Kannan, Lovász, and Simonovits, which predicts the approximate value of the Poincaré constants of convex bodies (a.k.a. inverse spectral gap of the Neumann Laplacian). More precisely if μ is a probability measure on some Euclidean space, one denotes by $C_P(\mu)$ (respectively, $C_P^{Lin}(\mu)$) the smallest constant C such that for all locally Lipschitz (respectively, linear) functions f, it holds

$$\mathrm{Var}_\mu(f) \le C \int |\nabla f|^2 d\mu.$$

Obviously $C_P^{Lin}(\mu) \le C_P(\mu)$, and the KLS conjecture predicts the existence of a universal constant c such that for any dimension n and any convex body $K \subset \mathbb{R}^n$,

$$C_P(\lambda_K) \le c\, C_P^{Lin}(\lambda_K),$$

where λ_K stands for the uniform probability measure on K. The conjecture turned out to be central in the understanding in high-dimension volume distributions of convex sets. We refer to, e.g., [1, 5, 6, 13, 14] for more background and references, and to [12] for a recent breakthrough. Kolesnikov and Milman have verified the conjecture for some Orlicz balls. We state next a simplified version of their full result on generalized Orlicz balls. Part of the simplification is unessential, as it amounts to reduce by dilation and translations to a convenient setting. A more significant simplification, compared to their work, is that we consider balls where all coordinates play the same role.

Theorem 5.1 ([13]) *Let $V : \mathbb{R} \to \mathbb{R}^+$ be a convex function with $V(0) = 0$ and such that $d\mu(x) = e^{-V(x)}dx$ is a probability measure. We also assume that the function $x \mapsto xV'(x)$, defined almost everywhere, belongs to the space $L^2(\mu)$. For each dimension $n \ge 1$, let*

$$\mathrm{Level}_n(V) := \left\{ E \geq 0;\ e^{-E}\mathrm{Vol}_n\left(B^n_{V/E}\right) \geq \frac{1}{e}\frac{n^n e^{-n}}{n!} \right\}.$$

Then there exists a constant c, which depends only on V (through $\|xV'(X)\|_{L^2(\mu)}$) such that for all $E \in \mathrm{Level}_n(V)$,

$$C_P(\lambda_{B^n_{V/E}}) \leq c\, C_P^{Lin}(\lambda_{B^n_{V/E}}).$$

Moreover, $\mathrm{Level}_n(V)$ *is an interval of length at most* $e\frac{n!e^n}{n^n} = e\sqrt{2\pi n}(1 + o(1))$ *as* $n \to \infty$*, and*

$$1 + n\int_{\mathbb{R}} V(x)e^{-V(x)}dx \in \mathrm{Level}_n(V).$$

We can prove more about the set $\mathrm{Level}_n(V)$ and in particular we show that its length is of order $\sqrt{n}$:

Proposition 5.2 *Let* $V : \mathbb{R} \to \mathbb{R}^+$ *be a Young function such that* $d\mu(x) = e^{-V(x)}dx$ *is a probability measure. Let* $m_1 = \int Ve^{-V}$ *be the average of V with respect to* μ*, and* σ_1^2 *its variance. For every* $\varepsilon \in (0, 1)$ *there exists an integer* $n_0 = n_0(V, \varepsilon)$ *depending on V such that for all* $n \geq n_0$,

$$\left[m_1 n - \sigma_1(1-\varepsilon)\sqrt{2n}\, ;\ m_1 n + \sigma_1(1-\varepsilon)\sqrt{2n}\right] \subset \mathrm{Level}_n(V).$$

Proof We apply Theorem 1.1, with $\Psi = V$ and $\lambda = 1$. With the notation of the theorem $\mu = \mu_1$ and $Z_1 = \int e^{-V} = 1$. We choose E of the following form: $E = m_1 n + \alpha\sigma_1\sqrt{n}$ with $|\alpha| \leq (1-\varepsilon)\sqrt{2}$. The theorem ensures that

$$\mathrm{Vol}\left(B^n_{V/E}\right) = \frac{e^E}{\sigma_1\sqrt{2\pi n}} e^{-\alpha^2/2}\left(1 + O\left(\frac{1}{\sqrt{n}}\right)\right),$$

where the $O(n^{-1/2})$ is uniform in $\alpha \in [-(1-\varepsilon)\sqrt{2}, (1-\varepsilon)\sqrt{2}]$. A sharp inequality due to Nguyen and Wang ensures that $\sigma_1^2 = \mathrm{Var}_{e^{-V}}(V) \leq 1$ (see [17, 22], [16] and for a short proof [9]). Therefore

$$e^{-E}\mathrm{Vol}\left(B^n_{V/E}\right) \geq \frac{1}{\sqrt{2\pi n}} e^{-(1-\varepsilon)^2}\left(1 + O\left(\frac{1}{\sqrt{n}}\right)\right),$$

whereas

$$\frac{1}{e}\frac{n^n e^{-n}}{n!} = \frac{e^{-1}}{\sqrt{2\pi n}}(1 + o(1)).$$

Hence for n large enough and for all α in the above interval $e^{-E}\mathrm{Vol}\big(B^n_{V/E}\big) \geq \frac{1}{e}\frac{n^n e^{-n}}{n!}$. □

Corollary 5.3 *Let $V : \mathbb{R} \to \mathbb{R}^+$ be a Young function such that $d\mu(x) = e^{-V(x)}dx$ is a probability measure. Let m_1 and σ_1^2 denote the average and the variance of V with respect to μ. We also assume that the function $x \mapsto xV'(x)$ belongs to the space $L^2(\mu)$. Let $\varepsilon \in (0, 1)$. Then there exists $c = c(V, \varepsilon)$ such that for all $n \geq 1$ and all $E \in \Big[m_1 n - \sigma_1(1-\varepsilon)\sqrt{2n}\,;\ m_1 n + \sigma_1(1-\varepsilon)\sqrt{2n}\Big]$,*

$$C_P(\lambda_{B^n_{V/E}}) \leq c\, C_P^{Lin}(\lambda_{B^n_{V/E}}).$$

Proof Combining the later proposition and theorem yields the result for $n \geq n_0(V, \varepsilon)$. In order to deal with smaller dimensions, we simply apply known dimension dependent bounds: e.g., Kannan, Lovász, and Simonovits [11] proved that $C_P(\lambda_K) \leq \kappa n C_P^{Lin}(\lambda_K)$ for all convex bodies K in $\mathbb{R}^n$, with κ a universal constant. □

6 Asymptotic Independence of Coordinates

A classical observation, going back to Maxwell, but also attributed to Borel and to Poincaré, states that for a fixed k, the law of the first k coordinates of a uniform random vector on the Euclidean sphere of $\mathbb{R}^n$, centered at the origin and of radius $\sqrt{n}$, tends to the law of k independent standard Gaussian random variables as n tends to infinity. Quantitative versions of this asymptotic independence property were given by Diaconis and Freedman [8], as well as a similar result for the unit sphere of the ℓ_1-norm, involving exponential variables in the limit. Extensions to random vectors distributed according to the cone measure on the surface of the unit ball B_p^n were given by Rachev and Rüschendorf [20], while Mogul'skiĭ [15] dealt with the case of the normalized surface measure. Explicit calculations, or the probabilistic representation put forward in [4], easily yield asymptotic independence results for the first k coordinates of a uniform vector on the set B_p^n itself, when k is fixed and n tend to infinity.

In this section we study marginals of a random vector $\xi^{(n)}$ uniformly distributed on B^n_{Ψ/E_n}, where E_n and n tend to ∞.

Let us start with the simple case when $E_n = mn$ for some $m > 0$, which can be written as $m = m_\lambda$ for some $\lambda > 0$. Let $k \geq 1$ be a fixed integer, then the density at $(x_1, \ldots, x_k) \in \mathbb{R}^k$ of the first k coordinates $(\xi_1^{(n)}, \ldots, \xi_k^{(n)})$ is equal to

$$\frac{\mathrm{Vol}_{n-k}\big(B^n_{\Phi/E_n} \cap \{y \in \mathbb{R}^n;\ y_i = x_i,\ \forall i \leq k\}\big)}{\mathrm{Vol}\big(B^n_{\Phi/E_n}\big)} = \frac{\mathrm{Vol}\big(B^{n-k}_{\Psi/(E_n - \sum_{i=1}^k \Psi(x_i))}\big)}{\mathrm{Vol}\big(B^n_{\Psi/E_n}\big)}$$

We apply Corollary 1.2 twice: once for the denominator, and once for the numerator after writing

$$m_\lambda n - \sum_{i\le k} \Psi(x_i) = m_\lambda (n-k) + \frac{m_\lambda k - \sum_{i\le k}\Psi(x_i)}{\sqrt{n-k}}\sqrt{n-k}.$$

We obtain that the above ratio is equivalent to

$$\frac{Z_\lambda^{n-k} e^{\lambda(E_n - \sum_{i\le k}\Psi(x_i))}}{\lambda\sigma_\lambda\sqrt{2\pi(n-k)}} \cdot \frac{\lambda\sigma_\lambda\sqrt{2\pi n}}{Z_\lambda^n e^{\lambda E_n}} \sim \frac{e^{-\lambda\sum_{i=1}^k \Psi(x_i)}}{Z_\lambda^k}.$$

Thus we have proved the convergence in distribution of $(\xi_1^{(n)}, \dots, \xi_k^{(n)})$ to $\mu_\lambda^{\otimes k}$ as n tends to infinity. In other words the first k coordinates of $\xi^{(n)}$ are asymptotically i.i.d. of law μ_λ. This is true for more general balls and for a number of coordinates going also to infinity:

Theorem 6.1 *Let $E_n = m_\lambda n + \alpha_n\sigma_\lambda\sqrt{n}$, where $(\alpha_n)_{n\ge1}$ is bounded. Let the random vector $\xi^{(n)}$ be uniformly distributed on B^n_{Ψ/E_n}. For any $k_n = o(\sqrt{n})$,*

$$\lim_{n\to\infty} d_{TV}\big((\xi_1^{(n)}, \dots, \xi_{k_n}^{(n)}), \mu_\lambda^{\otimes k_n}\big) = 0.$$

Proof Below, we simply write ξ_i for $\xi_i^{(n)}$. Recall that (X_i) are i.i.d. r.v.'s with the distribution μ_λ. Set $t_n := n^{1/4}k_n^{1/2}$ so that $t_n = o(\sqrt{n})$ and $k_n = o(t_n)$. The total variation distance between the law of $(\xi_1^{(n)}, \dots, \xi_{k_n}^{(n)})$ and $\mu_\lambda^{\otimes k_n}$ is

$$\begin{aligned}
&\int_{\mathbb{R}^{k_n}} \left| \frac{1}{\mathrm{Vol}(B^n_{\Psi/E_n})}\int_{\mathbb{R}^{n-k_n}} \mathbf{1}_{\{(x,y)\in B^n_{\Psi/E_n}\}}\,dy - \frac{1}{Z_\lambda^{k_n}} e^{-\lambda(\Psi(x_1)+\dots+\Psi(x_{k_n}))}\right| dx \\
&\le \int_{B^{k_n}_{\Psi/t_n}} \left| \frac{\mathrm{Vol}\big(B^{n-k_n}_{\Psi/(E_n-\sum_{i=1}^{k_n}\Psi(x_i))}\big)}{\mathrm{Vol}(B^n_{\Psi/E_n})} - \frac{1}{Z_\lambda^{k_n}} e^{-\lambda(\Psi(x_1)+\dots+\Psi(x_{k_n}))}\right| dx \\
&\quad + \mathbb{P}\big((\xi_1,\dots,\xi_{k_n}) \notin B^{k_n}_{\Psi/t_n}\big) + \mathbb{P}\big((X_1,\dots,X_{k_n}) \notin B^{k_n}_{\Psi/t_n}\big) \\
&= \int_0^{t_n} \left| \frac{\mathrm{Vol}\big(B^{n-k_n}_{\Psi/(E_n-t)}\big)}{\mathrm{Vol}(B^n_{\Psi/E_n})} - \frac{e^{-\lambda t}}{Z_\lambda^{k_n}}\right| \frac{d}{dt}\mathrm{Vol}(B^{k_n}_{\Psi/t})\,dt \qquad (6.1)\\
&\quad + \mathbb{P}\left(\sum_{i=1}^{k_n}\Psi(\xi_i) > t_n\right) + \mathbb{P}\left(\sum_{i=1}^{k_n}\Psi(X_i) > t_n\right).
\end{aligned}$$

By Markov's inequality,

$$\mathbb{P}\left(\sum_{i=1}^{k_n} \Psi(X_i) > t_n\right) \leq \frac{\mathbb{E}\big(\sum_{i=1}^{k_n} \Psi(X_i)\big)}{t_n} = \frac{k_n m_\lambda}{t_n} = o(1).$$

Similarly, and since by definition $\sum_{i=1}^{n} \Psi(\xi_i) \leq E_n$ and the ξ_i's are exchangeable

$$\mathbb{P}\left(\sum_{i=1}^{k_n} \Psi(\xi_i) > t_n\right) \leq \frac{\mathbb{E}\big(\sum_{i=1}^{k_n} \Psi(\xi_i)\big)}{t_n} \leq \frac{k_n E_n}{n t_n} = \frac{k_n m_\lambda}{t_n} = o(1).$$

In order to estimate (6.1), we use Theorem 1.1. Since $k_n = o(\sqrt{n})$ and $t_n = o(\sqrt{n})$, we know that $E_n - t = m_\lambda(n - k_n) + \beta_n \sigma_\lambda \sqrt{n - k_n}$, where

$$\beta_n := \alpha_n \sqrt{\frac{n}{n - k_n}} + \frac{m_\lambda k_n - t}{\sigma_\lambda \sqrt{n - k_n}}$$

is a bounded sequence such that $\beta_n - \alpha_n = o(1)$, both properties holding uniformly in $t \in [0, t_n]$. Therefore, Theorem 1.1 applied to $B^{n-k_n}_{\Psi/(E_n - t)}$ gives

$$\mathrm{Vol}\big(B^{n-k_n}_{\Psi/(E_n-t)}\big) = \frac{Z_\lambda^{n-k_n} e^{\lambda(E_n - t)}}{\lambda \sigma_\lambda \sqrt{2\pi (n - k_n)}} e^{-\beta_n^2/2}(1 + o(1))$$

uniformly in $t \in [0, t_n]$. On the other hand, Theorem 1.1 applied to B^n_{Ψ/E_n} yields

$$\mathrm{Vol}\big(B^n_{\Psi/E_n}\big) = \frac{Z_\lambda^n e^{\lambda E_n}}{\lambda \sigma_\lambda \sqrt{2\pi n}} e^{-\alpha_n^2/2}(1 + o(1)).$$

Combining the above two asymptotic expansions, we obtain

$$\frac{\mathrm{Vol}\big(B^{n-k_n}_{\Psi/(E_n-t)}\big)}{\mathrm{Vol}(B^n_{\Psi/E_n})} = \frac{e^{-\lambda t}}{Z_\lambda^{k_n}}(1 + o(1))$$

uniformly in $t \in [0, t_n]$. Therefore the term (6.1) equals

$$o(1) \int_0^{t_n} \frac{e^{-\lambda t}}{Z_\lambda^{k_n}} \frac{d}{dt} \mathrm{Vol}(B^{k_n}_{\Psi/t})\, dt = o(1) \cdot \mathbb{P}\left(\sum_{i=1}^{k_n} \Psi(X_i) \leq t_n\right) = o(1).$$

□

The next result gives the asymptotic distribution of a sort of distance to the boundary for high-dimensional Orlicz balls.

Theorem 6.2 *Let $E_n = m_\lambda n + \alpha_n \sigma_\lambda \sqrt{n}$, where $(\alpha_n)_{n\geq 1}$ is bounded. Let the random vector $\xi^{(n)}$ be uniformly distributed on B^n_{Ψ/E_n}. Then the following convergence in distribution occurs as n goes to infinity:*

$$\lambda \cdot \Big(E_n - \sum_{i=1}^{n} \Psi\big(\xi_i^{(n)}\big)\Big) \longrightarrow \mathcal{E}xp(1).$$

Proof Let $S_n := E_n - \sum_{i=1}^{n} \Psi\big(\xi_i^{(n)}\big) \geq 0$. For $t \geq 0$,

$$\mathbb{P}(S_n \geq t) = \mathbb{P}\left(\sum_{i=1}^{n} \Psi\big(\xi_i^{(n)}\big) \leq E_n - t\right) = \frac{\mathrm{Vol}\big(B^n_{\Psi/(E_n-t)}\big)}{\mathrm{Vol}\big(B^n_{\Psi/E_n}\big)}.$$

As before, Theorem 1.1 applied to B^n_{Ψ/E_n} yields

$$\mathrm{Vol}\big(B^n_{\Psi/E_n}\big) \sim \frac{Z_\lambda^n e^{\lambda E_n}}{\lambda \sigma_\lambda \sqrt{2\pi n}} e^{-\alpha_n^2/2},$$

whereas applied to $B^n_{\Psi/E_{n-t}}$ it gives

$$\mathrm{Vol}\big(B^n_{\Psi/E_n-t}\big) \sim \frac{Z_\lambda^n e^{\lambda(E_n-t)}}{\lambda \sigma_\lambda \sqrt{2\pi n}} e^{-\left(\alpha_n - \frac{t}{\sigma_\lambda \sqrt{n}}\right)^2/2}.$$

Taking the quotient gives $\lim_n \mathbb{P}(S_n \geq t) = e^{-\lambda t}$. □

7 Integrability of Linear Functionals

Linear functionals of uniform random vectors on convex bodies are well studied quantities. Their density function, known as the parallel section function, measures the volume of hyperplane sections in a given direction. We refer, e.g., to the book [5], and in particular to its sections 2.4 and 8.2 about the ψ_1 and ψ_2 properties, which describe uniform integrability features (exponential integrability for ψ_1, Gaussian type integrability for ψ_2). They can be expressed by upper bounds on the Laplace transform.

In this section, we deal with even Young functions Ψ, so that the corresponding sets B^n_Ψ are origin-symmetric, and actually unconditional. The forthcoming study is valid for any dimension, without taking limits, so we consider the dimension n fixed and write $\xi = (\xi_1, \ldots, \xi_n)$ for a uniform random vector on B^n_Ψ. We show that the arguments of [3] for ℓ_p^n unit balls extend to Orlicz balls.

Lemma 7.1 *Let $a \in \mathbb{R}^n$, and ξ be uniform on B^n_Ψ, then*

$$\mathbb{E}e^{\langle a,\xi\rangle} \leq \prod_{i=1}^{n} \mathbb{E}e^{a_i\xi_1}.$$

Proof Let $\varepsilon_1, \ldots, \varepsilon_n$ be i.i.d. random variables with $\mathbb{P}(\varepsilon_i = 1) = \mathbb{P}(\varepsilon_i = -1) = \frac{1}{2}$, and independent of ξ. Then by symmetry of Ψ, $(\varepsilon_1\xi_1, \ldots, \varepsilon_n\xi_n)$ has the same distribution as ξ. Hence,

$$\mathbb{E}e^{\langle a,\xi\rangle} = \mathbb{E}\prod_{i=1}^{n} e^{a_i\varepsilon_i\xi_i} = \mathbb{E}\left(\mathbb{E}\Big(\prod_{i=1}^{n} e^{a_i\varepsilon_i\xi_i}\,\Big|\,\xi\Big)\right) = \mathbb{E}\prod_{i=1}^{n}\cosh(a_i\xi_i).$$

Next by the subindependence property of coordinates, due to Pilipczuk and Wojtaszczyk [19], and using the symmetry again as well as exchangeability:

$$\mathbb{E}e^{\langle a,\xi\rangle} \leq \prod_{i=1}^{n}\mathbb{E}\cosh(a_i\xi_i) = \prod_{i=1}^{n}\mathbb{E}e^{a_i\xi_i} = \prod_{i=1}^{n}\mathbb{E}e^{a_i\xi_1}.$$

□

The above lemma shows that the Laplace transform of any linear functional $\langle a, \xi\rangle$ can be upper estimated using the Laplace transform of the first coordinate ξ_1. Therefore it is natural to study the law of ξ_1. For $t \in \mathbb{R}$ consider the section of B_Ψ^n:

$$S(t) := \{y \in \mathbb{R}^{n-1};\ (t, y) \in B_\Psi^n\}$$

and $f(t) := \mathrm{Vol}_{n-1}(S(t))$. Then $\mathbb{P}_{\xi_1}(dt) = f(t)dt/\mathrm{Vol}_n(B_\Psi^n)$. By the Brunn principle, f is a log-concave function. It is also even by symmetry of the ball, therefore it is non-increasing on $\mathbb{R}^+$. We observe that a slightly stronger property holds:

Lemma 7.2 *Let Ψ be an even Young function and $f(t) = Vol_n\big(\{y \in \mathbb{R}^{n-1};\ (t, y) \in B_\Psi^n\}\big)$. Then the function $\log f \circ \Psi^{-1}$ is concave and non-increasing on $\mathbb{R}^+$. Here Ψ^{-1} is the reciprocal function of the restriction of Ψ to $\mathbb{R}^+$.*

Proof Let $t, u \geq 0$. Let $a \in S(t)$ and $b \in S(u)$. Then by definition

$$\Psi(t) + \sum_{i=1}^{n-1}\Psi(a_i) \leq 1 \quad \text{and} \quad \Psi(u) + \sum_{i=1}^{n-1}\Psi(b_i) \leq 1.$$

Averaging these two inequalities and using the convexity of Ψ, we get for any $\theta \in (0, 1)$:

$$(1-\theta)\Psi(t) + \theta\Psi(u) + \sum_{i=1}^{n-1}\Psi\left((1-\theta)a_i + \theta b_i\right) \leq 1. \tag{7.1}$$

This can be rewritten as

$$(1-\theta)a+\theta b \in S\left(\Psi^{-1}\left((1-\theta)\Psi(t)+\theta\Psi(u)\right)\right).$$

Hence we have shown that

$$(1-\theta)S(t)+\theta S(u) \subset S\left(\Psi^{-1}\left((1-\theta)\Psi(t)+\theta\Psi(u)\right)\right),$$

and by the Brunn–Minkowski inequality, in multiplicative form

$$f(t)^{1-\theta} f(u)^{\theta} \leq f\left(\Psi^{-1}\left((1-\theta)\Psi(t)+\theta\Psi(u)\right)\right).$$

Note that if in (7.1) we had used convexity in the form $\Psi((1-\theta)t+\theta u) \leq (1-\theta)\Psi(t)+\theta\Psi(u)$, then we would have derived the Brunn principle from the Brunn–Minkowski inequality. □

The next result shows that Ψ is more convex than the square function, the corresponding Orlicz balls enjoy the ψ_2 property. This applies in particular to B_p^n for $p \geq 2$, a case which was treated in [3].

Theorem 7.3 *Let Ψ be an even Young function, such that $t > 0 \mapsto \Psi(\sqrt{t})$ is convex. Let ξ be a uniform random vector on B_Ψ^n. Then for all $a \in \mathbb{R}^n$,*

$$\mathbb{E}e^{\langle a,\xi\rangle} \leq \left(\mathbb{E}e^{\frac{|a|}{\sqrt{n}}\xi_1}\right)^n \leq e^{\frac{1}{2}\mathbb{E}\left(\langle a,\xi\rangle^2\right)}.$$

Proof Let $L_X(t) = \mathbb{E}e^{tX}$ denote the Laplace transform of a real valued random variable. Then with the notation of Lemma 7.2,

$$L_{\xi_1}(t) = \int e^{tu} f(u) \frac{du}{\mathrm{Vol}(B_\Psi^n)}.$$

Lemma 7.2 ensures that there exists a concave function c such that for all $u \geq 0$, $\log f(u) = c(\Psi(u))$. Note that c is also non-increasing on $\mathbb{R}^+$ since the section function f is. Hence

$$u \geq 0 \mapsto \log f(\sqrt{u}) = c(\Psi(\sqrt{u}))$$

is concave. Theorem 12 of [3] ensures that $t \geq 0 \mapsto \int_{\mathbb{R}} e^{u\sqrt{t}} f(u)\, du$ is log-concave. In other words,

$$t \geq 0 \mapsto \log L_{\xi_1}(\sqrt{t})$$

is concave.

From Lemma 7.1, using symmetry and the above concavity property

$$\mathbb{E}e^{\langle a,\xi\rangle} \le \prod_{i=1}^{n} L_{\xi_1}(a_i) = \prod_{i=1}^{n} L_{\xi_1}\left(\sqrt{a_i^2}\right) \le \left(L_{\xi_1}\left(\sqrt{\frac{1}{n}\sum_i a_i^2}\right)\right)^n = L_{\xi_1}\left(\frac{|a|}{\sqrt{n}}\right)^n.$$

To conclude we need the bound $L_{\xi_1}(t) \le e^{t^2\mathbb{E}(\xi_1^2)/2}$ (it follows from the fact that $t \ge 0 \mapsto \log L_{\xi_1}(\sqrt{t})$ is concave, hence upper bounded by its tangent application at 0, which is easily seen to be $t\mathbb{E}(\xi_1^2)/2$). We obtain

$$\mathbb{E}e^{\langle a,\xi\rangle} \le e^{\frac{1}{2}|a|^2\mathbb{E}(\xi_1^2)},$$

and we conclude using the symmetries of ξ since

$$\mathbb{E}\big(\langle a,\xi\rangle^2\big) = \sum_i a_i^2\mathbb{E}(\xi_i^2) + \sum_{i\ne j} a_i a_j \mathbb{E}(\xi_i\xi_j) = |a|^2\mathbb{E}(\xi_1^2).$$

□

Acknowledgments We are grateful to Emanuel Milman and Reda Chhaibi for useful discussions on related topics. We also thank Joscha Prochno for communicating his recent work to us.

References

1. D. Alonso-Gutiérrez, J. Bastero, *Approaching the Kannan-Lovász-Simonovits and Variance Conjectures*. Lecture Notes in Mathematics, vol. 2131 (Springer, Cham, 2015)
2. R.R. Bahadur, R. Ranga Rao, On deviations of the sample mean. Ann. Math. Stat. **31**, 1015–1027 (1960)
3. F. Barthe, A. Koldobsky, Extremal slabs in the cube and the Laplace transform. Adv. Math. **174**, 89–114 (2003)
4. F. Barthe, O. Guédon, S. Mendelson, A. Naor, A probabilistic approach to the geometry of the l_p^n-ball. Ann. Probab. **33**(2), 480–513 (2005)
5. S. Brazitikos, A. Giannopoulos, P. Valettas, B.-H. Vritsiou, *Geometry of Isotropic Convex Bodies*. Mathematical Surveys and Monographs, vol. 196 (American Mathematical Society, Providence, 2014)
6. Y. Chen, An almost constant lower bound of the isoperimetric coefficient in the KLS conjecture. Geom. Funct. Anal. **31**(1), 34–61 (2021)
7. H. Cramér, Sur un nouveau théorème-limite de la théorie des probabilités. Actual. Sci. Industr. **736**, 5–23 (1938) (Confér. internat. Sci. math. Univ. Genève. Théorie des probabilités. III: Les sommes et les fonctions de variables aléatoires.)
8. P. Diaconis, D. Freedman, A dozen de Finetti-style results in search of a theory. Ann. Inst. H. Poincaré Probab. Stat. **23**(2, suppl.), 397–423 (1987)
9. M. Fradelizi, M. Madiman, L. Wang, Optimal concentration of information content for log-concave densities. In *High Dimensional Probability VII*. Progress in Probability, vol. 71 (Springer, Cham, 2016), pp. 45–60

10. Z. Kabluchko, J. Prochno, The maximum entropy principle and volumetric properties of Orlicz balls. J. Math. Anal. Appl. **495**(1), 124687 (2021)
11. R. Kannan, L. Lovász, M. Simonovits, Isoperimetric problems for convex bodies and a localization lemma. Discret. Comput. Geom. **13**(3–4), 541–559 (1995)
12. B. Klartag, J. Lehec, Bourgain's slicing problem and KLS isoperimetry up to polylog. arXiv:2203.15551 (2022)
13. A.V. Kolesnikov, E. Milman, The KLS isoperimetric conjecture for generalized Orlicz balls. Ann. Probab. **46**(6), 3578–3615 (2018)
14. Y.T. Lee, S.S. Vempala, The Kannan-Lovász-Simonovits conjecture. In *Current Developments in Mathematics 2017* (International Press, Somerville, 2019), pp. 1–36
15. A.A. Mogul'skiĭ, De Finetti-type results for l_p. Sibirsk. Mat. Zh. **32**(4), 88–95, 228 (1991)
16. V.H. Nguyen, Dimensional variance inequalities of Brascamp-Lieb type and a local approach to dimensional Prékopa's theorem. J. Funct. Anal. **266**(2), 931–955 (2014)
17. V.H. Nguyen, *Inégalités fonctionnelles et convexité.* PhD thesis, Université Pierre et Marie Curie (Paris VI) (2013)
18. V.V. Petrov, *Sums of Independent Random Variables* (Springer, New York/Heidelberg, 1975). Translated from the Russian by A. A. Brown, Ergebnisse der Mathematik und ihrer Grenzgebiete, Band 82
19. M. Pilipczuk, J.O. Wojtaszczyk, The negative association property for the absolute values of random variables equidistributed on a generalized Orlicz ball. Positivity **12**(3), 421–474 (2008)
20. S.T. Rachev, L. Rüschendorf, Approximate independence of distributions on spheres and their stability properties. Ann. Probab. **19**(3), 1311–1337 (1991)
21. M.M. Rao, Z.D. Ren, *Theory of Orlicz Spaces*. Monographs and Textbooks in Pure and Applied Mathematics, vol. 146 (Marcel Dekker, Inc., New York, 1991)
22. L. Wang, *Heat capacity bound, energy fluctuations and convexity*. PhD thesis, Yale University (2014)

Entropic Isoperimetric Inequalities

Sergey G. Bobkov and Cyril Roberto

2020 Mathematics Subject Classification: 60E15; 94A17

1 Introduction

The entropic isoperimetric inequality asserts that

$$N(X)\, I(X) \geq 2\pi e\, n \tag{1.1}$$

for any random vector X in $\mathbb{R}^n$ with a smooth density. Here

$$N(X) = \exp\Big\{ -\frac{2}{n} \int p(x) \log p(x)\, dx \Big\} \quad \text{and} \quad I(X) = \int \frac{|\nabla p(x)|^2}{p(x)}\, dx$$

denote the Shannon entropy power and the Fisher information of X with density p, respectively (with integration with respect to Lebesgue measure dx on $\mathbb{R}^n$ which may be restricted to the supporting set $\mathrm{supp}(p) = \{x : p(x) > 0\}$).

This inequality was discovered by Stam [15] where it was treated in dimension one. It is known to hold in any dimension, and the standard normal distribution on $\mathbb{R}^n$ plays an extremal role in it. Later on, Costa and Cover [6] pointed out a remarkable analogy between (1.1) and the classical isoperimetric inequality relating the surface of an arbitrary body A in $\mathbb{R}^n$ to its volume $\mathrm{vol}_n(A)$. The terminology "isoperimetric inequality for entropies" goes back to Dembo, Costa, and Thomas [8].

S. G. Bobkov
School of Mathematics, University of Minnesota, Minneapolis, MN, USA
e-mail: bobkov@math.umn.edu

C. Roberto (✉)
Modal'x, UPL, Univ. Paris Nanterre, CNRS, Nanterre, France
e-mail: croberto@math.cnrs.fr

R. Adamczak et al. (eds.), *High Dimensional Probability IX*, Progress in Probability 80, https://doi.org/10.1007/978-3-031-26979-0_3

As Rényi entropies have become a focus of numerous investigations in the recent time, it is natural to explore more general relations of the form

$$N_\alpha(X)\, I(X) \geq c_{\alpha,n} \tag{1.2}$$

for the functional

$$N_\alpha(X) = \Big(\int p(x)^\alpha dx\Big)^{-\frac{2}{n(\alpha-1)}}. \tag{1.3}$$

It is desirable to derive (1.2) with optimal constants $c_{\alpha,n}$ independent of the density p, where $\alpha \in [0,\infty]$ is a parameter called the order of the Rényi entropy power $N_\alpha(X)$. Another representation

$$N_\alpha(X)^{-\frac{n}{2}} = \|p\|_{L^{\alpha-1}(p(x)\,dx)}$$

shows that N_α is non-increasing in α. This allows one to define the Rényi entropy power for the two extreme values by the monotonicity to be

$$\begin{aligned} N_\infty(X) &= \lim_{\alpha\to\infty} N_\alpha(X) &= \|p\|_\infty^{-\frac{2}{n}}, \\ N_0(X) &= \lim_{\alpha\to 0} N_\alpha(X) &= \mathrm{vol}_n(\mathrm{supp}(p))^{\frac{2}{n}}, \end{aligned} \tag{1.4}$$

where $\|p\|_\infty = \operatorname{ess\,sup} p(x)$. As a standard approach, one may also put $N_1(X) = \lim_{\alpha\downarrow 1} N_\alpha(X)$ which returns us to the usual definition of the Shannon entropy power $N_1(X) = N(X)$ under mild moment assumptions (such as $N_\alpha(X) > 0$ for some $\alpha > 1$).

Returning to (1.1)–(1.2), the following two natural questions arise.

Question 1 Given n, for which range $\mathfrak{A}_n$ of the values of α does (1.2) hold with some positive constant?

Question 2 What is the value of the optimal constant $c_{\alpha,n}$ and can the extremizers in (1.2) be described?

The entropic isoperimetric inequality (1.1) answers both questions for the order $\alpha = 1$ with an optimal constant $c_{1,n} = 2\pi e\, n$. As for the general order, let us first stress that, by the monotonicity of N_α with respect to α, the function $\alpha \mapsto c_{\alpha,n}$ is also non-increasing. Hence, the range in Question 1 takes necessarily the form $\mathfrak{A}_n = [0,\alpha_n)$ or $\mathfrak{A}_n = [0,\alpha_n]$ for some critical value $\alpha_n \in [0,\infty]$. The next assertion specifies these values.

Theorem 1.1 *We have*

$$\mathfrak{A}_n = \begin{cases} [0,\infty] & \text{for } n = 1, \\ [0,\infty) & \text{for } n = 2, \\ [0,\frac{n}{n-2}] & \text{for } n \geq 3. \end{cases}$$

Thus, in the one dimensional case there is no restriction on α (the range is full). In fact, this already follows from the elementary sub-optimal inequality

$$N_\infty(X) I(X) \geq 1, \tag{1.5}$$

implying that $c_{\alpha,1} \geq 1$ for all α. To see this, assume that $I(X)$ is finite, so that X has a (locally) absolutely continuous density p, thus differentiable almost everywhere. Since p is non-negative, any point $y \in \mathbb{R}$ such that $p(y) = 0$ is a local minimum, and necessarily $p'(y) = 0$ (as long as p is differentiable at y). Hence, applying the Cauchy inequality, we have

$$\begin{aligned}\int_{-\infty}^{\infty} |p'(y)|\, dy &= \int_{p(y)>0} \frac{|p'(y)|}{\sqrt{p(y)}} \sqrt{p(y)}\, dy \\ &\leq \left(\int_{p(y)>0} \frac{p'(y)^2}{p(y)}\, dy\right)^{1/2} \left(\int_{p(y)>0} p(y)\, dy\right)^{1/2} = \sqrt{I(X)}.\end{aligned}$$

It follows that p has a bounded total variation not exceeding $\sqrt{I(X)}$, so $p(x) \leq \sqrt{I(X)}$ for every $x \in \mathbb{R}$. This amounts to (1.5) according to (1.4) for $n = 1$.

Turning to Question 2, we will see that the optimal constants $c_{\alpha,1}$ together with the extremizers in (1.2) may be explicitly described in the one dimensional case for every α using the results due to Nagy [13]. Since the transformation of these results in the information-theoretic language is somewhat technical, we discuss this case in detail in the next three sections (Sects. 2, 3, and 4). Let us only mention here that

$$4 \leq c_{\alpha,1} \leq 4\pi^2,$$

where the inequalities are sharp for $\alpha = \infty$ and $\alpha = 0$, respectively, with extremizers

$$p(x) = \frac{1}{2} e^{-|x|} \quad \text{and} \quad p(x) = \frac{2}{\pi} \cos^2(x)\, 1_{\{|x| \leq \frac{\pi}{2}\}}.$$

The situation in higher dimensions is more complicated, and only partial answers to Question 2 will be given here. Anyway, in order to explore the behavior of the constants $c_{\alpha,n}$, one should distinguish between the dimensions $n = 2$ and $n \geq 3$ (which is also suggested by Theorem 1.1). In the latter case, these constants can be shown to satisfy

$$4\pi n(n-2) \left(\frac{\Gamma(\frac{n}{2})}{\Gamma(n)}\right)^{\frac{2}{n}} \leq c_{\alpha,n} \leq 4\pi^2 n, \qquad 0 \leq \alpha \leq \frac{n}{n-2},$$

where the left inequality is sharp and corresponds to the critical order $\alpha = \frac{n}{n-2}$. With respect to the growing dimension, these constants are asymptotically $2\pi e n + O(1)$, which exhibits nearly the same behavior as for the order $\alpha = 1$. However (which is rather surprising), the extremizers for the critical order exist for $n \geq 5$ only and are described as densities of the (generalized) Cauchy distributions on $\mathbb{R}^n$. We discuss these issues in Sect. 7, while Sect. 6 deals with dimension $n = 2$, where some description of the constants $c_{\alpha,2}$ will be given for the range $\alpha \in [\frac{1}{2}, \infty)$.

We end this introduction by giving an equivalent formulation of the isoperimetric inequalities (1.2) in terms of functional inequalities of Sobolev type. As was noticed by Carlen [5], in the classical case $\alpha = 1$, (1.1) is equivalent to the logarithmic Sobolev inequality of Gross [9], cf. also [4]. However, when $\alpha \neq 1$, a different class of inequalities should be involved. Namely, using the substitution $p = f^2/\int f^2$ (here and in the sequel integrals are understood with respect to the Lebesgue measure on $\mathbb{R}^n$), we have

$$N_\alpha(X) = \Big(\int f^{2\alpha}\Big)^{-\frac{2}{n(\alpha-1)}} \Big(\int f^2\Big)^{\frac{2\alpha}{n(\alpha-1)}}$$

and

$$I(X) = 4\int |\nabla f|^2 \Big/ \int f^2.$$

Therefore (provided that f is square integrable), (1.2) can be equivalently reformulated as a homogeneous analytic inequality

$$\Big(\int |f|^{2\alpha}\Big)^{\frac{2}{n(\alpha-1)}} \leq \frac{4}{c_{\alpha,n}} \int |\nabla f|^2 \Big(\int f^2\Big)^{\frac{\alpha(2-n)+n}{n(\alpha-1)}}, \tag{1.6}$$

where we can assume that f is smooth and has gradient ∇f (however, when speaking about extremizers, the function f should be allowed to belong to the Sobolev class $W_1^2(\mathbb{R}^n)$). Such inequalities were introduced by Moser [11, 12] in the following form

$$\Big(\int |f|^{2+\frac{4}{n}}\Big) \leq B_n \int |\nabla f|^2 \Big(\int f^2\Big)^{\frac{2}{n}}. \tag{1.7}$$

More precisely, (1.7) corresponds to (1.6) for the specific choice $\alpha = 1+\frac{2}{n}$. Here, the one dimensional case is covered by Nagy's paper with the optimal factor $B_1 = \frac{4}{\pi^2}$. This corresponds to $\alpha = 3$ and $n = 1$, and therefore $c_{3,1} = \pi^2$ which complements the picture depicted above. To the best of our knowledge, the best constants B_n for $n \geq 2$ are not known. However, using the Euclidean log-Sobolev inequality and the optimal Sobolev inequality, Beckner [2] proved that asymptotically $B_n \sim \frac{2}{\pi e n}$.

Both Moser's inequality (1.7) and (1.6) with a certain range of α enter the general framework of Gagliardo–Nirenberg's inequalities

$$\Big(\int |f|^r\Big)^{\frac{1}{r}} \leq \kappa_n(q,r,s)\Big(\int |\nabla f|^q\Big)^{\frac{\theta}{q}} \Big(\int |f|^s\Big)^{\frac{1-\theta}{s}} \tag{1.8}$$

with $1 \le q, r, s \le \infty$, $0 \le \theta \le 1$, and $\frac{1}{r} = \theta\,(\frac{1}{q} - \frac{1}{n}) + (1-\theta)\,\frac{1}{s}$. We will make use of the knowledge on Gagliardo–Nirenberg's inequalities to derive information on (1.2).

In the sequel, we denote by $\|f\|_r = (\int |f|^r)^{\frac{1}{r}}$ the L^r-norm of f with respect to the Lebesgue measure on $\mathbb{R}^n$ (and use this functional also in the case $0 < r < 1$).

2 Nagy's Theorem

In the next three sections we focus on dimension $n = 1$, in which case the entropic isoperimetric inequality (1.2) takes the form

$$N_\alpha(X)\, I(X) \ge c_{\alpha,1} \tag{2.1}$$

for the Rényi entropy

$$N_\alpha(X) = \left(\int p(x)^\alpha dx\right)^{-\frac{2}{\alpha-1}}$$

and the Fisher information

$$I(X) = \int \frac{p'(x)^2}{p(x)}\, dx = 4 \int \left(\frac{d}{dx}\sqrt{p(x)}\right)^2 dx.$$

In dimension one, our basic functional space is the collection of all (locally) absolutely continuous functions on the real line whose derivatives are understood in the Radon–Nikodym sense. We already know that (2.1) holds for all $\alpha \in [0, \infty]$.

According to (1.6), the family (2.1) takes now the form

$$\int |f|^{2\alpha} \le \left(\frac{4}{c_{\alpha,1}}\right)^{\frac{\alpha-1}{2}} \left(\int f'^2\right)^{\frac{\alpha-1}{2}} \left(\int f^2\right)^{\frac{\alpha+1}{2}} \tag{2.2}$$

when $\alpha > 1$, and

$$\int f^2 \le \left(\frac{4}{c_{\alpha,1}}\right)^{\frac{1-\alpha}{1+\alpha}} \left(\int f'^2\right)^{\frac{1-\alpha}{1+\alpha}} \left(\int |f|^{2\alpha}\right)^{\frac{2}{1+\alpha}} \tag{2.3}$$

when $\alpha \in (0, 1)$.

In fact, these two families of inequalities can be seen as sub-families of the following one, studied by Nagy [13],

$$\int |f|^{\gamma+\beta} \le D\left(\int |f'|^p\right)^{\frac{\beta}{pq}} \left(\int |f|^\gamma\right)^{1+\frac{\beta(p-1)}{pq}}$$

with

$$p > 1, \quad \beta, \gamma > 0, \quad q = 1 + \frac{\gamma(p-1)}{p}, \tag{2.4}$$

and some constants $D = D_{\gamma,\beta,p}$ depending on γ, β, and p, only. For such parameters, introduce the functions $y_{p,\gamma} = y_{p,\gamma}(t)$ defined for $t \geq 0$ by

$$y_{p,\gamma}(t) = \begin{cases} (1+t)^{\frac{p}{p-\gamma}} & \text{if } p < \gamma, \\ e^{-t} & \text{if } p = \gamma, \\ (1-t)^{\frac{p}{p-\gamma}} \, 1_{[0,1]}(t) & \text{if } p > \gamma. \end{cases}$$

To involve the parameter β, define additionally $y_{p,\gamma,\beta}$ implicitly as follows. Put $y_{p,\gamma,\beta}(t) = u$, $0 \leq u \leq 1$, with

$$t = \int_u^1 \left(s^\gamma(1-s^\beta)\right)^{-\frac{1}{p}} ds$$

if $p \leq \gamma$. If $p > \gamma$, then $y_{p,\gamma,\beta}(t) = u$, $0 \leq u \leq 1$, is the solution of the above equation for

$$t \leq t_0 = \int_0^1 (s^\gamma(1-s^\beta))^{-\frac{1}{p}} ds$$

and $y_{p,\gamma,\beta}(t) = 0$ for all $t > t_0$. With these notations, Nagy established the following result.

Theorem 2.1 ([13]) *Under the constraint* (2.4), *for any (locally) absolutely continuous function* $f : \mathbb{R} \to \mathbb{R}$,

(i)

$$\|f\|_\infty \leq \left(\frac{q}{2}\right)^{\frac{1}{q}} \left(\int |f'|^p\right)^{\frac{1}{pq}} \left(\int |f|^\gamma\right)^{\frac{p-1}{pq}}. \tag{2.5}$$

Moreover, the extremizers take the form $f(x) = a y_{p,\gamma}(|bx+c|)$ *with* a, b, c *constants* ($b \neq 0$).

(ii)

$$\int |f|^{\beta+\gamma} \leq \left(\frac{q}{2} H\left(\frac{q}{\beta}, \frac{p-1}{p}\right)\right)^{\frac{\beta}{q}} \left(\int |f'|^p\right)^{\frac{\beta}{pq}} \left(\int |f|^\gamma\right)^{1+\frac{\beta(p-1)}{pq}}, \tag{2.6}$$

where

$$H(u,v) = \frac{\Gamma(1+u+v)}{\Gamma(1+u)\,\Gamma(1+v)} \Big(\frac{u}{u+v}\Big)^u \Big(\frac{v}{u+v}\Big)^v, \quad u, v \geq 0.$$

Moreover, the extremizers take the form $f(x) = a y_{p,\gamma,\beta}(|bx+c|)$ *with* a, b, c *constants* ($b \neq 0$).

Here, Γ denotes the classical Gamma function, and we use the convention that $H(u,0) = H(0,v) = 1$ for $u, v \geq 0$. It was mentioned by Nagy that H is monotone in each variable. Moreover, since $H(u,1) = (1+\frac{1}{u})^{-u}$ is between 1 and $\frac{1}{e}$, one has $1 > H(u,v) > (1+\frac{1}{u})^{-u} > \frac{1}{e}$ for all $0 < v < 1$. This gives a two-sided bound

$$1 \geq H\Big(\frac{q}{\beta}, \frac{p-1}{p}\Big) > \Big(1+\frac{\beta}{q}\Big)^{-\frac{q}{\beta}} > \frac{1}{e}.$$

3 One Dimensional Isoperimetric Inequalities for Entropies

The inequalities (2.2) and (2.3) correspond to (2.6) with parameters

$$p = \gamma = q = 2, \ \beta = 2(\alpha - 1) \ \text{ in the case } \alpha > 1$$

and

$$p = 2, \ \beta = 2(1-\alpha), \ \gamma = 2\alpha, \ q = 1+\alpha \ \text{ in the case } \alpha \in (0,1),$$

respectively. Hence, as a corollary from Theorem 2.1, we get the following statement which solves Question 2 when $n = 1$. Note that, by Theorem 2.1, the extremal distributions (their densities p) in (2.1) are determined in a unique way up to non-degenerate affine transformations of the real line. So, it is sufficient to indicate just one specific extremizer for each admissible collection of the parameters. Recall the definition of the optimal constants $c_{\alpha,1}$ from (2.1).

Theorem 3.1

(i) *In the case* $\alpha = \infty$, *we have*

$$c_{\infty,1} = 4.$$

Moreover, the density $p(x) = \frac{1}{2} e^{-|x|}$ ($x \in \mathbb{R}$) *of the two-sided exponential distribution represents an extremizer in* (2.1).

(ii) *In the case* $1 < \alpha < \infty$, *we have*

$$c_{\alpha,1} = \frac{2\pi}{\alpha - 1} \Big(\frac{2}{\alpha+1}\Big)^{\frac{\alpha-3}{\alpha-1}} \Bigg(\frac{\Gamma(\frac{1}{\alpha-1})}{\Gamma(\frac{\alpha+1}{2(\alpha-1)})}\Bigg)^2.$$

Moreover, the density $p(x) = a\cosh(x)^{-\frac{2}{\alpha-1}}$ *with a normalization constant* $a = \frac{1}{\sqrt{\pi}}\frac{\Gamma(\frac{\alpha+1}{2(\alpha-1)})}{\Gamma(\frac{1}{\alpha-1})}$ *represents an extremizer in* (2.1).

(iii) *In the case* $0 < \alpha < 1$,

$$c_{\alpha,1} = \frac{2\pi}{1-\alpha}\Big(\frac{2}{1+\alpha}\Big)^{\frac{1+\alpha}{1-\alpha}}\left(\frac{\Gamma(\frac{1+\alpha}{2(1-\alpha)})}{\Gamma(\frac{1}{1-\alpha})}\right)^2.$$

Moreover, the density $p(x) = a\cos(x)^{\frac{2}{1-\alpha}}\,1_{[-\frac{\pi}{2},\frac{\pi}{2}]}(x)$ *with constant* $a = \frac{1}{\sqrt{\pi}}\frac{\Gamma(\frac{3-\alpha}{1-\alpha})}{\Gamma(\frac{3-\alpha}{2(3-\alpha)})}$ *represents an extremizer in* (2.1).

To prove the theorem, we need a simple technical lemma.

Lemma 3.2

(i) *Given* $a > 0$ *and* $t \ge 0$, *the (unique) solution* $y \in (0, 1]$ *to the equation* $\int_y^1 \frac{ds}{s\sqrt{1-s^a}} = t$ *is given by*

$$y = \Big[\cosh\Big(\frac{at}{2}\Big)\Big]^{-\frac{2}{a}}.$$

(ii) *Given* $a, b > 0$ *and* $c \in \mathbb{R}$, *we have*

$$\int_{-\infty}^{\infty} \cosh(|bx+c|)^{-a}\,dx = \frac{\sqrt{\pi}}{b}\frac{\Gamma(\frac{a}{2})}{\Gamma(\frac{a+1}{2})}.$$

(iii) *Given* $a \in (0, 1)$ *and* $u \in [0, 1]$, *we have*

$$\int_u^1 \frac{ds}{s^a\sqrt{1-s^{2(1-a)}}} = \frac{1}{1-a}\arccos(u^{1-a}).$$

Remark 3.3 Since $\Gamma(\frac{a+1}{2}) = \Gamma(m+\frac{1}{2}) = \frac{(2m)!}{4^m m!}\sqrt{\pi}$ for $a = 2m$ with an integer $m \ge 1$, for such particular values of a, we have

$$\int_{-\infty}^{\infty} \cosh(|bx+c|)^{-a}\,dx = \frac{1}{b}\cdot\frac{4^m m!\,(m-1)!}{(2m)!}.$$

Proof of Lemma **3.2** Changing the variable $u = \sqrt{1-s^a}$, we have

$$\int_y^1 \frac{ds}{s\sqrt{1-s^a}} = \frac{2}{a}\int_0^{\sqrt{1-y^a}}\frac{du}{1-u^2} = \frac{1}{a}\log\Big(\frac{1+\sqrt{1-y^a}}{1-\sqrt{1-y^a}}\Big).$$

Inverting this equality leads to the desired result of item (i).

For item (ii) we use the symmetry of the cosh-function together with the change of variables $u = bx + c$ and then $t = \sinh(u)^2$ to get

$$\begin{aligned}\int_{-\infty}^{\infty} \cosh(|bx+c|)^{-a}\, dx &= \frac{1}{b}\int_{-\infty}^{\infty} \cosh(|u|)^{-a}\, du \\ &= \frac{2}{b}\int_{0}^{\infty} \cosh(u)^{-a}\, du = \frac{1}{b}\int_{0}^{\infty} t^{-\frac{1}{2}}(1+t)^{-\frac{a+1}{2}}\, dt.\end{aligned}$$

To obtain the result, we need to perform a final change of variables $v = \frac{1}{1+t}$. This turns the last integral into

$$\int_0^1 (1-v)^{-\frac{1}{2}} v^{\frac{a}{2}-1}\, dv = B\Big(\frac{1}{2}, \frac{a}{2}\Big) = \sqrt{\pi}\,\frac{\Gamma(\frac{a}{2})}{\Gamma(\frac{a+1}{2})},$$

where we used the beta function $B(x, y) = \int_0^1 (1-v)^{x-1} v^{y-1}\, dv = \frac{\Gamma(x)\Gamma(y)}{\Gamma(x+y)}$, $x, y > 0$.

Finally, in item (iii), a change of variables leads to

$$\begin{aligned}\int_u^1 \frac{ds}{s^a\sqrt{1-s^{2(1-a)}}} &= \frac{1}{1-a}\int_u^1 \frac{ds^{1-a}}{\sqrt{1-s^{2(1-a)}}} \\ &= \frac{1}{1-a}\int_{u^{1-a}}^1 \frac{dv}{\sqrt{1-v^2}} = \frac{1}{1-a}\arccos(u^{1-a}).\end{aligned}$$

□

***Proof of Theorem* 3.1** When $\alpha = \infty$ as in the case (i), (2.2) with $\int f^2 = 1$ becomes

$$\|f\|_\infty \le \Big(\frac{4}{c_{\infty,1}}\int f'^2\Big)^{\frac{1}{4}}.$$

This corresponds to (2.5) with parameters $p = q = \gamma = 2$. Therefore, item (i) of Theorem 2.1 applies and leads to

$$\|f\|_\infty \le \Big(\int f'^2\Big)^{\frac{1}{4}},$$

that is, $c_{\infty,1} = 4$. Moreover, the extremizers in (2.5) are given by

$$f(x) = a y_{2,2}(|bx+c|) = a\, e^{-|bx+c|}, \quad b \ne 0,\ a, c \in \mathbb{R}.$$

But, the extremizers in (2.1) are of the form $p = f^2/\int f^2$ with f an extremizer in (2.5). The desired result then follows after a change of variables.

Next, let us turn to the case (ii), where $1 < \alpha < \infty$. Here (2.1) is equivalent to (2.2) and corresponds to (2.6) with $p = \gamma = q = 2$ and $\beta = 2(\alpha - 1)$. Therefore, by Theorem 2.1, $(\frac{4}{c_{\alpha,1}})^{\frac{\alpha-1}{2}} = H(\frac{1}{\alpha-1}, \frac{1}{2})^{\alpha-1}$, so that

$$c_{\alpha,1} = \frac{4}{H(\frac{1}{\alpha-1}, \frac{1}{2})^2} = 4\,\frac{\Gamma(1+\frac{1}{\alpha-1})^2\,\Gamma(\frac{3}{2})^2}{\Gamma(\frac{3}{2}+\frac{1}{\alpha-1})^2}\left(\frac{\frac{1}{\alpha-1}+\frac{1}{2}}{\frac{1}{\alpha-1}}\right)^{\frac{2}{\alpha-1}}\left(\frac{\frac{1}{\alpha-1}+\frac{1}{2}}{\frac{1}{2}}\right)$$

$$= \pi\left(\frac{\frac{1}{\alpha-1}}{\frac{\alpha+1}{2(\alpha-1)}}\right)^2\frac{\Gamma(\frac{1}{\alpha-1})^2}{\Gamma(\frac{\alpha+1}{2(\alpha-1)})^2}\left(\frac{\alpha+1}{2}\right)^{\frac{2}{\alpha-1}}\left(\frac{\alpha+1}{\alpha-1}\right),$$

where we used the identities $\Gamma(3/2) = \sqrt{\pi}/2$ and $\Gamma(1+z) = z\Gamma(z)$. This leads to the desired expression for $c_{\alpha,1}$.

As for extremizers, item (ii) of Theorem 2.1 applies and asserts that the equality cases in (2.2) are reached, up to numerical factors, for functions $f(x) = y(|bx+c|)$, with $b \neq 0$, $c \in \mathbb{R}$, and $y\colon [0,\infty) \to \mathbb{R}$ defined implicitly for $t \in [0,\infty)$ by $y(t) = u$, $0 \le u \le 1$, with

$$t = \int_u^1 \left(s^2(1 - s^{2(\alpha-1)})\right)^{-\frac{1}{2}} ds = \int_u^1 \frac{1}{s\sqrt{1-s^{2(\alpha-1)}}}\, ds.$$

Now, Lemma 3.2 provides the solution $y(t) = (\cosh((\alpha-1)\,t))^{-\frac{1}{\alpha-1}}$. Therefore, the extremizers in (2.2) are reached, up to numerical factors, for functions of the form

$$f(x) = (\cosh(|bx+c|))^{-\frac{1}{\alpha-1}}, \quad b \neq 0,\ c \in \mathbb{R}.$$

Similarly to the case (i), the extremizers in (2.1) are of the form $p = f^2/\int f^2$ with f an extremizer in (2.2). Therefore, by Lemma 3.2, with some $b > 0$ and $c \in \mathbb{R}$,

$$p(x) = \frac{\cosh(|bx+c|)^{-\frac{2}{\alpha-1}}}{\int \cosh(|bx+c|)^{-\frac{2}{\alpha-1}}\, dx} = \frac{b}{\sqrt{\pi}}\,\frac{\Gamma(\frac{\alpha+1}{2(\alpha-1)})}{\Gamma(\frac{1}{\alpha-1})}\cosh(bx+c)^{-\frac{2}{\alpha-1}}$$

as announced.

Finally, let us turn to item (iii), when $\alpha \in (0,1)$. As already mentioned, (2.1) is equivalent to (2.3) and therefore corresponds to (2.6) with $p = 2$, $\beta = 2(1-\alpha)$, $\gamma = 2\alpha$, and $q = 1+\alpha$. An application of Theorem (2.1) leads to the desired conclusion after some algebra (which we leave to the reader) concerning the explicit value of $c_{\alpha,1}$. In addition, the extremizers are of the form $p(x) = ay^2(|bx+c|)$, with a a normalization constant, $b \neq 0$, and $c \in \mathbb{R}$. Here $y = y(t)$ is defined implicitly by the equation

$$t = \int_y^1 \frac{1}{s^\alpha \sqrt{1 - s^{2(1-\alpha)}}}\, ds$$

for $t \le t_0 = \int_0^1 \frac{1}{s^\alpha \sqrt{1-s^{2(1-\alpha)}}} ds$ and $y(t) = 0$ for $t > t_0$. Item (iii) of Lemma 3.2 asserts that

$$t_0 = \frac{\pi}{2(1-\alpha)} \quad \text{and} \quad y(t) = \Big(\cos((1-\alpha)\, t)\Big)^{\frac{1}{1-\alpha}}\, 1_{[0, \frac{\pi}{2(1-\alpha)}]}(t).$$

This leads to the desired conclusion. □

4 Special Orders

As an illustration, here we briefly mention some explicit values of $c_{\alpha,1}$ and extremizers for specific values of the parameter α in the one dimensional entropic isoperimetric inequality

$$N_\alpha(X)\, I(X) \ge c_{\alpha,1}. \tag{4.1}$$

The order $\alpha = 0$ The limit in item (iii) of Theorem 3.1 leads to the optimal constant

$$c_{0,1} = \lim_{\alpha \to 0} c_{\alpha,1} = 4\pi^2.$$

Since all explicit expressions are continuous with respect to α, the limits of the extremizers in (2.1) for $\alpha \to 0$ represent extremizers in (2.1) for $\alpha = 0$. Therefore, the densities

$$p(x) = \frac{2b}{\pi} \cos^2(bx + c)\, 1_{[-\frac{\pi}{2}; \frac{\pi}{2}]}(bx + c), \qquad b > 0,\ c \in \mathbb{R},$$

are extremizers in (2.1) with $\alpha = 0$.

The order $\alpha = \frac{1}{2}$ Direct computation leads to $c_{\frac{1}{2},1} = (4/3)^3\pi^2$. Moreover, the extremizers in (2.1) are of the form

$$p(x) = \frac{8b}{3\pi} \cos^4(bx + c)\, 1_{[-\frac{\pi}{2}; \frac{\pi}{1}]}(bx + c), \qquad b > 0,\ c \in \mathbb{R}.$$

The order $\alpha = 1$ This case corresponds to Stam's isoperimetric inequality for entropies. Here $c_{1,1} = 2\pi e$, and, using the Stirling formula, one may notice that indeed

$$c_{1,1} = \lim_{\alpha \to 1} c_{\alpha,1} = 2\pi e.$$

Moreover, Gaussian densities can be obtained from the extremizers $p(x) = \cosh(bx + c)^{-\frac{2}{\alpha-1}}$ with $b = b'\sqrt{\alpha - 1}$, $c = c'\sqrt{\alpha - 1}$ in the limit as $\alpha \downarrow 1$. (Note that the limit $\alpha \uparrow 1$ would lead to the same conclusion.)

The order $\alpha = 2$ A direct computation leads to $c_{2,1} = 12$ with extremizers of the form

$$p(x) = \frac{b}{2\cosh^2(bx + c)}, \qquad b > 0,\ c \in \mathbb{R}.$$

In this case, the entropic isoperimetric inequality may equivalently be stated in terms of the Fourier transform $\hat{p}(t) = \int e^{itx} p(x)$, $t \in \mathbb{R}$, of the density p. Indeed, thanks to Plancherel's identity, we have

$$N_2(X)^{-1/2} = \int p^2 = \frac{1}{2\pi} \int |\hat{p}|^2.$$

Therefore, the (optimal) isoperimetric inequality for entropies yields the relation

$$\int |\hat{p}|^2 \leq \pi \sqrt{\frac{I(X)}{3}}$$

which is a global estimate on the L^2-norm of $\hat{p}$. In [18], Zhang derived the following pointwise estimate: If the random variable X with density p has finite Fisher information $I(X)$, then (see also [3] for an alternative proof)

$$|\hat{p}(t)| \leq \frac{I(X)}{I(X) + t^2}, \qquad t \in \mathbb{R}.$$

The latter leads to some bounds on $c_{2,1}$, namely

$$N_2(X)^{-1/2} = \frac{1}{2\pi} \int |\hat{p}|^2 \leq \frac{1}{2\pi} \int \frac{I(X)^2}{(I(X) + t^2)^2} dt = \frac{1}{2} \sqrt{I(X)}.$$

Hence $N_2(X) I(X) \geq 4$ that should be compared to $N_2(X) I(X) \geq 12$.

The order $\alpha = 3$ Then $c_{3,1} = \pi^2$, and the extremizers are of the form

$$p(x) = \frac{b}{\pi \cosh(bx + c)}, \qquad b > 0,\ c \in \mathbb{R}.$$

The order $\alpha = \infty$ From Theorem 3.1, $c_{\infty,1} = 4$, and the extremizers are of the form

$$p(x) = b\, e^{-|bx + c|}, \qquad b > 0,\ c \in \mathbb{R}.$$

5 Fisher Information in Higher Dimensions

In order to perform the transition from the entropic isoperimetric inequality (1.2) to the form of the Gagliardo–Nirenberg inequality such as (1.8) via the change of functions $p = f^2/\int f^2$ and back, and to justify the correspondence of the constants in the two types of inequalities, let us briefly fix some definitions and recall some approximation properties of the Fisher information. This is dictated by the observation that in general f in (1.8) does not need to be square integrable, and then p will not be defined as a probability density.

The Fisher information of a random vector X in $\mathbb{R}^n$ with density p may be defined by means of the formula

$$I(X) = I(p) = 4\int |\nabla\sqrt{p}|^2. \tag{5.1}$$

This functional is well-defined and finite if and only if $f = \sqrt{p}$ belongs to the Sobolev space $W_1^2(\mathbb{R}^n)$. There is the following characterization: A function f belongs to $W_1^2(\mathbb{R}^n)$, if and only if it belongs to $L^2(\mathbb{R}^n)$ and

$$\sup_{h\neq 0}\left[\frac{1}{|h|}\,\|f(x+h)-f(x)\|_2\right] < \infty.$$

In this case, there is a unique vector-function $g = (g_1,\dots,g_n)$ on $\mathbb{R}^n$ with components in $L^2(\mathbb{R}^n)$, called a weak gradient of f and denoted $g = \nabla f$, with the property that

$$\int gv = -\int f\nabla v \quad \text{for all } v \in C_0^\infty(\mathbb{R}^n). \tag{5.2}$$

As usual, $C_0^\infty(\mathbb{R}^n)$ denotes the class of all C^∞-smooth, compactly supported functions on $\mathbb{R}^n$. Still equivalently, there is a representative $\bar{f}$ of f which is absolutely continuous on almost all lines parallel to the coordinate axes and whose partial derivatives $\partial_{x_k}\bar{f}$ belong to $L^2(\mathbb{R}^n)$. In particular, $g_k(x) = \partial_{x_k}\bar{f}(x)$ for almost all $x \in \mathbb{R}^n$ (cf. [19], Theorems 2.1.6 and 2.1.4).

Applied to $f = \sqrt{p}$ with a probability density p on $\mathbb{R}^n$, the property that $f \in W_1^2(\mathbb{R}^n)$ ensures that p has a representative $\bar{p}$ which is absolutely continuous on almost all lines parallel to the coordinate axes and such that the functions $\partial_{x_k}\bar{p}/\sqrt{p}$ belong to $L^2(\mathbb{R}^n)$. Moreover,

$$I(p) = \sum_{k=1}^n \left\|\frac{\partial_{x_k}\bar{p}}{\sqrt{p}}\right\|_2^2.$$

Note that $W_1^2(\mathbb{R}^n)$ is a Banach space for the norm defined by

$$\|f\|_{W_1^2}^2 = \|f\|_2^2 + \|\nabla f\|_2^2$$
$$= \|f\|_2^2 + \|g_1\|_2^2 + \cdots + \|g_n\|_2^2 \qquad (g = \nabla f).$$

We use the notation $N_\alpha(X) = N_\alpha(p)$ when a random vector X has density p.

Proposition 5.1 *Given a (probability) density p on $\mathbb{R}^n$ such that $I(p)$ is finite, there exists a sequence of densities $p_k \in C_0^\infty(\mathbb{R}^n)$ satisfying as $k \to \infty$*

(a) $I(p_k) \to I(p)$, *and*
(b) $N_\alpha(p_k) \to N_\alpha(p)$ *for any* $\alpha \in (0, \infty)$, $\alpha \neq 1$.

Proof Let us recall two standard approximation arguments. Fix a non-negative function $\omega \in C_0^\infty(\mathbb{R}^n)$ supported in the closed unit ball $\bar{B}_n(0, 1) = \{x \in \mathbb{R}^n : |x| \le 1\}$ and such that $\int \omega = 1$, and put $\omega_\varepsilon(x) = \varepsilon^{-n}\omega(x/\varepsilon)$ for $\varepsilon > 0$. Given a locally integrable function f on $\mathbb{R}^n$, one defines its regularization (mollification) as the convolution

$$f_\varepsilon(x) = (f * \omega_\varepsilon)(x) = \int \omega_\varepsilon(x - y) f(y)\, dy$$
$$= \int f(x - \varepsilon y)\omega(y)\, dy, \qquad x \in \mathbb{R}^n. \tag{5.3}$$

It belongs to $C^\infty(\mathbb{R}^n)$, has gradient $\nabla f_\varepsilon = f * \nabla\omega_\varepsilon$, and is non-negative, when f is non-negative. From the definition it follows that, if $f \in L^2(\mathbb{R}^n)$, then

$$\|f_\varepsilon\|_2 \le \|f\|_2, \qquad \lim_{\varepsilon \to 0} \|f_\varepsilon - f\|_2 = 0.$$

Moreover, if $f \in W_1^2(\mathbb{R}^n)$, then, by (5.2)–(5.3), we have $\nabla f_\varepsilon = \nabla f * \omega_\varepsilon$. Hence

$$\|\nabla f_\varepsilon\|_2 \le \|\nabla f\|_2, \qquad \lim_{\varepsilon \to 0} \|\nabla f_\varepsilon - \nabla f\|_2 = 0,$$

so that

$$\|f_\varepsilon\|_{W_1^2} \le \|f\|_{W_1^2}, \quad \lim_{\varepsilon \to 0} \|f_\varepsilon - f\|_{W_1^2} = 0. \tag{5.4}$$

Thus, $C^\infty(\mathbb{R}^n) \cap W_1^2(\mathbb{R}^n)$ is dense in $W_1^2(\mathbb{R}^n)$.

To obtain (a), define $f = \sqrt{p}$. Given $\delta \in (0, \frac{1}{2})$, choose $\varepsilon > 0$ such that $\|f_\varepsilon - f\|_{W_1^2} < \delta$. Let us take a non-negative function $w \in C_0^\infty(\mathbb{R}^n)$ with $w(0) = 1$ and consider a sequence

$$u_l(x) = f_\varepsilon(x) w(x/l).$$

These functions belong to $C_0^\infty(\mathbb{R}^n)$, and by the Lebesgue dominated convergence theorem, $u_l \to f_\varepsilon$ in $W_1^2(\mathbb{R}^n)$ as $l \to \infty$. Hence

$$\|u - f\|_{W_1^2} < \delta$$

for some $u = u_l$, which implies

$$|\,\|u\|_2 - 1| = |\,\|u\|_2 - \|f\|_2| \le \|u - f\|_2 < \delta$$

and thus $\|u\|_2 > \frac{1}{2}$. As a result, the normalized function $\tilde{f} = u/\|u\|_2$ satisfies

$$\|\tilde{f} - f\|_{W_1^2} = \frac{\|u - \|u\|_2\, f\|_{W_1^2}}{\|u\|_2} \le \frac{\delta + \delta\|f\|_{W_1^2}}{\|u\|_2} < 4\delta\, \|f\|_{W_1^2},$$

where we used $\|f\|_{W_1^2} \ge \|f\|_2 = 1$. This gives

$$|\,\|\nabla \tilde{f}\|_2 - \|\nabla f\|_2| < 4\delta\, \|f\|_{W_1^2} \le 2\, \|f\|_{W_1^2}$$

and hence

$$\begin{aligned} \big|\,\|\nabla \tilde{f}\|_2^2 - \|\nabla f\|_2^2\,\big| &\le 4\delta\, \|f\|_{W_1^2} \big(\|\nabla \tilde{f}\|_2 + \|\nabla f\|_2\big) \\ &\le 4\delta\, \|f\|_{W_1^2} \big(2\, \|f\|_{W_1^2} + 2\, \|\nabla f\|_2\big) \\ &= 8\delta\, \big(\|f\|_{W_1^2}^2 + \|f\|_{W_1^2}\, \|\nabla f\|_2\big). \end{aligned}$$

Here $\|f\|_{W_1^2}^2 = 1 + I(p)$ and

$$\|f\|_{W_1^2}\, \|\nabla f\|_2 \le \frac{1}{2}\, \|f\|_{W_1^2}^2 + \frac{1}{2}\, \|\nabla f\|_2^2 \le \frac{1}{2} + I(p).$$

Eventually, the probability density $\tilde{p} = \tilde{f}^2$ satisfies

$$|I(\tilde{p}) - I(p)| \le 4\delta\, (3 + 4I(p)). \tag{5.5}$$

With $\delta = \delta_k \to 0$, we therefore obtain a sequence $p_k = \tilde{p}$ such that $I(p_k) \to I(p)$ as $k \to \infty$, thus proving (a).

Let us see that similar functions p_k may be used in (b) when

$$\int p(x)^\alpha\, dx = \int f(x)^{2\alpha}\, dx = \infty$$

which corresponds to the case where $N_\alpha(p) = 0$ for $\alpha > 1$ and $N_\alpha(p) = \infty$ for $0 < \alpha < 1$. Returning to the previously defined functions u_l, we observe that $\|u_l\|_{2\alpha} \to$

$\|f_\varepsilon\|_{2\alpha}$ as $l \to \infty$. Hence, it is sufficient to check that $\|f_\varepsilon\|_{2\alpha} \to \|f\|_{2\alpha} = \infty$ for some sequence $\varepsilon = \varepsilon_k \to 0$. Indeed, since $\|f\|_2 = 1$, the function f is locally integrable, implying that $f_\varepsilon(x) \to f(x)$ as $\varepsilon \to 0$ for almost all points $x \in \mathbb{R}$. This follows from (5.2) and the Lebesgue differentiation theorem which yields

$$|f_\varepsilon(x) - f(x)| \leq \int \omega_\varepsilon(x-y)\,|f(y) - f(x)|\,dy$$

$$\leq \|\omega\|_\infty\,\varepsilon^{-n} \int_{|y-x|<\varepsilon} |f(y) - f(x)|\,dy \to 0 \quad \text{a.e.}$$

Hence, by Fatou's lemma, $\|f\|_{2\alpha} \leq \liminf_{\varepsilon\to 0} \|f_\varepsilon\|_{2\alpha}$, and we are done.

Now, let us turn to the basic case where $\int p(x)^\alpha\,dx < \infty$, $\alpha \in (0,\infty)$. To prove (b), we borrow arguments from the proof of Theorem 2.3.2 in [19]. Consider a partition $\{w_i\}_{i=0}^\infty$ of unity of $\mathbb{R}^n$ subordinate to the covering $G_i = B_n(0, i+1) \setminus \bar{B}_n(0, i-1)$, in which $B_n(0,-1) = B_n(0,0) = \emptyset$. Every function w_i is supposed to be in $C_0^\infty(\mathbb{R}^n)$ with a support lying in G_i, to be non-negative, and all of them satisfy

$$\sum_{i=0}^{\infty} w_i(x) = 1, \quad x \in \mathbb{R}^n. \tag{5.6}$$

As before, let $f = \sqrt{p}$. Given $0 < \delta < \frac{1}{2}$, for each $i \geq 0$ choose $\varepsilon_i > 0$ small enough such that $(w_i f)_{\varepsilon_i}$ is still supported in G_i and

$$\|(w_i f)_{\varepsilon_i} - w_i f\|_{W_1^2} < 2^{-i-1}\delta. \tag{5.7}$$

The latter is possible due to the property (5.3) applied to $w_i f$.

By the integrability assumption on p, we have $\|w_i f\|_{2\alpha} < \infty$, implying

$$\|(w_i f)_\varepsilon - w_i f\|_{2\alpha} \to 0 \quad \text{as } \varepsilon \to 0 \tag{5.8}$$

as long as $2\alpha \geq 1$. Since $f \in L^2(\mathbb{R}^n)$, we similarly have $\|(w_i f)_\varepsilon - w_i f\|_2 \to 0$. The latter implies that (5.8) holds in the case $2\alpha < 1$ as well, since $w_i f$ is supported on a bounded set. Therefore, in addition to (5.7), we may require that

$$\int |(w_i f)_{\varepsilon_i} - w_i f|^{2\alpha}\,dx < (2^{-i-1}\delta)^{\max(2\alpha,1)}. \tag{5.9}$$

Now, by (5.6), $f(x) = \sum_{i=0}^\infty w_i(x) f(x)$, where the series contains only finitely many non-zero terms. More precisely,

$$f(x) = \sum_{i=0}^{m} w_i(x) f(x), \quad |x| < m+1.$$

Similarly, for the function $u(x) = \sum_{i=0}^{\infty}(w_i(x)f(x))_{\varepsilon_i}$, we have

$$u(x) = \sum_{i=0}^{m}(w_i(x)f(x))_{\varepsilon_i}, \qquad |x| < m+1.$$

This equality shows that u is non-negative and belongs to the class $C_0^\infty(\mathbb{R}^n)$. In addition, by (5.7),

$$\|u - f\|_{W_1^2} \leq \sum_{i=0}^{\infty} \|(w_i f)_{\varepsilon_i} - w_i f\|_{W_1^2} < \delta.$$

Hence

$$\|u - f\|_2 < \delta, \tag{5.10}$$

and repeating the arguments from the previous step, we arrive at the bound (5.5) for the density $\tilde{p} = \tilde{f}^2$ with $\tilde{f} = u/\|u\|_2$.

Next, if $\alpha \geq \frac{1}{2}$, by the triangle inequality in $L^{2\alpha}$, from (5.9) we also get $\|u - f\|_{2\alpha} < \delta$, so

$$|\,\|u\|_{2\alpha} - \|f\|_{2\alpha}| < \delta. \tag{5.11}$$

If $\alpha < \frac{1}{2}$, then, applying the inequality $(a_1+\dots+a_N)^{2\alpha} \leq a_1^{2\alpha}+\dots+a_N^{2\alpha}$ ($a_k \geq 0$), from (5.9) we deduce that

$$\int |u - f|^{2\alpha}\,dx \leq \sum_{i=1}^{\infty} \int |u - w_i f|^{2\alpha}\,dx < \delta.$$

This yields

$$\Big| \int u^{2\alpha}\,dx - \int f^{2\alpha}\,dx \Big| < \delta$$

and therefore, by Jensen's inequality,

$$|\,\|u\|_{2\alpha} - \|f\|_{2\alpha}| < (2\delta)^{1/(2\alpha)}. \tag{5.12}$$

In view of (5.10), inequalities similar to (5.11)–(5.12) hold also true for the function $\tilde{f} = u/\|u\|_2$ in place of u. Applying this with $\delta = \delta_k \to 0$, we obtain a sequence $\tilde{f}_k$ such that the probability densities $\tilde{p} = \tilde{f}^2$ satisfy $(a) - (b)$ for any $\alpha \neq 1$. □

Corollary 5.2 *For any $\alpha > 0$, $\alpha \neq 1$, the infimum*

$$\inf_{I(p)<\infty} \big[N_\alpha(p)I(p)\big]$$

may be restricted to the class of compactly supported, C^∞-smooth densities p on $\mathbb{R}^n$ with finite Fisher information.

6 Two Dimensional Isoperimetric Inequalities for Entropies

In this section we deal with dimension $n = 2$. As will be clarified, the entropic isoperimetric inequality

$$N_\alpha(X)I(X) \geq c_{\alpha,2} \tag{6.1}$$

holds true for any $\alpha \in [0, \infty)$ with a positive constant $c_{\alpha,2}$ and does not hold for $\alpha = \infty$ which answers Question 1 in the introduction. In addition, we will give a certain description of the optimal constants $c_{\alpha,2}$ in (6.1) for the range $\alpha \in [\frac{1}{2}, \infty)$, thus answering partially Question 2.

When $n = 2$, the family of inequalities (1.6) takes now the form

$$\Big(\int |f|^{2\alpha}\Big)^{\frac{1}{2\alpha}} \leq \Big(\frac{4}{c_{\alpha,2}}\Big)^{\frac{\alpha-1}{2\alpha}} \Big(\int |\nabla f|^2\Big)^{\frac{\theta}{2}} \Big(\int f^2\Big)^{\frac{1-\theta}{2}} \tag{6.2}$$

with $\theta = \frac{\alpha-1}{\alpha}$ when $\alpha > 1$, and

$$\Big(\int f^2\Big)^{\frac{1}{2}} \leq \Big(\frac{4}{c_{\alpha,2}}\Big)^{\frac{1-\alpha}{2}} \Big(\int |\nabla f|^2\Big)^{\frac{\theta}{2}} \Big(\int |f|^{2\alpha}\Big)^{\frac{1-\theta}{2\alpha}} \tag{6.3}$$

with $\theta = 1 - \alpha$ when $\alpha \in (0, 1)$.

Both inequalities enter the framework of Gagliardo–Nirenberg's inequality (1.8). The best constants and extremizers in (1.8) are not known for all admissible parameters. The most recent paper on this topic is due to Liu and Wang [10] (see references therein and historical comments). The case $q = s = 2$ in (1.8) that corresponds to (6.2) with $r = 2\alpha$ goes back to Weinstein [17] who related the best constants to the solutions of non-linear Schrödinger equations.

We present now part of the results of [10] that are useful for us. Since we will use them for any dimension $n \geq 2$, the next statement does not deal only with the case $n = 2$. Also, since all the inequalities of interest for us deal with the L^2-norm of the gradient only, we may restrict ourselves to $q = 2$ for simplicity, when (1.8) becomes

$$\Big(\int |f|^r\Big)^{\frac{1}{r}} \le \kappa_n(2,r,s)\Big(\int |\nabla f|^2\Big)^{\frac{\theta}{2}}\Big(\int |f|^s\Big)^{\frac{1-\theta}{s}} \tag{6.4}$$

with parameters satisfying $1 \le r, s \le \infty$, $0 \le \theta \le 1$, and $\frac{1}{r} = \theta(\frac{1}{2} - \frac{1}{n}) + (1-\theta)\,\frac{1}{s}$. This inequality may be restricted to the class of all smooth, compactly supported functions $f \ge 0$ on $\mathbb{R}^n$. Once (6.4) holds in $C_0^\infty(\mathbb{R}^n)$, this inequality is extended by a regularization and density arguments to the Sobolev space of functions $f \in L^s(\mathbb{R}^n)$ such that $|\nabla f| \in L^2(\mathbb{R}^n)$ (the gradients in this space are understood in a weak sense).

The next statement relates the optimal constant in (6.4) to the solutions of the ordinary non-linear equation

$$u''(t) + \frac{n-1}{t}u'(t) + u(t)^{r-1} = u(t)^{s-1} \tag{6.5}$$

on the positive half-axis. Put

$$\sigma = \begin{cases} \frac{n+2}{n-2} & \text{if } n \ge 3, \\ \infty & \text{if } n = 2. \end{cases}$$

We denote by $|x|$ the Euclidean norm of a vector $x \in \mathbb{R}^n$.

Theorem 6.1 ([10]) *In the range* $1 \le s < \sigma$, $s < r < \sigma + 1$,

$$\kappa_n(2,r,s) = \theta^{-\frac{\theta}{2}}(1-\theta)^{\frac{\theta}{2}-\frac{1}{r}} M_s^{-\frac{\theta}{n}}, \quad M_s = \int_{\mathbb{R}^n} u_{r,s}^s(|x|)\,dx,$$

where the functions $u_{r,s} = u_{r,s}(t)$ *are defined for* $t \ge 0$ *as follows.*

(i) *If* $s < 2$, *then* $u_{r,s}$ *is the unique positive decreasing solution to the equation* (6.5) *in* $0 < t < T$ *(for some* T*), satisfying* $u'(0) = 0$, $u(T) = u'(T) = 0$, *and* $u(t) = 0$ *for all* $t \ge T$.
(ii) *If* $s \ge 2$, *then* $u_{r,s}$ *is the unique positive decreasing solution to* (6.5) *in* $t > 0$, *satisfying* $u'(0) = 0$ *and* $\lim_{t\to\infty} u(t) = 0$.

Moreover, the extremizers in (6.4) *exist and have the form* $f(x) = au_{r,s}(|bx+c|)$ *with* $a \in \mathbb{R}$, $b \ne 0$, $c \in \mathbb{R}^n$.

Note that (6.2) corresponds to Gagliardo–Nirenberg's inequality (6.4) with $s = 2$, $r = 2\alpha$, and $\theta = \frac{\alpha-1}{\alpha}$ for $\alpha > 1$, while (6.3) with $\alpha \in [\frac{1}{2}, 1)$ corresponds to (6.4) with $r = 2$, $s = 2\alpha$, and $\theta = 1-\alpha$. Applying Corollary 5.2, we therefore conclude that

$$\kappa_2(2,r,s) = (4/c_{\alpha,2})^{\frac{\alpha-1}{2\alpha}} \quad \text{when } \alpha > 1,$$

$$\kappa_2(2,r,s) = (4/c_{\alpha,2})^{\frac{1-\alpha}{2}} \quad \text{when } \alpha \in [1/2, 1).$$

Together with Liu–Wang's theorem, we immediately get the following corollary, where we put as before

$$M_s = \int_{\mathbb{R}^2} u^s(|x|)\,dx = 2\pi \int_0^\infty u^s(t)\,t\,dt.$$

Corollary 6.2

(i) *For any $\alpha > 1$, we have*

$$c_{\alpha,2} = 4(\alpha - 1)\,\alpha^{-\frac{1}{\alpha-1}} M_2,$$

where M_2 is defined for the unique positive decreasing solution $u(t)$ on $(0,\infty)$ to the equation $u''(t) + \frac{u'(t)}{t} + u(t)^{2\alpha-1} = u(t)$ with $u'(0) = 0$ and $\lim_{t\to\infty} u(t) = 0$.

(ii) *For any $\alpha \in [\frac{1}{2}, 1)$, we have*

$$c_{\alpha,2} = 4(1-\alpha)\,\alpha^{\frac{\alpha}{1-\alpha}} M_{2\alpha},$$

where $M_{2\alpha}$ is defined for the unique positive decreasing solution $u(t)$ to $u''(t) + \frac{1}{t}u'(t) + u(t) = u(t)^{2\alpha-1}$ in $0 < t < T$ with $u'(0) = 0$, $u(T) = u'(T) = 0$, and $u(t) = 0$ for all $t \geq T$.

In both cases the extremizers in (6.1) represent densities of the form $p(x) = \frac{b}{M}u^2(|bx + c|)$, $x \in \mathbb{R}^2$, with $b > 0$ and $c \in \mathbb{R}^2$.

So far, we have seen that (6.1) holds for any $\alpha \in [1/2, \infty)$. Since, as observed in the introduction, $\alpha \mapsto c_{\alpha,n}$ is non-increasing, (6.1) holds also for $\alpha < 1/2$ and therefore for any $\alpha \in [0,\infty)$. Note that the case $\alpha = 1$, which is formally not contained in the results above, is the classical isoperimetry inequality for entropies (1.1). Let us now explain why (6.1) cannot hold for $\alpha = \infty$. The functional form for (6.1) should be the limit case of (6.2) as $\alpha \to \infty$, when it becomes

$$\|f\|_\infty^2 \leq D \int |\nabla f|^2\,dx \tag{6.6}$$

with $D = 4/c_{\infty,2}$. To see that (6.6) may not hold with any constant D, we reproduce Example 1.1.1 in [14]. Let, for $x \in \mathbb{R}^2$,

$$f(x) = \begin{cases} \log|\log|x|| & \text{if } |x| \leq 1/e, \\ 0 & \text{otherwise.} \end{cases}$$

Then, passing to radial coordinates, we have

$$\int |\nabla f|^2 = 2\pi \int_0^{1/e} \frac{dr}{r|\log r|^2} = 2\pi,$$

while f is not bounded. In fact, (6.6) is also violated for a sequence of smooth bounded approximations of f.

7 Isoperimetric Inequalities for Entropies in Dimension $n = 3$ and Higher

One may exhibit two different behaviors between $n = 3, 4$, and $n \geq 5$ in the entropic isoperimetric inequality

$$N_\alpha(X)I(X) \geq c_{\alpha,n}. \tag{7.1}$$

Let us rewrite the inequality (1.6) separately for the three natural regions, namely as

$$\Big(\int |f|^{2\alpha}\Big)^{\frac{1}{2\alpha}} \leq \Big(\frac{4}{c_{\alpha,n}}\Big)^{\frac{n(\alpha-1)}{4\alpha}} \Big(\int |\nabla f|^2\Big)^{\frac{\theta}{2}} \Big(\int f^2\Big)^{\frac{1-\theta}{2}} \tag{7.2}$$

with $\theta = \frac{n(\alpha-1)}{2\alpha}$ when $1 < \alpha \leq \frac{n}{n-2}$,

$$\Big(\int |f|^{2\alpha}\Big)^{\frac{\theta}{2\alpha}} \Big(\int f^2\Big)^{\frac{1-\theta}{2}} \leq \frac{2}{\sqrt{c_{\alpha,n}}} \Big(\int |\nabla f|^2\Big)^{\frac{1}{2}} \tag{7.3}$$

with $\theta = \frac{2\alpha}{n(\alpha-1)}$ when $\alpha > \frac{n}{n-2}$ (observe that $\theta \in (0,1)$ in this case), and finally

$$\Big(\int f^2\Big)^{\frac{1}{2}} \leq \Big(\frac{4}{c_{\alpha,n}}\Big)^{\frac{n(1-\alpha)}{2[\alpha(2-n)+n]}} \Big(\int |\nabla f|^2\Big)^{\frac{\theta}{2}} \Big(\int |f|^{2\alpha}\Big)^{\frac{1-\theta}{2\alpha}} \tag{7.4}$$

with $\theta = \frac{n(1-\alpha)}{\alpha(2-n)+n}$ when $\alpha \in (0,1)$.

Both (7.2) and (7.4) enter the framework of Gagliardo–Nirenberg's inequality (1.8). As for (7.3), we will show that such an inequality cannot hold. To that aim, we need to introduce the limiting case $\theta = 1$ in (7.2), which corresponds to $\alpha = \frac{n}{n-2}$. It amounts to the classical Sobolev inequality

$$\Big(\int |f|^{\frac{2n}{n-2}}\Big)^{\frac{n-2}{2n}} \leq S_n \Big(\int |\nabla f|^2\Big)^{\frac{1}{2}} \tag{7.5}$$

which is known to hold true with best constant

$$S_n = \frac{1}{\sqrt{\pi n(n-2)}} \Big(\frac{\Gamma(n)}{\Gamma(\frac{n}{2})}\Big)^{\frac{1}{n}}.$$

Moreover, the only extremizers in (7.5) have the form

$$f(x) = \frac{a}{(1 + b\,|x - x_0|^2)^{\frac{n-2}{2}}}, \qquad a \in \mathbb{R},\ b > 0,\ x_0 \in \mathbb{R}^n \tag{7.6}$$

(sometimes called the Barenblatt profile), see [1, 7, 16]. If $f \in L^2(\mathbb{R}^n)$ and $|\nabla f| \in L^2(\mathbb{R}^n)$, then, by (7.3), we would have that $f \in L^p(\mathbb{R}^n)$ with $p = 2\alpha > \frac{2n}{n-2}$ which contradicts the Sobolev embeddings. Therefore (7.3) cannot be true, so that (7.1) holds only for $\alpha \in [0, \frac{n}{n-2}]$.

As for the value of the best constant $c_{\alpha,n}$ in (7.1) and the form of the extremizers, we need to use again Theorem 6.1 which can, however, be applied only for $n \le 5$. As in Corollary 6.2, we adopt the notation

$$M_s = \int_{\mathbb{R}^n} u^s(|x|)\,dx$$

for a function u satisfying the non-linear ordinary differential equation

$$u''(t) + \frac{n-1}{t}u'(t) + u(t)^{2\alpha-1} = u(t), \qquad 0 < t < \infty, \tag{7.7}$$

or (in a different scenario)

$$u''(t) + \frac{n-1}{t}u'(t) + u(t) = u(t)^{2\alpha-1}, \qquad 0 < t < T. \tag{7.8}$$

Corollary 7.2 *Let* $3 \le n \le 5$.

(i) *For any* $1 < \alpha < \frac{n}{n-2}$, *we have*

$$c_{\alpha,n} = \frac{2n(\alpha-1)}{\alpha}\Big(\frac{2\alpha}{\alpha(2-n)+n}\Big)^{\frac{n(\alpha-1)-2}{n(\alpha-1)}} M_2^{\frac{2}{n}},$$

where M_2 *is defined for the unique positive decreasing solution* $u(t)$ *to* (7.7) *on* $(0,\infty)$ *with* $u'(0) = 0$ *and* $\lim_{t\to\infty} u(t) = 0$.

(ii) *For any* $\alpha \in [\frac{1}{2}, 1)$,

$$c_{\alpha,n} = 4\,\frac{n(1-\alpha)}{\alpha(2-n)+n}\Big(\frac{2\alpha}{\alpha(2-n)+n}\Big)^{\frac{2\alpha}{n(1-\alpha)}} M_{2\alpha}^{\frac{2}{n}}$$

where $M_{2\alpha}$ *is defined for the unique positive decreasing solution* $u(t)$ *to* (7.8) *with* $u'(0) = 0$, $u(T) = u'(T) = 0$, *and* $u(t) = 0$ *for all* $t \ge T$.

In both cases, the extremizers in (7.1) *are densities of the form* $p(x) = \frac{b}{M}u^2(|bx + c|)$, $x \in \mathbb{R}^n$, *with* $b > 0$ *and* $c \in \mathbb{R}^n$.

For the critical value of α, the picture is more complete but is different.

Corollary 7.3 *Let $n \geq 3$ and $\alpha = \frac{n}{n-2}$. Then*

$$c_{\alpha,n} = 4\pi n(n-2)\Big(\frac{\Gamma(\frac{n}{2})}{\Gamma(n)}\Big)^{\frac{2}{n}}.$$

(i) *For $n = 3$ and $n = 4$,* (7.1) *has no extremizers, i.e., there does not exist any density p for which equality holds in* (7.1) *with the optimal constant.*
(ii) *For $n \geq 5$, the extremizers in* (7.1) *exist and have the form*

$$p(x) = \frac{a}{(1 + b|x - x_0|^2)^{n-2}}, \quad a, b > 0, \ x_0 \in \mathbb{R}^n. \tag{7.9}$$

Remark 7.4 Recall that $c_{1,n} = 2\pi en$. Using the Stirling formula, it is easy to see that, for $\alpha = \frac{n}{n-2}$,

$$c_{\alpha,n} \sim 2\pi en - 2\pi e\,(2 + \log 2) + O\Big(\frac{1}{n}\Big) \quad \text{as } n \to \infty.$$

In particular, $c_{\alpha,n} \geq 2\pi en - c_0$ for all $0 \leq \alpha \leq \frac{n}{n-2}$ with some absolute constant $c_0 > 0$. To get a similar upper bound, it is sufficient to test (7.1) with $\alpha = 0$ on some specific probability distributions. In this case, this inequality becomes

$$\mathrm{vol}_n(\mathrm{supp}(p))^{\frac{2}{n}}\, I(X) \geq c_{0,n}. \tag{7.10}$$

Suppose that the random vector $X = (X_1, \dots, X_n)$ in $\mathbb{R}^n$ has independent components such that every X_k has a common density $w(s) = \frac{2}{\pi}\cos^2(s)$, $|s| \leq \frac{\pi}{2}$. As we already mentioned in Sect. 4, this one dimensional probability distribution appears as an extremal one in the entropic isoperimetric inequality (1.2) for the parameter $\alpha = 0$. The random vector X has density

$$p(x) = w(x_1) \dots w(x_n), \quad x = (x_1, \dots, x_n) \in \mathbb{R}^n,$$

so that

$$N_0(X) = N_0(X_1) = \pi^2, \quad I(X) = nI(X_1) = 4n.$$

Therefore, from (7.10) we may conclude that $c_{0,n} \leq 4\pi^2 n$.

***Proof of Corollaries* 7.2–7.3** The first corollary is obtained by a straight forward application of Theorem 6.1 with

$$s = 2,\ r = 2\alpha,\ \theta = \frac{n(\alpha - 1)}{2\alpha}, \quad \kappa_n(2, r, s) = \big(4/c_{\alpha,n}\big)^{\frac{n(\alpha-1)}{4\alpha}}$$

when $1 < \alpha < \frac{n}{n-2}$, and with

$$q = r = 2,\ s = 2\alpha,\ \theta = \frac{n(1-\alpha)}{\alpha(2-n)+n}, \qquad \kappa_n(2,r,s) = \big(4/c_{\alpha,n}\big)^{\frac{n(1-\alpha)}{2[\alpha(2-n)+n]}}$$

when $\alpha \in (0,1)$. Details are left to the reader.

For the second corollary, we first observe that (7.2) can be recast for $n \geq 3$ and $\alpha = \frac{n}{n-2}$ as

$$\Big(\int |f|^{\frac{2n}{n-2}}\Big)^{\frac{n-2}{2n}} \leq \Big(\frac{4}{c_{\alpha,n}}\Big)^{1/2}\Big(\int |\nabla f|^2\Big)^{\frac{1}{2}}. \tag{7.11}$$

Therefore $\frac{4}{c_{\alpha,n}} = S_n^2$ from which the explicit value of $c_{\alpha,n}$ follows (recalling Corollary 5.2).

Now, in order to analyze the question about the extremizers in (7.1), suppose that we have an equality in it for a fixed (probability) density p on $\mathbb{R}^n$. In particular, we should assume that the function $f = \sqrt{p}$ belongs to $W_1^2(\mathbb{R}^n)$. Rewriting (7.1) in terms of f, we then obtain an equality in (7.11), which is the same as (7.5). As mentioned earlier, this implies that f must be of the form (7.6), thus leading to (7.9). However, whether or not this function p is integrable depends on the dimension. Using polar coordinates, one immediately realizes that

$$\int \frac{dx}{(1+b|x-x_0|^2)^{n-2}}$$

has the same behavior as $\int_1^\infty \frac{1}{r^{n-3}}\,dr$. But, the latter integral converges only if $n \geq 5$. □

Acknowledgments Research of the first author was partially supported by the NSF grant DMS-2154001. Research of the second author was partially supported by the grants ANR-15-CE40-0020-03 – LSD – Large Stochastic Dynamics, ANR 11-LBX-0023-01 – Labex MME-DII and Fondation Simone et Cino del Luca in France. This research has been conducted within the FP2M federation (CNRS FR 2036).

References

1. T. Aubin, Problèmes isopérimétriques et espaces de Sobolev. J. Differ. Geom. **11**, 573–598 (1976)
2. W. Beckner, Asymptotic estimates for Gagliardo-Nirenberg embedding constants. Potential Anal. **17**, 253–266 (2002)
3. S.G. Bobkov, G.P. Chistyakov, F. Götze, Fisher information and the central limit theorem. Probab. Theory Relat. Fields **159**(1–2), 1–59 (2014)
4. S.G. Bobkov, N. Gozlan, C. Roberto, P.-M. Samson, Bounds on the deficit in the logarithmic Sobolev inequality. J. Funct. Anal. **267**(11), 4110–4138 (2014)

5. E.A. Carlen, Superadditivity of Fisher's information and logarithmic Sobolev inequalities. J. Funct. Anal. **101**(1), 194–211 (1991)
6. M.H.M. Costa, T.M. Cover, On the similarity of the entropy power inequality and the Brunn-Minkowski inequality. IEEE Trans. Inform. Theory **30**, 837–839 (1984)
7. M. Del Pino, J. Dolbeault, Best constants for Gagliardo-Nirenberg inequalities and applications to nonlinear diffusions. J. Math. Pures Appl. (9) **81**(9), 847–875 (2002)
8. A. Dembo, T.M. Cover, J. Thomas, Information theoretic inequalities. IEEE Trans. Inform. Theory **37**(6), 1501–1518 (1991)
9. L. Gross, Logarithmic Sobolev inequalities. Am. J. Math. **97**, 1061–1083 (1975)
10. J.-G. Liu, J. Wang, On the best constant for Gagliardo-Nirenberg interpolation inequalities. Preprint (2017). Available at http://arxiv.org/abs/1712.10208v1
11. J. Moser, On Harnack's theorem for elliptic differential equations. Commun. Pure Appl. Math. **14**, 577–591 (1961)
12. J. Moser, A Harnack inequality for parabolic differential equations. Commun. Pure Appl. Math. **17**, 101–134 (1964); correction in **20**, 231–236 (1967)
13. B. Nagy, Über integralungleichungen zwischen einer Funktion und ihrer Ableitung. Acta Univ. Szeged. Sect. Sci. Math. **10**, 64–74 (1941)
14. L. Saloff-Coste, *Aspects of Sobolev-Type Inequalities*, vol. 289 (Cambridge University Press, Cambridge/New York, 2002)
15. A.J. Stam, Some inequalities satisfied by the quantities of information of Fisher and Shannon. Inform. Control **2**, 101–112 (1959)
16. G. Talenti, Best constant in Sobolev inequality. Ann. Mat. Pura Appl. **110**, 353–372 (1976)
17. M.I. Weinstein, Nonlinear Schrödinger equations and sharp interpolation estimates. Commun. Math. Phys. **87**, 567–576 (1983)
18. Z. Zhang, Inequalities for characteristic functions involving Fisher information. C. R. Math. Acad. Sci. Paris **344**(5), 327–330 (2007)
19. W.P. Ziemer, *Weakly Differentiable Functions. Sobolev Spaces and Functions of Bounded Variation*. Graduate Texts in Mathematics, vol. 120 (Springer, New York, 1989), xvi+308pp

Transport Proofs of Some Functional Inverse Santaló Inequalities

Matthieu Fradelizi, Nathael Gozlan, and Simon Zugmeyer

2020 Mathematics Subject Classification: 52A20; 52A40; 60E15

1 Introduction

The classical Blaschke-Santaló inequality [27] gives the following sharp relation between the volume of a convex body K in $\mathbb{R}^n$ and the volume of its polar $K^* = \{y \in \mathbb{R}^n; x \cdot y \leq 1, \forall\, x \in K\}$: there exists $z \in \mathbb{R}^n$ such that $|K||(K-z)^*| \leq |B_2^n|^2$, where B_2^n denotes the Euclidean ball of radius one. Mahler [22] conjectured that the following optimal lower bound holds:

$$|K||K^*| \geq \frac{4^n}{n!},$$

for any centrally symmetric convex body K, with equality, for example, if K is a cube. Among general convex bodies K, the conjecture is that the lower bound should be reached for simplices. Both conjectures were proved by Mahler in dimension 2 [21], while the conjecture for symmetric bodies was established by Iriyeh and Shibata in dimension 3 [17] (see also [8]). The conjectures were proved for particular families of convex bodies like unconditional convex bodies [23, 30], zonoids [13, 25], bodies having symmetries [3, 18]. Bourgain and Milman [5] (see

M. Fradelizi
LAMA UMR8050, Univ Gustave Eiffel, Univ Paris Est Creteil, CNRS, Marne-la-Vallée, France
e-mail: matthieu.fradelizi@univ-eiffel.fr

N. Gozlan (✉)
Université Paris Cité, CNRS, MAP5, Paris, France
e-mail: nathael.gozlan@u-paris.fr

S. Zugmeyer
ENS de Lyon, UMPA UMR 5669, Lyon, France
e-mail: simon.zugmeyer@ens-lyon.fr

R. Adamczak et al. (eds.), *High Dimensional Probability IX*, Progress in Probability 80, https://doi.org/10.1007/978-3-031-26979-0_4

also [1, 2, 4, 15, 20, 24]) established an asymptotic form of the conjectures by proving that there exists a constant c such that $|K||K^*| \geq c^n/n!$.

Functional forms of the Mahler conjectures were proposed, where the convex bodies are replaced by log-concave functions and polar convex bodies by the Fenchel-Legendre transform. More precisely, it is conjectured that, for any convex function $V : \mathbb{R}^n \to \mathbb{R} \cup \{+\infty\}$ such that $0 < \int e^{-V}\, dx < +\infty$, it holds

$$\int e^{-V}\, dx \int e^{-V^*}\, dx \geq e^n,$$

where the Fenchel-Legendre transform of V is defined by

$$V^*(y) = \sup_{x \in \mathbb{R}^n} \{x \cdot y - V(x)\}, \qquad y \in \mathbb{R}^n.$$

If, in addition, V is even, it is conjectured that

$$\int e^{-V}\, dx \int e^{-V^*}\, dx \geq 4^n.$$

These functional forms were proved in dimension 1 in [9–11] and the even case was proved in dimension 2 in [12]. The inequality was proved for unconditional functions in [9, 10]. These conjectures are slightly stronger than Mahler's conjectures for sets, because the latter are implied by the former, whereas the inequality for sets must be true in any dimension for the functional inequality to hold, as proved in [10].

To present the class of Entropy-Transport inequalities considered in this work, we need to introduce some definitions and notations.

The set of all Borel probability measures on $\mathbb{R}^n$ will be denoted by $\mathcal{P}(\mathbb{R}^n)$. For $k \geq 1$, we will denote by $\mathcal{P}_k(\mathbb{R}^n)$ the subset of $\mathcal{P}(\mathbb{R}^n)$ of probability measures admitting a finite moment of order k. Recall that $\eta \in \mathcal{P}(\mathbb{R}^n)$ is called log-concave, if it admits a density with respect to the Lebesgue measure of the form e^{-V}, where $V : \mathbb{R}^n \to \mathbb{R} \cup \{+\infty\}$ is a lower semicontinuous convex function. The function V will be referred to as the potential of η. Note that we will not consider log-concave measures supported by a strict affine subspace of $\mathbb{R}^n$. The moment measure associated with a log-concave probability measure η with potential V is the measure $\nu = \nabla V \# \eta$ defined as the pushforward of η under the map ∇V: in other words, for any bounded measurable test functions, it holds

$$\int f(x)\, \nu(dx) = \int f(\nabla V(x))\, \eta(dx).$$

We recall that convex functions are differentiable Lebesgue almost everywhere, so that this definition makes sense. When η does not have full support, i.e., when $\text{supp}(\eta) \neq \mathbb{R}^n$, some extra regularity will be required at the boundary. We will say that a log-concave probability measure η, with potential V, has an essentially continuous density, if $e^{-V}(x) = 0$ for $\mathcal{H}_{n-1}$ almost all $x \in \partial\,\mathrm{Supp}(\eta)$, where $\mathrm{Supp}(\eta)$ denotes the support of η. Note that this terminology slightly differs from the one of [7] where it was the potential V that was called essentially continuous.

Definition 1.1 (Entropy-Transport inequality) We will say that the inequality $\mathrm{ET}_n(c)$ is satisfied for some constant $c > 0$ if, for all log-concave probability measures η_1, η_2 on $\mathbb{R}^n$ having essentially continuous densities, it holds

$$H(\eta_1) + H(\eta_2) \leq -n \log(ce^2) + \mathcal{T}(\nu_1, \nu_2), \tag{1}$$

where ν_1, ν_2 are the moment measures of η_1, η_2.
Similarly, we say that $\mathrm{ET}_{n,s}(c)$ is satisfied, if equation (1) holds for all log-concave measures η_1, η_2 that are also symmetric (i.e., such that $\nu_i(A) = \nu_i(-A)$ for all measurable sets A).

In the definition above, $H(\eta)$ denotes the relative entropy of η with respect to the Lebesgue measure (which is also equal to minus the Shannon entropy of η) and is defined by

$$H(\eta) = \int \log\left(\frac{d\eta}{dx}\right) d\eta.$$

The quantity $\mathcal{T}$ appearing in (1) is the so-called maximal correlation optimal transport cost, defined, for any $\mu_1, \mu_2 \in \mathcal{P}_1(\mathbb{R}^n)$, by

$$\mathcal{T}(\mu_1, \mu_2) = \inf_{f \in \mathcal{F}(\mathbb{R}^n)} \left\{ \int f \, d\mu_1 + \int f^* \, d\mu_2 \right\},$$

where $\mathcal{F}(\mathbb{R}^n)$ is the set of convex and lower semicontinuous functions $f : \mathbb{R}^n \to \mathbb{R} \cup \{+\infty\}$ which are proper (i.e., take at least one finite value). Since elements of $\mathcal{F}(\mathbb{R}^n)$ always admit affine lower bounds, note that $\int g \, d\mu_i$ makes sense in $\mathbb{R} \cup \{+\infty\}$ for all $g \in \mathcal{F}(\mathbb{R}^n)$, so that $\mathcal{T}(\mu_1, \mu_2)$ is well defined whenever $\mu_1, \mu_2 \in \mathcal{P}_1(\mathbb{R}^n)$. In the case where $\mu_1, \mu_2 \in \mathcal{P}_2(\mathbb{R}^n)$, it follows from the Kantorovich duality theorem [31] that

$$\mathcal{T}(\mu_1, \mu_2) = \sup_{X_1 \sim \mu_1, X_2 \sim \mu_2} \mathbb{E}[X_1 \cdot X_2] = \sup_{\pi \in \Pi(\mu_1, \mu_2)} \int x \cdot y \, \pi(dxdy),$$

where $\Pi(\mu_1, \mu_2)$ denotes the set of probability measures on $\mathbb{R}^n \times \mathbb{R}^n$ with marginals μ_1 and μ_2.

Definition 1.1 is motivated by a recent result of the second author [14], which states that inequality (1) is equivalent to the functional version of Mahler's conjecture (also called inverse Santaló inequality), as formulated by Klartag and Milman [19] and Fradelizi and Meyer [10] that we now recall.

Definition 1.2 (Inverse Santaló inequality) We will say that the inequality $\mathrm{IS}_n(c)$ is satisfied for some c, if for all functions $f \in \mathcal{F}(\mathbb{R}^n)$ such that both $\int e^{-f(x)} dx$ and

$\int e^{-f^*(x)}dx$ are positive, it holds

$$\int e^{-f(x)}\,dx \int e^{-f^*(x)}\,dx \geq c^n. \tag{2}$$

Similarly, we say that $\mathrm{IS}_{n,s}(c)$ is satisfied if equation (2) holds for all even functions in $\mathcal{F}(\mathbb{R}^n)$.

With this definition, the functional forms of Mahler's conjectures are $\mathrm{IS}_n(e)$ and $\mathrm{IS}_{n,s}(4)$.

Theorem 1.3 ([14]) *The inequality* $\mathrm{ET}_n(c)$ *(respectively,* $\mathrm{ET}_{n,s}(c)$*) is equivalent to* $\mathrm{IS}_n(c)$ *(respectively,* $\mathrm{IS}_{n,s}(c)$*).*

As shown in Theorem 1.2 of [14], inequalities $\mathrm{ET}_n(c)$ or $\mathrm{ET}_{n,s}(c)$ can be restated as improved versions of the Gaussian log-Sobolev inequality. In particular, the results of [9, 10] lead to sharp lower bounds on the deficit in the Gaussian log-Sobolev inequality for unconditional probability measures (see Theorem 1.4 of [14]).

The main contributions of the paper are the following. In Sect. 2 we give a new proof of the implication

$$\mathrm{ET}_n(c) \Rightarrow \mathrm{IS}_n(c),$$

and we show, in particular in Corollary 2.5, that only a restricted form of the inequality $\mathrm{ET}_n(c)$ is enough to get $\mathrm{IS}_n(c)$. This new proof significantly simplifies the proof given in [14]. Then, we prove in Sect. 3, using transport arguments together with correlation inequalities, that $\mathrm{ET}_1(e)$ and $\mathrm{ET}_{1,s}(4)$ are satisfied. In particular, this gives new and short proofs of the sharp functional Mahler conjecture in dimension 1. Finally, in Sect. 4, we propose a short proof of $\mathrm{IS}_{n,s}(4)$ when we restrict ourselves to unconditional functions, i.e., functions that are symmetric with respect to all coordinate hyperplanes, blending tools from this paper and the proof given in [10].

2 Entropy-Transport and Inverse Santaló Inequalities

2.1 From Entropy-Transport to Inverse Santaló Inequalities

The following result provides the key identity connecting the quantities appearing in the inverse functional inequalities to their dual transport-entropy counterparts.

Lemma 2.1 *Let* $V : \mathbb{R}^n \to \mathbb{R} \cup \{+\infty\}$ *be a convex function such that* $Z := \int e^{-V}\,dx \in (0, \infty)$ *and let* ν *be the moment measure of* $\eta(dx) = \frac{1}{Z}e^{-V}\,dx$*. Then, it holds*

$$-\log\left(\int e^{-V}\,dx\right) = \int -V^*\,d\nu + \mathcal{T}(\nu,\eta) + H(\eta). \tag{3}$$

Proof According to Proposition 7 of [7] and its proof, $V^* \in L^1(\nu)$ and $V \in L^1(\eta)$. We claim that

$$\mathcal{T}(\nu,\eta) = \int V^*\,d\nu + \int V\,d\eta = \int x\cdot\nabla V(x)\,\eta(dx). \tag{4}$$

Namely, if $f \in \mathcal{F}(\mathbb{R}^n)$, then

$$\begin{aligned}
\int f^*\,d\nu + \int f\,d\eta &= \int f^*(\nabla V(x)) + f(x)\,\eta(dx)\\
&\geq \int \nabla V(x)\cdot x\,\eta(dx)\\
&= \int V^*(\nabla V(x)) + V(x)\,\eta(dx)\\
&= \int V^*\,d\nu + \int V\,d\eta\\
&\geq \mathcal{T}(\nu,\eta).
\end{aligned}$$

Therefore, optimizing over $f \in \mathcal{F}(\mathbb{R}^n)$ gives (4). To conclude the proof of (3), just observe that

$$H(\eta) = -\log Z - \int V\,d\eta.$$

□

Note that we proved Lemma 2.1 for general convex V, but it turns out that the quantity $\mathcal{T}(\eta,\nu)$ appearing in the conclusion of this lemma simplifies when the measure η is smooth enough, as the following result shows.

Lemma 2.2 *For any essentially continuous log-concave probability measure $\eta \in \mathcal{P}(\mathbb{R}^n)$, its moment measure ν satisfies $\mathcal{T}(\eta,\nu) = n$.*

The proof is straightforward when V has full domain and is C^1-smooth and follows by an integration by part (see also Corollary 3 in [14], for general V with full domain). The proof of the general case is given in the appendix. In this case, equation (3) reduces to

$$-\log\left(\int e^{-V}\,dx\right) = \int -V^*\,d\nu + n + H(\eta).$$

In what follows, it will be convenient to introduce a particular class of potentials.

Definition 2.3 We denote by $\mathcal{V}(\mathbb{R}^n)$ the class of all convex functions $V : \mathbb{R}^n \to \mathbb{R}$ such that $V^* : \mathbb{R}^n \to \mathbb{R}$ (thus V, V^* are continuous and with full domain).

Thanks to Lemma 2.1, we can show the following.

Proposition 2.4 *Let $V \in \mathcal{V}(\mathbb{R}^n)$; denote by $\eta(dx) = \frac{1}{Z}e^{-V}\,dx$, $\eta^*(dx) = \frac{1}{Z^*}e^{-V^*}\,dx$, where Z, Z^* are the normalizing constants, and let ν, ν^* be the moment measures associated with η, η^*.*
If

$$H(\eta) + H(\eta^*) \le -n\log(ce^2) + \mathcal{T}(\nu, \nu^*), \tag{5}$$

then

$$\int e^{-V}\,dx \int e^{-V^*}\,dx \ge c^n.$$

Note that, according to, e.g., Lemma 4 in [14], if $V \in \mathcal{V}(\mathbb{R}^n)$ then $Z := \int e^{-V}\,dx$ and $Z^* := \int e^{-V^*}\,dx$ are both finite, and so the log-concave probability measures η and η^* are well defined.

Proof Applying Lemmas 2.1 and 2.2 to V and V^* yields

$$-\log\Big(\int e^{-V}\,dx\Big) = \int -V^*\,d\nu + \mathcal{T}(\nu,\eta) + H(\eta) = \int -V^*\,d\nu + n + H(\eta)$$

$$-\log\Big(\int e^{-V^*}\,dx\Big) = \int -V\,d\nu^* + n + H(\eta^*).$$

Adding these two identities yields

$$\begin{aligned}-\log\Big(\int e^{-V}\,dx \int e^{-V^*}\,dx\Big) &= -\Big(\int V^*\,d\nu + \int V\,d\nu^*\Big) + H(\eta) + H(\eta^*) + 2n\\ &\le -\mathcal{T}(\nu,\nu^*) + H(\eta) + H(\eta^*) + 2n\\ &\le -n\log(ce^2) + 2n = -\log(c^n),\end{aligned}$$

where the first inequality comes from the definition of $\mathcal{T}(\nu, \nu^*)$ and the second inequality from (5). □

Corollary 2.5 *Inequality $\mathrm{IS}_n(c)$ (respectively, $\mathrm{IS}_{n,s}(c)$) holds true as soon as for all $V \in \mathcal{V}(\mathbb{R}^n)$ (respectively, for all symmetric $V \in \mathcal{V}(\mathbb{R}^n)$)*

$$H(\eta) + H(\eta^*) \le -n\log(ce^2) + \mathcal{T}(\nu, \nu^*),$$

where $\eta(dx) = \frac{1}{Z}e^{-V}\,dx$, $\eta^*(dx) = \frac{1}{Z^*}e^{-V^*}\,dx$ with Z, Z^* the normalizing constants and where ν, ν^* are the moment measures associated with η, η^*.

Proof According to Proposition 2.4, it holds

$$\int e^{-V}\,dx \int e^{-V^*}\,dx \geq c^n,$$

for all $V \in \mathcal{V}(\mathbb{R}^n)$. Let $V \in \mathcal{F}(\mathbb{R}^n)$ be such that $0 < \int e^{-V}\,dx \int e^{-V^*}\,dx < \infty$. For all $k \geq 1$, consider

$$V_k(x) = V \,\square\, (k\frac{|\cdot|^2}{2})(x) + \frac{|x|^2}{2k}, \qquad x \in \mathbb{R}^n,$$

where $|\cdot|$ denotes the standard Euclidean norm on $\mathbb{R}^n$ and $\square$ is the infimum convolution operator, defined by

$$f \,\square\, g(x) = \inf\{f(y) + g(x-y) : y \in \mathbb{R}^n\}, \qquad x \in \mathbb{R}^n.$$

Since the infimum convolution leaves the class of convex functions stable, it is clear that V_k is still convex for all $k \geq 1$. It is also clear that V_k takes finite values on $\mathbb{R}^n$. Since $(f+g)^* = f^* \,\square\, g^*$ and (equivalently) $(f \,\square\, g)^* = f^* + g^*$ for all $f, g \in \mathcal{F}(\mathbb{R}^n)$, it is not difficult to check that

$$V_k^*(y) = (V^* + \frac{|\cdot|^2}{2k}) \,\square\, (k\frac{|\cdot|^2}{2})(y), \qquad y \in \mathbb{R}^n$$

and so V_k^* takes finite values on $\mathbb{R}^n$. In other words, $V_k \in \mathcal{V}(\mathbb{R}^n)$ for all $k \geq 1$. Since

$$V_k \geq V \,\square\, (k\frac{|\cdot|^2}{2}) \qquad \text{and} \qquad V_k^* \geq V^* \,\square\, (k\frac{|\cdot|^2}{2}),$$

one gets that

$$\int e^{-V\square(k\frac{|\cdot|^2}{2})}\,dx \int e^{-V^*\square(k\frac{|\cdot|^2}{2})}\,dx \geq \int e^{-V_k}\,dx \int e^{-V_k^*}\,dx \geq c^n.$$

Note that $V \,\square\, (k\frac{|\cdot|^2}{2})$ is the Moreau-Yosida approximation of V. In particular, it is well known that if $V \in \mathcal{F}(\mathbb{R}^n)$ then $V \,\square\, (k\frac{|\cdot|^2}{2})(x) \to V(x)$, for all $x \in \mathbb{R}^n$, as $k \to \infty$ (see, e.g., [12, Lemma 3.6]). Since $V \,\square\, (k\frac{|\cdot|^2}{2}) \geq V \,\square\, (\frac{|\cdot|^2}{2})$, it easily follows, from the dominated convergence theorem, that

$$\int e^{-V\square(k\frac{|\cdot|^2}{2})}\,dx \to \int e^{-V}\,dx,$$

as $k \to \infty$. Reasoning similarly for the other integral, one concludes that

$$\int e^{-V}\,dx \int e^{-V^*}\,dx \geq c^n,$$

which completes the proof. □

Remark 2.6 Note that the functions V_k and V_k^* are both continuously differentiable on $\mathbb{R}^n$. This follows from a well known regularizing property of the Moreau-Yosida approximation (see, e.g., [26, Theorem 26.3]). Therefore, the conclusion of Corollary 2.5 is still true if the Entropy-Transport inequality (5) is only assumed to hold for $V \in \mathcal{V}_1(\mathbb{R}^n)$, where $\mathcal{V}_1(\mathbb{R}^n)$ denotes the set of $V \in \mathcal{V}(\mathbb{R}^n)$ such that V and V^* are continuously differentiable.

2.2 Different Equivalent Formulations of Inverse Santaló Inequalities

The following result gathers different equivalent formulations of $\mathrm{IS}_n(c)$.

Theorem 2.7 *Let $c > 0$; the following statements are equivalent:*

(i) *The inequality $\mathrm{IS}_n(c)$ holds.*
(ii) *The inequality $\mathrm{ET}_n(c)$ holds.*
(iii) *For all $V \in \mathcal{V}(\mathbb{R}^n)$,*

$$H(\eta) + H(\eta^*) \leq -n\log(ce^2) + \mathcal{T}(\nu, \nu^*),$$

where η, η^* are the log-concave probability measures with respective potentials V, V^* and associated moment measures ν, ν^*.
(iv) *For all $V \in \mathcal{V}(\mathbb{R}^n)$,*

$$H(\eta) + H(\eta^*) \leq -n\log(ce^2) + \int V^*\, d\nu + \int V\, d\nu^*,$$

with the same notation as above.

The same equivalence is true for $\mathrm{IS}_{n,s}(c)$ and $\mathrm{ES}_{n,c}(c)$ assuming in (iii) and (iv) that $V \in \mathcal{V}(\mathbb{R}^n)$ is symmetric.

Proof $(i) \Rightarrow (ii)$ follows from Theorem 1.3 proved in [14].
$(ii) \Rightarrow (iii)$ is straightforward.
$(iii) \Rightarrow (iv)$ follows from the inequality $\mathcal{T}(\nu, \nu^*) \leq \int V^*\, d\nu + \int V\, d\nu^*$.
$(iv) \Rightarrow (i)$ follows from the proof of Proposition 2.4 and Corollary 2.5. □

Remark 2.8 Let us make some comments on Theorem 2.7.

(a) The proof of $(i) \Rightarrow (ii)$ given in [14] makes use of the following variational characterization of moment measures due to Cordero-Klartag [7] and Santambrogio [28]: a measure ν is the moment measure of a log-concave probability measure η with an essentially continuous density if and only if it is centered and not supported by an hyperplane; moreover, the measure η is the unique (up to translation) minimizer of the functional

$$\mathcal{P}_1(\mathbb{R}^n) \to \mathbb{R} \cup \{+\infty\} : \eta \mapsto \mathcal{T}(\nu, \eta) + H(\eta).$$

(b) In [14], the implication $(ii) \Rightarrow (i)$ has been established using the following duality formula: for all $V \in \mathcal{V}(\mathbb{R}^n)$ such that $\int e^{-V^*}\,dx > 0$, it holds

$$L(V) := -\log\left(\int e^{-V^*}\,dx\right) = \sup_{\nu \in \mathcal{P}_1(\mathbb{R}_n)} \left\{\int -V\,d\nu - K(\nu)\right\},$$

with $K(\nu) = \inf_{\eta \in \mathcal{P}_1(\mathbb{R}_n)}\{\mathcal{T}(\nu, \eta) + H(\eta)\}$, $\nu \in \mathcal{P}_1(\mathbb{R}^n)$. This equality, established in [14], shows that the functionals L and K are in convex duality. The route followed in the present paper, based on the key Lemma 2.1, turns out to be simpler and more direct.

(c) Let us finally highlight the fact that the equivalence of (iii) and (iv) is a bit surprising. Namely, for a fixed $V \in \mathcal{F}(\mathbb{R}^n)$, the formulation (iii) is in general strictly stronger than (iv), because the inequality $\mathcal{T}(\nu, \nu^*) \le \int V^*\,d\nu + \int V\,d\nu^*$ is strict in general. Indeed, equality here means that (V^*, V) is a couple of Kantorovich potentials between ν and ν^*. If ν has a density with respect to Lebesgue, this means that ∇V^* transports ν onto ν^* which is not true in general.

3 Proofs of Entropy-Transport Inequalities in Dimension 1

In this section, we show that inequalities $\mathrm{ET}_{1,s}(4)$ and $\mathrm{ET}_1(e)$ hold true. The reason why the case of dimension 1 is simple is that optimal transport maps for the cost $\mathcal{T}$ are given in an explicit form. Recall that the cumulative distribution function of $\mu \in \mathcal{P}(\mathbb{R})$ is the function

$$F_\mu(x) = \mu((-\infty, x]), \qquad x \in \mathbb{R}.$$

Its generalized inverse is the function denoted F_μ^{-1} defined by

$$F_\mu^{-1}(t) = \inf\{x : F_\mu(x) \ge t\}, \qquad t \in (0, 1).$$

Lemma 3.1 *Let $\eta_1, \eta_2 \in \mathcal{P}_1(\mathbb{R})$ be such that $\mathcal{T}(\eta_1, \eta_2)$ is finite. It holds*

$$\mathcal{T}(\eta_1, \eta_2) \ge \int_0^1 F_{\eta_1}^{-1}(x) F_{\eta_2}^{-1}(x)\,dx,$$

with equality if $\eta_1, \eta_2 \in \mathcal{P}_2(\mathbb{R})$. More generally, if $\nu_1 = S_1\#\eta_1$ and $\nu_2 = S_2\#\eta_2$ with $S_1, S_2 : \mathbb{R} \to \mathbb{R}$ two measurable maps, and if $\nu_1, \nu_2 \in \mathcal{P}_1(\mathbb{R})$ and satisfy that $\mathcal{T}(\nu_1, \nu_2)$ is finite, then

$$\mathcal{T}(\nu_1, \nu_2) \geq \int_0^1 S_1(F_{\eta_1}^{-1})(x) S_2(F_{\eta_2}^{-1})(x)\, dx.$$

Proof It is well known that, if X is uniformly distributed on $(0, 1)$, then $(F_{\eta_1}^{-1}(X), F_{\eta_2}^{-1}(X))$ is a coupling between η_1 and η_2 called the monotone coupling. Therefore, $(S_1(F_{\eta_1}^{-1}(X)), S_2(F_{\eta_2}^{-1}(X)))$ is a coupling between ν_1, ν_2. Suppose that $\mathcal{T}(\nu_1, \nu_2)$ is finite, then, if $f \in \mathcal{F}(\mathbb{R})$ is such that $f \in L^1(\nu_1)$ and $f^* \in L^1(\nu_2)$, Young's inequality yields

$$f(S_1(F_{\eta_1}^{-1}(X))) + f^*(S_2(F_{\eta_2}^{-1}(X))) \geq S_1(F_{\eta_1}^{-1}(X)) S_2(F_{\eta_2}^{-1}(X)).$$

Therefore, $[S_1(F_{\eta_1}^{-1}(X)) S_2(F_{\eta_2}^{-1}(X))]_+$ is integrable, and taking expectation, we get

$$\begin{aligned}\int_0^1 S_1(F_{\eta_1}^{-1}(x)) S_2(F_{\eta_2}^{-1}(x))\, dx &= \mathbb{E}[S_1(F_{\eta_1}^{-1}(X)) S_2(F_{\eta_2}^{-1}(X))] \\ &\leq \int f\, d\nu_1 + \int f^*\, d\nu_2.\end{aligned}$$

Optimizing over f gives the desired inequality. In the case where $S_1 = S_2 = \mathrm{Id}$ and η_1, η_2 have finite moments of order 2, then it is well known that the monotone coupling is optimal for W_2^2 (the square of the 2-Wasserstein distance), and so also for $\mathcal{T}$. □

Lemma 3.2 *The inequality* $\mathrm{ET}_1(c)$ *is satisfied as soon as for all concave functions* $f_1, f_2 : [0, 1] \to \mathbb{R}_+$ *such that* $f_1(0) = f_2(0) = f_1(1) = f_2(1) = 0$,

$$\int_0^1 \log(f_1 f_2)\, dx \leq -\log(e^2 c) + \int_0^1 f_1' f_2'\, dx. \tag{6}$$

Similarly, the inequality $\mathrm{ET}_{1,s}(c)$ *is satisfied as soon as inequality* (6) *holds for all functions* f_1, f_2 *that are also symmetric with respect to* $1/2$, *i.e.,* $f_i(x) = f_i(1 - x)$ *for all* $x \in [0, 1]$.

Proof Let $\eta_i(dx) = e^{-V_i}\, dx$, $i = 1, 2$ be two log-concave probability measures on $\mathbb{R}$ with essentially continuous densities. This latter condition means that, for some $-\infty \leq a_i < b_i \leq +\infty$, the convex function V_i takes finite values on (a_i, b_i), is $+\infty$ on $\mathbb{R} \setminus (a_i, b_i)$ and is such that $V_i(x) \to +\infty$ when $x \to a_i$ and $x \to b_i$.

To prove $\mathrm{ET}_1(c)$, one can assume that $\mathcal{T}(\nu_1, \nu_2)$ is finite, otherwise there is nothing to prove. Using Lemma 3.1 with $S_i = V_i'$, we see that the inequality

$$H(\eta_1) + H(\eta_2) \leq -\log(ce^2) + \int_0^1 V_1'(F_{\eta_1}^{-1}(x)) V_2'(F_{\eta_2}^{-1}(x))\, dx \tag{7}$$

implies $\mathrm{ET}_1(c)$. For $i = 1, 2$, define

$$f_i(x) = F'_{\eta_i} \circ F_{\eta_i}^{-1}(x) = \exp(-V_i \circ F_{\eta_i}^{-1}(x)), \qquad x \in (0, 1).$$

Note that, since F_{η_i} is strictly increasing and differentiable on (a_i, b_i), the function $F_{\eta_i}^{-1}$ is the regular inverse of the restriction of F_{η_i} to (a_i, b_i) and is also differentiable on $(0, 1)$. Since $F_{\eta_i}^{-1}(x) \to b_i$ as $x \to 1$ and $\exp(-V_i(y)) \to 0$ as $y \to b_i$, one sees that $f_i(x) \to 0$ as $x \to 1$. Similarly, $f_i(x) \to 0$ as $x \to 0$. Setting $f_i(0) = f_i(1) = 0$ thus provides a continuous extension of f_i to $[0, 1]$. The function f_i is moreover concave on $[0, 1]$. Indeed, denoting by f'_i and V'_i the left derivatives of f_i, V_i which are well defined everywhere on $(0, 1)$, we see that for all $x \in (0, 1)$,

$$f'_i(x) = (F'_{\eta_i} \circ F_{\eta_i}^{-1})'(x) = \frac{F''_{\eta_i} \circ F_{\eta_i}^{-1}(x)}{F'_{\eta_i} \circ F_{\eta_i}^{-1}(x)} = -V'_i(F_{\eta_i}^{-1}(x)).$$

So, f'_i is decreasing on $(0, 1)$, and thus f_i is concave. Finally, note that

$$H(\eta_1) + H(\eta_2) = \int_0^1 \log(f_1 f_2)\, dx$$

and

$$\int_0^1 V'_1(F_{\eta_1}^{-1}) V'_2(F_{\eta_2}^{-1})\, dx = \int_0^1 f'_1 f'_2\, dx,$$

so that inequality (7) becomes

$$\int_0^1 \log(f_1 f_2)\, dx \le -\log(e^2 c) + \int_0^1 f'_1 f'_2\, dx.$$

It is furthermore clear that whenever η_1, η_2 are symmetric, then f_1, f_2 are also symmetric with respect to $1/2$, which concludes the proof. □

Remark 3.3 The functions f_i are related to the isoperimetric profiles of the measures η_i in dimension 1. Moreover, there is a one to one correspondence between log-concave measures η and concave f on $(0, 1)$, see, for example, [6, Proposition A.1].

3.1 The One-Dimensional Symmetric Case

Theorem 3.4 *The inequality* $\mathrm{ET}_{1,s}(4)$ *is satisfied and the constant* 4 *is optimal.*

Proof Let f_1, f_2 be two concave functions on $[0, 1]$, equal to zero at 0 and 1, and symmetric with respect to $1/2$. Let us show that inequality (6) holds true with $c = 4$. It is enough to prove that

$$\int_0^{1/2} \log(f_1 f_2)\,dx \le -1 - \log(2) + \int_0^{1/2} f_1' f_2'\,dx.$$

We use the following classical correlation inequality: if $h, k : \mathbb{R} \to \mathbb{R}$ are two non-increasing functions (or non-decreasing), and if μ is a finite measure on $\mathbb{R}$, then

$$\int_{\mathbb{R}} h(x)\,\mu(dx) \int_{\mathbb{R}} k(x)\,\mu(dx) \le \mu(\mathbb{R}) \int_{\mathbb{R}} h(x)k(x)\,\mu(dx), \tag{8}$$

which follows from the integration of the inequality

$$(h(x) - h(y))(k(x) - k(y)) \ge 0.$$

As a result, since f_1' and f_2' are non-increasing, we get, for all $x \in [0, 1]$, that

$$f_1(x) f_2(x) = \int_0^x f_1'(t)\,dt \int_0^x f_2'(t)\,dt \le x \int_0^x f_1'(t) f_2'(t)\,dt. \tag{9}$$

For a later use, note that this inequality holds also even if f_1, f_2 are not symmetric. By symmetry, $f_1'(t) f_2'(t) \ge 0$ for all $t \in [0, 1/2]$, so we get

$$f_1(x) f_2(x) \le x \int_0^{1/2} f_1'(t) f_2'(t)\,dt, \qquad \forall x \in [0, 1/2].$$

Thus, after integrating,

$$\begin{aligned}
\int_0^{1/2} \log(f_1(x) f_2(x))\,dx &\le \int_0^{1/2} \log(x)\,dx + \frac{1}{2}\log\Big(\int_0^{1/2} f_1'(t) f_2'(t)\,dt\Big) \\
&\le \frac{1}{2}\log\Big(\frac{1}{2}\Big) - \frac{1}{2} + \int_0^{1/2} f_1'(t) f_2'(t)\,dt + \frac{1}{2}\log\Big(\frac{1}{2}\Big) - \frac{1}{2} \\
&= -1 - \log(2) + \int_0^{1/2} f_1'(t) f_2'(t)\,dt,
\end{aligned}$$

where we used the inequality $\log(x) - \log(1/2) \le 2x - 1$.

To see that this inequality is sharp, we can use the functions $f_1(x) = \min(x, 1 - x)$ and f_2 an approximation of the constant function equal to $1/2$. The optimal constant is reached at the limit. □

Remark 3.5 The choice $f_1(x) = \min(x, 1 - x)$ corresponds to the log-concave probability measure $\eta(dx) = e^{-|x|}\,dx/2$, the polar transform of which is the uniform probability measure on $[-1, 1]$. These densities are the equality case in the functional Mahler inequality [10]. However, the uniform probability measure on $[-1, 1]$ is not an admissible measure in our case, since it is not essentially continuous, thus the optimality is only reached at the limit.

Remark 3.6 Inequality (6) is also satisfied if we assume only one of the functions to be symmetric. Indeed, if f_2 is symmetric with respect to 1/2, define $\tilde{f_1}(x) = \frac{1}{2}(f(x) + f(1-x))$. On the one hand, using the concavity of the logarithm,

$$\begin{aligned}\int_0^1 \log(\tilde{f_1}(x) f_2(x))\, dx &= \int_0^1 \log \tilde{f_1}(x)\, dx + \int_0^1 \log f_2(x)\, dx \\ &\geq \frac{1}{2}\int_0^1 \log(f_1(x)) + \log(f_1(1-x))\, dx \\ &+ \int_0^1 \log f_2(x)\, dx \\ &= \int_0^1 \log f_1(x)\, dx + \log f_2(x)\, dx \\ &= \int_0^1 \log(f_1(x) f_2(x))\, dx,\end{aligned}$$

and on the other hand,

$$\int \tilde{f_1}' f_2'\, dx = \frac{1}{2}\int_0^1 f_1'(x) f_2'(x)\, dx - \frac{1}{2}\int_0^1 f_1'(x) f_2'(1-x)\, dx,$$

hence the claim, since $f_2'(x) = -f_2'(1-x)$ for all $x \in [0, 1]$.

3.2 *The One-Dimensional General Case*

Theorem 3.7 *The inequality* $\mathrm{ET}_1(e)$ *is satisfied and the constant* e *is sharp.*

Proof Let us show that, if $f_1, f_2 : [0, 1] \to \mathbb{R}^+$ are concave functions vanishing at 0 and 1, then

$$\int_0^1 \log(f_1 f_2)\, dx \leq -3 + \int_0^1 f_1' f_2'\, dx.$$

Just like before, it is enough to show that

$$\int_0^{1/2} \log(f_1 f_2)\, dx \leq -\frac{3}{2} + \int_0^{1/2} f_1' f_2'\, dx.$$

Applying the inequality $\log(b) \leq \log(a) + \frac{(b-a)}{a}$ to $b = f_1 f_2$ and $a = x(1-x)$, $x \in (0, 1)$, and using again the correlation inequality (9), we get

$$\int_0^{1/2} \log(f_1 f_2)\,dx \le \int_0^{1/2} \Big(\frac{f_1(x) f_2(x)}{x(1-x)} + \log(x(1-x)) - 1\Big)\,dx$$
$$\le -\frac{3}{2} + \int_0^{1/2} \frac{1}{1-x}\Big(\int_0^x f_1'(t) f_2'(t)dt\Big)\,dx$$
$$= -\frac{3}{2} + \int_0^{1/2} f_1'(t) f_2'(t) \log(2-2t)\,dt,$$

and Lemma 3.9 below concludes the proof of the inequality.

To see that the inequality is optimal, we choose for f_1 and f_2 approximations of the functions $x \mapsto x$ and $x \mapsto 1-x$, which of course are not admissible, since they are not zero on the boundary. It is a straightforward calculation to see that equality is reached at the limit. □

Remark 3.8 The function $f_1(x) = x$ is the isoperimetric profile of the log-concave probability measure $\nu(dx) = e^{-(1+x)} \mathbb{1}_{[-1,+\infty[}(x)\,dx$, which density is an equality case in the functional Mahler inequality [10].

Lemma 3.9 *Let $f, g : [0,1] \to \mathbb{R}_+$ be two concave functions vanishing at* 0 *and* 1. *The following inequality holds:*

$$\int_0^{1/2} f'(t) g'(t) \log(2-2t)\,dt \le \int_0^{1/2} f'(t) g'(t)\,dt. \tag{10}$$

Proof For $0 \le t \le 1/2$, we define $\varphi(t) = 1 - \log(2) - \log(1-t)$ and $\Phi(t) = \int_0^t \varphi(x)\,dx$. Notice that φ is increasing on $[0, 1/2]$ and $\varphi(0) = 1 - \log(2) > 0$, hence $\varphi > 0$ on $[0, 1/2]$. Let $u = f'$ and $v = g'$. The functions u and v are non-increasing and satisfy $\int_0^1 u\,dx = \int_0^1 v\,dx = 0$. Applying the correlation inequality (8) again, and integrating with respect to the measure with density φ on $[0, 1/2]$, we get

$$\int_0^{1/2} \varphi\,dx \int_0^{1/2} uv\varphi\,dx \ge \int_0^{1/2} u\varphi\,dx \int_0^{1/2} v\varphi\,dx.$$

Integrating by parts, one has

$$\int_0^{1/2} u\varphi\,dx = \Big[u\Phi\Big]_0^{1/2} + \int_0^{1/2} (-u')\Phi\,dx = u\left(\frac{1}{2}\right)\Phi\left(\frac{1}{2}\right) + \int_0^{1/2} (-u')\Phi\,dx.$$

A quick calculation shows that $\Phi(1/2) = 1 - \log(2) = \varphi(0)$. Since φ is increasing, it follows that $\Phi(x) \ge \varphi(0)x = \Phi(1/2)x$. Using this inequality, the fact that u is non-increasing and integrating again by parts, we get

$$\int_0^{1/2} (-u'(x))\Phi(x)\,dx \ge \Phi\left(\frac{1}{2}\right) \int_0^{1/2} (-u'(x))x\,dx = \Phi\left(\frac{1}{2}\right)\left(-\Big[u(x)x\Big]_0^{1/2} + \int_0^{1/2} u(x)\,dx\right).$$

Thus, using that u is non-increasing again, we get

$$\int_0^{1/2} u\varphi\, dx \geq \Phi\left(\frac{1}{2}\right)\left(\frac{1}{2}u\left(\frac{1}{2}\right) + \int_0^{1/2} u(x)\, dx\right) \geq \Phi\left(\frac{1}{2}\right)\int_0^1 u(x)\, dx = 0.$$

One also has $\int_0^{1/2} v\varphi\, dx \geq 0$, so we conclude that $\int_0^{1/2} uv\varphi\, dx \geq 0$, which establishes (10). □

4 Revisiting the Unconditional Case

Recall that a function $V : \mathbb{R}^n \to \mathbb{R}$ is called unconditional if

$$V(x_1, \ldots, x_n) = V(|x_1|, \ldots, |x_n|), \qquad \forall x \in \mathbb{R}^n.$$

The following result is due to Fradelizi and Meyer [9, 10].

Theorem 4.1 *Let* $V : \mathbb{R}^n \to \mathbb{R} \cup \{+\infty\}$ *be a convex unconditional function such that* $0 < \int_{\mathbb{R}^n} e^{-V}\, dx < \infty$ *then*

$$\int_{\mathbb{R}^n} e^{-V}\, dx \int_{\mathbb{R}^n} e^{-V^*}\, dx \geq 4^n. \tag{11}$$

Below, we show how Lemma 2.1 can be used to shorten the proof of [10]. More precisely, from Lemma 2.1 we quickly derive the inequality (13) below, which is the key step of the argument, and then the rest of the proof follows the same path as in [10].

Proof Reasoning as in the proof of Corollary 2.5, it is enough to prove (11) when V, V^* have full domain and are continuously differentiable on $\mathbb{R}^n$. Since V and V^* are unconditional, it is clear that (11) is equivalent to

$$\int_{\mathbb{R}^n_+} e^{-V}\, dx \int_{\mathbb{R}^n_+} e^{-V^*}\, dx \geq 1. \tag{12}$$

Let us prove (12) by induction on n.

- For $n = 1$, (12) follows from Theorem 3.4.
- Let $V : \mathbb{R}^n \to \mathbb{R}$, with $n \geq 2$, satisfying the assumption of the theorem. For all $t > 0$, let $a(t) = \int_{\mathbb{R}^n_+} e^{-tV}\, dx$ and $\eta_t(dx) = \frac{1}{a(t)} e^{-tV(x)} \mathbb{1}_{\mathbb{R}^n_+}(x)\, dx$, and let ν_t be the moment measure of η_t. Applying Lemma 2.1 to η_t, and then Jensen's inequality, we get

$$\begin{aligned} H(\eta_t) + n + \log a(t) &= \int (tV)^* d\nu_t = t \int V^*\left(\frac{x}{t}\right) \nu_t(dx) \\ &= t \int V^*(\nabla V)\, d\eta_t \geq tV^*\left(\int_{\mathbb{R}^n_+} \nabla V\, d\eta_t\right) \end{aligned}$$

for $t > 0$. Here, we have used that $\mathcal{T}(\nu, \eta) = \int_{\mathbb{R}^n_+} x \cdot \nabla V(x) e^{-V(x)}\, dx = n$ because the boundary terms in the integration by parts are 0. A simple integration by parts shows that, for all $t > 0$,

$$\int_{\mathbb{R}^n_+} \nabla V \, d\eta_t = \frac{G(t)}{ta(t)},$$

where $G(t) = (a_1(t), \dots, a_n(t))$ and $a_i(t) = \int_{\mathbb{R}^{n-1}_+} e^{-tV_i(x)}\, dx$, with

$$V_i(x) = V(x_1, \dots, x_{i-1}, 0, x_{i+1}, \dots, x_n), \qquad x \in \mathbb{R}^{n-1}_+.$$

Since $H(\eta_t) + \log a(t) = t\frac{a'(t)}{a(t)}$, we get

$$\frac{a'(t)}{a(t)} + \frac{n}{t} \geq V^*\left(\frac{G(t)}{ta(t)}\right), \qquad \forall t > 0. \tag{13}$$

Denoting $\alpha(t) = \int_{\mathbb{R}^n_+} e^{-tV^*}\, dx$ and $\Gamma(t) = (\alpha_1(t), \dots, \alpha_n(t))$, with $\alpha_i(t) = \int_{\mathbb{R}^{n-1}_+} e^{-t(V^*)_i(x)}\, dx$, a similar calculation gives

$$\frac{\alpha'(t)}{\alpha(t)} + \frac{n}{t} \geq V\left(\frac{\Gamma(t)}{t\alpha(t)}\right), \qquad \forall t > 0. \tag{14}$$

Adding (13) and (14) and applying Young's inequality gives, for all $t > 0$,

$$\begin{aligned}\frac{a'(t)}{a(t)} + \frac{\alpha'(t)}{\alpha(t)} + \frac{2n}{t} &\geq V^*\left(\frac{G(t)}{ta(t)}\right) + V\left(\frac{\Gamma(t)}{t\alpha(t)}\right) \geq \frac{G(t)}{ta(t)} \cdot \frac{\Gamma(t)}{t\alpha(t)} \\ &= \frac{1}{t^2 a(t)\alpha(t)} \sum_{i=1}^n a_i(t)\alpha_i(t).\end{aligned}$$

Note that for all $1 \leq i \leq n$, $(V_i)^* = (V^*)_i$ because V is non-decreasing with respect to each coordinate. By induction, for all $1 \leq i \leq n$ and $t > 0$,

$$t^{n-1} a_i(t)\alpha_i(t) = \int_{\mathbb{R}^{n-1}_+} e^{-tV_i}\, dx \int_{\mathbb{R}^{n-1}_+} e^{-(tV_i)^*}\, dx \geq 1.$$

Therefore, for all $t > 0$,

$$\frac{a'(t)}{a(t)} + \frac{\alpha'(t)}{\alpha(t)} + \frac{2n}{t} \geq \frac{n}{t^{n+1} a(t)\alpha(t)},$$

which amounts to

$$F'(t) \geq nt^{n-1},$$

with $F(t) = t^{2n}a(t)\alpha(t)$. Since $F(0) = 0$, one gets $F(1) \geq 1$, which is exactly (12).

□

Appendix: Proof of Lemma 2.2

For completeness' sake, we provide here the proof of Lemma 2.2, which mostly follows the arguments given in [7].

Proof Let $\eta(dx) = e^{-V(x)}\,dx$ be an essentially continuous probability measure, and $\nu = \nabla V \# \eta$ its moment measure. Recall that we want to prove that $\mathcal{T}(\eta, \nu) = n$. As established in Lemma 2.1, the maximal correlation is given by

$$\mathcal{T}(\mu, \nu) = \int x \cdot \nabla V(x) e^{-V(x)}\,dx.$$

Assuming everything is smooth, an integration by parts immediately proves that

$$\mathcal{T}(\mu, \nu) = \int \mathrm{div}(x) e^{-V(x)}\,dx - \int_{\partial \,\mathrm{dom}\, V} x \cdot n_{\mathrm{dom}\, V}(x) e^{-V(x)}\,d\mathcal{H}_{n-1}(x) = n,$$

since $e^{-V(x)} = 0$ for $\mathcal{H}_{n-1}$-almost all $x \in \partial \,\mathrm{dom}\, V$. In the general case, however, V is only Lipschitz on the interior of its domain. Thus, let us choose x_0 in the interior of the domain of V. According to [7, Lemma 4],

$$\int \nabla V(x) e^{-V(x)}\,dx = 0$$

by essential continuity, and thus

$$\mathcal{T}(\mu, \nu) = \int x \cdot \nabla V(x) e^{-V(x)}\,dx = \int (x - x_0) \cdot \nabla V(x) e^{-V(x)}\,dx.$$

Convexity of V implies that the function $x \mapsto (x - x_0) \cdot \nabla V(x)$ is bounded from below by some constant (which is, of course, integrable against η), and so, if $(K_N)_{N \in \mathbb{N}}$ is an increasing sequence of compact sets such that $\bigcup_N K_N = \mathrm{dom}\, V$,

$$\int (x - x_0) \cdot \nabla V(x) e^{-V(x)}\,dx = \lim_{N \to \infty} \int_{K_N} (x - x_0) \cdot \nabla V(x) e^{-V(x)}\,dx.$$

For $N \in \mathbb{N}$, with $N > \min V$, the sets $\{V \leq N\}$ are convex, closed because of lower semicontinuity, with nonempty interior since $\int e^{-V} > 0$, bounded since $\lim_{|x|\to+\infty} V(x) = +\infty$ and strictly increasing by the essential continuity of e^{-V}. Since convex bodies may be approximated by smooth convex bodies (see [16, Lemma 2.3.2]), we can find a sequence (K_N) of smooth convex bodies such that

$$\{V \leq N\} \subset K_N \subset \{V \leq 2N\}$$

for all $N > \min V$. It is clear that then $\bigcup_N K_N = \operatorname{dom} V$. Since K_N is smooth, and V is Lipschitz on K_N, the divergence theorem applies:

$$\int_{K_N} (x - x_0) \cdot \nabla V(x) e^{-V(x)}\, dx = \int_{K_N} \operatorname{div}(x) e^{-V(x)}\, dx - \int_{\partial K_N} n_{K_N}(x) \cdot (x - x_0) e^{-V(x)}\, d\mathcal{H}_{n-1}(x),$$

where $n_{K_N}(x)$ is the outer normal vector to K_N at x. Clearly,

$$\lim_{N\to\infty} \int_{K_N} \operatorname{div}(x) e^{-V(x)}\, dx = n \lim_{N\to+\infty} \eta(K_N) = n,$$

and we will show that the second term converges towards zero. To that end, note that since $e^{-V(x)}$ is integrable, there exist constants $a > 0$ and b such that $V(x) \geq a|x| + b$. As an immediate consequence, for all $N > b$, the sublevel set $\{V \leq N\}$ is included in the ball of center 0 and of radius $R_N = (N - b)/a$. Hence, whenever N is large enough so that $x_0 \in K_N$,

$$\left|\int_{\partial K_N} n_{K_N}(x) \cdot (x - x_0) e^{-V(x)}\, d\mathcal{H}_{n-1}(x)\right| \leq \int_{\partial K_N} |x - x_0| e^{-V(x)}\, d\mathcal{H}_{n-1}(x) \leq 2R_{2N}\, e^{-N} \mathcal{H}_{n-1}(\partial K_N).$$

Finally, if K, L are two convex bodies such that $K \subset L$, then $\mathcal{H}_{n-1}(\partial K) \leq \mathcal{H}_{n-1}(\partial L)$ (see [29, (5.25)]), and so $\mathcal{H}_{n-1}(\partial K_N) \leq R_{2N}^{n-1} \mathcal{H}_{n-1}(\mathbb{S}^{n-1})$, which is enough to conclude that

$$\left|\int_{\partial K_N} n_{K_N}(x) \cdot (x - x_0) e^{-V(x)}\, d\mathcal{H}_{n-1}(x)\right| \leq p(N) e^{-N},$$

where p is some polynomial, which proves our claim. □

Acknowledgments The authors were supported by a grant of the Simone and Cino Del Duca foundation. The third author has benefited from a post doctoral position funded by the Simone and Cino Del Duca foundation. This research has been conducted within the FP2M federation (CNRS

FR 2036). The authors warmly thank Anne Estrade, the head of MAP 5, for making the meetings that led to this publication possible in the midst of the epidemic situation. The authors would like to thank the anonymous referee for helpful comments.

References

1. B. Berndtsson, Bergman kernels for Paley-wiener spaces and Nazarov's proof of the Bourgain-Milman theorem. Pure Appl. Math. Q. **18**(2), 395–409 (2022)
2. B. Berndtsson, Complex integrals and Kuperberg's proof of the Bourgain-Milman theorem. Adv. Math. **388** (2021). https://doi.org/10.1016/j.aim.2021.107927
3. F. Barthe, M. Fradelizi, The volume product of convex bodies with many hyperplane symmetries. Am. J. Math. **135**(2), 311–347 (2013)
4. Z. Blocki, A lower bound for the Bergman kernel and the Bourgain-Milman inequality, in *Geometric Aspects of Functional Analysis*. Lecture Notes in Mathematics, vol. 2116 (Springer, Cham, 2014), pp. 53–63
5. J. Bourgain, V.D. Milman, New volume ratio properties for convex symmetric bodies in $\mathbf{R}^n$. Invent. Math. **88**(2), 319–340 (1987)
6. S. Bobkov, Extremal properties of half-spaces for log-concave distributions. Ann. Probab. **24**(1), 35–48 (1996)
7. D. Cordero-Erausquin, B. Klartag, Moment measures. J. Funct. Anal. **268**(12), 3834–3866 (2015)
8. M. Fradelizi, A. Hubard, M. Meyer, E. Roldán-Pensado, A. Zvavitch, Equipartitions and Mahler volumes of symmetric convex bodies. Am. J. Math. **144**(5), 1201–1219 (2022). https://doi.org/10.1353/ajm.2022.0027
9. M. Fradelizi, M. Meyer, Increasing functions and inverse Santaló inequality for unconditional functions. Positivity **12**(3), 407–420 (2008)
10. M. Fradelizi, M. Meyer, Some functional inverse Santaló inequalities. Adv. Math. **218**(5), 1430–1452 (2008)
11. M. Fradelizi, M. Meyer, Functional inequalities related to Mahler's conjecture. Monatsh. Math. **159**(1–2), 13–25 (2010)
12. M. Fradelizi, E. Nakhle, The functional form of Mahler conjecture for even log-concave functions in dimension 2. Int. Math. Res. Not. (2022). https://doi.org/10.1093/imrn/rnac120
13. Y. Gordon, M. Meyer, S. Reisner, Zonoids with minimal volume-product—a new proof. Proc. Am. Math. Soc. **104**(1), 273–276 (1988)
14. N. Gozlan, The deficit in the Gaussian Log-Sobolev inequality and inverse Santaló inequalities. Int. Math. Res. Not. IMRN **17**, 12940–12983 (2022)
15. A. Giannopoulos, G. Paouris, B.H. Vritsiou, The isotropic position and the reverse Santaló inequality. Israel J. Math. **203**(1), 1–22 (2014)
16. L. Hörmander, *Notions of Convexity*. Modern Birkhäuser Classics (Birkhäuser, Boston, 2007). Reprint of the 1994 edition
17. H. Iriyeh, M. Shibata, Symmetric Mahler's conjecture for the volume product in the 3-dimensional case. Duke Math. J. **169**(6), 1077–1134 (2020)
18. H. Iriyeh, M. Shibata, Minimal volume product of three dimensional convex bodies with various discrete symmetries. Discrete Comput. Geom. **68**(3), 738–773 (2022). https://doi.org/10.1007/s00454-021-00357-6
19. B. Klartag, V.D. Milman, Geometry of log-concave functions and measures. Geom. Dedicata **112**, 169–182 (2005)
20. G. Kuperberg, From the Mahler conjecture to Gauss linking integrals. Geom. Funct. Anal. **18**(3), 870–892 (2008)
21. K. Mahler, Ein Minimalproblem für konvexe Polygone. Mathematica (Zutphen) (1939)

22. K. Mahler, Ein Übertragungsprinzip für konvexe Körper. Časopis Pěst. Mat. Fys. **68**, 93–102 (1939)
23. M. Meyer, Une caractérisation volumique de certains espaces normés de dimension finie. Israel J. Math. **55**(3), 317–326 (1986)
24. F. Nazarov, The Hörmander proof of the Bourgain-Milman theorem, in *Geometric Aspects of Functional Analysis*. Lecture Notes in Mathematics, vol. 2050 (Springer, Heidelberg, 2012), pp. 335–343
25. S. Reisner, Zonoids with minimal volume-product. Math. Z. **192**(3), 339–346 (1986)
26. R.T. Rockafellar, *Convex Analysis* (Princeton University Press, Princeton, 1997)
27. L.A. Santaló, An affine invariant for convex bodies of n-dimensional space. Port. Math. **8**, 155–161 (1949)
28. F. Santambrogio, Dealing with moment measures via entropy and optimal transport. J. Funct. Anal. **271**(2), 418–436 (2016)
29. R. Schneider, *Convex Bodies: The Brunn-Minkowski Theory*. Encyclopedia of Mathematics and Its Applications, vol. 151, expanded edition (Cambridge University Press, Cambridge, 2014)
30. J. Saint-Raymond, Sur le volume des corps convexes symétriques, in *Initiation Seminar on Analysis: G. Choquet-M. Rogalski-J. Saint-Raymond, 20th Year: 1980/1981*. Publications mathématiques de l'Université Pierre et Marie Curie, vol. 46, Exp. No. 11 (University of Paris VI, Paris, 1981)
31. C. Villani, *Optimal Transport. Old and New*, vol. 338 (Springer, Berlin, 2009)

Tail Bounds for Sums of Independent Two-Sided Exponential Random Variables

Jiawei Li and Tomasz Tkocz

2020 Mathematics Subject Classification: 60E15; 60G50

1 Introduction

Concentration inequalities establish conditions under which random variables are *close* to their *typical values* (such as the expectation or median) and provide quantitative probabilistic bounds. Their significance cannot be overestimated, both across probability theory and in applications in related areas (see [1, 2]). Particularly, such inequalities often concern sums of independent random variables.

Let $X_1, \ldots, X_n$ be independent exponential random variables, each with mean 1. Consider their weighted sum $S = \sum_{i=1}^n a_i X_i$ with some positive weights $a_1, \ldots, a_n$. Janson in [11] showed the following concentration inequality: for every $t > 1$,

$$\frac{1}{2e\alpha} \exp\big(-\alpha(t-1)\big) \leq \mathbb{P}(S \geq t\mathbb{E}S) \leq \frac{1}{t} \exp\big(-\alpha(t-1-\log t)\big), \tag{1}$$

where $\alpha = \frac{\mathbb{E}S}{\max_{i \leq n} a_i}$ (in fact, he derived (1) from its analogue for the geometric distribution). Note that as $t \to \infty$, the lower and upper bounds are of the same order $e^{-\alpha t + o(t)}$. Moreover, $e^{-\alpha t} = \mathbb{P}\left(X_1 > t\frac{\mathbb{E}S}{\max_{i \leq n} a_i}\right)$. In words, the asymptotic

TT's research supported in part by NSF grant DMS-1955175.

J. Li
School of Mathematics, University of Edinburgh, Edinburgh, UK
e-mail: jiawei.li@ed.ac.uk

T. Tkocz (✉)
Carnegie Mellon University, Pittsburgh, PA, USA
e-mail: ttkocz@math.cmu.edu

R. Adamczak et al. (eds.), *High Dimensional Probability IX*, Progress in Probability 80, https://doi.org/10.1007/978-3-031-26979-0_5

behaviour of the tail of the sum S is the same as that of one summand carrying the largest weight.

The goal of this short note is to exhibit that the same behaviour holds for sums of two-sided exponentials (Laplace). Our main result reads as follows.

Theorem 1 *Let $X_1, \dots, X_n$ be independent standard two-sided exponential random variables (i.e., with density $\frac{1}{2}e^{-|x|}$, $x \in \mathbb{R}$). Let $S = \sum_{i=1}^n a_i X_i$ with $a_1, \dots, a_n$ be positive. For every $t > 1$,*

$$\frac{1}{57}\frac{1}{\sqrt{\alpha t}}\exp\left(-\alpha t\right) \le \mathbb{P}\left(S > t\sqrt{\mathrm{Var}(S)}\right) \le \exp\left(-\frac{\alpha^2}{2}h\left(\frac{2t}{\alpha}\right)\right), \tag{2}$$

where $\alpha = \frac{\sqrt{\mathrm{Var}(S)}}{\max_{i\le n} a_i} = \frac{\sqrt{2\sum_{i=1}^n a_i^2}}{\max_{i\le n} a_i}$, $h(u) = \sqrt{1+u^2} - 1 - \log\frac{1+\sqrt{1+u^2}}{2}$, $u > 0$.

In (2), as $t \to \infty$, the lower and the upper bounds are of the same order, $e^{-\alpha t + o(t)}$ (plainly, $h(u) = u + o(u)$).

Our proof of Theorem 1 presented in Sect. 2 is based on an observation that two-sided exponentials are Gaussian mixtures, allowing to leverage (1) (this idea has recently found numerous uses in convex geometry, see [4, 5, 15]). In Sect. 3, we provide further generalisations of Janson's inequality (1) to certain nonnegative distributions, which also allows to extend Theorem 1 to a more general framework. We finish in Sect. 4 with several remarks (for instance, we deduce from (2) a formula for moments of S).

2 Proof of Theorem 1

For the upper bound, we begin with a standard Chernoff-type calculation. Denote $\sigma = \sqrt{\mathrm{Var}(S)} = \sqrt{2\sum a_i^2}$. For $\theta > 0$, we have

$$\mathbb{P}\left(S \ge t\sigma\right) \le e^{-\theta t\sigma}\mathbb{E}e^{\theta S}$$

and

$$\mathbb{E}e^{\theta S} = \prod \mathbb{E}e^{\theta a_i X_i} = \prod \frac{1}{1-\theta^2 a_i^2} = \exp\left\{-\sum \log(1-\theta^2 a_i^2)\right\},$$

for $\theta < \frac{1}{a_*}$, $a_* = \max_{i\le n} a_i$. By convexity,

$$-\sum \log(1-\theta^2 a_i^2) \le -\sum \frac{a_i^2}{a_*^2}\log(1-\theta^2 a_*^2),$$

so changing θ to θ/a_*, for every $0 < \theta < 1$, we have

$$\mathbb{P}(S \geq t\sigma) \leq \exp\left\{-\theta t\alpha - \frac{\alpha^2}{2}\log(1-\theta^2)\right\}$$
$$= \exp\left\{-\frac{\alpha^2}{2}\left(\frac{2t}{\alpha}\theta + \log(1-\theta^2)\right)\right\},$$

where $\alpha = \frac{\sigma}{a_*}$. Optimising over θ and using

$$\sup_{\theta\in(0,1)}\left(\theta u + \log(1-\theta^2)\right) = \sqrt{1+u^2} - 1 - \log\frac{1+\sqrt{1+u^2}}{2}, \qquad u > 0$$

gives the upper bound in (2) and thus finishes the argument.

For the lower bound, we shall use that a standard two-sided exponential random variable with density $\frac{1}{2}e^{-|x|}$, $x \in \mathbb{R}$, has the same distribution as $\sqrt{2Y}G$, where Y is an exponential random variable with mean 1 and G is a standard Gaussian random variable independent of Y (this follows, for instance, by checking that the characteristic functions are the same; see also a remark following Lemma 23 in [5]). This and the fact that sums of independent Gaussians are Gaussian justify the following claim, central to our argument.

Proposition 2 *The sum* $S = \sum_{i=1}^n a_i X_i$ *has the same distribution as* $(2\sum_{i=1}^n a_i^2 Y_i)^{1/2} G$ *with* $Y_1, \ldots, Y_n$ *being independent mean* 1 *exponential random variables, independent of the standard Gaussian* G.

Recall $\alpha = \frac{\sigma}{\max a_i}$. Fix $t > 1$. By Proposition 2, for $\theta > 0$, we have

$$\mathbb{P}(S \geq t\sigma) = \mathbb{P}\left(\sqrt{2\sum a_i^2 Y_i}G \geq t\sigma\right) \geq \mathbb{P}\left(\sqrt{2\sum a_i^2 Y_i}\right.$$
$$\left.\geq \sqrt{\theta t\sigma^2},\ G \geq \sqrt{\theta^{-1}t}\right)$$
$$= \mathbb{P}\left(\sum a_i^2 Y_i \geq \frac{1}{2}\theta t\sigma^2\right)\mathbb{P}\left(G \geq \sqrt{\theta^{-1}t}\right).$$

Case 1. $t \geq \alpha$. With hindsight, choose $\theta = \frac{1}{\alpha}$. Applying (1) to the first term yields

$$\mathbb{P}\left(\sum a_i^2 Y_i \geq \frac{1}{2}\theta t\sigma^2\right) = \mathbb{P}\left(\sum a_i^2 Y_i \geq \frac{t}{\alpha}\sum a_i^2\right)$$
$$\geq \frac{1}{e\alpha^2}\exp\left\{-\frac{\alpha^2}{2}\left(\frac{t}{\alpha} - 1\right)\right\}.$$

For the second term we use a standard bound on the Gaussian tail,

$$\begin{aligned}\mathbb{P}(G > u) &\geq \frac{1}{\sqrt{2\pi}} \frac{u}{u^2+1} e^{-u^2/2}, && u > 0,\\ &\geq \frac{1}{2\sqrt{2\pi}} \frac{1}{u} e^{-u^2/2}, && u \geq 1 \end{aligned} \tag{3}$$

and as $\theta^{-1}t = \alpha t \geq \sqrt{2}$, (3) applies in our case. Combining the above estimates gives

$$\mathbb{P}(S \geq t\sigma) \geq \frac{\exp(\alpha^2/2)}{2\sqrt{2\pi}e\alpha^2} \frac{1}{\sqrt{\alpha t}} \exp\big(-\alpha t\big) \geq \frac{1}{4\sqrt{2\pi}} \frac{1}{\sqrt{\alpha t}} \exp\big(-\alpha t\big),$$

where in the last inequality we use that $\inf_{x>1} \frac{1}{x} e^{x/2} = \frac{e}{2}$.

Case 2. $t \leq \alpha$. With hindsight, choose $\theta = \frac{1}{t}$. Then

$$\mathbb{P}\left(\sum a_i^2 Y_i \geq \frac{1}{2}\theta t\sigma^2\right) = \mathbb{P}\left(\sum a_i^2 Y_i \geq \sum a_i^2\right).$$

In order to lower-bound the last expression, we use a standard Paley-Zygmund type inequality (see, e.g. Lemma 3.2 in [17]).

Lemma 3 *Let* $Z_1, \ldots, Z_n$ *be independent mean* 0 *random variables such that* $\mathbb{E}Z_i^4 \leq C(\mathbb{E}Z_i^2)^2$ *for all* $1 \leq i \leq n$ *for some constant* $C \geq 1$. *Then for* $Z = Z_1 + \cdots + Z_n$,

$$\mathbb{P}(Z \geq 0) \geq \frac{1}{16^{1/3}\max\{C, 3\}}.$$

Proof We can assume that $\mathbb{P}(Z = 0) < 1$. Since Z has mean 0,

$$\mathbb{E}|Z| = 2\mathbb{E}Z\mathbf{1}_{Z\geq 0} \leq 2(\mathbb{E}Z^4)^{1/4}\mathbb{P}(Z \geq 0)^{3/4}.$$

Moreover, by Hölder's inequality, $\mathbb{E}|Z| \geq \frac{(\mathbb{E}Z^2)^{3/2}}{(\mathbb{E}Z^4)^{1/2}}$, so

$$\mathbb{P}(Z \geq 0) \geq 16^{-1/3}\frac{(\mathbb{E}Z^2)^2}{\mathbb{E}Z^4}.$$

Using independence, $\mathbb{E}Z_i = 0$ and the assumption $\mathbb{E}Z_i^4 \leq C(\mathbb{E}Z_i^2)^2$, we have

$$\begin{aligned}\mathbb{E}Z^4 = \sum_{i=1}^n \mathbb{E}Z_i^4 + 6\sum_{i<j} \mathbb{E}Z_i^2\mathbb{E}Z_j^2 &\leq \max\{C, 3\}\left(\sum_{i=1}^n (\mathbb{E}Z_i^2)^2 + 2\sum_{i<j}\mathbb{E}Z_i^2\mathbb{E}Z_j^2\right)\\ &= \max\{C, 3\}(\mathbb{E}Z^2)^2.\end{aligned}$$

□

Take $Z_i = a_i(Y_i - 1)$. We have, $\mathbb{E}(Y_i - 1)^2 = 1$, $\mathbb{E}(X_i - \gamma)^4 = 9$. Thus we can apply Lemma 3 with $C = 9$ and obtain

$$\mathbb{P}\left(\sum a_i^2 Y_i \geq \sum a_i^2\right) \geq \frac{1}{9 \cdot 16^{1/3}}. \tag{4}$$

By (3),

$$\mathbb{P}\left(G \geq \sqrt{\theta^{-1}t}\right) = \mathbb{P}(G \geq t) \geq \frac{1}{2\sqrt{2\pi}} \frac{1}{t} e^{-t^2/2} \geq \frac{1}{2\sqrt{2\pi}} \frac{1}{\sqrt{\alpha t}} e^{-\alpha t/2},$$

where in the last inequality we use that in this case $t \leq \sqrt{\alpha t}$. Moreover, since $\alpha t \geq \sqrt{2}$, $e^{-\alpha t/2} \geq e^{1/\sqrt{2}} e^{-\alpha t}$. Thus,

$$\mathbb{P}(S \geq t\sigma) \geq \frac{e^{1/\sqrt{2}}}{18 \cdot 16^{1/3}\sqrt{2\pi}} \frac{1}{\sqrt{\alpha t}} \exp\big(-\alpha t\big) > \frac{1}{57} \frac{1}{\sqrt{\alpha t}} \exp\big(-\alpha t\big).$$

Combining Case 1 and 2 finishes the proof of the lower bound in (2) and thus the proof of Theorem 1 is complete. □

3 Generalisations

In this section, we provide general tail bounds for weighted sums of nonnegative random variables which for certain distributions allow to capture the same behaviour as featured in Janson's inequality (1), viz. asymptotically the sum has the same tail as the summand carrying the largest weight.

Theorem 4 *Let $X_1, \ldots, X_n$ be i.i.d. nonnegative random variables, $\mu = \mathbb{E}X_1$. Let $S = \sum_{i=1}^n a_i X_i$ with $a_1, \ldots, a_n$ positive. For every $t > 1$,*

$$\mathbb{P}(S \geq \mathbb{E}S)\, r((t-1)\alpha\mu) \leq \mathbb{P}(S > t\mathbb{E}S) \leq \exp\{-\alpha I(\mu t)\}, \tag{5}$$

where $\alpha = \frac{\sum_{i=1}^n a_i}{\max_{i \leq n} a_i}$, for $v > 0$,

$$r(v) = \inf_{u>0} \frac{\mathbb{P}(X_1 > u + v)}{\mathbb{P}(X_1 > u)} \tag{6}$$

and for $t > 0$,

$$I(t) = \sup_{\theta>0} \left(t\theta - \log \mathbb{E}e^{\theta X_1}\right). \tag{7}$$

Before presenting the proof, we look at the example of the exponential and gamma distribution.

3.1 Examples

When the X_i are exponential rate 1 random variables, $I(t) = t - 1 - \log t$, $r(v) = e^{-v}$, $\mathbb{P}(S \geq \mathbb{E}S) \geq \frac{1}{9 \cdot 16^{1/3}}$ (see (4)) and we obtain

$$\frac{1}{9 \cdot 16^{1/3}} e^{-\alpha(t-1)} \leq \mathbb{P}(S > t\mathbb{E}S) \leq e^{-\alpha(t-1-\log t)}.$$

Comparing with (1), the extra factor $\frac{1}{t}$ in the upper bound was obtained in [11] through rather delicate computations for the moment generating function specific for the exponential distribution. Since $\alpha \geq 1$, our lower bound up to a universal constant recovers the one from (1) (improves on it as long as $\alpha > 9 \cdot 16^{1/3}/(2e)$ and is worse otherwise). Along the same lines, for the gamma distribution with parameter $\gamma > 0$ (i.e., with density $\Gamma(\gamma)^{-1} x^{\gamma-1} e^{-x}$, $x > 0$), we have $\mu = \gamma$, $I(t\mu) = \gamma(t - 1 - \log t)$ and with some extra work,

$$r_\gamma(v) = \begin{cases} \frac{1}{2\Gamma(\gamma)} \min\{v^{\gamma-1}, 1\} e^{-v}, & 0 < \gamma < 1, \\ e^{-v}, & \gamma \geq 1. \end{cases}$$

Moreover, via Lemma 3, $\mathbb{P}(S \geq \mathbb{E}S) > \frac{1}{3 \cdot 16^{1/3}(1+2\gamma^{-1})}$. Then (5) yields

$$\frac{1}{3 \cdot 16^{1/3}(1 + 2\gamma^{-1})} r_\gamma\big(\alpha\gamma(t-1)\big) \leq \mathbb{P}(S > t\mathbb{E}S) \leq \exp\big(-\alpha\gamma(t - 1 - \log t)\big). \tag{8}$$

In particular, $\mathbb{P}(S > t\mathbb{E}S) = \exp\{-\alpha\gamma t + o(t)\}$ as $t \to \infty$. It would perhaps be interesting to find a larger class of distributions for which the upper and lower bounds from (5) are asymptotically tight. For more precise results involving the variance of S for weighted sums of independent Gamma random variables (not necessarily with the same parameter), we refer to Theorem 2.57 in [1].

3.2 Proof of Theorem 4: The Upper Bound

For the log-moment generating function $\psi : \mathbb{R} \to (-\infty, \infty]$,

$$\psi(u) = \log \mathbb{E}e^{uX_1}, \qquad u \in \mathbb{R},$$

we have $\psi(0) = 0$, ψ is convex (by Hölder's inequality). Thus, by the monotonicity of slopes of convex functions,

$$\mathbb{R} \ni u \mapsto \frac{\psi(u)}{u} \text{ is nondecreasing.} \tag{9}$$

This is what Janson's proof specified to the case of exponentials relies on. We turn to estimating the tails (using of course Chernoff-type bounds). Fix $t > 1$. For $\theta > 0$, we have

$$\mathbb{P}(S \geq t\mathbb{E}S) = \mathbb{P}\left(e^{\theta S} \geq e^{\theta t \mathbb{E}S}\right) \leq e^{-\theta t\mathbb{E}S}\mathbb{E}e^{\theta S} = e^{-\theta t\mathbb{E}S}\prod_{i=1}^{n}\mathbb{E}e^{\theta a_i X_i}$$

$$= \exp\left\{-\theta t\mathbb{E}S + \sum_{i=1}^{n}\psi(\theta a_i)\right\}.$$

Let $a_* = \max_{i \leq n} a_i$. Thanks to (9),

$$\sum_{i=1}^{n}\psi(\theta a_i) = \sum_{i=1}^{n}(\theta a_i)\frac{\psi(\theta a_i)}{\theta a_i} \leq \sum_{i=1}^{n}(\theta a_i)\frac{\psi(\theta a_*)}{\theta a_*}$$

$$= \frac{\sum_{i=1}^{n} a_i}{a_*}\psi(\theta a_*) = \alpha\psi(\theta a_*),$$

where we set $\alpha = \frac{\sum_{i=1}^{n} a_i}{a_*}$. Note $\mathbb{E}S = \mu\sum a_i = \mu\alpha a_*$. We obtain

$$\mathbb{P}(S \geq t\mathbb{E}S) \leq \exp\{-\theta t\mathbb{E}S + \alpha\psi(\theta a_*)\} = \exp\{-\alpha(t\mu\theta a_* - \psi(\theta a_*))\},$$

so optimising over θ gives the upper bound of (5). □

3.3 *Proof of Theorem 4: The Lower Bound*

We follow a general idea from [11]. The whole argument is based on the following simple lemma.

Lemma 5 *Suppose X and Y are independent random variables and Y is such that $\mathbb{P}(Y \geq u + v) \geq r(v)\mathbb{P}(Y \geq u)$ for all $u \in \mathbb{R}$ and $v > 0$, for some function $r(v)$. Then $\mathbb{P}(X + Y \geq u + v) \geq r(v)\mathbb{P}(X + Y \geq u)$ for all $u \in \mathbb{R}$ and $v > 0$.*

Proof By independence, conditioning on X, we get

$$\mathbb{P}(X + Y \geq u + v) = \mathbb{E}_X\mathbb{P}_Y(Y \geq u - X + v) \geq r(v)\mathbb{E}_X\mathbb{P}_Y(Y \geq u - X)$$

$$= r(v)\mathbb{P}(X + Y \geq u).$$

□

Let $S = \sum_{i=1}^{n} a_i X_i$ be the weighted sum of i.i.d. random variables and without loss of generality let us assume $a_1 = \max_{i \leq n} a_i$. Fix $t > 1$. We write $S = S' + a_1 X_1$, with $S' = \sum_{i=2}^{n} a_i X_i$. Note that the definition of function r from (6) remains unchanged if the infimum is taken over all $u \in \mathbb{R}$ (since X_1 is nonnegative). Thus Lemma 5 gives

$$\mathbb{P}(S \geq t\mathbb{E}S) = \mathbb{P}(S \geq \mathbb{E}S + (t-1)\mathbb{E}S) \geq r\left((t-1)\frac{\mathbb{E}S}{a_1}\right)\mathbb{P}(S \geq \mathbb{E}S),$$

as desired. □

4 Further Remarks

4.1 Moments

The upper bound from (2) allows us to recover precise estimates for moments (a special case of Gluskin and Kwapień results from [8]), with a straightforward proof. Here and throughout, $\|a\|_p = (\sum_{i=1}^n |a_i|^p)^{1/p}$ denotes the p-norm of a sequence $a = (a_1, \ldots, a_n)$, $p > 0$, and $\|a\|_\infty = \max_{i \leq n} |a_i|$.

Theorem 6 (Gluskin and Kwapień, [8]) *Under the assumptions of Theorem 1, for every $p \geq 2$,*

$$\frac{\sqrt{2e}}{\sqrt{2e}+1}\left(p\|a\|_\infty + \sqrt{p}\|a\|_2\right) \leq \left(\mathbb{E}|S|^p\right)^{1/p} \leq 4\sqrt{2}\left(p\|a\|_\infty + \sqrt{p}\|a\|_2\right). \quad (10)$$

Proof For the upper bound, letting $\tilde{S} = \frac{S}{\sqrt{\mathrm{Var}(S)}}$ and using (2), we get

$$\mathbb{E}|\tilde{S}|^p = \int_0^\infty pt^{p-1}\mathbb{P}\left(|\tilde{S}| > t\right)\mathrm{d}t \leq \int_0^1 pt^{p-1}\mathrm{d}t + 2\int_1^\infty pt^{p-1}\exp\left(-\frac{\alpha^2}{2}h\left(\frac{2t}{\alpha}\right)\right)\mathrm{d}t.$$

We check that as u increases, $h(u)$ behaves first quadratically, then linearly. More precisely,

$$h(u) \geq \frac{1}{5}u^2, \quad u \in (0, \sqrt{2}), \qquad h(u) \geq \frac{1}{4}u, \quad u \in (\sqrt{2}, \infty). \quad (11)$$

Thus the second integral $\int_1^\infty \ldots \mathrm{d}t$ can be upper bounded by (recall that $\mathrm{Var}(S) = 2\|a\|_2^2$, $\frac{\alpha}{\sqrt{2}} = \frac{\|a\|_2}{\|a\|_\infty} > 1$),

$$\int_1^{\alpha/\sqrt{2}} pt^{p-1}\exp\left(-\frac{\alpha^2}{2}\frac{1}{5}\left(\frac{2t}{\alpha}\right)^2\right)\mathrm{d}t + \int_{\alpha/\sqrt{2}}^\infty pt^{p-1}\exp\left(-\frac{\alpha^2}{2}\frac{1}{4}\frac{2t}{\alpha}\right)\mathrm{d}t$$

$$\leq \int_0^\infty pt^{p-1} \exp\left(-\frac{2}{5}t^2\right) \mathrm{d}t + \int_0^\infty pt^{p-1} \exp\left(-\frac{1}{4}\alpha t\right) \mathrm{d}t$$
$$= \left(\frac{5}{2}\right)^{p/2} \Gamma\left(\frac{p}{2}+1\right) + \left(\frac{4}{\alpha}\right)^p \Gamma(p+1).$$

Using $\Gamma(x+1) \leq x^x$, $x \geq 1$, yields

$$\left(\mathbb{E}|S|^p\right)^{1/p} = \sqrt{2}\|a\|_2 \left(\mathbb{E}|\tilde{S}|^p\right)^{1/p} \leq \sqrt{2}\|a\|_2 \left(1 + 2\left(\frac{5p}{4}\right)^{p/2} + 2\left(\frac{4p}{\alpha}\right)^p\right)^{1/p}$$
$$\leq 4\sqrt{2}(p\|a\|_\infty + \sqrt{p}\|a\|_2).$$

For the lower bound, suppose $a_1 = \|a\|_\infty$. Then, by independence and Jensen's inequality,

$$\mathbb{E}|S|^p \geq \mathbb{E}\big|a_1 X_1 + \mathbb{E}(a_2 X_2 + \cdots + a_n X_n)\big|^p = a_1^p \mathbb{E}|X_1|^p = a_1^p \Gamma(p+1).$$

Using $\Gamma(x+1)^{1/x} \geq x/e$, $x > 0$ (Stirling's formula, [10]), this gives

$$(\mathbb{E}|S|^p)^{1/p} \geq \frac{p}{e}\|a\|_\infty.$$

On the other hand, by Proposition 2, and Jensen's inequality,

$$\mathbb{E}|S|^p = \mathbb{E}\left(2\sum a_i^2 Y_i\right)^{p/2} \mathbb{E}|G|^p \geq \left(2\sum a_i^2\right)^{p/2} \mathbb{E}|G|^p.$$

Using $\mathbb{E}|G|^p \geq (p/e)^{p/2}$, $p \geq 1$ (again, by, e.g., Stirling's approximation), we obtain

$$(\mathbb{E}|S|^p)^{1/p} \geq \sqrt{\frac{2}{e}}\sqrt{p}\|a\|_2.$$

Combining gives

$$(\mathbb{E}|S|^p)^{1/p} \geq \max\left\{\frac{1}{e}p\|a\|_\infty, \sqrt{\frac{2}{e}}\sqrt{p}\|a\|_2\right\} \geq \frac{\sqrt{2e}}{\sqrt{2e}+1}\left(p\|a\|_\infty + \sqrt{p}\|a\|_2\right),$$

which finishes the proof. □

Remark 7 Using Markov and Payley–Zygmund type inequalities, it is possible to recover two-sided tail bounds from moment estimates (like (10)), but incurring loss of (universal) constants in the exponents, as it is done in, e.g., [8], or [9].

4.2 Upper Bounds on Upper Tails from S-Inequalities

Let S be as in (1). The upper bound in (1) for $t = 1$ is trivial, whereas as a result of Lemma 3, viz. (4), we obtain $\mathbb{P}(S \geq \mathbb{E}S) \in (\frac{1}{24}, \frac{23}{24})$, where the upper bound $\frac{23}{24}$ is obtained by applying Lemma 3 to $-Z$. Letting $a > 0$ be such that $\mathbb{P}(S \geq \mathbb{E}S) = \mathbb{P}(X_1 \geq a) = e^{-a}$, by the S-inequality for the two-sided product exponential measure and the set $\{x \in \mathbb{R}^n, \sum a_i |x_i| \leq \mathbb{E}S\}$ (Theorem 2 in [13]), we obtain that for every $t \geq 1$,

$$\mathbb{P}(S \geq t\mathbb{E}S) \leq \mathbb{P}(X_1 \geq ta) = e^{-at} \leq \left(\frac{23}{24}\right)^t. \tag{12}$$

This provides an improvement of (1) for small enough t (of course the point of (1) is that it is optimal for large t). The same can be said about the upper bound in (8) for $\gamma \geq 1$ (in view of (4) and the results from [14] for gamma distributions with parameter $\gamma \geq 1$). Complimentary to such concentration bounds are small ball probability estimates and anti-concentration phenomena, typically treating, however, the regime of $t = O(1/\mathbb{E}S)$ (under our normalisation). We refer, for instance, to the comprehensive survey [16] of Nguyen and Vu, as well as the recent work [12] of Li and Madiman for further results and references. Specific reversals of (12) concerning the exponential measure can be found, e.g., in [5] (Corollary 15), [18] (Proposition 3.4), [19] ((5.5) and Theorem 5.7).

4.3 Heavy-Tailed Distributions

Janson's as well as this paper's techniques strongly rely on Chernoff-type bounds involving exponential moments to establish the *largest-weight* summand tail asymptotics from (1) or (2). Interestingly, when the exponential moments do *not* exist, i.e., for heavy-tailed distributions, under some natural additional assumptions (subexponential distributions), a different phenomenon occurs: in the simplest case of i.i.d. summands, we have

$$\mathbb{P}(X_1 + \cdots + X_n > t) = (1 + o(1))\mathbb{P}\left(\max_{i \leq n} X_i > t\right) \qquad \text{as } t \to \infty,$$

often called the single big jump or catastrophe principle. We refer to the monograph [7] (Chapters 3.1 and 5.1), as well as the papers [3] and [6] for extensions including weighted sums and continuous time, respectively.

4.4 Theorem 1 in a More General Framework

A careful inspection of the proof of Theorem 1 shows that thanks to Theorem 2.57 from [1] (or the simpler but weaker bound (8)), the former can be extended to the case where the X_i have the same distribution as $\sqrt{Y_i}G_i$ with the Y_i being i.i.d. gamma random variables and the G_i independent standard Gaussian. For simplicity, we have decided to present it for the symmetric exponentials.

Acknowledgments We are indebted to an anonymous referee for many valuable comments, leading in particular to the remarks in Sects. 4.2 and 4.3.

References

1. B. Bercu, B. Delyon, E. Rio, *Concentration Inequalities for Sums and Martingales*. SpringerBriefs in Mathematics (Springer, Cham, 2015)
2. S. Boucheron, G. Lugosi, P. Massart, *Concentration Inequalities. A Nonasymptotic Theory of Independence. With a Foreword by Michel Ledoux* (Oxford University Press, Oxford, 2013)
3. Y. Chen, K.W. Ng, Q. Tang, Weighted sums of subexponential random variables and their maxima. Adv. Appl. Probab. **37**(2), 510–522 (2005)
4. A. Eskenazis, On extremal sections of subspaces of L_p. Discret. Comput. Geom. **65**(2), 489–509 (2021)
5. A. Eskenazis, P. Nayar, T. Tkocz, Gaussian mixtures: entropy and geometric inequalities. Ann. Probab. **46**(5), 2908–2945 (2018)
6. S. Foss, T. Konstantopoulos, S. Zachary, Discrete and continuous time modulated random walks with heavy-tailed increments. J. Theor. Probab. **20**(3), 581–612 (2007)
7. S. Foss, D. Korshunov, S. Zachary, *An Introduction to Heavy-Tailed and Subexponential Distributions*, 2nd edn. Springer Series in Operations Research and Financial Engineering (Springer, New York, 2013)
8. E.D. Gluskin, S. Kwapień, Tail and moment estimates for sums of independent random variables with logarithmically concave tails. Studia Math. **114**(3), 303–309 (1995)
9. P. Hitczenko, S. Montgomery-Smith, A note on sums of independent random variables, in *Advances in Stochastic Inequalities (Atlanta, GA, 1997)*. Contemp. Math., vol. 234, Amer. Math. Soc., Providence, RI (1999), pp.69–73
10. G. Jameson, A simple proof of Stirling's formula for the gamma function. Math. Gaz. **99**(544), 68–74 (2015)
11. S. Janson, Tail bounds for sums of geometric and exponential variables. Stat. Probab. Lett. **135**, 1–6 (2018)
12. J. Li, M. Madiman, A combinatorial approach to small ball inequalities for sums and differences. Comb. Probab. Comput. **28**(1), 100–129 (2019)
13. P. Nayar, T. Tkocz, The unconditional case of the complex S-inequality. Israel J. Math. **197**(1), 99–106 (2013)
14. P. Nayar, T. Tkocz, S-inequality for certain product measures. Math. Nachr. **287**(4), 398–404 (2014)
15. P. Nayar, T. Tkocz, On a convexity property of sections of the cross-polytope. Proc. Am. Math. Soc. **148**(3), 1271–1278 (2020)
16. H.H. Nguyen, V.H. Vu, Small ball probability, inverse theorems, and applications, in *Erdös Centennial*. Bolyai Society Mathematical Studies, vol. 25 (János Bolyai Mathematical Society, Budapest, 2013), pp. 409–463

17. K. Oleszkiewicz, Precise moment and tail bounds for Rademacher sums in terms of weak parameters. Israel J. Math. **203**(1), 429–443 (2014)
18. G. Paouris, P. Valettas, A Gaussian small deviation inequality for convex functions. Ann. Probab. **46**(3), 1441–1454 (2018)
19. G. Paouris, P. Valettas, Variance estimates and almost Euclidean structure. Adv. Geom. **19**(2), 165–189 (2019)

Boolean Functions with Small Second-Order Influences on the Discrete Cube

Krzysztof Oleszkiewicz

2020 Mathematics Subject Classification: 60E15; 42C10

1 Introduction

Throughout the chapter, n stands for an integer greater than 1, and we use the standard notation $[n] := \{1, 2, \ldots, n\}$. We equip the discrete cube $\{-1, 1\}^n = \{-1, 1\}^{[n]}$ with the normalized counting (uniform probability) measure $\mu_n = (\frac{1}{2}\delta_{-1} + \frac{1}{2}\delta_1)^{\otimes n}$. Let $\mathbb{E}$ denote the expectation with respect to this measure, and let $r_1, r_2, \ldots, r_n$ be the standard Rademacher functions on the discrete cube, i.e., the coordinate projections $r_i(x) = x_i$ for $x \in \{-1, 1\}^n$ and $i \in [n]$. Furthermore, for $A \subseteq [n]$, we define the Walsh functions by $w_A = \prod_{i \in A} r_i$, with $w_\emptyset \equiv 1$.

The Walsh functions $(w_A)_{A \subseteq [n]}$ form a complete orthonormal system in $L^2(\{-1, 1\}^n, \mu_n)$. Thus, every $f : \{-1, 1\}^n \to \mathbb{R}$ admits a unique Walsh–Fourier expansion $f = \sum_{A \subseteq [n]} \hat{f}(A) w_A$, whose coefficients are given by

$$\hat{f}(A) = \langle f, w_A \rangle = \mathbb{E}[f \cdot w_A] = \frac{1}{2^n} \sum_{x \in \{-1,1\}^n} f(x) w_A(x).$$

For $p \geq 1$, we abbreviate $\|f\|_{L^p(\{-1,1\}^n, \mu_n)}$ to $\|f\|_p$.

In a standard way, for $i \in [n]$, we define the influence of the i-th coordinate on the function f by

$$I_i = I_i(f) := \sum_{A \subseteq [n]:\, i \in A} \left(\hat{f}(A)\right)^2 = \sum_{A \subseteq [n]:\, i \in A} \hat{f}^2(A),$$

K. Oleszkiewicz (✉)
Institute of Mathematics, University of Warsaw, Warsaw, Poland
e-mail: koles@mimuw.edu.pl

R. Adamczak et al. (eds.), *High Dimensional Probability IX*, Progress in Probability 80, https://doi.org/10.1007/978-3-031-26979-0_6

and given distinct integers $i, j \in [n]$, we define (following, up to a minor modification, the notation of Kevin Tanguy's paper [6]) the influence of the couple (i, j) on the function f by

$$I_{i,j} = I_{i,j}(f) := \sum_{A\subseteq[n]:\, i,j\in A} \hat{f}^2(A).$$

Note that if f is Boolean, i.e., $\{-1, 1\}$-valued, then $I_i, I_{i,j} \le 1$, since, by the orthonormality of the Walsh–Fourier system, $\sum_{A\subseteq[n]} \hat{f}^2(A) = \mathbb{E}[f^2] = 1$.

2 Main Results

Theorem 2.1 *There exists a universal constant $C > 0$ with the following property. Let $n \ge 2$ be an integer. Assume that $f : \{-1, 1\}^n \to \{-1, 1\}$ satisfies the bound $I_{i,j}(f) \le \alpha n^{-2} \ln^2 n$ for all $1 \le i < j \le n$ for some $\alpha > 0$. Then $\hat{f}^2(\emptyset) \ge 1 - C\alpha$, or there exists exactly one $i \in [n]$ such that $\hat{f}^2(\{i\}) \ge 1 - C\alpha n^{-1} \ln n$.*

We postpone the proof of Theorem 2.1 till Sect. 4. This theorem says that if a Boolean function f on the discrete cube has uniformly small second-order influences, then it has to be close to one of the functions $1, -1$, or very close to one of the dictatorship/antidictatorship functions $r_1, -r_1, \dots, r_n, -r_n$. Thus, it may be viewed as a modified version of two classical theorems: the KKL theorem of Kahn, Kalai, and Linial, [4], which says that if a Boolean function f on the discrete cube has influences I_i uniformly bounded from above by $\alpha n^{-1} \ln n$, then $\hat{f}^2(\emptyset) \ge 1 - C\alpha$, where $C > 0$ is some universal constant, and the FKN theorem of Friedgut, Kalai, and Naor, [2], which says that if a Boolean function on the discrete cube is close—in a certain sense, different from the assumptions of Theorem 2.1—to an affine function, then it must be close to one of the constant functions or to one of the dictatorship/antidictatorship functions.

While the numerical value of the constant C that can be deduced from the proof of Theorem 2.1 is quite large (with some additional effort, many numerical constants in the proofs can be improved, though perhaps at the cost of clarity), it is not difficult to obtain a reasonable estimate in the case of α close to zero.

Theorem 2.2 *There exists a bounded function $C : (0, \infty) \to (0, \infty)$ such that $\lim_{\alpha\to 0^+} C(\alpha) = 4$ and with the following property. Let $n \ge 2$ be an integer. Assume that a function $f : \{-1, 1\}^n \to \{-1, 1\}$ satisfies the bound $\max_{1\le i<j\le n} I_{i,j}(f) \le \alpha n^{-2} \ln^2 n$ for some $\alpha > 0$. Then $\hat{f}^2(\emptyset) \ge 1 - C(\alpha)\alpha$, or there exists exactly one $i \in [n]$ such that $\hat{f}^2(\{i\}) \ge 1 - C(\alpha)\alpha n^{-1} \ln n$.*

3 Auxiliary Notation and Tools

For any $g : \{-1,1\}^n \to \mathbb{R}$ and $t \geq 0$, we define $P_t g : \{-1,1\}^n \to \mathbb{R}$ by

$$P_t g = \sum_{A \subseteq [n]} e^{-|A|t} \hat{g}(A) w_A.$$

$(P_t)_{t\geq 0}$ is called the heat semigroup on the discrete cube. By the classical hypercontractive inequality of Bonami [1], $\|P_t g\|_2 \leq \|g\|_{1+e^{-2t}}$.

Given $f : \{-1,1\}^n \to \mathbb{R}$ and $i \in [n]$, define the discrete partial derivative $D_i f : \{-1,1\}^n \to \mathbb{R}$ of f by $(D_i f)(x) = \big(f(x) - f(x^i)\big)/2$, where

$$x^i = (x_1, \dots, x_{i-1}, -x_i, x_{i+1}, \dots, x_n) \text{ for } x \in \{-1,1\}^n.$$

Given distinct integers $i, j \in [n]$, let $D_{i,j} = D_i \circ D_j$, so that $D_{i,j} f = D_i(D_j f)$. Note that

$$D_i f = \sum_{A \subseteq [n]:\, i \in A} \hat{f}(A) w_A \text{ and } D_{i,j} f = \sum_{A \subseteq [n]:\, i,j \in A} \hat{f}(A) w_A,$$

so that $I_i(f) = \|D_i f\|_2^2$ and $I_{i,j}(f) = \|D_{i,j} f\|_2^2$.

A slightly different partial derivative operator ∂_i is defined by the formula $(\partial_i f)(x) = \big(f(x^{i\to 1}) - f(x^{i\to -1})\big)/2$, where

$$x^{i\to\varepsilon} = (x_1, \dots, x_{i-1}, \varepsilon, x_{i+1}, \dots, x_n) \text{ for } x \in \{-1,1\}^n,\ \varepsilon \in \{-1,1\}.$$

Again, given distinct integers $i, j \in [n]$, let $\partial_{i,j} = \partial_i \circ \partial_j$, i.e., $\partial_{i,j} f = \partial_i(\partial_j f)$. One easily checks that $D_i f = r_i \cdot \partial_i f$, and thus, $D_{i,j} f = r_i r_j \cdot \partial_{i,j} f$ for all functions f. The Rademacher functions r_i and r_j are $\{-1,1\}$-valued, so $\|D_i f\|_p = \|\partial_i f\|_p$ and $\|D_{i,j} f\|_p = \|\partial_{i,j} f\|_p$ for every $p \geq 1$. Furthermore, $P_t D_{i,j} f = e^{-2t} r_i r_j \cdot P_t \partial_{i,j} f$ (this identity is an easy consequence of the same equality for Walsh functions that is easy to verify), and thus, $e^{2t} \|P_t D_{i,j} f\|_p = \|P_t \partial_{i,j} f\|_p$ for all $p \geq 1$, $t \geq 0$, and $f : \{-1,1\}^n \to \mathbb{R}$.

Lemma 3.1 *For every $f : \{-1,1\}^n \to \mathbb{R}$, there is*

$$\sum_{A \subseteq [n]:\, |A| \geq 2} \hat{f}^2(A) = 4 \sum_{i,j:\, 1 \leq i < j \leq n} \int_0^\infty (e^{2t} - 1) \|P_t D_{i,j} f\|_2^2 \, dt.$$

Proof Since $P_t D_{i,j} f = \sum_{A \subseteq [n]:\, i,j \in A} e^{-|A|t} \hat{f}(A) w_A$, we have

$$\sum_{i,j:\, i<j} \|P_t D_{i,j} f\|_2^2 = \sum_{i,j:\, i<j} \sum_{A \subseteq [n]:\, i,j \in A} e^{-2|A|t} \hat{f}^2(A)$$

$$= \sum_{A \subseteq [n]:\, |A| \geq 2} \sum_{i,j \in A:\, i<j} e^{-2|A|t} \hat{f}^2(A) = \sum_{A \subseteq [n]:\, |A| \geq 2} \binom{|A|}{2} e^{-2|A|t} \hat{f}^2(A),$$

so that

$$\sum_{i,j:\, 1 \leq i < j \leq n} \int_0^\infty (e^{2t} - 1) \| P_t D_{i,j} f \|_2^2 \, dt$$

$$= \sum_{A \subseteq [n]:\, |A| \geq 2} \frac{|A|(|A|-1)}{2} \hat{f}^2(A) \cdot \int_0^\infty (e^{2t} - 1) e^{-2|A|t} \, dt.$$

It remains to note that, for $k > 1$,

$$\int_0^\infty (e^{2t} - 1) e^{-2kt} \, dt = \frac{1}{2(k-1)} - \frac{1}{2k} = \frac{1}{2k(k-1)}.$$

□

Lemma 3.2 *For every $f : \{-1, 1\}^n \to \{-1, 1\}$ and integers $1 \leq i < j \leq n$, there is*

$$\int_0^\infty (e^{2t} - 1) \| P_t D_{i,j} f \|_2^2 \, dt \leq \frac{2 I_{i,j}}{\ln^2(2/I_{i,j})},$$

where $I_{i,j}$ denotes the influence of the pair (i, j) on the function f.

Proof Bonami's hypercontractive bound applied to $g = \partial_{i,j} f$ yields

$$e^{4t} \| P_t D_{i,j} f \|_2^2 = \| P_t \partial_{i,j} f \|_2^2 \leq \| \partial_{i,j} f \|_{1+e^{-2t}}^2 = \left(\mathbb{E}\left[|\partial_{i,j} f|^{1+e^{-2t}} \right] \right)^{\frac{2}{1+e^{-2t}}}.$$

Note that $\partial_{i,j} f$ is $\{-1, -1/2, 0, 1/2, 1\}$-valued because f is Boolean. Since $|w|^{1+e^{-2t}} \leq 2^{1-e^{-2t}} \cdot w^2$ for every $w \in \{-1, -1/2, 0, 1/2, 1\}$, we also have

$$\mathbb{E}\left[|\partial_{i,j} f|^{1+e^{-2t}} \right] \leq 2^{1-e^{-2t}} \cdot \mathbb{E}\left[(\partial_{i,j} f)^2 \right] = 2^{1-e^{-2t}} \cdot \| \partial_{i,j} f \|_2^2 = 2^{1-e^{-2t}} \cdot I_{i,j},$$

therefore,

$$\| P_t D_{i,j} f \|_2^2 \leq e^{-4t} \cdot 4^{\frac{1-e^{-2t}}{1+e^{-2t}}} \cdot I_{i,j}^{\frac{2}{1+e^{-2t}}} = \frac{(1-u)^2}{(1+u)^2} \cdot 4^u \cdot I_{i,j}^{1+u},$$

where we use the change of variables $u = \frac{1-e^{-2t}}{1+e^{-2t}} \in [0, 1]$, i.e., $t = \frac{\ln(1+u) - \ln(1-u)}{2}$, and thus $\frac{dt}{du} = \frac{1}{2}\left(\frac{1}{1+u} + \frac{1}{1-u} \right) = \frac{1}{(1-u)(1+u)}$. Hence,

$$\int_0^\infty (e^{2t}-1)\|P_t D_{i,j} f\|_2^2\,dt \le \int_0^1 \frac{2u}{1-u}\cdot\frac{(1-u)^2}{(1+u)^2}\cdot 4^u \cdot I_{i,j}^{1+u}\cdot\frac{du}{(1-u)(1+u)}$$

$$= 2\int_0^1 \frac{4^u\,u}{(1+u)^3} I_{i,j}^{1+u}\,du \le 2I_{i,j}\int_0^\infty u\left(\frac{I_{i,j}}{2}\right)^u du = \frac{2I_{i,j}}{\ln^2(2/I_{i,j})}.$$

We have used the fact that, by the convexity of the exponential function, $2^u \le 1+u$ for $u \in [0, 1]$. □

Lemma 3.3 *For every* $f : \{-1,1\}^n \to \mathbb{R}$ *and* $i \in [n]$*, there is*

$$I_i(f) - \hat{f}^2(\{i\}) = \sum_{A\subseteq[n]:\,|A|\ge 2,\, i\in A} \hat{f}^2(A) = 2\sum_{j\in[n]\setminus\{i\}} \int_0^\infty e^{2t}\,\|P_t D_{i,j} f\|_2^2\,dt.$$

Proof Since $P_t D_{i,j} f = \sum_{A\subseteq[n]:\,i,j\in A} e^{-|A|t}\hat{f}(A)w_A$, we have

$$\sum_{j\in[n]\setminus\{i\}} \|P_t D_{i,j} f\|_2^2 = \sum_{j\in[n]\setminus\{i\}}\ \sum_{A\subseteq[n]:\,i,j\in A} e^{-2|A|t}\,\hat{f}^2(A)$$

$$= \sum_{A\subseteq[n]:\,|A|\ge 2,\, i\in A}\ \sum_{j\in A\setminus\{i\}} e^{-2|A|t}\,\hat{f}^2(A) = \sum_{A\subseteq[n]:\,|A|\ge 2,\, i\in A} (|A|-1)\,e^{-2|A|t}\,\hat{f}^2(A),$$

so that

$$\sum_{j\in[n]\setminus\{i\}} \int_0^\infty e^{2t}\,\|P_t D_{i,j} f\|_2^2\,dt = \sum_{A\subseteq[n]:\,|A|\ge 2,\, i\in A} (|A|-1)\,\hat{f}^2(A)\int_0^\infty e^{2t}\,e^{-2|A|t}\,dt$$

$$= \sum_{A\subseteq[n]:\,|A|\ge 2,\, i\in A} (|A|-1)\,\hat{f}^2(A)\,\frac{1}{2|A|-2} = \frac{1}{2}\sum_{A\subseteq[n]:\,|A|\ge 2,\, i\in A} \hat{f}^2(A).$$

□

Lemma 3.4 *For every* $f : \{-1,1\}^n \to \{-1,1\}$ *and integers* $1 \le i < j \le n$*, there is*

$$\int_0^\infty e^{2t}\,\|P_t D_{i,j} f\|_2^2\,dt \le \frac{I_{i,j}}{\ln(1/I_{i,j})},$$

where $I_{i,j}$ *denotes the influence of the pair* (i, j) *on the function* f.

Proof In the same way as in the proof of Lemma 3.2, we obtain the bound

$$\|P_t D_{i,j} f\|_2^2 \le e^{-4t}\cdot 4^{\frac{1-e^{-2t}}{1+e^{-2t}}}\cdot I_{i,j}^{\frac{2}{1+e^{-2t}}} = \frac{(1-u)^2}{(1+u)^2}\cdot 4^u\cdot I_{i,j}^{1+u},$$

where again $u = \frac{1-e^{-2t}}{1+e^{-2t}} \in [0,1]$, $t = \frac{\ln(1+u)-\ln(1-u)}{2}$, and $\frac{dt}{du} = \frac{1}{(1-u)(1+u)}$. Hence,

$$\int_0^\infty e^{2t}\,\|P_t D_{i,j} f\|_2^2\,dt \le \int_0^1 \frac{1+u}{1-u}\cdot\frac{(1-u)^2}{(1+u)^2}\cdot 4^u\cdot I_{i,j}^{1+u}\cdot\frac{du}{(1-u)(1+u)}$$

$$= \int_0^1 \frac{4^u}{(1+u)^2}\,I_{i,j}^{1+u}\,du \le \int_0^\infty I_{i,j}^{1+u}\,du = \frac{I_{i,j}}{\ln(1/I_{i,j})}.$$

We have once more used the fact that $2^u \le 1+u$ for $u \in [0,1]$. □

Lemma 3.5 *Let $z \in [0,1/4)$ and $0 \le x \le y \le x^2 + z$. Then either $y \le 2z$ or $x \ge 1-2z$.*

Proof Note that $\left(\frac{1}{4}-z\right)^{1/2} \ge \frac{1}{2} - 2z$ for $z \in [0,1/4]$, so it suffices to prove that $y \le x_1$ or $x \ge x_2$, where

$$x_1 = \frac{1}{2} - \left(\frac{1}{4}-z\right)^{1/2} \quad\text{and}\quad x_2 = \frac{1}{2} + \left(\frac{1}{4}-z\right)^{1/2}.$$

This is easy: since $x \le x^2 + z$, we have $x \in (-\infty, x_1] \cup [x_2,\infty)$. Thus, if $x < x_2$, then $x \in [0, x_1]$, which in turn implies that

$$y \le x^2 + z \le x_1^2 + z = x_1.$$

□

We will also make use of the following slightly more precise observation.

Remark 3.6 Let $z \in [0,1/4)$ and $0 \le x \le y \le x^2 + z$. Then either $y \le z + 4z^2$ or $x \ge 1 - z - 4z^2$. To prove this, it suffices to replace in the proof of Lemma 3.5 the $\left(\frac{1}{4}-z\right)^{1/2} \ge \frac{1}{2} - 2z$ bound by a stronger one, $\left(\frac{1}{4}-z\right)^{1/2} \ge \frac{1}{2} - z - 4z^2$, which is also satisfied for all $z \in [0,1/4]$.

Lemma 3.7 *For every $f : \{-1,1\}^n \to \{-1,1\}$ and $i \in [n]$, the influence I_i of the i-th coordinate on the function f satisfies the inequality*

$$|\hat f(\{i\})| \le I_i = |\hat f(\{i\})|^2 + \sum_{A\subseteq[n]:\,|A|\ge 2,\, i\in A} \hat f^2(A).$$

Proof Using the triangle inequality and the fact that $\partial_i f$ is $\{-1,0,1\}$-valued, so that $|\partial_i f| \equiv (\partial_i f)^2$, we arrive at

$$|\hat f(\{i\})| = |\mathbb{E}[\partial_i f]| \le \mathbb{E}[|\partial_i f|] = \mathbb{E}\big[(\partial_i f)^2\big] = I_i.$$

□

4 Proof of the Main Results

Proof of Theorem 2.1 Let us define a positive constant κ by

$$\kappa = \min\left(\inf_{n\geq 2} \frac{n}{16\ln n}, \inf_{n\geq 2} \frac{n}{\ln^2 n}\right),$$

and let $C = \max\left(\kappa^{-1}, 20\right)$. For $\alpha \geq \kappa$, the assertion of the theorem holds true in a trivial way; therefore, we may and will assume that $0 < \alpha < \kappa$.

Since $\alpha < n/\ln^2 n$ and $I_{i,j} \leq \alpha n^{-2} \ln^2 n$, we have

$$\ln(2/I_{i,j}) \geq \ln(1/I_{i,j}) > \ln n,$$

and thus also

$$\frac{I_{i,j}}{\ln^2(2/I_{i,j})} \leq \frac{\alpha}{n^2} \quad \text{and} \quad \frac{I_{i,j}}{\ln(1/I_{i,j})} \leq \frac{\alpha \ln n}{n^2}$$

for all $1 \leq i < j \leq n$, so that

$$\sum_{i,j:\, 1\leq i<j\leq n} \frac{I_{i,j}}{\ln^2(2/I_{i,j})} \leq \binom{n}{2} \cdot \frac{\alpha}{n^2} \leq \frac{\alpha}{2}$$

and, for every $i \in [n]$,

$$\sum_{j\in[n]\setminus\{i\}} \frac{I_{i,j}}{\ln(1/I_{i,j})} \leq (n-1) \cdot \frac{\alpha \ln n}{n^2} \leq \frac{\alpha \ln n}{n}.$$

Using Lemma 3.1 and Lemma 3.2, we arrive at

$$\sum_{A\subseteq[n]:\, |A|\geq 2} \hat{f}^2(A) \leq 4\alpha,$$

and from Lemma 3.3 and Lemma 3.4, we obtain, for every $i \in [n]$,

$$\sum_{A\subseteq[n]:\, |A|\geq 2,\, i\in A} \hat{f}^2(A) \leq 2\alpha n^{-1} \ln n.$$

Applying Lemma 3.5 to $x = |\hat{f}(\{i\})|$, $y = I_i(f)$, and $z = 2\alpha n^{-1} \ln n < 1/4$, we deduce from Lemma 3.7 that, for every $i \in [n]$, either $I_i(f) \leq 4\alpha n^{-1} \ln n$ or $|\hat{f}(\{i\})| \geq 1 - 4\alpha n^{-1} \ln n$.

Case 1: $I_i(f) \leq 4\alpha n^{-1} \ln n$ **for all** $i \in [n]$. Then, by Lemma 3.7, for all $i \in [n]$, we have $|\hat{f}(\{i\})| \leq 4\alpha n^{-1} \ln n$, so that

$$\sum_{i\in[n]} \hat{f}^2(\{i\}) \leq n \cdot 16\alpha^2 n^{-2} \ln^2 n < 16\alpha,$$

and therefore,

$$\hat{f}^2(\emptyset) = 1 - \sum_{A\subseteq[n]:\,|A|=1} \hat{f}^2(A) - \sum_{A\subseteq[n]:\,|A|\geq 2} \hat{f}^2(A) \geq 1 - 16\alpha - 4\alpha \geq 1 - C\alpha.$$

Case 2: $|\hat{f}(\{i\})| \geq 1 - 4\alpha n^{-1} \ln n$ **for some** $i \in [n]$. Then we obviously have $\hat{f}^2(\{i\}) \geq 1 - 8\alpha n^{-1} \ln n > 1/2$, which implies that there is exactly one such $i \in [n]$ (recall that $\sum_{A\subseteq[n]} \hat{f}^2(A)=1$). It remains to note that $C > 8$. □

Proof of Theorem 2.2 It suffices to repeat the proof of Theorem 2.1, using Remark 3.6 instead of Lemma 3.5 when α is close to zero. □

Remark 4.1 Throughout the chapter, we have restricted our interest to the uniform estimate assumption. However, it is easy to see that the assumption is used in the proof only via the

$$\alpha \geq \max\left(\sum_{i,j:\,1\leq i<j\leq n} \frac{2I_{i,j}}{\ln^2(2/I_{i,j})},\ \frac{n}{\ln n} \cdot \max_{i\in[n]} \sum_{j\in[n]\setminus\{i\}} \frac{I_{i,j}}{\ln(1/I_{i,j})} \right)$$

condition, allowing for significant extensions. In particular, if a function $f : \{-1, 1\}^n \to \{-1, 1\}$ satisfies

$$\max_{1\leq i<j\leq n} I_{i,j}(f) \leq n^{-\gamma} \text{ and } \max_{i\in[n]} \sum_{j\in[n]\setminus\{i\}} I_{i,j}(f) \leq \beta \frac{\ln^2 n}{n}$$

for some constants $\beta > 0$ and $\gamma \in (0, 1]$, then the proof works for $\alpha = \beta\gamma^{-2}$, so that $\hat{f}^2(\emptyset) \geq 1 - C\beta\gamma^{-2}$, or there exists a unique $i \in [n]$ such that $\hat{f}^2(\{i\}) \geq 1 - C\beta\gamma^{-2}n^{-1} \ln n$, where $C > 0$ is the universal constant from Theorem 2.1.

5 Alternative Proof

Here we present another proof of Theorem 2.1, based on a quite different approach. We will prove the following more general statement.

Theorem 5.1 *Let* $n \geq 2$. *For a function* $f : \{-1, 1\}^n \to \{-1, 1\}$, *let*

$$\theta = \max_{i\in[n]} \sum_{j\in[n]\setminus\{i\}} I_{i,j}(f).$$

Assume $\theta \le 1/25$. Then either there exists exactly one $i \in [n]$ such that

$$\left|\hat{f}(\{i\})\right| \ge 1 - \frac{8\theta}{\ln(1/\theta)}$$

or

$$\max_{k\in[n]} I_k(f) \le \frac{4\theta}{\ln(1/\theta)}.$$

For a function $\varphi : \{-1,1\}^n \to \mathbb{R}$, let us denote by $I(\varphi)$ its total influence,

$$I(\varphi) = \sum_{i\in[n]} I_i(\varphi).$$

Since $P_0\varphi \equiv \varphi$, Bonami's bound $\forall_{t\ge 0}\ \|P_t\varphi\|_2 \le \|\varphi\|_{1+\mathrm{e}^{-2t}}$ yields

$$\frac{\mathrm{d}}{\mathrm{d}t}\|P_t\varphi\|_2\Big|_{t=0+} \le \frac{\mathrm{d}}{\mathrm{d}t}\|\varphi\|_{1+\mathrm{e}^{-2t}}\Big|_{t=0+},$$

which amounts to the classical logarithmic Sobolev inequality,

$$2\,I(\varphi) \ge \mathrm{Ent}(\varphi^2) = \mathbb{E}\big[\varphi^2 \ln\left(\varphi^2\right)\big] - \mathbb{E}\big[\varphi^2\big]\ln\mathbb{E}\big[\varphi^2\big].$$

If the function φ is $\{0,1\}$-valued, then $\varphi^2 \equiv \varphi$ and $\varphi^2\ln\left(\varphi^2\right) \equiv 0$. In this well-known way, we derive a weak functional version of the edge-isoperimetric inequality:

Lemma 5.2 *For every function $h : \{-1,1\}^n \to \{0,1\}$, there is*

$$I(h) \ge \frac{1}{2}\cdot\mathbb{E}[h]\cdot\ln(1/\mathbb{E}[h]) = \frac{1}{2}\cdot\mathbb{P}(h=1)\cdot\ln\big(1/\mathbb{P}(h=1)\big).$$

Remark 5.3 A slightly stronger version of this inequality is known to hold true, with ln replaced by $\log_2$. It can be deduced from Harper's solution [3] of the edge-isoperimetric problem for the discrete cube (this result has also been proved by Lindsey, Bernstein, and Hart).

Corollary 5.4 *For each $n \ge 2$ and every function $g : \{-1,1\}^n \to \{-1,0,1\}$ satisfying $I(g) \le 1/25$, there exists $\eta \in \{-1,0,1\}$ such that*

$$\mathbb{P}(g \ne \eta) \le 4\cdot I(g)/\ln\big(1/I(g)\big).$$

Proof Since $\mathbb{P}(g=-1) + \mathbb{P}(g=0) + \mathbb{P}(g=1) = 1$, certainly there exists $\eta \in \{-1,0,1\}$ for which $\mathbb{P}(g=\eta) \ge 1/3$, so that $\mathbb{P}(g\ne\eta) \le 2/3$. We will prove that the bound of Corollary 5.4 holds true for this η. Let us define $h : \{-1,1\}^n \to \{0,1\}$ by $h = 1_{g\ne\eta}$. Then, for every $x \in \{-1,1\}^n$ and every $i \in [n]$,

$$|\partial_i h(x)| = \left|h\big(x^{i\to 1}\big) - h\big(x^{i\to -1}\big)\right|/2 = \left|1_{g\neq\eta}\big(x^{i\to 1}\big) - 1_{g\neq\eta}\big(x^{i\to -1}\big)\right|/2$$

$$= \frac{1}{2}\left|1_{\{-1,0,1\}\setminus\{\eta\}}\Big(g\big(x^{i\to 1}\big)\Big) - 1_{\{-1,0,1\}\setminus\{\eta\}}\Big(g\big(x^{i\to -1}\big)\Big)\right|$$

$$\leq \left|g\big(x^{i\to 1}\big) - g\big(x^{i\to -1}\big)\right|/2 = |\partial_i g(x)|,$$

because $1_{\{-1,0,1\}\setminus\{\eta\}} : \{-1, 0, 1\} \to \{0, 1\}$ is 1-Lipschitz. Therefore,

$$I(h) = \sum_{i\in[n]} I_i(h) = \sum_{i\in[n]} \|\partial_i h\|_2^2 \leq \sum_{i\in[n]} \|\partial_i g\|_2^2 = \sum_{i\in[n]} I_i(g) = I(g),$$

so that, by Lemma 5.2,

$$\mathbb{P}(h = 1) \leq 2I(h)/\ln\big(1/\mathbb{P}(h = 1)\big) \leq 2I(g)/\ln\big(1/\mathbb{P}(h = 1)\big)$$

$$= 2I(g)/\ln\big(1/\mathbb{P}(g \neq \eta)\big) \leq 2I(g)/\ln(3/2) \leq 5I(g) \leq \sqrt{I(g)}.$$

Thus, applying Lemma 5.2 again, we get

$$\mathbb{P}(g \neq \eta) = \mathbb{P}(h = 1) \leq \frac{2I(h)}{\ln\big(1/\mathbb{P}(h = 1)\big)} \leq \frac{2I(g)}{\ln\big(1/\sqrt{I(g)}\big)} = 4 \cdot \frac{I(g)}{\ln\big(1/I(g)\big)}.$$

□

Proof of Theorem 5.1 Let $i \in [n]$. Since the function f is $\{-1, 1\}$-valued, its i-th partial derivative $\partial_i f$ is $\{-1, 0, 1\}$-valued. Note that

$$I(\partial_i f) = \sum_{j\in[n]} I_j(\partial_i f) = \sum_{j\in[n]\setminus\{i\}} \|\partial_j(\partial_i f)\|_2^2 = \sum_{j\in[n]\setminus\{i\}} I_{i,j}(f) \leq \theta \leq 1/25.$$

Applying Corollary 5.4 to $g = \partial_i f$ and using the fact that the function $(0, 1) \ni x \mapsto x/\ln(1/x)$ is increasing, we prove the existence of $\eta_i \in \{-1, 0, 1\}$ such that $\mathbb{P}(\partial_i f \neq \eta_i) \leq 4\theta/\ln(1/\theta)$. Obviously, this η_i is unique, because $4\theta/\ln(1/\theta) \leq 0.16/\ln(25) < 1/20$, so that $\mathbb{P}(\partial_i f = \eta_i) > 1/2$.

Observe that if $\partial_i f(x) = 1$ for some $x = (x_1, x_2, \ldots, x_n) \in \{-1, 1\}^n$, then $f\big(x^{i\to 1}\big) = 1$ and $f\big(x^{i\to -1}\big) = -1$, i.e., $f(x) = x_i$. In other words, $\{f \neq r_i\} \subseteq \{\partial_i f \neq 1\}$. In a similar way, we prove $\{f \neq -r_i\} \subseteq \{\partial_i f \neq -1\}$. Thus, if there exists $i \in [n]$ for which $\eta_i \in \{-1, 1\}$, then

$$\mathbb{P}(f \neq \eta_i r_i) \leq 4\theta/\ln(1/\theta),$$

and since

$$\hat{f}(\{i\}) = \mathbb{E}[f \cdot r_i] = \mathbb{P}(f = r_i) - \mathbb{P}(f \neq r_i) = \mathbb{P}(f \neq -r_i) - \mathbb{P}(f \neq r_i)$$

$$= 1 - 2\,\mathbb{P}(f \neq r_i) = 2\,\mathbb{P}(f \neq -r_i) - 1,$$

we arrive at

$$\left|\hat{f}(\{i\})\right| \geq 1 - 8\theta/\ln(1/\theta),$$

as desired. The uniqueness of i follows from $\hat{f}^2(\{i\}) > (9/10)^2 > 1/2$.

It remains to consider the case $\eta_1 = \eta_2 = \ldots = \eta_n = 0$, in which

$$I_i(f) = \mathbb{E}\left[(\partial_i f)^2\right] = \mathbb{E}[1_{\partial_i f \neq 0}] = \mathbb{P}(\partial_i f \neq 0) \leq 4\theta/\ln(1/\theta)$$

for each $i \in [n]$, as desired. Since $\left|\hat{f}(\{i\})\right| \leq \sqrt{I_i(f)}$ and $4\theta/\ln(1/\theta) < 1/20$, the assertion of Theorem 5.1 is an exclusive disjunction, as stated. □

Theorem 2.1 easily follows from Theorem 5.1 by considering $\theta = \alpha n^{-1}\ln^2 n$ (and using the KKL theorem to derive the $\hat{f}^2(\emptyset) \geq 1 - C\alpha$ estimate from the upper bound on $\max_{i\in[n]} I_i(f)$, as explained in Sect. 2). If $\alpha \leq 1/25$, then also $\theta \leq 1/25$, and by taking $C \geq 25$, we make the case $\alpha > 1/25$ trivial.

For the standard tribes function $T : \{-1, 1\}^n \to \{-1, 1\}$, with disjoint tribes of size $\sim \log_2(n/\ln n)$ each, one easily checks that $I_{i,j}(T) \simeq n^{-2}\ln^2 n$ if i and j belong to different tribes, but $I_{i,j}(T) \simeq n^{-1}\ln n$ if i and j belong to the same tribe—Remark (1) on page 703 of [6] is not correct. Thus, T cannot serve as an example showing the essential optimality of Theorem 2.1, though to some extent it does the job for Theorem 5.1, with $\theta \simeq n^{-1}\ln^2 n$ and $I_i(T) \simeq |\hat{T}(\{i\})| \simeq n^{-1}\ln n \simeq \theta/\ln(1/\theta)$ uniformly for all $i \in [n]$.

This is complemented by the example of $V : \{-1, 1\}^n \to \{-1, 1\}$ defined by

$$V(x) = x_1 \cdot \left(1 - 2\prod_{k=2}^{n}\left(\frac{1+x_k}{2}\right)\right).$$

The function V satisfies the assumptions of Theorem 5.1 with $\theta \simeq n \cdot 2^{-n}$, i.e., $\theta/\ln(1/\theta) \simeq 2^{-n}$, whereas $\hat{V}(\{1\}) = 1 - 4 \cdot 2^{-n}$ and $I_1(V) = 1$.

Very recently, Tomasz Przybyłowski [5] established a nice counterpart of Theorem 2.1 for influences of orders higher than 2. Following an advice of Peter Keevash, he also provided an example demonstrating the essential optimality of Theorem 2.1.

Acknowledgments Research supported by the National Science Centre, Poland, project number 2015/18/A/ST1/00553.

References

1. A. Bonami, Étude des coefficients Fourier des fonctions de $L^p(G)$. Ann. Inst. Fourier **20**, 335–402 (1970)
2. E. Friedgut, G. Kalai, A. Naor, Boolean functions whose Fourier transform is concentrated on the first two levels. Adv. Appl. Math. **29**, 427–437 (2002)

3. L. Harper, Optimal assignments of numbers to vertices. J. Soc. Ind. Appl. Math. **12**, 131–135 (1964)
4. J. Kahn, G. Kalai, N. Linial, The influence of variables on Boolean functions, in *Proceedings of 29th Annual Symposium on Foundations of Computer Science*, vol. 62 (Computer Society Press, Washington, 1988), pp. 68–80
5. T. Przybyłowski, Influences of higher orders in the Boolean analysis. Master's thesis, University of Warsaw (2022)
6. K. Tanguy, Talagrand inequality at second order and application to Boolean analysis. J. Theoret. Probab. **33**, 692–714 (2020)

Some Notes on Concentration for α-Subexponential Random Variables

Holger Sambale

2020 Mathematics Subject Classification: 60E15; 60F10; 46E30; 46N30

1 Introduction

The aim of this note is to compile a number of smaller results that extend some classical as well as more recent concentration inequalities for bounded or sub-Gaussian random variables to random variables with heavier (but still exponential type) tails. In detail, we shall consider random variables X that satisfy

$$\mathbb{P}(|X - \mathbb{E}X| \geq t) \leq 2\exp(-t^\alpha / C_{1,\alpha}^\alpha) \tag{1.1}$$

for any $t \geq 0$, some $\alpha \in (0, 2]$, and a suitable constant $C_{1,\alpha} > 0$. Such random variables are sometimes called α-subexponential (for $\alpha = 2$, they are sub-Gaussian) or sub-Weibull(α) (cf. [23, Definition 2.2]).

There are several equivalent reformulations of (1.1), e. g., in terms of L^p norms:

$$\|X\|_{L^p} \leq C_{2,\alpha} p^{1/\alpha} \tag{1.2}$$

for any $p \geq 1$. Another characterization is that these random variables have finite Orlicz norms of order α, i. e.,

$$C_{3,\alpha} := \|X\|_{\Psi_\alpha} := \inf\{t > 0 \colon \mathbb{E}\exp((|X|/t)^\alpha) \leq 2\} < \infty. \tag{1.3}$$

If $\alpha < 1$, $\|\cdot\|_{\Psi_\alpha}$ is actually a quasi-norm; however, many norm-like properties (such as a triangle-type inequality) can nevertheless be recovered up to α-dependent

H. Sambale (✉)
Faculty of Mathematics, Bielefeld University, Bielefeld, Germany
e-mail: hsambale@math.uni-bielefeld.de

R. Adamczak et al. (eds.), *High Dimensional Probability IX*, Progress in Probability 80, https://doi.org/10.1007/978-3-031-26979-0_7

constants (see, e. g., [12, Appendix A]). In fact, $C_{1,\alpha}$, $C_{2,\alpha}$, and $C_{3,\alpha}$ can be chosen such that they only differ by a constant α-dependent factor.

Note that α-subexponential random variables have log-convex (if $\alpha \le 1$) or log-concave (if $\alpha \ge 1$) tails, i. e., $t \mapsto -\log \mathbb{P}(|X| \ge t)$ is convex or concave, respectively. For log-convex or log-concave measures, two-sided L^p norm estimates for polynomial chaos (and as a consequence, concentration bounds) have been established over the last 25 years. In the log-convex case, results of this type have been derived for linear forms in [17] and for forms of any order in [12, 21]. For log-concave measures, starting with linear forms again in [10], important contributions have been made in [3, 24, 25, 27].

In this note, we mainly present four different results for functions of α-subexponential random variables: a Hanson–Wright-type inequality in Sect. 2, a version of the convex concentration inequality in Sect. 3, a uniform Hanson–Wright inequality in Sect. 4, and finally a convex concentration inequality for simple random tensors in Sect. 5. These results are partly based on and generalize recent research, e. g., [20] and [42]. In fact, they partially build upon each other: for instance, in the proofs of Sect. 5, we apply results both from Sects. 2 and 3. A more detailed discussion is provided in each of the sections.

Finally, let us introduce some conventions that we will use in this chapter.

Notations. If $X_1, \ldots, X_n$ is a sequence of random variables, we denote by $X = (X_1, \ldots, X_n)$ the corresponding random vector. Moreover, we shall need the following types of norms throughout the paper:

- The norms $\|x\|_p := (\sum_{i=1}^n |x_i|^p)^{1/p}$ for $x \in \mathbb{R}^n$
- The L^p norms $\|X\|_{L^p} := (\mathbb{E}|X|^p)^{1/p}$ for random variables X (cf. (1.2))
- The Orlicz (quasi-)norms $\|X\|_{\Psi_\alpha}$ as introduced in (1.3)
- The Hilbert–Schmidt and operator norms $\|A\|_{\mathrm{HS}} := (\sum_{i,j} a_{ij}^2)^{1/2}$, $\|A\|_{\mathrm{op}} := \sup\{\|Ax\|_2 \colon \|x\|_2 = 1\}$ for matrices $A = (a_{ij})$

The constants appearing in this chapter (typically denoted C or c) may vary from line to line. Without subscript, they are assumed to be absolute, and if they depend on α (only), we shall write C_α or c_α.

2 A Generalized Hanson–Wright Inequality

Arguably, the most famous concentration result for quadratic form is the Hanson–Wright inequality, which first appeared in [16]. We may state it as follows: assuming $X_1, \ldots, X_n$ are centered, independent random variables satisfying $\|X_i\|_{\Psi_2} \le K$ for any i and $A = (a_{ij})$ is a symmetric matrix, we have for any $t \ge 0$

$$\mathbb{P}\big(|X^T AX - \mathbb{E}X^T AX| \ge t\big) \le 2\exp\Big(-\frac{1}{C}\min\Big(\frac{t^2}{K^4\|A\|_{\mathrm{HS}}^2}, \frac{t}{K^2\|A\|_{\mathrm{op}}}\Big)\Big).$$

For a modern proof, see [33], and for various developments, cf. [2, 4, 18, 43].

In this note, we provide an extension of the Hanson–Wright inequality to random variables with bounded Orlicz norms of any order $\alpha \in (0, 2]$. This complements the results in [12], where the case of $\alpha \in (0, 1]$ was considered, while for $\alpha = 2$, we get back the actual Hanson–Wright inequality.

Theorem 2.1 *For any $\alpha \in (0, 2]$, let $X_1, \ldots, X_n$ be independent, centered random variables such that $\|X_i\|_{\Psi_\alpha} \leq K$ for any i and $A = (a_{ij})$ be a symmetric matrix. Then, for any $t \geq 0$,*

$$\mathbb{P}\big(|X^T AX - \mathbb{E}X^T AX| \geq t\big) \leq 2\exp\Big(-\frac{1}{C_\alpha}\min\Big(\frac{t^2}{K^4\|A\|_{\mathrm{HS}}^2}, \Big(\frac{t}{K^2\|A\|_{\mathrm{op}}}\Big)^{\frac{\alpha}{2}}\Big)\Big).$$

Theorem 2.1 generalizes and implies a number of inequalities for quadratic forms in α-subexponential random variables (in particular for $\alpha = 1$) that are spread throughout the literature. For a detailed discussion, see [12, Remark 1.7]. Note that it is possible to sharpen the tail estimate given by Theorem 2.1, cf., e. g., [12, Corollary 1.4] for $\alpha \in (0, 1]$ or [3, Theorem 3.2] for $\alpha \in [1, 2]$ (in fact, the proof of Theorem 2.1 works by evaluating the family of norms used therein). The main benefit of Theorem 2.1 is that it uses norms that are easily calculable and in many situations already sufficient for applications.

Before we give the proof of Theorem 2.1, let us briefly mention that for the standard Hanson–Wright inequality, a number of selected applications can be found in [33]. Some of them were generalized to α-subexponential random variables with $\alpha \leq 1$ in [12], and it is no problem to extend these proofs to any order $\alpha \in (0, 2]$ using Theorem 2.1. Here, we just focus on a single example that yields a concentration result for the Euclidean norm of a linear transformation of a vector X having independent components with bounded Orlicz norms around the Hilbert–Schmidt norm of the transformation matrix. This is a variant and extension of [12, Proposition 2.1] and will be applied in Sect. 5.

Proposition 2.2 *Let $X_1, \ldots, X_n$ be independent, centered random variables such that $\mathbb{E}X_i^2 = 1$ and $\|X_i\|_{\Psi_\alpha} \leq K$ for some $\alpha \in (0, 2]$ and let $B \neq 0$ be an $m \times n$ matrix. For any $t \geq 0$, we have*

$$\mathbb{P}(|\|BX\|_2 - \|B\|_{\mathrm{HS}}| \geq tK^2\|B\|_{\mathrm{op}}) \leq 2\exp(-t^\alpha/C_\alpha). \tag{2.1}$$

In particular, for any $t \geq 0$, it holds

$$\mathbb{P}(|\|X\|_2 - \sqrt{n}| \geq tK^2) \leq 2\exp(-t^\alpha/C_\alpha). \tag{2.2}$$

For the proofs, let us recall some elementary relations that we will use throughout the paper to adjust the constants in the tail bounds we derive.

Adjusting constants. For any two constants $C_1 > C_2 > 1$, we have for all $r \geq 0$ and $C > 0$

$$C_1 \exp(-r/C) \le C_2 \exp\Big(-\frac{\log(C_2)}{C\log(C_1)} r\Big) \tag{2.3}$$

whenever the left-hand side is smaller or equal to 1 (cf., e. g., [35, Eq. (3.1)]). Moreover, for any $\alpha \in (0, 2)$, any $\gamma > 0$, and all $t \ge 0$, we may always estimate

$$\exp(-(t/C)^2) \le 2\exp(-(t/C')^\alpha), \tag{2.4}$$

using $\exp(-s^2) \le \exp(1 - s^\alpha)$ for any $s > 0$ and (2.3). More precisely, we may choose $C' := C/\log^{1/\alpha}(2)$. Note that strictly speaking, the range of $t/C \le 1$ is not covered by (2.3); however, in this case (in particular, choosing C' as suggested), both sides of (2.4) are at least 1 anyway so that the right-hand side still provides a valid upper bound for any probability.

Let us now turn to the proof of Theorem 2.1. In what follows, we actually show that for any $p \ge 2$,

$$\|X^T AX - \mathbb{E}X^T AX\|_{L^p} \le C_\alpha K^2\big(p^{1/2}\|A\|_{\mathrm{HS}} + p^{2/\alpha}\|A\|_{\mathrm{op}}\big). \tag{2.5}$$

From here, Theorem 2.1 follows by standard means (cf. [34, Proof of Theorem 3.6]). Moreover, we may restrict ourselves to $\alpha \in (1, 2]$, since the case of $\alpha \in (0, 1]$ has been proven in [12].

Proof of Theorem 2.1 First we shall treat the off-diagonal part of the quadratic form. Let $w_i^{(1)}, w_i^{(2)}$ be independent (of each other as well as of the X_i) symmetrized Weibull random variables with scale 1 and shape α, i. e., $w_i^{(j)}$ are symmetric random variables with $\mathbb{P}(|w_i^{(j)}| \ge t) = \exp(-t^\alpha)$. In particular, the $w_i^{(j)}$ have logarithmically concave tails.

Using standard decoupling and symmetrization arguments (cf. [8, Theorem 3.1.1 & Lemma 1.2.6]) as well as [3, Theorem 3.2] in the second inequality, for any $p \ge 2$, it holds

$$\begin{aligned}\|\sum_{i\neq j} a_{ij}X_iX_j\|_{L^p} &\le C_\alpha K^2\|\sum_{i\neq j} a_{ij}w_i^{(1)}w_j^{(2)}\|_{L^p}\\ &\le C_\alpha K^2(\|A\|^{\mathcal{N}}_{\{1,2\},p} + \|A\|^{\mathcal{N}}_{\{\{1\},\{2\}\},p}),\end{aligned} \tag{2.6}$$

where the norms $\|A\|^{\mathcal{N}}_{\mathcal{J},p}$ are defined as in [3]. Instead of repeating the general definitions, we will only focus on the case we need in our situation. Indeed, for the symmetric Weibull distribution with parameter α, we have (again, in the notation of [3]) $N(t) = t^\alpha$, and so for $\alpha \in (1, 2]$, it follows that $\hat{N}(t) = \min(t^2, |t|^\alpha)$. Hence, the norms can be written as follows:

$$\|A\|^{\mathcal{N}}_{\{1,2\},p} = 2\sup\Big\{\sum_{i,j} a_{ij}x_{ij} : \sum_{i=1}^n \min\Big(\sum_j x_{ij}^2, \big(\sum_j x_{ij}^2\big)^{\alpha/2}\Big) \le p\Big\},$$

$$\|A\|^{\mathcal{N}}_{\{\{1\},\{2\}\},p} = \sup\Big\{\sum_{i,j} a_{ij}x_i y_j : \sum_{i=1}^n \min(x_i^2, |x_i|^\alpha) \le p, \sum_{j=1}^n \min(y_j^2, |y_j|^\alpha) \le p\Big\}.$$

Before continuing with the proof, we next introduce a lemma that will help to rewrite the norms in a more tractable form. □

Lemma 2.3 *For any $p \ge 2$, define*

$$I_1(p) := \Big\{x = (x_{ij}) \in \mathbb{R}^{n\times n} : \sum_{i=1}^n \min\Big(\Big(\sum_{j=1}^n x_{ij}^2\Big)^{\alpha/2}, \sum_{j=1}^n x_{ij}^2\Big) \le p\Big\},$$

$$I_2(p) := \Big\{x_{ij} = z_i y_{ij} \in \mathbb{R}^{n\times n} : \sum_{i=1}^n \min(|z_i|^\alpha, z_i^2) \le p, \max_{i=1,\dots,n} \sum_{j=1}^n y_{ij}^2 \le 1\Big\}.$$

Then $I_1(p) = I_2(p)$.

Proof The inclusion $I_1(p) \supseteq I_2(p)$ is an easy calculation, and the inclusion $I_1(p) \subseteq I_2(p)$ follows by defining $z_i = \|(x_{ij})_j\|$ and $y_{ij} = x_{ij}/\|(x_{ij})_j\|$ (or 0, if the norm is zero). □

Proof of Theorem 2.1*, continued* For brevity, for any matrix $A = (a_{ij})$, let us write $\|A\|_m := \max_{i=1,\dots,n}(\sum_{j=1}^n a_{ij}^2)^{1/2}$. Note that clearly, $\|A\|_m \le \|A\|_{\mathrm{op}}$.

Now, fix some vector $z \in \mathbb{R}^n$ such that $\sum_{i=1}^n \min(|z_i|^\alpha, z_i^2) \le p$. The condition also implies

$$p \ge \sum_{i=1}^n |z_i|^\alpha \mathbb{1}_{\{|z_i|>1\}} + \sum_{i=1}^n z_i^2 \mathbb{1}_{\{|z_i|\le 1\}} \ge \max\Big(\sum_{i=1}^n z_i^2 \mathbb{1}_{\{|z_i|\le 1\}}, \sum_{i=1}^n |z_i| \mathbb{1}_{\{|z_i|>1\}}\Big),$$

where in the second step we used $\alpha \in [1,2]$ to estimate $|z_i|^\alpha \mathbb{1}_{\{|z_i|>1\}} \ge |z_i|\mathbb{1}_{\{|z_i|>1\}}$. So, given any z and y satisfying the conditions of $I_2(p)$, we can write

$$\begin{aligned}
\Big|\sum_{i,j} a_{ij} z_i y_{ij}\Big| &\le \sum_{i=1}^n |z_i| \Big(\sum_{j=1}^n a_{ij}^2\Big)^{1/2}\Big(\sum_{j=1}^n y_{ij}^2\Big)^{1/2} \le \sum_{i=1}^n |z_i|\Big(\sum_{j=1}^n a_{ij}^2\Big)^{1/2} \\
&\le \sum_{i=1}^n |z_i| \mathbb{1}_{\{|z_i|\le 1\}} \Big(\sum_{j=1}^n a_{ij}^2\Big)^{1/2} + \sum_{i=1}^n |z_i| \mathbb{1}_{\{|z_i|>1\}} \Big(\sum_{j=1}^n a_{ij}^2\Big)^{1/2} \\
&\le \|A\|_{\mathrm{HS}} \Big(\sum_{i=1}^n z_i^2 \mathbb{1}_{\{|z_i|\le 1\}}\Big)^{1/2} + \|A\|_m \sum_{i=1}^n |z_i| \mathbb{1}_{\{|z_i|>1\}}.
\end{aligned}$$

So, this yields

$$\|A\|^{\mathcal{N}}_{\{1,2\},p} \le 2p^{1/2}\|A\|_{\mathrm{HS}} + 2p\|A\|_m \le 2p^{1/2}\|A\|_{\mathrm{HS}} + 2p\|A\|_{\mathrm{op}}. \tag{2.7}$$

As for $\|A\|^{\mathcal{N}}_{\{\{1\},\{2\}\},p}$, we can use the decomposition $z = z_1 + z_2$, where $(z_1)_i = z_i \mathbb{1}_{\{|z_i|>1\}}$ and $z_2 = z - z_1$, and obtain

$$\begin{aligned}\|A\|^{\mathcal{N}}_{\{\{1\},\{2\}\},p} &\le \sup\Big\{\sum_{ij} a_{ij}(x_1)_i(y_1)_j : \|x_1\|_\alpha \le p^{1/\alpha}, \|y_1\|_\alpha \le p^{1/\alpha}\Big\}\\ &\quad + 2\sup\Big\{\sum_{ij} a_{ij}(x_1)_i(y_2)_j : \|x_1\|_\alpha \le p^{1/\alpha}, \|y_2\|_2 \le p^{1/2}\Big\}\\ &\quad + \sup\Big\{\sum_{ij} a_{ij}(x_2)_i(y_2)_j : \|x_2\|_2 \le p^{1/2}, \|y_2\|_2 \le p^{1/2}\Big\}\\ &= p^{2/\alpha}\sup\{\ldots\} + 2p^{1/\alpha+1/2}\sup\{\ldots\} + p\|A\|_{\mathrm{op}}\end{aligned}$$

(in the braces, the conditions $\|\cdot\|_\beta \le p^{1/\beta}$ have been replaced by $\|\cdot\|_\beta \le 1$). Clearly, since $\|x_1\|_\alpha \le 1$ implies $\|x_1\|_2 \le 1$ (and the same for y_1), all of the norms can be upper bounded by $\|A\|_{\mathrm{op}}$, i. e., we have

$$\|A\|^{\mathcal{N}}_{\{\{1\},\{2\}\},p} \le (p^{2/\alpha} + 2p^{1/\alpha+1/2} + p)\|A\|_{\mathrm{op}} \le 4p^{2/\alpha}\|A\|_{\mathrm{op}}, \tag{2.8}$$

where the last inequality follows from $p \ge 2$ and $1/2 \le 1/\alpha \le 1 \le (\alpha+2)/(2\alpha) \le 2/\alpha$.

Combining the estimates (2.6), (2.7), and (2.8) yields

$$\Big\|\sum_{i,j} a_{ij}X_iX_j\Big\|_{L^p} \le C_\alpha K^2\big(2p^{1/2}\|A\|_{\mathrm{HS}} + 6p^{2/\alpha}\|A\|_{\mathrm{op}}\big).$$

To treat the diagonal terms, we use Corollary 6.1 in [12], as X_i^2 are independent and satisfy $\|X_i^2\|_{\Psi_{\alpha/2}} \le K^2$, so that it yields

$$\mathbb{P}\Big(\Big|\sum_{i=1}^n a_{ii}(X_i^2 - \mathbb{E}X_i^2)\Big| \ge t\Big) \le 2\exp\Big(-\frac{1}{C_\alpha K^2}\min\Big(\frac{t^2}{\sum_{i=1}^n a_{ii}^2}, \Big(\frac{t}{\max_{i=1,\ldots,n}|a_{ii}|}\Big)^{\alpha/2}\Big)\Big).$$

Now it is clear that $\max_{i=1,\ldots,n}|a_{ii}| \le \|A\|_{\mathrm{op}}$ and $\sum_{i=1}^n a_{ii}^2 \le \|A\|_{\mathrm{HS}}^2$. In particular,

$$\Big\|\sum_{i=1}^n a_{ii}(X_i^2 - \mathbb{E}X_i^2)\Big\|_{L^p} \le C_\alpha K^2(p^{1/2}\|A\|_{\mathrm{HS}} + p^{2/\alpha}\|A\|_{\mathrm{op}}).$$

The claim (2.5) now follows from Minkowski's inequality. □

Finally, we prove Proposition 2.2.

Proof of Proposition 2.2 It suffices to prove (2.1) for matrices satisfying $\|B\|_{\mathrm{HS}} = 1$, as otherwise we set $\widetilde{B} = B\|B\|_{\mathrm{HS}}^{-1}$ and use the equality

$$\{|\|BX\|_2 - \|B\|_{\mathrm{HS}}| \geq \|B\|_{\mathrm{op}} t\} = \{|\|\widetilde{B}X\|_2 - 1| \geq \|\widetilde{B}\|_{\mathrm{op}} t\}.$$

Now let us apply Theorem 2.1 to the matrix $A := B^T B$. An easy calculation shows that $\mathrm{trace}(A) = \mathrm{trace}(B^T B) = \|B\|_{\mathrm{HS}}^2 = 1$, so that we have for any $t \geq 0$

$$\begin{aligned}
\mathbb{P}\big(|\|BX\|_2 - 1| \geq t\big) &\leq \mathbb{P}\big(|\|BX\|_2^2 - 1| \geq \max(t, t^2)\big) \\
&\leq 2\exp\Big(- \frac{1}{C_\alpha} \min\Big(\frac{\max(t, t^2)^2}{K^4\|B\|_{\mathrm{op}}^2}, \Big(\frac{\max(t, t^2)}{K^4\|B\|_{\mathrm{op}}^2}\Big)^{\alpha/2}\Big)\Big) \\
&\leq 2\exp\Big(- \frac{1}{C_\alpha} \min\Big(\frac{t^2}{K^4\|B\|_{\mathrm{op}}^2}, \Big(\frac{t^2}{K^4\|B\|_{\mathrm{op}}^2}\Big)^{\alpha/2}\Big)\Big) \\
&\leq 2\exp\Big(- \frac{1}{C_\alpha} \Big(\frac{t}{K^2\|B\|_{\mathrm{op}}}\Big)^{\alpha}\Big).
\end{aligned}$$

Here, the first step follows from $|z - 1| \leq \min(|z^2 - 1|, |z^2 - 1|^{1/2})$ for $z \geq 0$, in the second step, we have used the estimates $\|A\|_{\mathrm{HS}}^2 \leq \|B\|_{\mathrm{op}}^2 \|B\|_{\mathrm{HS}}^2 = \|B\|_{\mathrm{op}}^2$ and $\|A\|_{\mathrm{op}} \leq \|B\|_{\mathrm{op}}^2$, and moreover, the fact that since $\mathbb{E}X_i^2 = 1$, $K \geq C_\alpha > 0$ (cf., e. g., [12, Lemma A.2]), while the last step follows from (2.4) and (2.3). Setting $t = K^2 s\|B\|_{\mathrm{op}}$ for $s \geq 0$ finishes the proof of (2.1). Finally, (2.2) follows by taking $m = n$ and $B = I$. □

3 Convex Concentration for Random Variables with Bounded Orlicz Norms

Assume $X_1, \ldots, X_n$ are independent random variables each taking values in some bounded interval $[a, b]$. Then, by convex concentration as established in [19, 29, 38], for every convex 1-Lipschitz function $f : [a, b]^n \to \mathbb{R}$,

$$\mathbb{P}(|f(X) - \mathbb{E}f(X)| > t) \leq 2\exp\Big(- \frac{t^2}{2(b - a)^2}\Big) \tag{3.1}$$

for any $t \geq 0$ (see, e. g., [36, Corollary 3]).

While convex concentration for bounded random variables is by now standard, there is less literature for unbounded random variables. In [31], a Martingale-type approach is used, leading to a result for functionals with stochastically bounded increments. The special case of suprema of unbounded empirical processes was treated in [1, 28, 40]. Another branch of research, begun in [29] and continued, e. g., in [5, 13–15, 36, 37], is based on functional inequalities (such as Poincaré or log-Sobolev inequalities) restricted to convex functions and weak transport-entropy inequalities. In [20, Lemma 1.8], a generalization of (3.1) for sub-Gaussian random variables ($\alpha = 2$) was proven, which we may extend to any order $\alpha \in (0, 2]$.

Proposition 3.1 *Let $X_1, \ldots, X_n$ be independent random variables, $\alpha \in (0, 2]$ and $f\colon \mathbb{R}^n \to \mathbb{R}$ convex and 1-Lipschitz. Then, for any $t \geq 0$,*

$$\mathbb{P}(|f(X) - \mathbb{E}f(X)| > t) \leq 2\exp\Big(-\frac{t^\alpha}{C_\alpha \|\max_i |X_i|\|_{\Psi_\alpha}^\alpha}\Big).$$

In particular,

$$\|f(X) - \mathbb{E}f(X)\|_{\Psi_\alpha} \leq C_\alpha \|\max_i |X_i|\|_{\Psi_\alpha}. \tag{3.2}$$

Note that the main results of the following two sections can be regarded as applications of Proposition 3.1. If f is separately convex only (i. e., convex is every coordinate with the other coordinates being fixed), it is still possible to prove a corresponding result for the upper tails. Indeed, it is no problem to modify the proof below accordingly, replacing (3.1) by [7, Theorem 6.10]. Moreover, note that $\|\max_i |X_i|\|_{\Psi_\alpha}$ cannot be replaced by $\max_i \||X_i|\|_{\Psi_\alpha}$ (a counterexample for $\alpha = 2$ is provided in [20]). In general, the Orlicz norm of $\max_i |X_i|$ will be of order $(\log n)^{1/\alpha}$ (cf. Lemma 5.6).

Proof of Proposition 3.1 Following the lines of the proof of [20, Lemma 3.5], the key step is a suitable truncation that goes back to [1]. Indeed, write

$$X_i = X_i 1_{\{|X_i| \leq M\}} + X_i 1_{\{|X_i| > M\}} =: Y_i + Z_i \tag{3.3}$$

with $M := 8\mathbb{E}\max_i |X_i|$ (in particular, $M \leq C_\alpha \|\max_i |X_i|\|_{\Psi_\alpha}$, cf. [12, Lemma A.2]), and let $Y = (Y_1, \ldots, Y_n)$, $Z = (Z_1, \ldots, Z_n)$. By the Lipschitz property of f,

$$\begin{aligned} &\mathbb{P}(|f(X) - \mathbb{E}f(X)| > t) \\ &\leq \mathbb{P}(|f(Y) - \mathbb{E}f(Y)| + |f(X) - f(Y)| + |\mathbb{E}f(Y) - \mathbb{E}f(X)| > t) \\ &\leq \mathbb{P}(|f(Y) - \mathbb{E}f(Y)| + \|Z\|_2 + \mathbb{E}\|Z\|_2 > t), \end{aligned} \tag{3.4}$$

and hence, it suffices to bound the terms in the last line.

Applying (3.1) to Y and using (2.4) and (2.3), we obtain

$$\mathbb{P}(|f(Y) - \mathbb{E}f(Y)| > t) \leq 2\exp\Big(-\frac{t^\alpha}{C_\alpha^\alpha \|\max_i |X_i|\|_{\Psi_\alpha}^\alpha}\Big). \tag{3.5}$$

Furthermore, below we will show that

$$\| \|Z\|_2 \|_{\Psi_\alpha} \leq C_\alpha \| \max_i |X_i| \|_{\Psi_\alpha}. \tag{3.6}$$

Hence, for any $t \geq 0$,

$$\mathbb{P}(\|Z\|_2 \geq t) \leq 2 \exp\Big(- \frac{t^\alpha}{C_\alpha^\alpha \| \max_i |X_i| \|_{\Psi_\alpha}^\alpha} \Big), \tag{3.7}$$

and by [12, Lemma A.2],

$$\mathbb{E}\|Z\|_2 \leq C_\alpha \| \max_i |X_i| \|_{\Psi_\alpha}. \tag{3.8}$$

Temporarily writing $K := C_\alpha \| \max_i |X_i| \|_{\Psi_\alpha}$, where C_α is large enough so that (3.5), (3.7), and (3.8) hold, (3.4) and (3.8) yield

$$\mathbb{P}(|f(X) - \mathbb{E}f(X)| > t) \leq \mathbb{P}(|f(Y) - \mathbb{E}f(Y)| + \|Z\|_2 > t - K)$$

if $t \geq K$. Using subadditivity and invoking (3.5) and (3.7), we obtain

$$\mathbb{P}(|f(X) - \mathbb{E}f(X)| > t) \leq 4 \exp\Big(- \frac{(t-K)^\alpha}{(2K)^\alpha} \Big) \leq 4 \exp\Big(- \frac{t^\alpha}{c_\alpha (2K)^\alpha} \Big),$$

where the last step holds for $t \geq K + \delta$ for some $\delta > 0$. This bound extends trivially to any $t \geq 0$ (if necessary, by a suitable change of constants). Finally, the constant in front of the exponential may be adjusted to 2 by (2.3), which finishes the proof.

It remains to show (3.6). To this end, recall the Hoffmann-Jørgensen inequality (cf. [30, Theorem 6.8]) in the following form: if $W_1, \ldots, W_n$ are independent random variables, $S_k := W_1 + \ldots + W_k$, and $t \geq 0$ is such that $\mathbb{P}(\max_k |S_k| > t) \leq 1/8$, then

$$\mathbb{E} \max_k |S_k| \leq 3\mathbb{E} \max_i |W_i| + 8t.$$

In our case, we set $W_i := Z_i^2$, $t = 0$, and note that by Chebyshev's inequality,

$$\mathbb{P}(\max_i Z_i^2 > 0) = \mathbb{P}(\max_i |X_i| > M) \leq \mathbb{E} \max_i |X_i|/M = 1/8,$$

and consequently, recalling that $S_k = Z_1^2 + \ldots + Z_k^2$,

$$\mathbb{P}(\max_k |S_k| > 0) \leq \mathbb{P}(\max_i Z_i^2 > 0) \leq 1/8.$$

Thus, together with [12, Lemma A.2], we obtain

$$\mathbb{E}\|Z\|_2^2 \le 3\mathbb{E}\max_i Z_i^2 \le C_\alpha \|\max_i Z_i^2\|_{\Psi_{\alpha/2}}.$$

Now it is easy to see that $\|\max_i Z_i^2\|_{\Psi_{\alpha/2}} \le \|\max_i |X_i|\|_{\Psi_\alpha}^2$, so that altogether we arrive at

$$\mathbb{E}\|Z\|_2^2 \le C_\alpha \|\max_i |X_i|\|_{\Psi_\alpha}^2. \tag{3.9}$$

Furthermore, by [30, Theorem 6.21], if $W_1, \ldots, W_n$ are independent random variables with zero mean and $\alpha \in (0, 1]$,

$$\|\sum_{i=1}^n W_i\|_{\Psi_\alpha} \le C_\alpha(\|\sum_{i=1}^n W_i\|_{L^1} + \|\max_i |W_i|\|_{\Psi_\alpha}).$$

In our case, we consider $W_i = Z_i^2 - \mathbb{E}Z_i^2$ and $\alpha/2$ (instead of α). Together with the previous arguments (in particular, (3.9)) and [12, Lemma A.3], this yields

$$\begin{aligned}\|\sum_{i=1}^n (Z_i^2 - \mathbb{E}Z_i^2)\|_{\Psi_{\alpha/2}} &\le C_\alpha(\mathbb{E}|\|Z\|_2^2 - \mathbb{E}\|Z\|_2^2| + \|\max_i |Z_i^2 - \mathbb{E}Z_i^2|\|_{\Psi_{\alpha/2}})\\ &\le C_\alpha(\mathbb{E}\|Z\|_2^2 + \|\max_i Z_i^2\|_{\Psi_{\alpha/2}}) \le C_\alpha \|\max_i |X_i|\|_{\Psi_\alpha}^2.\end{aligned}$$

Combining this with [12, Lemma A.3] and (3.9), we arrive at (3.6). □

4 Uniform Tail Bounds for First- and Second-Order Chaos

In this section, we discuss bounds for the tails of the supremum of certain chaos-type classes of functions. Even if we are particularly interested in quadratic forms, i. e., uniform Hanson–Wright inequalities, let us first consider linear forms.

Let $X_1, \ldots, X_n$ be independent random variables, let $\alpha \in (0, 2]$, and let $\{a_{i,t} : i = 1, \ldots, n, t \in \mathcal{T}\}$ be a compact set of real numbers, where $\mathcal{T}$ is some index set. Consider $g(X) := \sup_{t\in\mathcal{T}} \sum_{i=1}^n a_{i,t}X_i$. Clearly, g is convex and has Lipschitz constant $D := \sup_{t\in\mathcal{T}}(\sum_{i=1}^n a_{i,t}^2)^{1/2}$. Therefore, applying Proposition 3.1, we immediately obtain that for any $t \ge 0$,

$$\mathbb{P}(|g(X) - \mathbb{E}g(X)| \ge t) \le 2\exp\Big(-\frac{t^\alpha}{C_\alpha D^\alpha \|\max_i |X_i|\|_{\Psi_\alpha}^\alpha}\Big). \tag{4.1}$$

For bounded random variables, corresponding tail bounds can be found, e. g., in [32, Eq. (14)], and choosing $\alpha = 2$, we get back this result up to constants.

Our main aim is to derive a second-order analogue of (4.1), i. e., a uniform Hanson–Wright inequality. A pioneering result in this direction (for Rademacher variables) can be found in [39]. Later results include [2] (which requires the so-called concentration property), [22], [9], and [11] (certain classes of weakly dependent random variables). In [20], a uniform Hanson–Wright inequality for sub-Gaussian random variables was proven. We may show a similar result for random variables with bounded Orlicz norms of any order $\alpha \in (0, 2]$.

Theorem 4.1 *Let $X_1, \ldots, X_n$ be independent, centered random variables and $K := \|\max_i |X_i|\|_{\Psi_\alpha}$, where $\alpha \in (0, 2]$. Let $\mathcal{A}$ be a compact set of real symmetric $n \times n$ matrices, and let $f(X) := \sup_{A \in \mathcal{A}}(X^T AX - \mathbb{E}X^T AX)$. Then, for any $t \geq 0$,*

$$\mathbb{P}(f(X) - \mathbb{E}f(X) \geq t) \leq 2\exp\Big(- \frac{1}{C_\alpha K^\alpha} \min \Big(\frac{t^\alpha}{(\mathbb{E}\sup_{A \in \mathcal{A}} \|AX\|_2)^\alpha}, \frac{t^{\alpha/2}}{\sup_{A \in \mathcal{A}} \|A\|_{\mathrm{op}}^{\alpha/2}}\Big)\Big).$$

For $\alpha = 2$, this gives back [20, Theorem 1.1] (up to constants and a different range of t). Comparing Theorems 4.1 to 2.1, we note that instead of a sub-Gaussian term, we obtain an α-subexponential term (which can be trivially transformed into a sub-Gaussian term for $t \leq \mathbb{E}\sup_{A \in \mathcal{A}} \|AX\|_2$, but this does not cover the complete α-subexponential regime). Moreover, Theorem 4.1 only gives a bound for the upper tails. Therefore, if $\mathcal{A}$ just consists of a single matrix, Theorem 2.1 is stronger. These differences have technical reasons.

To prove Theorem 4.1, we shall follow the basic steps of [20] and modify those where the truncation comes in. Let us first repeat some tools and results. In the sequel, for a random vector $W = (W_1, \ldots, W_n)$, we shall denote

$$f(W) := \sup_{A \in \mathcal{A}} (W^T AW - g(A)), \tag{4.2}$$

where $g \colon \mathbb{R}^{n \times n} \to \mathbb{R}$ is some function. Moreover, if A is any matrix, we denote by $\mathrm{Diag}(A)$ its diagonal part (regarded as a matrix with zero entries on its off-diagonal). The following lemma combines [20, Lemmas 3.2 & 3.5].

Lemma 4.2

(1) *Assume the vector W has independent components that satisfy $W_i \leq K$ a.s. Then, for any $t \geq 1$, we have*

$$f(W) - \mathbb{E}f(W) \leq C\big(K(\mathbb{E}\sup_{A \in \mathcal{A}} \|AW\|_2 + \mathbb{E}\sup_{A \in \mathcal{A}} \|\mathrm{Diag}(A)W\|_2)\sqrt{t} + K^2 \sup_{A \in \mathcal{A}} \|A\|_{\mathrm{op}} t\big)$$

with probability at least $1 - e^{-t}$.

(2) *Assuming the vector W has independent (but not necessarily bounded) components with mean zero, we have*

$$\mathbb{E} \sup_{A\in\mathcal{A}} \|\mathrm{Diag}(A)W\|_2 \le C\mathbb{E} \sup_{A\in\mathcal{A}} \|AW\|_2.$$

From now on, let X be the random vector from Theorem 4.1, and recall the truncated random vector Y that we introduced in (3.3) (and the corresponding "remainder" Z). Then, Lemma 4.2 (1) for $f(Y)$ with $g(A) = \mathbb{E}X^TAX$ yields

$$f(Y) - \mathbb{E}f(Y) \le C\big(M(\mathbb{E} \sup_{A\in\mathcal{A}} \|AY\|_2 + \mathbb{E} \sup_{A\in\mathcal{A}} \|\mathrm{Diag}(A)\|_2)t^{1/\alpha} + M^2t^{2/\alpha} \sup_{A\in\mathcal{A}} \|A\|_{\mathrm{op}}\big) \tag{4.3}$$

with probability at least $1 - e^{-t}$ (actually, (4.3) even holds with $\alpha = 2$, but in the sequel we will have to use the weaker version given above anyway). Here we recall that $M \le C_\alpha \|\max_i |X_i|\|_{\Psi_\alpha}$.

To prove Theorem 4.1, it remains to replace the terms involving the truncated random vector Y by the original vector X. First, by Proposition 3.1 and since $\sup_{A\in\mathcal{A}} \|AX\|_2$ is $\sup_{A\in\mathcal{A}} \|A\|_{\mathrm{op}}$-Lipschitz, we obtain

$$\mathbb{P}(\sup_{A\in\mathcal{A}} \|AX\|_2 > \mathbb{E} \sup_{A\in\mathcal{A}} \|AX\|_2 + C_\alpha \|\max_i|X_i|\|_{\Psi_\alpha} \sup_{A\in\mathcal{A}} \|A\|_{\mathrm{op}}t^{1/\alpha}) \le 2e^{-t}. \tag{4.4}$$

Moreover, by (3.8),

$$|\mathbb{E} \sup_{A\in\mathcal{A}} \|AY\|_2 - \mathbb{E} \sup_{A\in\mathcal{A}} \|AX\|_2| \le C_\alpha \|\max_i|X_i|\|_{\Psi_\alpha} \sup_{A\in\mathcal{A}} \|A\|_{\mathrm{op}}. \tag{4.5}$$

Next we estimate the difference between the expectations of $f(X)$ and $f(Y)$.

Lemma 4.3 *We have*

$$|\mathbb{E}f(Y) - \mathbb{E}f(X)| \le C_\alpha\big(\|\max_i|X_i|\|_{\Psi_\alpha}\mathbb{E} \sup_{A\in\mathcal{A}} \|AX\|_2 + \|\max_i|X_i|\|_{\Psi_\alpha}^2 \sup_{A\in\mathcal{A}} \|A\|_{\mathrm{op}}\big).$$

Proof First note that

$$\begin{aligned} f(X) &= \sup_{A\in\mathcal{A}} (Y^TAY - \mathbb{E}X^TAX + Z^TAX + Z^TAY) \\ &\le \sup_{A\in\mathcal{A}} (Y^TAY - \mathbb{E}X^TAX) + \sup_{A\in\mathcal{A}} |Z^TAX| + \sup_{A\in\mathcal{A}} |Z^TAY| \\ &\le f(Y) + \|Z\|_2 \sup_{A\in\mathcal{A}} \|AX\|_2 + \|Z\|_2 \sup_{A\in\mathcal{A}} \|AY\|_2. \end{aligned}$$

The same holds if we reverse the roles of X and Y. As a consequence,

$$|f(X) - f(Y)| \le \|Z\|_2 \sup_{A\in\mathcal{A}} \|AX\|_2 + \|Z\|_2 \sup_{A\in\mathcal{A}} \|AY\|_2, \tag{4.6}$$

and thus, taking expectations and applying Hölder's inequality,

$$|\mathbb{E}f(X) - \mathbb{E}f(Y)| \le (\mathbb{E}\|Z\|_2^2)^{1/2}((\mathbb{E} \sup_{A\in\mathcal{A}} \|AX\|_2^2)^{1/2} + (\mathbb{E} \sup_{A\in\mathcal{A}} \|AY\|_2^2)^{1/2}). \tag{4.7}$$

We may estimate $(\mathbb{E}\|Z\|_2^2)^{1/2}$ using (3.9). Moreover, by related arguments as in (3.8), from (4.4), we get that

$$\mathbb{E} \sup_{A\in\mathcal{A}} \|AX\|_2^2 \le C_\alpha((\mathbb{E} \sup_{A\in\mathcal{A}} \|AX\|_2)^2 + \|\max_i |X_i|\|_{\Psi_\alpha}^2 \sup_{A\in\mathcal{A}} \|A\|_{\mathrm{op}}^2).$$

Arguing similarly and using (4.5), the same bound also holds for $(\mathbb{E} \sup_{A\in\mathcal{A}} \|AY\|_2^2)^{1/2}$. Taking roots and plugging everything into (4.7) complete the proof. □

Finally, we prove the central result of this section.

Proof of Theorem 4.1 First, it immediately follows from Lemma 4.3 that

$$\mathbb{E}f(Y) \le \mathbb{E}f(X) + C_\alpha\big(\|\max_i |X_i|\|_{\Psi_\alpha} \mathbb{E} \sup_{A\in\mathcal{A}} \|AX\|_2 + \|\max_i |X_i|\|_{\Psi_\alpha}^2 \sup_{A\in\mathcal{A}} \|A\|_{\mathrm{op}}\big). \tag{4.8}$$

Moreover, by (4.5) and Lemma 4.2 (2),

$$\begin{aligned}\mathbb{E} \sup_{A\in\mathcal{A}} \|AY\|_2 + \mathbb{E} \sup_{A\in\mathcal{A}} \|\mathrm{Diag}(A)Y\|_2 \le C_\alpha(&\mathbb{E} \sup_{A\in\mathcal{A}} \|AX\|_2 \\ &+ \|\max_i |X_i|\|_{\Psi_\alpha} \sup_{A\in\mathcal{A}} \|A\|_{\mathrm{op}}).\end{aligned} \tag{4.9}$$

Finally, it follows from (4.6), (4.4), and (4.5) that

$$\begin{aligned}|f(X) - f(Y)| &\le \|Z\|_2 \sup_{A\in\mathcal{A}} \|AX\|_2 + \|Z\|_2 \sup_{A\in\mathcal{A}} \|AY\|_2 \\ &\le C_\alpha(\|Z\|_2 \mathbb{E} \sup_{A\in\mathcal{A}} \|AX\|_2 + \|Z\|_2 \|\max_i |X_i|\|_{\Psi_\alpha} \sup_{A\in\mathcal{A}} \|A\|_{\mathrm{op}} t^{1/\alpha})\end{aligned}$$

with probability at least $1 - 4e^{-t}$ for all $t \ge 1$. Using (3.7), it follows that

$$\begin{aligned}|f(X) - f(Y)| \le C_\alpha(&\|\max_i |X_i|\|_{\Psi_\alpha} \mathbb{E} \sup_{A\in\mathcal{A}} \|AX\|_2 t^{1/\alpha} \\ &+ \|\max_i |X_i|\|_{\Psi_\alpha}^2 \sup_{A\in\mathcal{A}} \|A\|_{\mathrm{op}} t^{2/\alpha})\end{aligned} \tag{4.10}$$

with probability at least $1 - 6e^{-t}$ for all $t \ge 1$. Combining (4.8), (4.9), and (4.10) and plugging into (4.3) thus yield that with probability at least $1 - 6e^{-t}$ for all $t \ge 1$,

$$\begin{aligned}f(X) - \mathbb{E}f(X) &\le C_\alpha(\|\max_i|X_i|\|_{\Psi_\alpha}\mathbb{E}\sup_{A\in\mathcal{A}}\|AX\|_2 t^{1/\alpha}\\ &\quad + \|\max_i|X_i|\|_{\Psi_\alpha}^2 \sup_{A\in\mathcal{A}}\|A\|_{\mathrm{op}}t^{2/\alpha})\\ &=: C_\alpha(at^{1/\alpha} + bt^{2/\alpha}).\end{aligned}$$

If $u \ge \max(a, b)$, it follows that

$$\mathbb{P}(f(X) - \mathbb{E}f(X) \ge u) \le 6\exp\Big(-\frac{1}{C_\alpha}\min\Big(\Big(\frac{u}{a}\Big)^\alpha, \Big(\frac{u}{b}\Big)^{\alpha/2}\Big)\Big).$$

By standard means (a suitable change of constants, using (2.3)), this bound may be extended to any $u \ge 0$, and the constant may be adjusted to 2. □

5 Random Tensors

By a *simple random tensor*, we mean a random tensor of the form

$$X := X_1 \otimes \cdots \otimes X_d = (X_{1,i_1}\cdots X_{d,i_d})_{i_1,\dots,i_d} \in \mathbb{R}^{n^d}, \tag{5.1}$$

where all X_k are independent random vectors in $\mathbb{R}^n$ whose coordinates are independent, centered random variables with variance one. Concentration results for random tensors (typically for polynomial-type functions) have been shown in [6, 12, 26], for instance.

Recently, in [42], new and interesting concentration bounds for simple random tensors were shown. In comparison to previous work, these inequalities focus on *small* values of t, e. g., a regime where sub-Gaussian tail decay holds. Moreover, in contrast to previous papers, [42] provides constants with optimal dependence on d. One of these results is the following convex concentration inequality: assuming that n and d are positive integers, $f: \mathbb{R}^{n^d} \to \mathbb{R}$ is convex and 1-Lipschitz, and the X_{ij} are bounded a.s., then for any $t \in [0, 2n^{d/2}]$,

$$\mathbb{P}(|f(X) - \mathbb{E}f(X)| > t) \le 2\exp\Big(-\frac{t^2}{Cdn^{d-1}}\Big), \tag{5.2}$$

where $C > 0$ only depends on the bound of the coordinates. Using Theorem 2.1 and Proposition 3.1, we may extend this result to unbounded random variables as follows:

Theorem 5.1 *Let $n, d \in \mathbb{N}$ and $f: \mathbb{R}^{n^d} \to \mathbb{R}$ be convex and 1-Lipschitz. Consider a simple random tensor $X := X_1 \otimes \cdots \otimes X_d$ as in (5.1). Fix $\alpha \in [1, 2]$, and assume that $\|X_{i,j}\|_{\Psi_\alpha} \le K$. Then, for any $t \in [0, c_\alpha n^{d/2}(\log n)^{1/\alpha}/K]$,*

$$\mathbb{P}(|f(X) - \mathbb{E}f(X)| > t) \le 2\exp\Big(-\frac{1}{C_\alpha}\Big(\frac{t}{d^{1/2}n^{(d-1)/2}(\log n)^{1/\alpha}K}\Big)^\alpha\Big).$$

On the other hand, if $\alpha \in (0, 1)$, *then, for any* $t \in [0, c_\alpha n^{d/2}(\log n)^{1/\alpha} d^{1/\alpha-1/2}/K]$,

$$\mathbb{P}(|f(X) - \mathbb{E}f(X)| > t) \le 2\exp\Big(-\frac{1}{C_\alpha}\Big(\frac{t}{d^{1/\alpha}n^{(d-1)/2}(\log n)^{1/\alpha}K}\Big)^\alpha\Big).$$

The logarithmic factor stems from the Orlicz norm of $\max_i |X_i|$ in Proposition 3.1. For a slightly sharper version that includes the explicit dependence on these norms (and also gives back (5.2) for bounded random variables and $\alpha = 2$), see (5.12) in the proof of Theorem 5.1. We believe that Theorem 5.1 is non-optimal for $\alpha < 1$ as we would expect a bound of the same type as for $\alpha \in [1, 2]$. However, a key difference in the proofs is that in the case of $\alpha \ge 1$ we can make use of moment-generating functions. This is clearly not possible if $\alpha < 1$, so that less subtle estimates must be invoked instead.

For the proof of Theorem 5.1, we first adapt some preliminary steps and compile a number of auxiliary lemmas whose proofs are deferred to the appendix. As a start, we need some additional characterizations of α-subexponential random variables via the behavior of the moment-generating functions:

Proposition 5.2 *Let* X *be a random variable and* $\alpha \in (0, 2]$. *Then, the properties* (1.1), (1.2), *and* (1.3) *are equivalent to*

$$\mathbb{E}\exp(\lambda^\alpha |X|^\alpha) \le \exp(C_{4,\alpha}^\alpha \lambda^\alpha) \tag{5.3}$$

for all $0 \le \lambda \le 1/C_{4,\alpha}$. *If* $\alpha \in [1, 2]$ *and* $\mathbb{E}X = 0$, *then the above properties are moreover equivalent to*

$$\mathbb{E}\exp(\lambda X) \le \begin{cases} \exp(C_{5,\alpha}^2 \lambda^2) & \text{if } |\lambda| \le 1/C_{5,\alpha} \\ \exp(C_{5,\alpha}^{\alpha/(\alpha-1)} |\lambda|^{\alpha/(\alpha-1)}) & \text{if } |\lambda| \ge 1/C_{5,\alpha} \text{ and } \alpha > 1. \end{cases} \tag{5.4}$$

The parameters $C_{i,\alpha}$, $i = 1, \ldots, 5$, *can be chosen such that they only differ by constant* α*-dependent factors. In particular, we can take* $C_{i,\alpha} = c_{i,\alpha}\|X\|_{\Psi_\alpha}$.

To continue, note that $\|X\|_2 = \prod_{i=1}^d \|X_i\|_2$. A key step in the proofs of [42] is a maximal inequality that simultaneously controls the tails of $\prod_{i=1}^k \|X_i\|_2$, $k = 1, \ldots, d$, where the X_i have independent sub-Gaussian components, i.e., $\alpha = 2$. Generalizing these results to any order $\alpha \in (0, 2]$ is not hard. The following preparatory lemma extends [42, Lemma 3.1]. Note that in the proof (given in the appendix again), we apply Proposition 2.2.

Lemma 5.3 *Let* $X_1, \ldots, X_d \in \mathbb{R}^n$ *be independent random vectors with independent, centered coordinates such that* $\mathbb{E}X_{i,j}^2 = 1$ *and* $\|X_{i,j}\|_{\Psi_\alpha} \le K$ *for some* $\alpha \in (0, 2]$. *Then, for any* $t \in [0, 2n^{d/2}]$,

$$\mathbb{P}\Big(\prod_{i=1}^{d}\|X_i\|_2 > n^{d/2}+t\Big) \le 2\exp\Big(-\frac{1}{C_\alpha}\Big(\frac{t}{K^2d^{1/2}n^{(d-1)/2}}\Big)^\alpha\Big).$$

To control all $k = 1, \dots, d$ simultaneously, we need a generalized version of the maximal inequality [42, Lemma 3.2] that we state next.

Lemma 5.4 *Let $X_1, \dots, X_d \in \mathbb{R}^n$ be independent random vectors with independent, centered coordinates such that $\mathbb{E}X_{i,j}^2 = 1$ and $\|X_{i,j}\|_{\Psi_\alpha} \le K$ for some $\alpha \in (0, 2]$. Then, for any $u \in [0, 2]$,*

$$\mathbb{P}\Big(\max_{1\le k\le d} n^{-k/2}\prod_{i=1}^{k}\|X_i\|_2 > 1+u\Big) \le 2\exp\Big(-\frac{1}{C_\alpha}\Big(\frac{n^{1/2}u}{K^2d^{1/2}}\Big)^\alpha\Big).$$

The following Martingale-type bound is directly taken from [42]:

Lemma 5.5 ([42], Lemma 4.1) *Let $X_1, \dots X_d$ be independent random vectors. For each $k = 1, \dots, d$, let $f_k = f_k(X_k, \dots, X_d)$ be an integrable real-valued function and $\mathcal{E}_k$ be an event that is uniquely determined by the vectors $X_k, \dots, X_d$. Let $\mathcal{E}_{d+1}$ be the entire probability space. Suppose that for every $k = 1, \dots, d$, we have*

$$\mathbb{E}_{X_k}\exp(f_k) \le \pi_k$$

for every realization of $X_{k+1}, \dots, X_d$ in $\mathcal{E}_{k+1}$. Then, for $\mathcal{E} := \mathcal{E}_2 \cap \dots \cap \mathcal{E}_d$, we have

$$\mathbb{E}\exp(f_1 + \ldots + f_d)1_{\mathcal{E}} \le \pi_1 \cdots \pi_d.$$

Finally, we need a bound for the Orlicz norm of $\max_i |X_i|$.

Lemma 5.6 *Let $X_1, \dots, X_n$ be independent, centered random variables such that $\|X_i\|_{\Psi_\alpha} \le K$ for any i and some $\alpha > 0$. Then,*

$$\|\max_i |X_i|\|_{\Psi_\alpha} \le C_\alpha K \max\Big\{\Big(\frac{\sqrt{2}+1}{\sqrt{2}-1}\Big)^{1/\alpha}, (\log n)^{1/\alpha}\Big(\frac{2}{\log 2}\Big)^{1/\alpha}\Big\}.$$

Here, we may choose $C_\alpha = \max\{2^{1/\alpha-1}, 2^{1-1/\alpha}\}$.

Note that for $\alpha \ge 1$, [8, Proposition 4.3.1] provides a similar result. However, we are also interested in the case of $\alpha < 1$ in the present note. The condition $\mathbb{E}X_i = 0$ in Lemma 5.6 can easily be removed only at the expense of a different absolute constant.

We are now ready to prove Theorem 5.1.

Proof of Theorem 5.1 We shall adapt the arguments from [42]. First let

$$\mathcal{E}_k := \Big\{ \prod_{i=k}^{d} \|X_i\|_2 \le 2n^{(d-k+1)/2} \Big\}, \qquad k = 1, \ldots, d,$$

and let $\mathcal{E}_{d+1}$ be the full space. It then follows from Lemma 5.4 for $u = 1$ that

$$\mathbb{P}(\mathcal{E}) \ge 1 - 2\exp\Big(-\frac{1}{C_\alpha}\Big(\frac{n^{1/2}}{K^2 d^{1/2}}\Big)^{\alpha}\Big), \tag{5.5}$$

where $\mathcal{E} := \mathcal{E}_2 \cap \cdots \cap \mathcal{E}_d$.

Now fix any realization $x_2, \ldots, x_d$ of the random vectors $X_2, \ldots, X_d$ in $\mathcal{E}_2$, and apply Proposition 3.1 to the function $f_1(x_1)$ given by $x_1 \mapsto f(x_1, \ldots x_d)$. Clearly, f_1 is convex, and since

$$|f(x \otimes x_2 \otimes \cdots \otimes x_d) - f(y \otimes x_2 \otimes \cdots \otimes x_d)| \le \|x-y\|_2 \prod_{i=2}^{d} \|x_i\|_2 \le \|x-y\|_2 2n^{(d-1)/2},$$

we see that it is $2n^{(d-1)/2}$-Lipschitz. Hence, it follows from (3.2) that

$$\|f - \mathbb{E}_{X_1} f\|_{\Psi_\alpha(X_1)} \le c_\alpha n^{(d-1)/2} \|\max_j |X_{1,j}|\|_{\Psi_\alpha} \tag{5.6}$$

for any $x_2, \ldots, x_d$ in $\mathcal{E}_2$, where $\mathbb{E}_{X_1}$ denotes taking the expectation with respect to X_1 (which, by independence, is the same as conditionally on $X_2, \ldots, X_d$).

To continue, fix any realization $x_3, \ldots, x_d$ of the random vectors $X_3, \ldots, X_d$ that satisfy $\mathcal{E}_3$ and apply Proposition 3.1 to the function $f_2(x_2)$ given by $x_2 \mapsto \mathbb{E}_{X_1} f(X_1, x_2, \ldots, x_d)$. Again, f_2 is a convex function, and since

$$\begin{aligned}
&|\mathbb{E}_{X_1} f(X_1 \otimes x \otimes x_3 \otimes \ldots \otimes x_d) - \mathbb{E}_{X_1} f(X_1 \otimes y \otimes x_3 \otimes \ldots \otimes x_d)| \\
&\le \mathbb{E}_{X_1} \|X_1 \otimes (x-y) \otimes x_3 \otimes \ldots \otimes x_d\|_2 \le (\mathbb{E}\|X_1\|_2^2)^{1/2} \|x-y\|_2 \prod_{i=3}^{d} \|x_i\|_2 \\
&\le \sqrt{n}\|x-y\|_2 \cdot 2n^{(d-2)/2} = \|x-y\|_2 \cdot 2n^{(d-1)/2},
\end{aligned}$$

f_2 is $2n^{(d-1)/2}$-Lipschitz. Applying (3.2), we thus obtain

$$\|\mathbb{E}_{X_1} f - \mathbb{E}_{X_1, X_2} f\|_{\Psi_\alpha(X_2)} \le c_\alpha n^{(d-1)/2} \|\max_j |X_{2,j}|\|_{\Psi_\alpha} \tag{5.7}$$

for any $x_3, \ldots, x_d$ in $\mathcal{E}_3$. Iterating this procedure, we arrive at

$$\|\mathbb{E}_{X_1, \ldots, X_{k-1}} f - \mathbb{E}_{X_1, \ldots, X_k} f\|_{\Psi_\alpha(X_k)} \le c_\alpha n^{(d-1)/2} \|\max_j |X_{k,j}|\|_{\Psi_\alpha} \tag{5.8}$$

for any realization $x_{k+1}, \ldots, x_d$ of $X_{k+1}, \ldots, X_d$ in $\mathcal{E}_{k+1}$.

We now combine (5.8) for $k = 1, \ldots, d$. To this end, we write

$$\Delta_k := \Delta_k(X_k, \ldots, X_d) := \mathbb{E}_{X_1,\ldots,X_{k-1}} f - \mathbb{E}_{X_1,\ldots,X_k} f$$

and apply Proposition 5.2. Here we have to distinguish between the cases where $\alpha \in [1, 2]$ and $\alpha \in (0, 1)$. If $\alpha \geq 1$, we use (5.4) to arrive at a bound for the moment-generating function. Writing $M_k := \|\max_j |X_{k,j}|\|_{\Psi_\alpha}$, we obtain

$$\mathbb{E} \exp(\lambda \Delta_k) \leq \begin{cases} \exp((c_\alpha n^{(d-1)/2} M_k)^2 \lambda^2) \\ \exp((c_\alpha n^{(d-1)/2} M_k)^{\alpha/(\alpha-1)} |\lambda|^{\alpha/(\alpha-1)}) \end{cases}$$

for all $x_{k+1}, \ldots, x_d$ in $\mathcal{E}_{k+1}$, where the first line holds if $|\lambda| \leq 1/(c_\alpha n^{(d-1)/2} M_k)$ and the second one if $|\lambda| \geq 1/(c_\alpha n^{(d-1)/2} M_k)$ and $\alpha > 1$. For the simplicity of presentation, temporarily assume that $c_\alpha n^{(d-1)/2} = 1$ (alternatively, replace M_k by $c_\alpha n^{(d-1)/2} M_k$ in the following arguments) and that $M_1 \leq \ldots \leq M_d$. Using Lemma 5.5, we obtain

$$\begin{aligned} &\mathbb{E} \exp(\lambda(f - \mathbb{E} f)) 1_{\mathcal{E}} = \mathbb{E} \exp(\lambda(\Delta_1 + \cdots + \Delta_d)) 1_{\mathcal{E}} \\ &\leq \exp((M_1^2 + \ldots + M_k^2)\lambda^2 + (M_{k+1}^{\alpha/(\alpha-1)} + \ldots + M_d^{\alpha/(\alpha-1)})|\lambda|^{\alpha/(\alpha-1)}) \end{aligned}$$

for $|\lambda| \in [1/M_{k+1}, 1/M_k]$, where we formally set $M_0 := 0$ and $M_{d+1} := \infty$. In particular, setting $M := (M_1^2 + \ldots + M_d^2)^{1/2}$, we have

$$\mathbb{E} \exp(\lambda(f - \mathbb{E} f)) 1_{\mathcal{E}} \leq \exp(M^2 \lambda^2)$$

for all $|\lambda| \leq 1/M_d = 1/(\max_k M_k)$. Furthermore, for $\alpha > 1$, it is not hard to see that

$$(M_1^2 + \ldots + M_k^2)\lambda^2 + (M_{k+1}^{\alpha/(\alpha-1)} + \ldots + M_d^{\alpha/(\alpha-1)})|\lambda|^{\alpha/(\alpha-1)} \leq M^{\alpha/(\alpha-1)} |\lambda|^{\alpha/(\alpha-1)}$$

If $|\lambda| \in [1/M_{k+1}, 1/M_k]$ for some $k = 0, 1, \ldots, d-1$ or $|\lambda| \in [1/M, 1/M_d]$ for $k = d$. Indeed, by monotonicity (divide by λ^2 and compare the coefficients), it suffices to check this for $\lambda = 1/M_{k+1}$ or $\lambda = 1/M$ if $k = d$. The cases of $k = 0$ and $k = d$ follow by simple calculations. In the general case, set $x^2 = (M_1^2 + \ldots + M_{k+1}^2)/M_{k+1}^2$ and $y^{\alpha/(\alpha-1)} = (M_{k+2}^{\alpha/(\alpha-1)} + \ldots + M_d^{\alpha/(\alpha-1)})/M_{k+1}^{\alpha/(\alpha-1)}$. Clearly, $(x^2 + y^{\alpha/(\alpha-1)})^{(\alpha-1)/\alpha} \leq (x^2 + y^2)^{1/2}$ since $x \geq 1$ and $\alpha/(\alpha-1) \geq 2$. Moreover, $y^2 \leq (M_{k+2}^2 + \ldots + M_d^2)/M_{k+1}^2$, which proves the inequality. Altogether, inserting the factor $c_\alpha n^{(d-1)/2}$ again, we therefore obtain

$$\mathbb{E} \exp(\lambda(f - \mathbb{E} f)) 1_{\mathcal{E}} = \mathbb{E} \exp(\lambda(\Delta_1 + \cdots + \Delta_d)) 1_{\mathcal{E}}$$

$$\leq \begin{cases} \exp((c_\alpha n^{(d-1)/2})^2 M^2 \lambda^2) \\ \exp((c_\alpha n^{(d-1)/2})^{\alpha/(\alpha-1)} M^{\alpha/(\alpha-1)} |\lambda|^{\alpha/(\alpha-1)}), \end{cases} \tag{5.9}$$

where the first line holds if $|\lambda| \leq 1/(c_\alpha n^{(d-1)/2} M)$ and the second one if $|\lambda| \geq 1/(c_\alpha n^{(d-1)/2} M)$ and $\alpha > 1$.

On the other hand, if $\alpha < 1$, we use (5.3). Together with Lemma 5.5 and the subadditivity of $|\cdot|^\alpha$ for $\alpha \in (0, 1)$, this yields

$$\begin{aligned} &\mathbb{E}\exp(\lambda^\alpha |f - \mathbb{E}f|^\alpha) 1_{\mathcal{E}} \leq \mathbb{E}\exp(\lambda^\alpha (|\Delta_1|^\alpha + \cdots + |\Delta_d|^\alpha)) 1_{\mathcal{E}} \\ &\leq \exp((c_\alpha n^{(d-1)/2})^\alpha (M_1^\alpha + \cdots + M_d^\alpha) \lambda^\alpha) \end{aligned} \tag{5.10}$$

for $\lambda \in [0, 1/(c_\alpha n^{(d-1)/2} \max_k M_k)]$.

To finish the proof, first consider $\alpha \in [1, 2]$. Then, for any $\lambda > 0$, we have

$$\begin{aligned} \mathbb{P}(f - \mathbb{E}f > t) &\leq \mathbb{P}(\{f - \mathbb{E}f > t\} \cap \mathcal{E}) + \mathbb{P}(\mathcal{E}^c) \\ &\leq \mathbb{P}(\exp(\lambda(f - \mathbb{E}f)) 1_{\mathcal{E}} > \exp(\lambda t)) + \mathbb{P}(\mathcal{E}^c) \\ &\leq \exp\Big(-\Big(\frac{t}{c_\alpha n^{(d-1)/2} M}\Big)^\alpha\Big) + 2\exp\Big(-\frac{1}{C_\alpha}\Big(\frac{n^{1/2}}{K^2 d^{1/2}}\Big)^\alpha\Big), \end{aligned} \tag{5.11}$$

where the last step follows by standard arguments (similarly as in the proof of Proposition 5.2 given in the appendix), using (5.9) and (5.5). Now, assume that $t \leq c_\alpha n^{d/2} M/(K^2 d^{1/2})$. Then, the right-hand side of (5.11) is dominated by the first term (possibly after adjusting constants), so that we arrive at

$$\mathbb{P}(f - \mathbb{E}f > t) \leq 3\exp\Big(-\frac{1}{C_\alpha}\Big(\frac{t}{n^{(d-1)/2} M}\Big)^\alpha\Big).$$

The same arguments hold if f is replaced by $-f$. Adjusting constants by (2.3), we obtain that for any $t \in [0, c_\alpha n^{d/2} M/(K^2 d^{1/2})]$,

$$\mathbb{P}(|f(X) - \mathbb{E}f(X)| > t) \leq 2\exp\Big(-\frac{1}{C_\alpha}\Big(\frac{t}{n^{(d-1)/2} M}\Big)^\alpha\Big). \tag{5.12}$$

Now it remains to note that by Lemma 5.6, we have

$$\|\max_j |X_{i,j}|\|_{\Psi_\alpha} \leq C_\alpha (\log n)^{1/\alpha} \max_j \|X_{i,j}\|_{\Psi_\alpha} \leq C_\alpha (\log n)^{1/\alpha} K.$$

If $\alpha \in (0, 1)$, similarly to (5.11), using (5.10), (5.5) and Proposition 5.2,

$$\mathbb{P}(|f - \mathbb{E}f| > t) \leq \mathbb{P}(\{|f - \mathbb{E}f| > t\} \cap \mathcal{E}) + \mathbb{P}(\mathcal{E}^c)$$

$$\leq 2\exp\Big(-\Big(\frac{t}{c_\alpha n^{(d-1)/2}M_\alpha}\Big)^\alpha\Big) + 2\exp\Big(-\frac{1}{C_\alpha}\Big(\frac{n^{1/2}}{K^2 d^{1/2}}\Big)^\alpha\Big),$$

where $M_\alpha := (M_1^\alpha + \ldots + M_d^\alpha)^{1/\alpha}$. The rest follows as above. □

Appendix A

Proof of Proposition 5.2 The equivalence of (1.1), (1.2), (1.3), and (5.3) is easily seen by directly adapting the arguments from the proof of [41, Proposition 2.5.2]. To see that these properties imply (5.4), first note that since in particular $\|X\|_{\Psi_1} < \infty$, the bound for $|\lambda| \leq 1/C'_{5,\alpha}$ directly follows from [41], Proposition 2.7.1 (e). To see the bound for large values of $|\lambda|$, we infer that by the weighted arithmetic–geometric mean inequality (with weights $\alpha - 1$ and 1),

$$y^{(\alpha-1)/\alpha} z^{1/\alpha} \leq \frac{\alpha-1}{\alpha} y + \frac{1}{\alpha} z$$

for any $y, z \geq 0$. Setting $y := |\lambda|^{\alpha/(\alpha-1)}$ and $z := |x|^\alpha$, we may conclude that

$$\lambda x \leq \frac{\alpha-1}{\alpha}|\lambda|^{\alpha/(\alpha-1)} + \frac{1}{\alpha}|x|^\alpha$$

for any $\lambda, x \in \mathbb{R}$. Consequently, using (5.3), assuming $C_{4,\alpha} = 1$, for any $|\lambda| \geq 1$,

$$\begin{aligned}\mathbb{E}\exp(\lambda X) &\leq \exp\big(\frac{\alpha-1}{\alpha}|\lambda|^{\alpha/(\alpha-1)}\big)\mathbb{E}\exp(|X|^\alpha/\alpha)\\ &\leq \exp\big(\frac{\alpha-1}{\alpha}|\lambda|^{\alpha/(\alpha-1)}\big)\exp(1/\alpha) \leq \exp(|\lambda|^{\alpha/(\alpha-1)}).\end{aligned}$$

This yields (5.4) for $|\lambda| \geq 1/C''_{5,\alpha}$. The claim now follows by taking $C_{5,\alpha} := \max(C'_{5,\alpha}, C''_{5,\alpha})$.

Finally, starting with (5.4), assuming $C_{5,\alpha} = 1$, let us check (1.1). To this end, note that for any $\lambda > 0$,

$$\mathbb{P}(X \geq t) \leq \exp(-\lambda t)\mathbb{E}\exp(\lambda X) \leq \exp(-\lambda t + \lambda^2 \mathbb{1}_{\{\lambda \leq 1\}} + \lambda^{\alpha/(\alpha-1)}\mathbb{1}_{\{\lambda > 1\}}).$$

Now choose $\lambda := t/2$ if $t \leq 2$, $\lambda := ((\alpha-1)t/\alpha)^{\alpha-1}$ if $t \geq \alpha/(\alpha-1)$, and $\lambda := 1$ if $t \in (2, \alpha/(\alpha-1))$. This yields

$$\mathbb{P}(X \geq t) \leq \begin{cases}\exp(-t^2/4) & \text{if } t \leq 2,\\ \exp(-(t-1)) & \text{if } t \in (2, \alpha/(\alpha-1)),\\ \exp(-\frac{(\alpha-1)^{\alpha-1}}{\alpha^\alpha}t^\alpha) & \text{if } t \geq \alpha/(\alpha-1).\end{cases}$$

Now use (2.3), (2.4), and the fact that $\exp(-(t-1)) \le \exp(-t^\alpha/C_\alpha^\alpha)$ for any $t \in (2, \alpha/(\alpha-1))$. It follows that

$$\mathbb{P}(X \ge t) \le 2\exp(-t^\alpha/C'^\alpha_{1,\alpha})$$

for any $t \ge 0$. The same argument for $-X$ completes the proof. □

Proof of Lemma 5.3 By the arithmetic and geometric means inequality and since $\mathbb{E}\|X_i\|_2 \le \sqrt{n}$, for any $s \ge 0$,

$$\begin{aligned}\mathbb{P}\Big(\prod_{i=1}^d \|X_i\|_2 > (\sqrt{n}+s)^d\Big) &\le \mathbb{P}\Big(\frac{1}{d}\sum_{i=1}^d (\|X_i\|_2 - \sqrt{n}) > s\Big)\\ &\le \mathbb{P}\Big(\frac{1}{d}\sum_{i=1}^d (\|X_i\|_2 - \mathbb{E}\|X_i\|_2) > s\Big).\end{aligned} \tag{A.1}$$

Moreover, by (2.2) and [12, Corollary A.5],

$$\big\| \|X_i\|_2 - \mathbb{E}\|X_i\|_2 \big\|_{\Psi_\alpha} = \big\| \|X_i\|_2 - \sqrt{n} - (\mathbb{E}\|X_i\|_2 - \sqrt{n}) \big\|_{\Psi_\alpha} \le C_\alpha K^2$$

for any $i = 1, \ldots, d$. On the other hand, if $Y_1, \ldots, Y_d$ are independent centered random variables with $\|Y_i\|_{\Psi_\alpha} \le M$, we have

$$\begin{aligned}\mathbb{P}\Big(\frac{1}{d}\Big|\sum_{i=1}^d Y_i\Big| \ge s\Big) &\le 2\exp\Big(-\frac{1}{C_\alpha}\min\Big(\Big(\frac{s\sqrt{d}}{M}\Big)^2, \Big(\frac{s\sqrt{d}}{M}\Big)^\alpha\Big)\Big)\\ &\le 2\exp\Big(-\frac{1}{C_\alpha}\Big(\frac{s\sqrt{d}}{M}\Big)^\alpha\Big).\end{aligned}$$

Here, the first estimate follows from [10] ($\alpha > 1$) and [17] ($\alpha \le 1$), while the last step follows from (2.4). As a consequence, (A.1) can be bounded by $2\exp(-s^\alpha d^{\alpha/2}/(K^{2\alpha}C_\alpha))$.

For $u \in [0, 2]$ and $s = u\sqrt{n}/2d$, we have $(\sqrt{n}+s)^d \le n^{d/2}(1+u)$. Plugging in, we arrive at

$$\mathbb{P}\Big(\prod_{i=1}^d \|X_i\|_2 > n^{d/2}(1+u)\Big) \le 2\exp\Big(-\frac{1}{C_\alpha}\Big(\frac{n^{1/2}u}{K^2 d^{1/2}}\Big)^\alpha\Big).$$

Now set $u := t/n^{d/2}$. □

Proof of Lemma 5.4 Let us first recall the partition into "binary sets" that appears in the proof of [42, Lemma 3.2]. Here we assume that $d = 2^L$ for some $L \in \mathbb{N}$ (if not, increase d). Then, for any $\ell \in \{0, 1, \ldots, L\}$, we consider the partition $\mathcal{I}_\ell$ of $\{1, \ldots, d\}$ into 2^ℓ successive (integer) intervals of length $d_\ell := d/2^\ell$ that we call "binary intervals." It is not hard to see that for any $k = 1, \ldots, d$, we can partition $[1, k]$ into binary intervals of different lengths such that this partition contains at most one interval of each family $\mathcal{I}_\ell$.

Now it suffices to prove that

$$\mathbb{P}\Big(\exists 0 \le \ell \le L, \exists I \in \mathcal{I}_\ell \colon \prod_{i\in I} \|X_i\|_2 > (1 + 2^{-\ell/4}u)n^{d_\ell/2}\Big) \le 2\exp\Big(-\frac{1}{C_\alpha}\Big(\frac{n^{1/2}u}{K^2d^{1/2}}\Big)^\alpha\Big)$$

(cf. Step 3 of the proof of [42, Lemma 3.2], where the reduction to this case is explained in detail). To this end, for any $\ell \in \{0, 1, \ldots, L\}$, any $I \in \mathcal{I}_\ell$, and $d_\ell := |I| = d/2^\ell$, we apply Lemma 5.3 for d_ℓ and $t := 2^{-\ell/4}n^{d_\ell/2}u$. This yields

$$\mathbb{P}\Big(\prod_{i\in I} \|X_i\|_2 > (1 + 2^{-\ell/4}u)n^{d_\ell/2}\Big) \le 2\exp\Big(-\frac{1}{C_\alpha}\Big(\frac{n^{1/2}u}{2^{\ell/4}K^2d_\ell^{1/2}}\Big)^\alpha\Big)$$
$$= 2\exp\Big(-\frac{1}{C_\alpha}\Big(2^{\ell/4}\frac{n^{1/2}u}{K^2d^{1/2}}\Big)^\alpha\Big).$$

Altogether, we arrive at

$$\mathbb{P}\Big(\exists \ell \in \{0, 1, \ldots, L\}, \exists I \in \mathcal{I}_\ell \colon \prod_{i\in I} \|X_i\|_2 > (1 + 2^{-\ell/4}u)n^{d_\ell/2}\Big)$$
$$\le \sum_{\ell=0}^{L} 2^\ell \cdot 2\exp\Big(-\frac{1}{C_\alpha}\Big(2^{\ell/4}\frac{n^{1/2}u}{K^2d^{1/2}}\Big)^\alpha\Big). \tag{A.2}$$

We may now assume that $(n^{1/2}u/(K^2d^{1/2}))^\alpha/C_\alpha \ge 1$ (otherwise the bound in Lemma 5.4 gets trivial by adjusting C_α). Using the elementary inequality $ab \ge (a + b)/2$ for all $a, b \ge 1$, we arrive at

$$2^{\ell\alpha/4}\frac{1}{C_\alpha}\Big(\frac{n^{1/2}u}{K^2d^{1/2}}\Big)^\alpha \ge \frac{1}{2}\Big(2^{\ell\alpha/4} + \frac{1}{C_\alpha}\Big(\frac{n^{1/2}u}{K^2d^{1/2}}\Big)^\alpha\Big).$$

Using this in (A.2), we obtain the upper bound

$$2\exp\Big(-\frac{1}{2C_\alpha}\Big(\frac{n^{1/2}u}{K^2d^{1/2}}\Big)^\alpha\Big)\sum_{\ell=0}^{L} 2^\ell \exp(-2^{\ell\alpha/4-1}) \le c_\alpha \exp\Big(-\frac{1}{2C_\alpha}\Big(\frac{n^{1/2}u}{K^2d^{1/2}}\Big)^\alpha\Big).$$

By (2.3), we can assume $c_\alpha = 2$. □

To prove Lemma 5.6, we first present a number of lemmas and auxiliary statements. In particular, recall that if $\alpha \in (0, \infty)$, then for any $x, y \in (0, \infty)$,

$$c_\alpha(x^\alpha + y^\alpha) \le (x + y)^\alpha \le \widetilde{c}_\alpha(x^\alpha + y^\alpha), \tag{A.3}$$

where $c_\alpha := 2^{\alpha-1} \wedge 1$ and $\widetilde{c}_\alpha := 2^{\alpha-1} \vee 1$. Indeed, if $\alpha \le 1$, using the concavity of the function $x \mapsto x^\alpha$, it follows by standard arguments that $2^{\alpha-1}(x^\alpha + y^\alpha) \le (x+y)^\alpha \le x^\alpha + y^\alpha$. Likewise, for $\alpha \ge 1$, using the convexity of $x \mapsto x^\alpha$, we obtain $x^\alpha + y^\alpha \le (x + y)^\alpha \le 2^{\alpha-1}(x^\alpha + y^\alpha)$.

Lemma A.1 *Let $X_1, \dots, X_n$ be independent, centered random variables such that $\|X_i\|_{\Psi_\alpha} \le 1$ for some $\alpha > 0$. Then, if $Y := \max_i |X_i|$ and $c := (c_\alpha^{-1} \log n)^{1/\alpha}$, we have*

$$\mathbb{P}(Y \ge c + t) \le 2\exp(-c_\alpha t^\alpha)$$

with c_α as in (A.3).

Proof We have

$$\begin{aligned}\mathbb{P}(Y \ge c + t) &\le n\mathbb{P}(|X_i| \ge c + t) \le 2n\exp(-(c + t)^\alpha)\\ &\le 2n\exp(-c_\alpha(t^\alpha + c^\alpha)) = 2\exp(-c_\alpha t^\alpha),\end{aligned}$$

where we have used (A.3) in the next-to-last step. □

Lemma A.2 *Let $Y \ge 0$ be a random variable that satisfies*

$$\mathbb{P}(Y \ge c + t) \le 2\exp(-t^\alpha)$$

for some $c \ge 0$ and any $t \ge 0$. Then,

$$\|Y\|_{\Psi_\alpha} \le \widetilde{c}_\alpha^{1/\alpha} \max\Big\{\Big(\frac{\sqrt{2}+1}{\sqrt{2}-1}\Big)^{1/\alpha}, c\Big(\frac{2}{\log 2}\Big)^{1/\alpha}\Big\}$$

with $\widetilde{c}_\alpha$ as in (A.3).

Proof By (A.3) and monotonicity, we have $Y^\alpha \le \widetilde{c}_\alpha((Y - c)_+^\alpha + c^\alpha)$, where $x_+ := \max(x, 0)$. Thus,

$$\begin{aligned}\mathbb{E}\exp\Big(\frac{Y^\alpha}{s^\alpha}\Big) &\le \exp\Big(\frac{\widetilde{c}_\alpha c^\alpha}{s^\alpha}\Big)\mathbb{E}\exp\Big(\frac{\widetilde{c}_\alpha(Y - c)_+^\alpha}{s^\alpha}\Big)\\ &= \exp\Big(\frac{c^\alpha}{t^\alpha}\Big)\mathbb{E}\exp\Big(\frac{(Y - c)_+^\alpha}{t^\alpha}\Big) =: I_1 \cdot I_2,\end{aligned}$$

where we have set $t := s\widetilde{c}_\alpha^{-1/\alpha}$. Obviously, $I_1 \le \sqrt{2}$ if $t \ge c(1/\log\sqrt{2})^{1/\alpha}$. As for I_2, we have

$$\begin{aligned} I_2 &= 1 + \int_1^\infty \mathbb{P}((Y-c)_+ \geq t(\log y)^{1/\alpha})dy \\ &\leq 1 + 2\int_1^\infty \exp(-t^\alpha \log y)dy = 1 + 2\int_1^\infty \frac{1}{y^{t^\alpha}}dy \leq \sqrt{2} \end{aligned}$$

if $t \geq ((\sqrt{2}+1)/(\sqrt{2}-1))^{1/\alpha}$. Therefore, $I_1 I_2 \leq 2$ if $t \geq \max\{((\sqrt{2}+1)/(\sqrt{2}-1))^{1/\alpha}, c(2/\log 2)^{1/\alpha}\}$, which finishes the proof. □

Having these lemmas at hand, the proof of Lemma 5.6 is easily completed.

Proof of Lemma 5.6 The random variables $\hat{X}_i := X_i/K$ obviously satisfy the assumptions of Lemma A.1. Hence, setting $Y := \max_i |\hat{X}_i| = K^{-1}\max_i |X_i|$,

$$\mathbb{P}(c_\alpha^{1/\alpha} Y \geq (\log n)^{1/\alpha} + t) \leq 2\exp(-t^\alpha).$$

Therefore, we may apply Lemma A.2 to $\hat{Y} := c_\alpha^{1/\alpha} K^{-1} \max_i |X_i|$. This yields

$$\|\hat{Y}\|_{\Psi_\alpha} \leq \widetilde{c}_\alpha^{1/\alpha} \max\Big\{\Big(\frac{\sqrt{2}+1}{\sqrt{2}-1}\Big)^{1/\alpha}, (\log n)^{1/\alpha}\Big(\frac{2}{\log 2}\Big)^{1/\alpha}\Big\},$$

i. e., the claim of Lemma 5.6, where we have set $C := (\widetilde{c}_\alpha c_\alpha^{-1})^{1/\alpha}$. □

Acknowledgments This work was supported by the German Research Foundation (DFG) via CRC 1283 "*Taming uncertainty and profiting from randomness and low regularity in analysis, stochastics and their applications.*" The author would moreover like to thank Arthur Sinulis for carefully reading this paper and many fruitful discussions and suggestions.

References

1. R. Adamczak, A tail inequality for suprema of unbounded empirical processes with applications to Markov chains. Electron. J. Probab. **13**(34), 1000–1034 (2008)
2. R. Adamczak, A note on the Hanson-Wright inequality for random vectors with dependencies. Electron. Commun. Probab. **20**(72), 13 (2015)
3. R. Adamczak, R. Latała, Tail and moment estimates for chaoses generated by symmetric random variables with logarithmically concave tails. Ann. Inst. Henri Poincaré Probab. Stat. **48**(4), 1103–1136 (2012)
4. R. Adamczak, R. Latała, R. Meller, Hanson-Wright inequality in Banach spaces. Ann. Inst. Henri Poincaré Probab. Stat. **56**(4), 2356–2376 (2020)
5. R. Adamczak, M. Strzelecki, On the convex Poincaré inequality and weak transportation inequalities. Bernoulli **25**(1), 341–374 (2019)
6. R. Adamczak, P. Wolff, Concentration inequalities for non-Lipschitz functions with bounded derivatives of higher order. Probab. Theory Relat. Fields **162**(3–4), 531–586 (2015)
7. S. Boucheron, G. Lugosi, P. Massart, *Concentration Inequalities* (Oxford University Press, Oxford, 2013). A nonasymptotic theory of independence, With a foreword by Michel Ledoux

8. V.H. de la Peña, E. Giné, *Decoupling*. Probability and Its Applications (New York) (Springer, New York, 1999)
9. L.H. Dicker, M.A. Erdogdu, Flexible results for quadratic forms with applications to variance components estimation. Ann. Stat. **45**(1), 386–414 (2017)
10. E.D. Gluskin, S. Kwapień, Tail and moment estimates for sums of independent random variables with logarithmically concave tails. Studia Math. **114**(3), 303–309 (1995)
11. F. Götze, H. Sambale, A. Sinulis, Concentration inequalities for bounded functionals via log-Sobolev-type inequalities. J. Theor. Probab. **34**(3), 1623–1652 (2021)
12. F. Götze, H. Sambale, A. Sinulis, Concentration inequalities for polynomials in α-sub-exponential random variables. Electron. J. Probab. **26**(48), 22 (2021)
13. N. Gozlan, C. Roberto, P.-M. Samson, From dimension free concentration to the Poincaré inequality. Calc. Var. Partial Differ. Equ. **52**(3–4), 899–925 (2015)
14. N. Gozlan, C. Roberto, P.-M. Samson, Y. Shu, P. Tetali, Characterization of a class of weak transport-entropy inequalities on the line. Ann. Inst. Henri Poincaré Probab. Stat. **54**(3), 1667–1693 (2018)
15. N. Gozlan, C. Roberto, P.-M. Samson, P. Tetali, Kantorovich duality for general transport costs and applications. J. Funct. Anal. **273**(11), 3327–3405 (2017)
16. D.L. Hanson, F.T. Wright, A bound on tail probabilities for quadratic forms in independent random variables. Ann. Math. Stat. **42**, 1079–1083 (1971)
17. P. Hitczenko, S.J. Montgomery-Smith, K. Oleszkiewicz, Moment inequalities for sums of certain independent symmetric random variables. Studia Math. **123**(1), 15–42 (1997)
18. D. Hsu, S.M. Kakade, T. Zhang, A tail inequality for quadratic forms of sub-Gaussian random vectors. Electron. Commun. Probab. **17**(52), 6 (2012)
19. W.B. Johnson, G. Schechtman, Remarks on Talagrand's deviation inequality for Rademacher functions, in *Functional Analysis (Austin, TX, 1987/1989)*. Lecture Notes in Mathematics, vol. 1470 (Springer, Berlin, 1991), pp. 72–77
20. Y. Klochkov, N. Zhivotovskiy, Uniform Hanson–Wright type concentration inequalities for unbounded entries via the entropy method. Electron. J. Probab. **25**(22), 30 (2020)
21. K. Kolesko, R. Latała, Moment estimates for chaoses generated by symmetric random variables with logarithmically convex tails. Stat. Probab. Lett. **107**, 210–214 (2015)
22. F. Krahmer, S. Mendelson, H. Rauhut, Suprema of chaos processes and the restricted isometry property. Commun. Pure Appl. Math. **67**(11), 1877–1904 (2014)
23. A.K. Kuchibhotla, A. Chakrabortty, Moving Beyond Sub-Gaussianity in High-Dimensional Statistics: Applications in Covariance Estimation and Linear Regression. arXiv preprint, 2018
24. R. Latała, Tail and moment estimates for sums of independent random vectors with logarithmically concave tails. Studia Math. **118**(3), 301–304 (1996)
25. R. Latała, Tail and moment estimates for some types of chaos. Studia Math. **135**(1), 39–53 (1999)
26. R. Latała, Estimates of moments and tails of Gaussian chaoses. Ann. Probab. **24**(6), 2315–2331 (2006)
27. R. Latała, R. Łochowski, Moment and tail estimates for multidimensional chaos generated by positive random variables with logarithmically concave tails, in *Stochastic Inequalities and Applications*. Progress in Probability, vol. 56 (Birkhäuser, Basel, 2003), pp. 77–92
28. J. Lederer, S. van de Geer, New concentration inequalities for suprema of empirical processes. Bernoulli **20**(4), 2020–2038 (2014)
29. M. Ledoux, On Talagrand's deviation inequalities for product measures. ESAIM Probab. Stat. **1**, 63–87 (1995/1997)
30. M. Ledoux, M. Talagrand, *Probability in Banach Spaces: Isoperimetry and Processes*. Ergebnisse der Mathematik und ihrer Grenzgebiete (3) [Results in Mathematics and Related Areas (3)], vol. 23 (Springer, Berlin, 1991)
31. A. Marchina, Concentration inequalities for separately convex functions. Bernoulli **24**(4A), 2906–2933 (2018)
32. P. Massart, Some applications of concentration inequalities to statistics. Ann. Fac. Sci. Toulouse Math. (6) **9**(2), 245–303 (2000)

33. M. Rudelson, R. Vershynin, Hanson-Wright inequality and sub-Gaussian concentration. Electron. Commun. Probab. **18**(82), 9 (2013)
34. H. Sambale, A. Sinulis, Logarithmic Sobolev inequalities for finite spin systems and applications. Bernoulli **26**(3), 1863–1890 (2020)
35. H. Sambale, A. Sinulis, Modified log-Sobolev inequalities and two-level concentration. ALEA Lat. Am. J. Probab. Math. Stat. **18**, 855–885 (2021)
36. P.-M. Samson, Concentration of measure inequalities for Markov chains and Φ-mixing processes. Ann. Probab. **28**(1), 416–461 (2000)
37. P.-M. Samson, Concentration inequalities for convex functions on product spaces, in *Stochastic Inequalities and Applications*. Progress in Probability, vol. 56 (Birkhäuser, Basel, 2003), pp. 33–52
38. M. Talagrand, An isoperimetric theorem on the cube and the Khintchine-Kahane inequalities. Proc. Am. Math. Soc. **104**(3), 905–909 (1988)
39. M. Talagrand, New concentration inequalities in product spaces. Invent. Math. **126**(3), 505–563 (1996)
40. S. van de Geer, J. Lederer, The Bernstein-Orlicz norm and deviation inequalities. Probab. Theory Relat. Fields **157**(1–2), 225–250 (2013)
41. R. Vershynin, *High-Dimensional Probability*. Cambridge Series in Statistical and Probabilistic Mathematics, vol. 47 (Cambridge University Press, Cambridge, 2018)
42. R. Vershynin, Concentration inequalities for random tensors. Bernoulli **26**(4), 3139–3162 (2020)
43. V.H. Vu, K. Wang, Random weighted projections, random quadratic forms and random eigenvectors. Random Struct. Algorithm **47**(4), 792–821 (2015)

Part II
Limit Theorems

Limit Theorems for Random Sums of Random Summands

David Grzybowski

2020 Mathematics Subject Classification: 60G50; 60F05

1 Introduction and Statement of Results

This chapter employs concentration of measure to prove limit theorems for sums of randomly chosen random numbers. This is a particular example of a quenched limit theorem. A quenched limit theorem involves two sources of randomness: a random environment X and a random object conditioned on that environment. In our case, we have a collection of random numbers and an independent selection of a subset of those numbers. The numbers, or weights, are the components of a random n-dimensional vector X and the random subset σ of size m. Our results describe the limiting distribution of $\sum_{a\in\sigma} \frac{X_a - \mathbb{E}X_a}{\sqrt{m}}$, or something similar, where the weights are taken as fixed, so that randomness only comes from σ.

It is worth distinguishing our situation from the classical problem concerning the distribution of $\sum_{i=1}^{N} X_i$, where both N and the $\{X_i\}_{i=1}^{\infty}$ are independent random variables. Here we are interested in the case that the summands are already sampled from their respective distributions, and we do not specify a fixed order of summation. Thus, we consider the distribution of the sum conditioned on the summands and choose the summands randomly. We also take N to be deterministic and consider the limiting behavior of the distribution as N diverges to ∞.

In [5], the authors address the corner-growth setting: X is indexed by elements of $\square_{N,M} = \{(i, j) : 1 \leq i \leq N, 1 \leq j \leq M\}$ with $M = \lfloor \xi N \rfloor, \xi > 0$. Notionally, one traverses this grid from $(1, 1)$ to (N, M), and so σ is chosen to represent an

D. Grzybowski (✉)
Department of Mathematics, Applied Mathematics, and Statistics, Case Western Reserve University, Cleveland, OH, USA
e-mail: drg67@case.edu

R. Adamczak et al. (eds.), *High Dimensional Probability IX*, Progress in Probability 80, https://doi.org/10.1007/978-3-031-26979-0_8

up-right path, i.e., $\sigma = \{(i_k, j_k) : k = 1, \ldots, M+N-1\}$ with $(i_1, j_1) = (1, 1)$, $(i_{N+M-1}, j_{N+M-1}) = (N, M)$, and $(i_{k+1} - i_k, j_{k+1} - j_k)$ is either $(1, 0)$ or $(0, 1)$. They provide moment conditions for a quenched limit theorem when σ is chosen according to three different schemes:

1. The up-right path σ is chosen from those proceeding from $(1, 1)$ to (N, M) without additional restriction.
2. As 1., but the up-right path σ is also specified to pass through points $(\lfloor \zeta_i N \rfloor, \lfloor \xi_i M \rfloor)$ for a finite set of numbers, $0 < \zeta_1 < \cdots < \zeta_k < 1$ and $0 < \xi_1 < \cdots < \xi_k < \xi$.
3. As 1., but $M = N$ and the up-right path σ is also specified to avoid a central square with side $(\lfloor \beta N \rfloor)$, $\beta \in (0, 1)$.

See the sources cited in [5] for more information on the corner-growth setting.

This chapter generalizes the concentration result used in [5] and proves limit theorems in both the corner-growth setting and other settings. In particular, that paper is concerned only with independent weights, whereas we cover certain forms of dependency among the weights, and we present results for weak convergence in probability as well as almost-sure convergence in distribution. We also extend to cases with heavy-tailed weights. Beyond the corner-growth setting, we prove a version of Hoeffding's combinatorial central limit theorem and results related to the empirical distribution of a large random sample.

Definition 1.1 Let $\{\mu_n\}_{n=1}^{\infty}$ be a sequence of random probability measures. We say $\{\mu_n\}_{n=1}^{\infty}$ **converges weakly in probability (WIP)** to μ if

$$\mathbb{P}\left[\left|\int f d\mu_n - \int f d\mu\right| > \epsilon\right] \to 0$$

for every $\epsilon > 0$ and every test function $f \in \mathcal{C}$, where $\mathcal{C}$ is a class of functions such that if $\int f d\nu_n \to \int f d\nu$ for every $f \in \mathcal{C}$, then $\nu_n \xrightarrow{D} \nu$.

We typically take $\mathcal{C}$ to be a more restricted class of functions than the bounded, continuous functions used to define convergence in distribution. In particular, we will here take $\mathcal{C}$ to be convex 1-Lipschitz functions.[1]

Our first three results concern the corner-growth setting. The following is a small extension of the theorems from [5] and involves both weak convergence in probability and $\mathbb{P}_X$-a.s. convergence:

Theorem 1.2 *Under each of the three methods for sampling σ described above, suppose the weights are independent with mean 0, variance 1, and $\mathbb{E}|X_a|^p \leq K < \infty$. If $p > 8$, then*

[1] The sufficiency of convergence for convex 1-Lipschitz functions to establish convergence in distribution can be demonstrated in several ways. One such way is to approximate power functions by convex Lipschitz functions. For x^{2n}, we can find an approximating function using Lemma 5 in [13]. For x^{2n+1}, we approximate $x^{2n+1}\mathbb{1}_{x \geq 0}$ and $x^{2n+1}\mathbb{1}_{x \leq 0}$ separately. See [5], [6], and [12] for alternative arguments.

$$\frac{1}{\sqrt{M+N-1}} \sum_{(i,j)\in\sigma} X_{i,j} \xrightarrow{WIP} \mathcal{N}(0,1), \ as\ N \to \infty,$$

and if $p > 12$*, then*

$$\frac{1}{\sqrt{M+N-1}} \sum_{(i,j)\in\sigma} X_{i,j} \xrightarrow{D} \mathcal{N}(0,1), \ as\ N \to \infty,$$

$\mathbb{P}_X$*-almost surely.*

The next result concerns dependent weights: the uniform distributions on the sphere and on two simplices.

Theorem 1.3 *Under each of the three methods for sampling* σ *described above, suppose* w *has the uniform distribution on…:*

1. $\sqrt{NM}S^{NM-1}$.
2. $\Delta_{(MN,=)} = \{x \in \mathbb{R}_+^{MN} : x_1 + \cdots + x_{MN} = MN\}$.
3. $\Delta_{(MN,\leq)} = \{x \in \mathbb{R}_+^{MN} : x_1 + \cdots + x_{MN} \leq MN\}$.

Then,

$$\frac{1}{\sqrt{M+N-1}} \sum_{(i,j)\in\sigma} \left(X_{i,j} - \mathbb{E}X_{i,j}\right) \xrightarrow{D} \mathcal{N}(0,1), \ as\ N \to \infty,$$

$\mathbb{P}_X$*-almost surely.*

For the above cases, we can offer a physical interpretation: $\sqrt{NM}S^{NM-1}$ and $\Delta_{(MN,=)}$ are level sets of specific functions of the weights, meaning that some notion of "energy" is constant. In $\Delta_{(MN,\leq)}$, this "energy" is merely bounded. Notably, this "energy" is constrained for the whole system, not merely for the particularly chosen summands.

Concentration is loose for stable vectors, but we can prove the following:

Theorem 1.4 *In each of the three situations above, suppose the weights are distributed as follows. Let* Y *be* α*-stable,* $\alpha > \frac{3}{2}$*, with distribution* ν_α*. Define* $\ell_k = \{(i,j) : i + j - 1 = k\}$*. Let the components of* X *all be independent, and let all the components indexed by elements of* ℓ_k *have the same distribution as* $k^{-\tau}Y$*. If* $\tau > 2$*, then*

$$\frac{1}{(2N-1)^{1/\gamma}} \sum_{(i,j)\in\sigma} X_{i,j} \xrightarrow{WIP} \nu_\infty,$$

and if $\tau > 3$*, then*

$$\frac{1}{(2N-1)^{1/\gamma}} \sum_{(i,j)\in\sigma} X_{i,j} \xrightarrow{D} \nu_\infty,$$

$\mathbb{P}_X$*-almost surely, where* ν_∞ *has characteristic exponent*

$$\psi(t) = -\kappa^\alpha \lim_{n\to\infty} \sum_{k=1}^{n} \frac{k^{-\alpha\tau}}{n^{\alpha/\gamma}} |t|^\alpha \left(1 - \iota\beta \operatorname{sign}(t) \tan \frac{\pi\alpha}{2}\right),$$

κ *and* β *are constants determined by* ν_α*, and* $\gamma = \frac{\alpha}{1-\alpha\tau}$.

Our version of Hoeffding's combinatorial central limit theorem is as follows.

Theorem 1.5 *Let* X_N *be an* $N \times N$ *array of independent random variables* $X_{i,j}$, $1 \le i, j \le N$*, with mean* 0*, variance* 1*, and* $\mathbb{E}|X_{i,j}|^p < K$ *for all* i, j*. Let* π *be a permutation of* $(1, \ldots, N)$ *chosen uniformly and* $S_N = \frac{1}{\sqrt{N}} \sum_{i=1}^{N} X_{i,\pi(i)}$*. If* $p > 4$*, then*

$$S_N \xrightarrow{WIP} \mathcal{N}(0,1), \text{ as } N \to \infty,$$

and if $p > 6$*, then*

$$S_N \xrightarrow{D} \mathcal{N}(0,1), \text{ as } N \to \infty,$$

$\mathbb{P}_X$*-almost surely.*

Hoeffding proved his combinatorial central limit theorem in 1951 [8] for deterministic weights, and it has been refined and proved in many ways since. Random weights were apparently first considered in [7]. In [2], a version is proved with random weights using concentration of measure and Stein's method.

When we select finitely many numbers, we have the following result:

Theorem 1.6 *Let* $X = \{X_n\}_{n=1}^{N}$ *be a sample of i.i.d. random variables with common distribution* μ *such that* $\mathbb{E}|X_1|^p < \infty$ *for some* $p > 2$*. If* $\sigma \subset \{1, \ldots, N\}$ *is chosen uniformly from subsets of size* m*, with* m *fixed, then*

$$\sum_{j\in\sigma} X_j \xrightarrow{WIP} \mu^{*m}, \text{ as } N \to \infty.$$

If $\mathbb{E}|X_1|^p < \infty$ *for some* $p > 4$*, then*

$$\sum_{j\in\sigma} X_j \xrightarrow{D} \mu^{*m}, \text{ as } N \to \infty,$$

$\mathbb{P}_X$*-almost surely.*

When $m = 1$, this is a form of the Glivenko–Cantelli theorem. In addition, we have the following result for the uniform distributions on the sphere and two simplices.

Theorem 1.7 *Let X be uniformly distributed on $K \subset \mathbb{R}^N$ and σ be chosen uniformly from subsets of $\{1, \ldots, N\}$ of size m, which is fixed. If $K = \ldots$:*

1. $\sqrt{N}S^{N-1}$, *then*

$$m^{-1/2} \sum_{j \in \sigma} X_j \xrightarrow{D} \mathcal{N}(0, 1) \text{ as } N \to \infty$$

2. $\Delta_{N,=}$ *or* $\Delta_{N,\leq}$, *then*

$$\sum_{j \in \sigma} X_j \xrightarrow{D} \Gamma_{m,1} \text{ as } N \to \infty$$

$\mathbb{P}_X$-almost surely, where $\Gamma_{m,1}$ is the Gamma distribution with shape parameter m and scale parameter 1.

2 Concentration and Convergence

2.1 General Concentration

We employ a classical result from Talagrand [16] as stated in [15]:

Theorem 2.1 *Let $X = (X_1, \ldots, X_n)$ be a random vector with independent components such that for all $1 \leq i \leq n$, $|X_i| \leq 1$-almost surely, and let $f : \mathbb{R} \to \mathbb{R}$ be a convex 1-Lipschitz function. Then for all $t > 0$,*

$$\mathbb{P}[|f(X) - \mathbb{E}f(X)| > t] \leq Ce^{-ct^2},$$

where $C, c > 0$ are absolute constants.

Talagrand's theorem is an example of sub-Gaussian concentration (SGC):

$$\mathbb{P}\big[|f(X) - \mathbb{E}f(X)| > t\big] \leq Ce^{-ct^2}$$

for a specified class of 1-Lipschitz functions f. C and c are not dependent on dimension, but they do vary depending on the distribution of X. Examples of distributions for X satisfying SGC include i.i.d. components satisfying a log-Sobolev inequality, independent bounded components (for convex 1-Lipschitz functions, per Theorem 2.1), and the uniform distribution on a sphere. As this last example shows, this property does not require independence among the components of X, but it does require all their moments to be finite.

We will use SGC in the form of Theorem 2.1 as a component in our general concentration result and to prove a similar result for dependent weights. In addition, we also have subexponential concentration (SEC). The inequality becomes

$$\mathbb{P}\big[|f(X) - \mathbb{E}f(X)| > t\big] \leq Ce^{-ct},$$

for a specified class of 1-Lipschitz functions f. SEC is proved for the two simplices $\Delta_{(N,=)}$ and $\Delta_{(N,\leq)}$ in [1] and [14].

We now present our new concentration lemma, which generalizes the result from [5]. As setting, suppose Σ is a set with n elements equipped with independent weights $\{X_a : a \in \Sigma\}$ with mean zero and $\mathbb{E}|X_a|^p \leq K < \infty$ for some $p > 1$. Let σ be a random subset of Σ with m elements chosen independently of X. We will specify the distribution of σ in applications. $\mathbb{P}_X$ and $\mathbb{P}_\sigma$ are the respective marginals. For a test function $f : \mathbb{R} \to \mathbb{R}$, we set

$$\int f d\mu_X = \mathbb{E}_\sigma f\left(m^{-1/\alpha} \sum_{a\in\sigma} X_a\right)$$

for some $0 < \alpha \leq \infty$. Also set $L = \left(\sum_{a\in\Sigma} \mathbb{P}_\sigma(a \in \sigma)^2\right)^{1/2}$. Finally, recall the 1-Wasserstein distance between probability measures μ and ν:

$$d_W(\mu, \nu) = \sup_{|f|_L \leq 1} \left| \int f d\mu - \int f d\nu \right|,$$

where $|f|_L$ is the Lipschitz constant of f.

Lemma 2.2 *In the setting described above, there exist absolute constants $C, c > 0$ such that for any s, t, R, $0 < \alpha \leq \infty$, any probability measure ν, and any convex 1-Lipschitz function $f : \mathbb{R} \to \mathbb{R}$:*

1. *If $1 \leq p < 2$, then*

$$\mathbb{P}_X\left[\left| \int f d\mu_X - \int f d\nu \right| \geq D + \frac{LKn}{m^{1/\alpha} R^{p-1}} + s + t\right] \leq \frac{LKn}{m^{1/\alpha} R^{p-1} s} + C\exp\left[-c\frac{m^{2/\alpha} t^2}{L^2 R^2}\right]$$

2. *If $\mathbb{E}X_a^2 = 1$ for all a and $2 \leq p$, then*

$$\mathbb{P}_X\left[\left| \int f d\mu_X - \int f d\nu \right| \geq D + \frac{L\sqrt{Kn}}{m^{1/\alpha} \sqrt{R^{p-2}}} + s + t\right] \leq \frac{L^2 Kn}{m^{2/\alpha} R^{p-2} s^2} + C\exp\left[-c\frac{m^{2/\alpha} t^2}{L^2 R^2}\right]$$

where $D = \max_{\sigma} d_W(\rho_\sigma, \nu)$ *and* ρ_σ *is the distribution of* $m^{-1/\alpha}\sum_{a\in\sigma} X_a$ *conditioned on* σ.

Proof First assume $\alpha < \infty$. We take the same approach as in the proof of the concentration lemma of [5]. For a fixed $R > 0$, we define the truncations $X_a^{(R)} = X_a \mathbb{1}_{|X_a|\le R}$ and denote the distribution of $m^{-1/\alpha}\sum_{a\in\sigma} X_a^{(R)}$ conditioned on X as $\mu_X^{(R)}$. We split the integral

$$\left|\int f d\mu_X - \int f d\nu\right| \le \left|\int f d\mu_X - \int f d\mu_X^{(R)}\right| \tag{2.1}$$

$$+\left|\int f d\mu_X^{(R)} - \mathbb{E}_X \int f d\mu_X^{(R)}\right| \tag{2.2}$$

$$+\left|\mathbb{E}_X \int f d\mu_X^{(R)} - \mathbb{E}_X \int f d\mu_X\right| \tag{2.3}$$

$$+\left|\mathbb{E}_X \int f d\mu_X - \int f d\nu\right|. \tag{2.4}$$

Each of these items can be bounded individually, either absolutely or with high probability. The easiest is (2.4), which is bounded absolutely by Fubini's theorem.

$$\begin{aligned}\left|\mathbb{E}_X \int f d\mu_X - \int f d\nu\right| =& \left|\mathbb{E}_X \mathbb{E}_\sigma f\left(m^{-1/\alpha}\sum_{a\in\sigma} X_a\right) - \int f d\nu\right| \\ \le& \mathbb{E}_\sigma \left|\mathbb{E}_X f\left(m^{-1/\alpha}\sum_{a\in\sigma} X_a\right) - \int f d\nu\right| \\ \le& \max_\sigma d_W(\rho_\sigma, \nu).\end{aligned}$$

(2.2) is next. As in [5], we can set

$$F(X) = \int f d\mu_X = \mathbb{E}_\sigma f\left(m^{-1/\alpha}\sum_{a\in\sigma} X_a\right),$$

which is convex and Lipschitz. For $X, X' \in \mathbb{R}^\Sigma$,

$$\begin{aligned}|F(X) - F(X')| \le& \mathbb{E}_\sigma \left| f\left(m^{-1/\alpha}\sum_{a\in\sigma} X_a\right) - f\left(m^{-1/\alpha}\sum_{a\in\sigma} X'_a\right)\right| \\ \le& m^{-1/\alpha}\mathbb{E}_\sigma \left|\sum_{a\in\sigma} X_a - \sum_{a\in\sigma} X'_a\right| \\ \le& m^{-1/\alpha}\mathbb{E}_\sigma \sum_{a\in\sigma} |X_a - X'_a| \\ \le& m^{-1/\alpha} L ||X - X'||_2,\end{aligned}$$

so F is $\frac{L}{m^{1/\alpha}}$-Lipschitz. Applying Theorem 2.1 gives

$$\mathbb{P}_X\left[\left|\int f d\mu_X^{(R)} - \mathbb{E}_X \int f d\mu_X^{(R)}\right| \geq t\right] \leq C \exp\left[-c\frac{m^{2/\alpha}t^2}{L^2R^2}\right].$$

For the other terms, the situation differs depending on case 1. or case 2. In either case, we use the Lipschitz estimate

$$\begin{aligned}\left|\int f d\mu_X - \int f d\mu_X^{(R)}\right| &\leq \frac{L}{m^{1/\alpha}}||X - X^{(R)}||_2 \\ &= \frac{L}{m^{1/\alpha}}\left(\sum_{a\in\Sigma} X_a^2 \mathbb{1}_{|X_a|>R}\right)^{1/2}.\end{aligned}$$

In case 2., we can use Hölder's and Chebyshev's inequalities to give

$$\begin{aligned}\mathbb{E}_X\left(\sum_{a\in\Sigma} X_a^2 \mathbb{1}_{|X_a|>R}\right)^{1/2} &\leq \left(\mathbb{E}_X|X_a|^p\right)^{2/p}\left(\mathbb{P}_X[|X_a| > R]\right)^{1-2/p} \\ &\leq \frac{\mathbb{E}_X|X_a|^p}{R^{p(1-2/p)}} \\ &\leq \frac{KL}{R^{p-2}}.\end{aligned}$$

By Markov's and Chebyshev's inequalities, then

$$\mathbb{P}_X\left[\sum_{a\in\Sigma} X_a^2 \mathbb{1}_{|X_a|>R} \geq u\right] \leq \frac{nK}{uR^{p-2}},$$

from which we have an absolute bound on (2.3)

$$\left|\mathbb{E}_X \int f d\mu_X - \mathbb{E}_X \int f d\mu_X^{(R)}\right| \leq \frac{L\sqrt{Kn}}{m^{1/\alpha}\sqrt{R^{p-2}}}$$

and a high-probability bound on (2.1)

$$\mathbb{P}_X\left[\left|\int f d\mu_X - \int f d\mu_X^{(R)}\right| \geq s\right] \leq \frac{nKL^2}{m^{2/\alpha}R^{p-2}s^2}.$$

Combining these bounds gives the result.

In case 1., the weights are not guaranteed to have finite variance, so we need to use

$$\left(\sum_{a\in\Sigma} X_a^2 \mathbb{1}_{|X_a|>R}\right)^{1/2} \le \sum_{a\in\Sigma} |X_a|\mathbb{1}_{|X_a|>R},$$

and again by Hölder's and Chebyshev's inequalities,

$$\begin{aligned}\mathbb{E}_X|X_a|\mathbb{1}_{|X_a|>R} \le& (\mathbb{E}_X|X_a|^p)^{1/p}\mathbb{P}_X(|X_a|>R)^{(p-1)/p}\\ \le& (\mathbb{E}_X|X_a|^p)^{1/p}\left(\frac{\mathbb{E}_X|X_a|^p}{R^p}\right)^{(p-1)/p}\\ =& \frac{\mathbb{E}_X|X_a|^p}{R^{p-1}}\\ =& \frac{K}{R^{p-1}}.\end{aligned}$$

From this result and Markov's inequality,

$$\mathbb{P}_X\left[\sum_{a\in\Sigma} |X_a|\mathbb{1}_{|X_a|>R} \ge u\right] \le \frac{Kn}{uR^{p-1}},$$

so that

$$\left|\mathbb{E}_X \int f d\mu_X - \mathbb{E}_X \int f d\mu_X^{(R)}\right| \le \frac{LKn}{m^{1/\alpha}R^{p-1}}$$

and

$$\mathbb{P}_X\left[\left|\int f d\mu_X - \int f d\mu_X^{(R)}\right| \ge s\right] \le \frac{LKn}{m^{1/\alpha}R^{p-1}s}.$$

For $\alpha = \infty$, this corresponds to the case of the sums not being rescaled at all. The result in this case follows by observing that removing the term $m^{-1/\alpha}$ altogether does not change the validity of the above argument. □

The next lemma covers the SGC and SEC cases. For this, we recall the Bounded Lipschitz distance between two probability measures μ and ν:

$$d_{BL}(\mu,\nu) = \sup_{|f|_{BL}\le 1}\left|\int f d\mu - \int f d\nu\right|,$$

where $|f|_{BL} = \max\{||f||_\infty, |f|_L\}$. It is worth noting that the bounded Lipschitz distance metrizes weak convergence.

Lemma 2.3 *Let the assumptions and notation be as in Lemma 2.2, except that X no longer needs to have independent components, f no longer needs to be convex, and $|f|_{BL} \le 1$. If for bounded Lipschitz functions X has, with constants C and c,...:*

1. SGC, then

$$\mathbb{P}_X\left[\left|\int f d\mu_X - \int f d\nu\right| > E + t\right] \leq C \exp\left(-c\frac{m^{2/\alpha}t^2}{L^2}\right)$$

2. SEC, then

$$\mathbb{P}_X\left[\left|\int f d\mu_X - \int f d\nu\right| > E + t\right] \leq C \exp\left(-c\frac{m^{1/\alpha}t}{L}\right)$$

where $E = \max_\sigma d_{BL}(\rho_\sigma, \nu)$ *and* ρ_σ *is the distribution of* $m^{-1/\alpha}\sum_{a\in\sigma} X_a$ *conditioned on* σ.

Proof The proof is the same as for Lemma 2.2 except that the truncation step is not necessary. We have

$$\left|\int f d\mu_X - \int f d\nu\right| \leq \left|\int f d\mu_X - \mathbb{E}_X \int f d\mu_X\right| + \left|\mathbb{E}_X \int f d\mu_X - \int f d\nu\right|.$$

The first term is bounded with high probability by SGC or SEC, and the second is bounded according to Fubini's theorem. Since we have specified $|f|_{BL} \leq 1$, we can use the Bounded Lipschitz distance instead of the Wasserstein distance. □

For a further extension, let us first recall the definition of stable distributions.

Definition 2.4 A random variable X has an **α-stable distribution** ν_α, $0 < \alpha < 2$, if its characteristic function $\phi_X(t) = \mathbb{E}e^{\iota X t} = e^{\psi(t)}$ where

$$\psi(t) = -\kappa^\alpha |t|^\alpha \left(1 - \iota\beta \operatorname{sign}(t) \tan\frac{\pi\alpha}{2}\right), \tag{2.5}$$

for some $\beta \in [-1, 1]$ and $\kappa > 0$. For a d-dimensional **α-stable random vector**, the characteristic exponent is

$$\psi(t) = -t \int_{S^{d-1}} \int_0^\infty d\lambda(\xi) \mathbb{1}_{B(r\xi,0)} e^{\iota tx} - 1 - \iota tx \mathbb{1}_{|t|<1} r^{-(\alpha+1)} dr d\lambda(\xi) \tag{2.6}$$

with λ a finite positive measure on S^{d-1}. The double integral can be rewritten as a single integral with respect to a measure called the Lévy measure.

Case 1. of Lemma 2.2 covers stable weights and weights in the domain of attraction of a stable distribution, but unfortunately, it is insufficient for proving convergence of any kind, nor are most concentration results for infinitely divisible vectors any better. The following is one useful result, however, from [11].

Theorem 2.5 *For* $\alpha > \frac{3}{2}$, *let* X *be a* d*-dimensional* α*-stable random vector with characteristic exponent as in Definition 2.4. For any* 1*-Lipschitz function* $f : \mathbb{R}^d \to \mathbb{R}$,

$$\mathbb{P}[|f(X) - \mathbb{E}f(X)| \geq x] \leq \frac{K\lambda(S^{d-1})}{x^{\alpha}}$$

for every x such that

$$x^{\alpha} \geq K_{\alpha}\lambda(S^{d-1})$$

where K is an absolute constant and K_{α} is a constant depending only on α.

Further concentration results can be found in [9], [10], and [11]. From this concentration inequality, we can prove a result leading to Theorem 1.4. Notice that here we distinguish between the index of stability for the vectors and the exponent used in rescaling the sum.

Lemma 2.6 *With the same assumptions and notation as in Lemma 2.2 except that X is an α'-stable random vector with $\alpha' > \frac{3}{2}$, $d = n$ and Lévy measure as in (2.6) and $\alpha < \infty$. For any 1-Lipschitz function $f : \mathbb{R} \to \mathbb{R}$,*

$$\mathbb{P}_X\left[\left|\int f d\mu_X - \int f d\nu\right| > D + t\right] \leq \frac{KL^{\alpha}\lambda(S^{d-1})}{mt^{\alpha}}$$

for any t satisfying

$$t^{\alpha'} \geq L^{\alpha}K_{\alpha'}\lambda(S^{d-1})m^{-1/\alpha}.$$

Proof The proof is the same as for Lemma 2.3 except using Theorem 2.5 instead of SGC. □

2.2 Convergence Conditions

In this section, we apply the concentration results of Lemmas 2.2, 2.3, and 2.6 to prove convergence results from which the main results will follow as simple corollaries. For this purpose, we assume an infinite family of random vectors X_N with entries indexed by finite index sets Σ_N, from which we take subsets $\sigma_N \subset \Sigma_N$, all in turn indexed by $N \in \mathbb{N}$.

We will assume here $n = n(N) = |\Sigma_N| \sim N^{\eta}$, $m = m(N) = |\sigma_N| \sim N^{\mu}$, and $L = L(N) \sim N^{\lambda}$ where $\eta > 0$, $\mu \geq 0$, and λ are constants. Our result establishes the necessary relationships between α, p, η, μ, and λ for convergence theorems. The truncation parameter R offers a measure of freedom, so we will assume $R(N) \sim N^{\rho}$, $\rho \geq 0$. Naturally, $\mu \leq \eta$.

Lemma 2.7 *Let the setting be as in Lemma 2.2 and described above. For $\alpha < \infty$, if we have $\lambda < \frac{\mu}{\alpha}$ and $p > \frac{\alpha\eta}{\mu-\alpha\lambda}$, then*

$$\mathbb{P}_{X_N}\left[\left|\int f d\mu_{X_N} - \int f d\nu\right| \geq D_N + o(1)\right] \to 0, \text{ as } N \to \infty,$$

where D_N is the value D in Lemma 2.2 for X_N, and if additionally $p > \frac{\alpha(\eta+1)}{\mu-\alpha\lambda}$, then

$$\sum_{N=1}^{\infty} \mathbb{P}_{X_N}\left[\left|\int f d\mu_{X_N} - \int f d\nu\right| > D_N + M_{N,i} + o(1)\right] < \infty,$$

where $i = 1, 2$ stands for the case in Lemma 2.2: in case 1., $M_{N,1} = \frac{L(N)Kn(N)}{m(N)^{1/\alpha}R(N)^{p-1}}$, and in case 2., $M_{N,2} = \frac{L(N)\sqrt{Kn(N)}}{m(N)^{1/\alpha}\sqrt{R(N)^{p-2}}}$.

The same results apply when $\alpha = \infty$ with moment conditions $p > \frac{\eta}{-\lambda}$ and $p > \frac{\eta+1}{-\lambda}$ provided $0 > \lambda$.

Proof In case 1., take $s = o(1)$, $t = o(1)$, and $R \sim N^{\rho}$. Using Lemma 2.2, the first convergence requires

$$\lambda + \eta - \frac{\mu}{\alpha} - \rho(p-1) < 0 \tag{2.7}$$

and

$$\frac{2\mu}{\alpha} - 2\lambda - 2\rho > 0. \tag{2.8}$$

The second condition implies $\lambda < \frac{\mu}{\alpha}$. Taking $\rho = \frac{\mu}{\alpha} - \lambda - \epsilon > 0$ with $\epsilon > 0$ gives $p > \frac{\alpha(\eta-\epsilon)}{\mu-\alpha\lambda-\alpha\epsilon}$. We pass to the limit $\epsilon \to 0^+$ to give the appropriate lower bound for p.

For summability, the first condition becomes

$$\lambda + \eta - \frac{\mu}{\alpha} - \rho(p-1) < -1, \tag{2.9}$$

which now gives $p > \frac{\alpha(\eta+1)}{\mu-\alpha\lambda}$.

In case 2., equation (2.8) is the same, but (2.7) and (2.9) are replaced with

$$2\lambda + \eta - \frac{2\mu}{\alpha} - \rho(p-2) < 0 \text{ or } -1,$$

respectively, which, interestingly, gives the same conditions as in case 1.

For $\alpha = \infty$, the same process gives the result, simply removing the m terms. The condition on λ is required so that $\rho > 0$. □

When Lemma 2.3 applies, convergence is easier.

Lemma 2.8 *Let the setting be as in Lemma 2.7 except that f need not be convex and $|f|_{BL} \leq 1$ and X_N has SGC or SEC for all N. If $\frac{\mu}{\alpha} > \lambda$, then*

$$\mathbb{P}_{X_N}\left[\left|\int f d\mu_{X_N} - \int f d\nu\right| \geq E_N + o(1)\right] \to 0, \text{ as } N \to \infty,$$

and

$$\sum_{N=1}^{\infty} \mathbb{P}_{X_N}\left[\left|\int f d\mu_{X_N} - \int f d\nu\right| > E_N + o(1)\right] < \infty.$$

Proof This follows from Lemma 2.3 in the same way as Lemma 2.7 follows from Lemma 2.2. □

Finally, we have a convergence result from Lemma 2.6.

Lemma 2.9 *Let the setting be as in Lemma 2.6 and the vectors X_N be α'-stable with $\alpha' > \frac{3}{2}$ and characteristic exponents as in Definition 2.4. Suppose further $\lambda_N(S^{n(N)-1}) = O(N^{-\tau})$, $\tau > 0$. If $\alpha\lambda - \tau - \mu < 0$, then*

$$\mathbb{P}_{X_N}\left[\left|\int f d\mu_{X_N} - \int f d\nu\right| \geq D_N + o(1)\right] \to 0, \text{ as } N \to \infty,$$

and if $\alpha\lambda - \tau - \mu < -1$, then

$$\sum_{N=1}^{\infty} \mathbb{P}_{X_N}\left[\left|\int f d\mu_{X_N} - \int f d\nu\right| > D_N + o(1)\right] < \infty.$$

Proof This follows from Lemma 2.6 in the same way that Lemma 2.7 follows from Lemma 2.3. □

3 Proofs of Main Results

Proving the main theorems requires only a little bit more than the results from the previous section. The first desideratum is for $\limsup_{N\to\infty} D_N = 0$ or $\limsup_{N\to\infty} E_N = 0$, which is guaranteed by appropriate choice of target measure ν. The second is a bound on λ, which we recall to be such that $L = L(N) \sim N^{\lambda}$. In Section 3 of [5], the authors prove the following:

Lemma 3.1 *In the corner-growth setting, in each of the three cases listed in Section 1, $\lambda < \frac{1}{4}$.*

This allows us to prove all our results in the corner-growth setting.

Proof of Theorems 1.2, 1.3, *and,* 1.4 Theorem 1.2 follows from Lemma 2.7, the Borel–Cantelli lemma, Lemma 3.1, and a central limit theorem along the lines of [4]. Notice that such a theorem requires bounded 3rd absolute moments, which are guaranteed by our conditions.

Theorem 1.3 for the sphere follows from Lemma 2.8 (on account of SGC), Lemma 3.1, and the following reasoning. For 1., follows from a result from [3]: the authors prove a convergence result in the total variation distance (Inequality (1))

$$d_{TV}(\mathcal{L}(X_\sigma), \mathcal{L}(Z)) \leq \frac{2(N+M+2)}{NM-N-M-2}, \tag{3.1}$$

where X_σ is the vector of weights indexed by σ and Z is a vector of i.i.d. standard Gaussian random variables. Since $d_{BL}(\mu, \nu) \leq d_{TV}(\mu, \nu)$, the triangle inequality gives the following:

$$\begin{aligned}\left|\int f d\mu_X - \int f d\nu\right| \leq & \left|\int f d\mu_X - \int f\Big((N+M-1)^{-1/2}\Big(\sum_{i=1}^{N+M-1} Z_i\Big)\Big) d\mathcal{L}(Z)\right| - \\ & \left|\int f\Big((N+M-1)^{-1/2}\Big(\sum_{i=1}^{N+M-1} Z_i\Big)\Big) d\mathcal{L}(Z) - \int f d\nu\right| \\ \leq & d_{BL}(\mathcal{L}(X_\sigma), \mathcal{L}(Z)) + d_{BL}(\mathcal{L}((N+M-1)^{-1/2}(Z_1+\cdots+Z_{N+M-1}), \nu),\end{aligned} \tag{3.2}$$

where $\nu \sim \mathcal{N}(0, 1)$. The second term on the right-hand side of (3.2) is 0, so (3.1) gives the result. For 2., the proof is the same except that we use (3.4) from [3], which gives a similar bound on the total variation distance between the coordinates of the simplex and a vector of i.i.d. 1-exponential random variables, and $d_{BL}(\mathcal{L}((N+M-1)^{-1/2}(x_1+\cdots+x_{N+M-1}), \nu) \to 0$ by the classical central limit theorem. Similar reasoning gives the result for 3.

Theorem 1.4 follows from Lemma 2.9, the Borel–Cantelli lemma, Lemma 3.1, and the following reasoning. As a preliminary, observe that $\sum_{k=1}^{N} k^{-\gamma} = O(N^{1-\gamma})$. Next, observe that by construction $(2N-1)^{-1/\gamma} \sum_{(i,j)\in\sigma} X_{(i,j)}$ has the same distribution, ν_N, regardless of σ, and that distribution converges to ν_∞, again by construction. To confirm $d_W(\nu_N, \nu_\infty)$ converges to 0, for any 1-Lipschitz function $f : \mathbb{R} \to \mathbb{R}$,

$$\begin{aligned}\left|\int f d\nu_N - \int f d\nu_\infty\right| \leq & \mathbb{E}||X_N - X_\infty|| \\ \leq & \mathbb{E}|X|\left(\lim_{n\to\infty}\sum_{k=1}^{n}\frac{k^{-\tau}}{n^{1/\gamma}} - \sum_{k=1}^{n}\frac{k^{-\tau}}{n^{1/\gamma}}\right),\end{aligned}$$

where X_N and X_∞ have distributions ν_N and ν_∞, respectively.

The final step is to show that under the specified conditions, $\lambda(S^{2N-2})$ satisfies Lemma 2.9. By construction,

$$\begin{aligned}\lambda(S^{2N-2}) =& 1+\sum_{n=2}^{N}\sum_{k=N}^{2N-2} 2k^{-\tau}+(2N-1)^{-\tau}\\ \leq& 2\sum_{n=1}^{N} N^{1-\tau}\\ \leq& 2N^{2-\tau},\end{aligned}$$

so the conditions satisfy Lemma 2.9. □

In the case of σ being uniformly distributed over subsets of size m, we can compute λ exactly.

Lemma 3.2 *When σ is chosen uniformly from subsets of size m, $\lambda = \mu - \frac{\eta}{2}$.*

Proof In this case, the probability that any particular element of Σ is in σ is $\frac{\binom{n-1}{m-1}}{\binom{n}{m}} = \frac{m}{n}$, so $L^2 = \frac{m^2}{n} \sim N^{2\mu-\eta}$. Thus, $\lambda = \mu - \frac{\eta}{2}$. □

Proof of Theorems 1.6 *and* 1.7 Theorem 1.6 follows from Lemma 2.7, the Borel–Cantelli lemma, and Lemma 3.2, taking $\alpha = \infty, \eta = 1$, and $\mu = 0$, since by construction $D = 0$.

Theorem 1.7 follows from Lemmas 2.8, 3.2, and the same reasoning as in the proof of Theorem 1.3. For the case of the simplex, recall that the m entries in the sum are approximately independent 1-exponential random variables, the sum of which has a Gamma distribution with shape parameter m and scale parameter 1. □

Another simple corollary of Lemma 3.2 vaguely connected to the corner-growth setting comes from taking $\alpha = \eta = 2$ and $\mu = 1$.

Corollary 3.3 *Let the setting be as in Lemma 2.4 with $\alpha = \eta = 2$ and $\mu = 1$, and σ is chosen uniformly from subsets of Σ of size m. If $p > 4$, then*

$$\frac{1}{\sqrt{M+N-1}}\sum_{(i,j)\in\sigma} X_{i,j} \xrightarrow{WIP} \mathcal{N}(0,1), \text{ as } N\to\infty,$$

and if $p > 6$, then

$$\frac{1}{\sqrt{M+N-1}}\sum_{(i,j)\in\sigma} X_{i,j} \xrightarrow{D} \mathcal{N}(0,1), \text{ as } N\to\infty,$$

$\mathbb{P}_X$-almost surely.

In the setting for the combinatorial central limit theorem, we again can compute λ exactly.

Lemma 3.4 *When X is an $N \times N$ rectangular array of numbers, so that $\Sigma = \{(i, j) : 1 \leq i, j \leq N\}$, with σ chosen uniformly from subsets each containing exactly one number from each row, $\lambda = 0$.*

Proof For fixed i, the probability that $(i, j) \in \sigma$ is exactly the same regardless of j, so it is $\frac{1}{N}$. Thus, $L^2(N) = \left(\frac{1}{N^2}\right)N^2 = 1$, so $\lambda = 0$. □

Proof of Theorem 1.5 Take $\alpha = \eta = 2$ and $\mu = 1$. The result follows from Lemma 2.7, the Borel–Cantelli lemma, and Lemma 3.4. □

Acknowledgments This chapter is part of the author's Ph.D. dissertation written under the supervision of Mark W. Meckes. The author would like to thank Professor Meckes for his helpful comments and discussions and the anonymous referee for suggestions that improved the exposition.

References

1. J. Arias-de-Reyna, R. Villa, The uniform concentration of measure phenomenon in ℓ_p^n $(1 \leq p \leq 2)$. *Geometric Aspects of Functional Analysis*. Lecture Notes in Mathematics, vol. 1745 (Springer, Berlin/Heidelberg, 2000), pp. 13–18
2. L.H.Y. Chen, X. Fang, On the error bound in a combinatorial central limit theorem. Bernoulli **21**(1), 335–359 (2015)
3. P. Diaconis, D. Freedman, A Dozen de Finetti-style results in search of a theory. Annales de l'IHP Probabilités et statistiques **23**(S2), 397–423 (1987)
4. C.-G. Esseen, On mean central limit theorems. Kungl. Tekn. Högsk. Handl. Stockholm **121**, 31 (1958)
5. H.C. Gromoll, M.W. Meckes, L. Petrov, Quenched central limit theorem in a corner growth setting. Electron. Commun. Probab. **23**(101), 1–12 (2018)
6. A. Guionnet, O. Zeitouni, Concentration of the spectral measure for large matrices. Electron. Commun. Probab. **5**, 119–136 (2000)
7. S.T. Ho, L.H.Y. Chen, An L_p bound for the remainder in a combinatorial central limit theorem. Ann. Probab. **6**, 231–249 (1978)
8. W. Hoeffding, A combinatorial central limit theorem. Ann. Math. Stat. **22**(4), 558–566 (1951)
9. C. Houdré, P. Marchal, Median, concentration and fluctuations for Lévy processes. Stoch. Process. Appl. **118**, 852–863 (2008)
10. C. Houdré, P. Marchal, On the concentration of measure phenomenon for stable and related random vectors. Ann. Probab. **32**(2), 1496–1508 (2004)
11. C. Houdré, Remarks on deivation inequalities for functions of infinitely divisible random vectors. Ann. Probab. **30**(3), 1223–1237 (2002)
12. M. Meckes, Some results on random circulant matrices, in *High Dimensional Probability V*, vol. 5 (Institute of Mathematical Statistics, Beachwood, OH, 2009), pp. 213–224
13. M. Meckes, S. Szarek, Concentration for noncommutative polynomials in random matrices. Proc. Am. Math. Soc. **140**(5), 1803–1813 (2012)
14. G. Schechtman, An editorial comment on the preceeding paper, in *Geometric Aspects of Functional Analysis*. Lecture Notes in Mathematics, vol. 1745 (Springer, Berlin/Heidelberg, 2000), pp. 19–20
15. G. Schechtman, Concentration results and applications, in *Handbook of the Geometry of Banach Spaces*, vol. 2 (North-Holland, Amsterdam, 2003), pp. 1603–1634
16. M. Talagrand, Concentration of measure and isoperimetric inequalities in product spaces. Publications Mathematiques de l'IHES **81**(1), 73–205 (1995)

A Note on Central Limit Theorems for Trimmed Subordinated Subordinators

David M. Mason

2020 Mathematics Subject Classification: 60F05; 60G51

1 Introduction

Ipsen et al [3] and Mason [7] have proved under general conditions that a trimmed subordinator satisfies a *self-standardized* central limit theorem [CLT]. One of their basic tools was a classic representation for subordinators (e.g., Rosiński [9]). Ipsen et al [3] used conditional characteristic function methods to prove their CLT, whereas Mason [7] applied a powerful normal approximation result for standardized infinitely divisible random variables by Zaitsev [12]. In this note, we shall examine self-standardized CLTs for trimmed subordinated subordinators. It turns out that there are two ways to trim a subordinated subordinator. One way leads to CLTs for the usual trimmed subordinator treated in [3] and [7], and a second way to a closely related *subordinated trimmed subordinator* and CLTs for it.

We begin by describing our setup and establishing some basic notation. Let $V = (V(t), t \geq 0)$ and $X = (X(t), t \geq 0)$ be independent 0 drift subordinators with Lévy measures Λ_V and Λ_X on $\mathbb{R}^+ = (0, \infty)$, respectively, with *tail function* $\overline{\Lambda}_V(x) = \Lambda_V((x, \infty))$, respectively, $\overline{\Lambda}_X(x) = \Lambda_X((x, \infty))$, defined for $x > 0$, satisfying

$$\overline{\Lambda}_V(0+) = \overline{\Lambda}_X(0+) = \infty. \tag{1}$$

For $u > 0$, let $\varphi_V(u) = \sup\{x : \overline{\Lambda}_V(x) > u\}$, where $\sup \varnothing := 0$. In the same way, define φ_X.

Remark 1 Observe that we always have

D. M. Mason (✉)
Department of Applied Economics and Statistics, University of Delaware, Newark, DE, USA
e-mail: davidm@udel.edu

R. Adamczak et al. (eds.), *High Dimensional Probability IX*, Progress in Probability 80, https://doi.org/10.1007/978-3-031-26979-0_9

$$\varphi_V(u) \to 0, \text{ as } u \to \infty.$$

Moreover, whenever $\overline{\Lambda}_V(0+) = \infty$, we have

$$\varphi_V(u) > 0 \text{ for all } u > 0.$$

For details, see Remark 1 of Mason [7]. The same statement holds for φ_X.

Recall that the Lévy measure Λ_V of a subordinator V satisfies

$$\int_0^1 x\Lambda_V(\mathrm{d}x) < \infty, \text{ equivalently, for all } y > 0, \int_y^\infty \varphi_V(x)\,\mathrm{d}x < \infty.$$

The subordinator V has Laplace transform defined for $t \geq 0$ by

$$E\exp(-\theta V(t)) = \exp(-t\Phi_V(\theta)),\ \theta \geq 0,$$

where

$$\Phi_V(\theta) = \int_0^\infty (1 - \exp(-\theta v))\,\Lambda_V(\mathrm{d}v),$$

which can be written after a change of variable to

$$= \int_0^\infty (1 - \exp(-\theta\varphi_V(u)))\,\mathrm{d}u.$$

In the same way, we define the Laplace transform of X.

Consider the subordinated subordinator process

$$W = (W(t) = V(X(t)), t \geq 0). \tag{2}$$

Applying Theorem 30.1 and Theorem 30.4 of Sato [11], we get that the process W is a 0 drift subordinator W with Lévy measure Λ_W defined for Borel subsets B of $(0, \infty)$ by

$$\Lambda_W(B) = \int_0^\infty P\{V(y) \in B\}\,\Lambda_X(\mathrm{d}y), \tag{3}$$

with Lévy tail function

$$\overline{\Lambda}_W(x) = \Lambda_W((x, \infty)),\ \text{for } x > 0.$$

Remark 2 Notice that (1) implies

$$\overline{\Lambda}_W(0+) = \infty.$$

To see this, we have by (3) that

$$\overline{\Lambda}_W(0+) = \lim_{n\to\infty}\int_0^\infty P\left\{V(y)\in\left(\frac{1}{n},\infty\right)\right\}\Lambda_X(dy).$$

Now $\overline{\Lambda}_V(0+) = \infty$ implies that for all $y > 0$, $P\{V(y)\in(0,\infty)\} = 1$. Hence by the monotone convergence theorem,

$$\lim_{n\to\infty}\int_0^\infty P\left\{V(y)\in\left(\frac{1}{n},\infty\right)\right\}\Lambda_X(dy) = \overline{\Lambda}_X(0+) = \infty.$$

For later use, we note that W has Laplace transform defined for $t \geq 0$ by

$$E\exp(-\theta W(t)) = \exp(-t\Phi_W(\theta)),\ \theta \geq 0,$$

where

$$\Phi_W(\theta) = \int_0^\infty \left(1-e^{-\theta x}\right)\Lambda_W(dx)$$

$$= \int_0^\infty\int_0^\infty \left(1-e^{-\theta x}\right)P(V(y)\in dx)\Lambda_X(dy)$$

$$= \int_0^\infty \left(1-e^{y\Phi_V(\theta)}\right)\Lambda_X(dy).$$

Definition 30.2 of Sato [11] calls the transformation of V into W given by $W(t) = V(X(t))$ subordination by the subordinator X, which is sometimes called the directing process.

2 Two Methods of Trimming W

In order to talk about trimming W, we must first discuss the ordered jump sequences of V, X, and W. For any $t > 0$, denote the ordered jump sequence $m_V^{(1)}(t) \geq m_V^{(2)}(t) \geq \cdots$ of V on the interval $[0,t]$. Let $\omega_1, \omega_2, \ldots$ be i.i.d. exponential random variables with parameter 1, and for each $n \geq 1$, let $\Gamma_n = \omega_1 + \ldots + \omega_n$. It is well-known that for each $t > 0$,

$$\left(m_V^{(r)}(t)\right)_{r\geq 1} \stackrel{D}{=} \left(\varphi_V\left(\frac{\Gamma_r}{t}\right)\right)_{r\geq 1}, \tag{4}$$

and hence for each $t > 0$,

$$V(t)=\sum_{r=1}^{\infty} m_V^{(r)}(t) \stackrel{\mathrm{D}}{=} \sum_{r=1}^{\infty} \varphi_V\left(\frac{\Gamma_r}{t}\right) =: \widetilde{V}(t). \tag{5}$$

See, for instance, equation (1.3) in IMR [3] and the references therein. It can also be inferred from a general representation for subordinators due to Rosiński [9].

In the same way, we define for each $t>0$, $\left(m_X^{(r)}(t)\right)_{r\geq 1}$ and $\left(m_W^{(r)}(t)\right)_{r\geq 1}$, and we see that the analogs of the distributional identity (4) hold with $m_V^{(r)}$ and φ_V replaced by $m_X^{(r)}$ and φ_X , respectively, $m_W^{(r)}$ and φ_W. Recalling (2), observe that for all $t>0$,

$$W(t)=\sum_{0<s\leq t} \Delta W(s)=V(X(t))=\sum_{0<s\leq X(t)} \Delta V(s). \tag{6}$$

From (6) and the version of (4) with $m_V^{(r)}$ and φ_V replaced by $m_W^{(r)}$ and φ_W, we have for each $t>0$

$$W(t)=\sum_{r=1}^{\infty} m_W^{(r)}(t) \stackrel{\mathrm{D}}{=} \sum_{r=1}^{\infty} \varphi_W\left(\frac{\Gamma_r}{t}\right) =: \widetilde{W}(t).$$

Let V, X and $(\Gamma_r)_{r\geq 1}$ be independent. In particular, V is independent of

$$\left\{\left(m_X^{(r)}(y)\right)_{r\geq 1}, y>0\right\} \text{ and } (\Gamma_r)_{r\geq 1}.$$

Next consider for each $t>0$

$$\left(m_V^{(r)}(X(t))\right)_{r\geq 1}.$$

Note that conditioned on $X(t)=y$

$$\left(m_V^{(r)}(X(t))\right)_{r\geq 1} \stackrel{\mathrm{D}}{=} \left(m_V^{(r)}(y)\right)_{r\geq 1}.$$

Therefore, using (4), we get for each $t>0$

$$\left(m_V^{(r)}(X(t))\right)_{r\geq 1} \stackrel{\mathrm{D}}{=} \left(\varphi_V\left(\frac{\Gamma_r}{X(t)}\right)\right)_{r\geq 1},$$

and thus by (5),

$$V(X(t))=\sum_{r=1}^{\infty} m_V^{(r)}(X(t)) \stackrel{\mathrm{D}}{=} \sum_{r=1}^{\infty} \varphi_V\left(\frac{\Gamma_r}{X(t)}\right) =: \widetilde{V}(X(t)).$$

Here are two methods of trimming $W(t) = V(X(t))$.

Method I For each $t > 0$, trim $W(t) = V(X(t))$ based on the ordered jumps of V on the interval $(0, X(t)]$. In this case, for each $t > 0$ and $k \geq 1$, define the kth trimmed version of $V(X(t))$

$$V^{(k)}(X(t)) := V(X(t)) - \sum_{r=1}^{k} m_V^{(r)}(X(t)),$$

which we will call the *subordinated trimmed subordinator process.* We note that

$$V^{(k)}(X(t)) \stackrel{\mathrm{D}}{=} \widetilde{V}(X(t)) - \sum_{r=1}^{k} \varphi_V\left(\frac{\Gamma_r}{X(t)}\right) =: \widetilde{V}^{(k)}(X(t)).$$

Method II For each $t > 0$, trim $W(t)$ based on the ordered jumps of W on the interval $(0, t]$. In this case, for each $t > 0$ and $k \geq 1$, define the kth trimmed version of $W(t)$

$$W^{(k)}(t) := W(t) - \sum_{r=1}^{k} m_W^{(r)}(t)$$

$$\stackrel{\mathrm{D}}{=} \widetilde{W}(t) - \sum_{r=1}^{k} \varphi_W\left(\frac{\Gamma_r}{t}\right) =: \widetilde{W}^{(k)}(t).$$

Remark 3 Notice that in method I trimming for each $t > 0$, we treat $V(X(t))$ as the subordinator V randomly evaluated at $X(t)$, whereas in method II trimming we consider $W = V(X)$ as the *subordinator*, which results when the subordinator V is randomly time changed by the subordinator X.

Remark 4 Though for each $t > 0$, $V(X(t)) = W(t)$, typically we cannot conclude that for each $t > 0$ and $k \geq 1$

$$V^{(k)}(X(t)) \stackrel{\mathrm{D}}{=} W^{(k)}(t).$$

This is because it is not necessarily true that

$$\left(m_V^{(r)}(X(t))\right)_{r\geq 1} \stackrel{\mathrm{D}}{=} \left(m_W^{(r)}(t)\right)_{r\geq 1}.$$

See the example in Appendix 1.

3 Self-Standardized CLTs for W

3.1 Self-Standardized CLTs for Method I Trimming

Set $V^{(0)}(t) := V(t)$, and for any integer $k \geq 1$, consider the trimmed subordinator

$$V^{(k)}(t) := V(t) - m_V^{(1)}(t) - \cdots - m_V^{(k)}(t),$$

which on account of (4) says for any integer $k \geq 0$ and $t > 0$

$$V^{(k)}(t) \stackrel{\mathrm{D}}{=} \sum_{i=k+1}^{\infty} \varphi_V\left(\frac{\Gamma_i}{t}\right) =: \widetilde{V}^{(k)}(t). \tag{7}$$

Let T be a strictly positive random variable independent of

$$\left\{\left(m_V^{(r)}(t)\right)_{r\geq 1}, t > 0\right\} \text{ and } (\Gamma_r)_{r\geq 1}. \tag{8}$$

Clearly, by (4), (7), and (8), we have for any integer $k \geq 0$

$$V^{(k)}(T) \stackrel{\mathrm{D}}{=} \widetilde{V}^{(k)}(T).$$

Set for any $y > 0$

$$\mu_V(y) := \int_y^{\infty} \varphi_V(x)\,\mathrm{d}x \text{ and } \sigma_V^2(y) := \int_y^{\infty} \varphi_V^2(x)\,\mathrm{d}x.$$

We see by Remark 1 that (1) implies that

$$\sigma_V^2(y) > 0 \text{ for all } y > 0.$$

Throughout these notes, Z denotes a standard normal random variable. We shall need the following formal extension of Theorem 1 of Mason [7]. Its proof is nearly exactly the same as the proof of the Mason [7] version, and just replace the sequence of positive constants $\{t_n\}_{n\geq 1}$ in the proof of Theorem 1 of Mason [7] by $\{T_n\}_{n\geq 1}$. The proof of Theorem 1 of Mason [7] is based on a special case of Theorem 1.2 of Zaitsev [12], which we state in the digression below. Here is our self-standardized CLT for method I trimmed subordinated subordinators.

Theorem 1 *Assume that* $\overline{\Lambda}_V(0+) = \infty$. *For any sequence of positive integers* $\{k_n\}_{n\geq 1}$ *and sequence of strictly positive random variables* $\{T_n\}_{n\geq 1}$ *independent of* $(\Gamma_k)_{k\geq 1}$ *satisfying*

$$\frac{\sqrt{T_n}\sigma_V\left(\Gamma_{k_n}/T_n\right)}{\varphi_V\left(\Gamma_{k_n}/T_n\right)} \xrightarrow{\mathrm{P}} \infty, \text{ as } n \to \infty,$$

we have uniformly in x, as $n \to \infty$,

$$\left| P\left\{ \frac{\widetilde{V}^{(k_n)}(T_n) - T_n\mu_V\left(\Gamma_{k_n}/T_n\right)}{\sqrt{T_n}\sigma_V\left(\Gamma_{k_n}/T_n\right)} \le x|\Gamma_{k_n}, T_n \right\} - P\{Z \le x\} \right| \xrightarrow{\mathrm{P}} 0,$$

which implies as $n \to \infty$

$$\frac{\widetilde{V}^{(k_n)}(T_n) - T_n\mu_V\left(\Gamma_{k_n}/T_n\right)}{\sqrt{T_n}\sigma_V\left(\Gamma_{k_n}/T_n\right)} \xrightarrow{\mathrm{D}} Z. \tag{9}$$

The remainder of this subsection will be devoted to examining a couple of special cases of the following example of Theorem 1.

Example For each $0 < \alpha < 1$, let $V_\alpha = (V_\alpha(t), t \ge 0)$ be an α-stable process with Laplace transform defined for $\theta > 0$ by

$$E\exp(-\theta V_\alpha(t)) = \exp\left(-t\int_0^\infty (1-\exp(-\theta x))\,\alpha\Gamma(1-\alpha)\,x^{-1-\alpha}\mathrm{d}x\right)$$

$$= \exp\left(-t\int_0^\infty \left(1-\exp(-\theta c_\alpha u^{-1/\alpha})\right)\mathrm{d}u\right) = \exp\left(-t\theta^\alpha\right), \tag{10}$$

where

$$c_\alpha = 1/\Gamma^{1/\alpha}(1-\alpha).$$

(See Example 24.12 of Sato [11].) Note that for V_α,

$$\varphi V_\alpha(x) =: \varphi_\alpha(x) = c_\alpha x^{-1/\alpha}\mathbf{1}_{\{x>0\}}.$$

We record that for each $t > 0$

$$V_\alpha(t) \stackrel{\mathrm{D}}{=} \widetilde{V}_\alpha(t) := c_\alpha \sum_{i=1}^\infty \left(\frac{\Gamma_i}{t}\right)^{-1/\alpha}. \tag{11}$$

For any $t > 0$, denote the ordered jump sequence $m_\alpha^{(1)}(t) \ge m_\alpha^{(2)}(t) \ge \ldots$ of V_α on the interval $[0, t]$. Consider the kth trimmed version of $V_\alpha(t)$ defined for each integer $k \ge 1$

$$V_\alpha^{(k)}(t) = V_\alpha(t) - m_\alpha^{(1)}(t) - \cdots - m_\alpha^{(k)}(t), \tag{12}$$

which for each $t > 0$

$$\stackrel{\mathrm{D}}{=} \widetilde{V}_{\alpha}^{(k)}(t) := c_{\alpha} \sum_{i=1}^{\infty} \left(\frac{\Gamma_{k+i}}{t} \right)^{-1/\alpha}. \tag{13}$$

In this example, for ease of notation, write for each $0 < \alpha < 1$ and $y > 0$, $\mu_{V_\alpha}(y) = \mu_\alpha(y)$ and $\sigma^2_{V_\alpha}(y) = \sigma^2_\alpha(y)$. With this notation, we get that

$$\mu_\alpha(y) = \int_y^\infty c_\alpha v^{-1/\alpha} \mathrm{d}v = \frac{c_\alpha \alpha}{1-\alpha} y^{1-1/\alpha}$$

and

$$\sigma^2_\alpha(y) = \int_y^\infty c^2_\alpha v^{-2/\alpha} \mathrm{d}v = \frac{c^2_\alpha \alpha}{2-\alpha} y^{1-2/\alpha}.$$

From (13), we have that for any $k \geq 1$ and $T > 0$

$$\frac{\widetilde{V}_{\alpha}^{(k)}(T) - T\mu_\alpha\left(\frac{\Gamma_k}{T}\right)}{T^{1/2}\sigma_\alpha\left(\frac{\Gamma_k}{T}\right)} = \frac{\sum_{i=1}^{\infty} (\Gamma_{k+i})^{-1/\alpha} - \frac{\alpha}{1-\alpha}\Gamma_k^{1-1/\alpha}}{\sqrt{\frac{\alpha}{2-\alpha}}\Gamma_k^{1/2-1/\alpha}}. \tag{14}$$

Notice that

$$\frac{\sqrt{T}\sigma_\alpha\left(\frac{\Gamma_k}{T}\right)}{\varphi_\alpha(\frac{\Gamma_k}{T})} = (\Gamma_k)^{1/2}\sqrt{\frac{\alpha}{2-\alpha}}. \tag{15}$$

Clearly by (15) for any sequence of positive integers $\{k_n\}_{n\geq 1}$ converging to infinity and sequence of strictly positive random variables $\{T_n\}_{n\geq 1}$ independent of $(\Gamma_k)_{k\geq 1}$,

$$\frac{\sqrt{T_n}\sigma_\alpha\left(\Gamma_{k_n}/T_n\right)}{\varphi_\alpha\left(\Gamma_{k_n}/T_n\right)} = \left(\Gamma_{k_n}\right)^{1/2}\sqrt{\frac{\alpha}{2-\alpha}} \stackrel{\mathrm{P}}{\to} \infty, \text{ as } n \to \infty.$$

Hence, by rewriting (9) in the above notation, we have by Theorem 1 that as $n \to \infty$

$$\frac{\widetilde{V}_{\alpha}^{(k_n)}(T_n) - T_n\mu_\alpha\left(\frac{\Gamma_{k_n}}{T_n}\right)}{T_n^{1/2}\sigma_\alpha\left(\frac{\Gamma_{k_n}}{T_n}\right)} \stackrel{\mathrm{D}}{\to} Z. \tag{16}$$

Digression To make the presentation of our Example more self-contained, we shall show in this digression how a special case of Theorem 1.2 of Zaitsev [12] can be used to give a direct proof of (16).

It is pointed out in Mason [7] that Theorem 1.2 of Zaitsev [12] implies the following normal approximation. Let Y be an infinitely divisible mean 0 and variance 1 random variable with Lévy measure Λ and Z be a standard normal random variable. Assume that the support of Λ is contained in a closed interval $[-\tau, \tau]$ with $\tau > 0$; then for universal positive constants C_1 and C_2 for any $\lambda > 0$ all $x \in \mathbb{R}$

$$P\{Z \leq x - \lambda\} - C_1 \exp\left(-\frac{\lambda}{C_2 \tau}\right) \leq P\{Y \leq x\}$$

$$\leq P\{Z \leq x + \lambda\} + C_1 \exp\left(-\frac{\lambda}{C_2 \tau}\right). \tag{17}$$

We shall show how to derive (16) from (17). Note that

$$\frac{\sum_{i=1}^{\infty} (\Gamma_{k+i})^{-1/\alpha} - \frac{\alpha}{1-\alpha}\Gamma_k^{1-1/\alpha}}{\sqrt{\frac{\alpha}{2-\alpha}}\Gamma_k^{1/2-1/\alpha}} \stackrel{D}{=} \frac{\sum_{i=1}^{\infty}\left(1+\frac{\Gamma_i'}{\Gamma_k}\right)^{-1/\alpha} - \frac{\alpha}{1-\alpha}\Gamma_k}{\sqrt{\frac{\alpha}{2-\alpha}}\Gamma_k^{1/2}}, \tag{18}$$

where $\left(\Gamma_i'\right)_{i\geq 1} \stackrel{D}{=} (\Gamma_i)_{i\geq 1}$ and is independent of $(\Gamma_i)_{i\geq 1}$. Let $Y_\alpha = (Y_\alpha(y), y \geq 0)$ be the subordinator with Laplace transform defined for each $y > 0$ and $\theta \geq 0$, by

$$E \exp(-\theta Y_\alpha(y)) = \exp\left(-y\int_0^1 (1-\exp(-\theta x))\,\alpha x^{-\alpha-1}\mathrm{d}x\right)$$

$$=: \exp\left(-y\int_0^1 (1-\exp(-\theta x))\,\Lambda_\alpha(\mathrm{d}x)\right). \tag{19}$$

Observe that the Lévy measure Λ_α of Y_α has Lévy tail function on $(0, \infty)$

$$\overline{\Lambda}_\alpha(x) = \left(x^{-\alpha} - 1\right)\mathbf{1}_{\{0<x\leq 1\}}$$

with φ function

$$\varphi_{Y_\alpha}(u) = (1+u)^{-1/\alpha}\,\mathbf{1}_{\{u>0\}}.$$

Thus from (5), for each $y > 0$,

$$Y_\alpha(y) \stackrel{D}{=} \sum_{i=1}^{\infty}\left(1+\frac{\Gamma_i'}{y}\right)^{-1/\alpha}.$$

Also, we find by differentiating the Laplace transform of $Y_\alpha(y)$ that for each $y > 0$

$$EY_\alpha(y) = \frac{\alpha y}{1-\alpha} =: \beta_\alpha y \text{ and } Var Y_\alpha(y) = \frac{\alpha y}{2-\alpha} =: \gamma_\alpha^2 y, \tag{20}$$

and hence,

$$Z_\alpha(y) := \frac{Y_\alpha(y) - \beta_\alpha y}{\gamma_\alpha \sqrt{y}}$$

is a mean 0 and variance 1 infinitely divisible random variable whose Lévy measure has support contained in the closed interval $[-\tau(y), \tau(y)]$, where

$$\tau(y) = 1/\left(\gamma_\alpha \sqrt{y}\right). \tag{21}$$

Thus by (17) for universal positive constants C_1 and C_2 for any $\lambda > 0$ all $x \in \mathbb{R}$ and $\lambda > 0$,

$$P\{Z \le x - \lambda\} - C_1 \exp\left(-\frac{\lambda}{C_2 \tau(y)}\right) \le P\{Z_\alpha(y) \le x\}$$

$$\le P\{Z \le x + \lambda\} + C_1 \exp\left(-\frac{\lambda}{C_2 \tau(y)}\right). \tag{22}$$

Clearly, since $\left(\Gamma_i'\right)_{i \ge 1} \stackrel{D}{=} (\Gamma_i)_{i \ge 1}$ and $\left(\Gamma_i'\right)_{i \ge 1}$ is independent of $\left(\Gamma_{k_n}\right)_{n \ge 1}$, we conclude by (22) and (21) that

$$P\{Z \le x - \lambda\} - C_1 \exp\left(-\frac{\lambda \gamma_\alpha \sqrt{\Gamma_{k_n}}}{C_2}\right) \le P\left\{Z_\alpha\left(\Gamma_{k_n}\right) \le x | \Gamma_{k_n}\right\}$$

$$\le P\{Z \le x + \lambda\} + C_1 \exp\left(-\frac{\lambda \gamma_\alpha \sqrt{\Gamma_{k_n}}}{C_2}\right). \tag{23}$$

Now by the arbitrary choice of $\lambda > 0$, we get from (23) that uniformly in x, as $k_n \to \infty$,

$$\left| P\left\{\frac{Y_\alpha\left(\Gamma_{k_n}\right) - \beta_\alpha \Gamma_{k_n}}{\gamma_\alpha \sqrt{\Gamma_{k_n}}} \le x | \Gamma_{k_n}\right\} - P\{Z \le x\} \right| \stackrel{P}{\to} 0.$$

This implies as $n \to \infty$

$$\frac{Y_\alpha\left(\Gamma_{k_n}\right) - \beta_\alpha \Gamma_{k_n}}{\gamma_\alpha \sqrt{\Gamma_{k_n}}} \stackrel{D}{\to} Z. \tag{24}$$

Since the identity (14) holds for any $k \ge 1$ and $T > 0$, (16) follows from (18) and (24). Of course, there are other ways to establish (24). For instance, (24) can be shown to be a consequence of Anscombe's Theorem for Lévy processes. For details, see Appendix 2.

Remark 5 For any $0 < \alpha < 1$ and $k \geq 1$, the random variable $Y_\alpha(\Gamma_k)$ has Laplace transform

$$E \exp(-\theta Y_\alpha(\Gamma_k)) = \left(1 + \int_0^1 (1 - \exp(-\theta x))\, \Lambda_\alpha(\mathrm{d}x)\right)^{-k}, \theta \geq 0.$$

It turns out that for any $t > 0$

$$Y_\alpha(\Gamma_k) \stackrel{\mathrm{D}}{=} V_\alpha^{(k)}(t) / m_\alpha^{(k)}(t),$$

where $V_\alpha^{(k)}(t)$ and $m_\alpha^{(k)}(t)$ are as in (12). See Theorem 1.1 (i) of Kevei and Mason [6]. Also refer to page 1979 of Ipsen et al [4].

Next we give two special cases of our example, which we shall return to in the next subsection when we discuss self-standardized CLTs for method II trimming.

Special Case 1: Subordination of Two Independent Stable Subordinators
For $0 < \alpha_1, \alpha_2 < 1$, let V_{α_1}, respectively V_{α_2}, be an α_1-stable process, respectively an α_2-stable process, with a Laplace transform of the form (10). Assume that V_{α_1} and V_{α_2} are independent. Set for $t \geq 0$

$$W(t) = V_{\alpha_1}\left(V_{\alpha_2}(t)\right)$$

and

$$W = (W(t), t \geq 0).$$

One finds that for each $t \geq 0$

$$W(t) = V_{\alpha_1}\left(V_{\alpha_2}(t)\right) = \sum_{0<s\leq V_{\alpha_2}(t)} \Delta V_{\alpha_1}(s).$$

Moreover, W is a stationary independent increment process, and for each $t \geq 0$ and $\theta \geq 0$,

$$\begin{aligned} E \exp(-\theta W(t)) &= E \exp\left(-V_{\alpha_2}(t)\, \theta^{\alpha_1}\right) \\ &= \exp\left(-t\theta^{\alpha_1 \alpha_2}\right). \end{aligned} \tag{25}$$

This says that W is the $\alpha_1\alpha_2$-stable subordinator $V_{\alpha_1\alpha_2}$ with Laplace transform (25). (See Example 30.5 on page 202 of Sato [11].) Thus for each $t \geq 0$ and $\theta \geq 0$,

$$E \exp(-\theta W(t)) = E \exp\left(-\theta V_{\alpha_1\alpha_2}(t)\right). \tag{26}$$

Therefore, with $c(\alpha_1\alpha_2) = \frac{1}{\Gamma^{1/(\alpha_1\alpha_2)}(1-\alpha_1\alpha_2)}$, we get

$$c(\alpha_1\alpha_2)\sum_{i=1}^{\infty}\left(\frac{\Gamma_i}{t}\right)^{-1/(\alpha_1\alpha_2)} =: \widetilde{V}_{\alpha_1\alpha_2}(t),$$

which by (11), (25), and (26) for each fixed $t > 0$ is

$$\stackrel{\mathrm{D}}{=} V_{\alpha_1}\left(V_{\alpha_2}(t)\right).$$

Here we get that for any sequence of positive integers $\{k_n\}_{n\geq 1}$ converging to infinity and sequence of positive constants $\{s_n\}_{n\geq 1}$, by setting $T_n = V_{\alpha_2}(s_n)$, for $n \geq 1$, we have by (16) that as $n \to \infty$

$$\frac{\widetilde{V}_{\alpha_1}^{(k_n)}\left(V_{\alpha_2}(s_n)\right) - V_{\alpha_2}(s_n)\,\mu_{\alpha_1}\left(\frac{\Gamma_{k_n}}{V_{\alpha_2}(s_n)}\right)}{\sqrt{V_{\alpha_2}(s_n)}\sigma_{\alpha_1}\left(\frac{\Gamma_{k_n}}{V_{\alpha_2}(s_n)}\right)} \stackrel{\mathrm{D}}{\to} Z.$$

Special Case 2: Mittag-Leffler Process
For each $0 < \alpha < 1$, let V_α be the α-stable process with Laplace transform (10). Now independent of V, let $X = (X(s),\ s \geq 0)$ be the standard Gamma process, i.e., X is a zero drift subordinator with density for each $s > 0$

$$f_{X(s)}(x) = \frac{1}{\Gamma(s)}x^{s-1}e^{-x}, \text{ for } x > 0,$$

mean and variance

$$EX(s) = s \text{ and } VarX(s) = s,$$

and Laplace transform for $\theta \geq 0$

$$E\exp(-\theta X(s)) = (1+\theta)^{-s},$$

which after a little computation is

$$= \exp\left[-s\int_0^\infty (1-\exp(-\theta x))\,x^{-1}e^{-x}\mathrm{d}x\right].$$

Notice that X has Lévy density

$$\rho(x) = x^{-1}e^{-x}, \text{ for } x > 0.$$

(See Applebaum [1] pages 54–55.)

Consider the subordinated process

$$W = (W(s) := V_\alpha(X(s)), s \geq 0).$$

Applying Theorem 30.1 and Theorem 30.4 of Sato [11], we see that W is a drift 0 subordinator with Laplace transform

$$E\exp(-\theta W(s)) = E\exp(-V(X(s)))$$

$$= E\exp\left(-X(s)\theta^\alpha\right) = \left(1+\theta^\alpha\right)^{-s}$$

$$= \exp\left[-s\int_0^\infty \left(1-\exp\left(-\theta^\alpha y\right)\right) y^{-1}e^{-y}\mathrm{d}y\right],\ \theta \geq 0.$$

It has Lévy measure Λ_W defined for Borel subsets B of $(0, \infty)$, by

$$\Lambda_W(B) = \int_0^\infty P\{V_\alpha(y) \in B\} y^{-1}e^{-y}\mathrm{d}y.$$

In particular, it has Lévy tail function

$$\overline{\Lambda}_W(x) = \int_0^\infty P\{V(y) \in (x, \infty)\} y^{-1}e^{-y}\mathrm{d}y, \text{ for } x > 0.$$

For later use, we note that

$$\int_0^\infty \left(1-e^{-\theta x}\right)\Lambda_W(\mathrm{d}x) = \int_0^\infty\int_0^\infty \left(1-e^{-\theta x}\right)P_{V_\alpha(y)}(\mathrm{d}x)\,ay^{-1}e^{-by}\mathrm{d}y$$

$$= \int_0^\infty \left(1-e^{y\theta^\alpha}\right)y^{-1}e^{-y}\mathrm{d}y.$$

Such a process W is called the Mittag-Leffler process. See, e.g., Pillai [8].

By Theorem 4.3 of Pillai [8] for each $s > 0$, the exact distribution function $F_{\alpha,s}(x)$ of $W(s)$ is for $x \geq 0$

$$F_{\alpha,s}(x) = \sum_{r=0}^\infty (-1)^r \frac{\Gamma(s+r)\,x^{\alpha(s+r)}}{\Gamma(s)\,r!\Gamma(1+\alpha(s+r))},$$

which says that for each $s > 0$ and $x \geq 0$

$$P\{W(s) \leq x\} = P\{V_\alpha(X(s)) \leq x\}$$

$$= P\left\{\widetilde{V}_\alpha(X(s)) \leq x\right\} = F_{\alpha,s}(x).$$

In this special case, for any sequence of positive integers $\{k_n\}_{n\geq 1}$ converging to infinity and sequence of positive constants $\{s_n\}_{n\geq 1}$, by setting $T_n = X(s_n)$, for $n \geq 1$, we get by (16) that as $n \to \infty$

$$\frac{\widetilde{V}_{\alpha}^{(k_n)}\left(X\left(s_n\right)\right) - X\left(s_n\right)\mu_{\alpha}\left(\Gamma_{k_n}/X\left(s_n\right)\right)}{\sqrt{X\left(s_n\right)}\sigma_{\alpha}\left(\Gamma_{k_n}/X\left(s_n\right)\right)} \xrightarrow{\mathrm{D}} Z.$$

3.2 Self-Standardized CLTs for Method II Trimming

Let W be a subordinator of the form (2). Set for any $y > 0$

$$\mu_W(y) := \int_y^{\infty} \varphi_W(x)\,\mathrm{d}x \text{ and } \sigma_W^2(y) := \int_y^{\infty} \varphi_W^2(x)\,\mathrm{d}x.$$

We see by Remarks 1 and 2 that (1) implies that

$$\sigma_W^2(y) > 0 \text{ for all } y > 0.$$

For easy reference for the reader, we state here a version of Theorem 1 of Mason [7] stated in terms of a self-standardized CLT for the method II trimmed subordinated subordinator W.

Theorem 2 *Assume that $\overline{\Lambda}_W(0+) = \infty$. For any sequence of positive integers $\{k_n\}_{n\geq 1}$ and sequence of positive constants $\{t_n\}_{n\geq 1}$ satisfying*

$$\frac{\sqrt{t_n}\sigma_W\left(\Gamma_{k_n}/t_n\right)}{\varphi_W\left(\Gamma_{k_n}/t_n\right)} \xrightarrow{\mathrm{P}} \infty, \text{ as } n \to \infty,$$

we have uniformly in x, as $n \to \infty$,

$$\left| P\left\{ \frac{\widetilde{W}^{(k_n)}(t_n) - t_n\mu_W\left(\Gamma_{k_n}/t_n\right)}{\sqrt{t_n}\sigma_W\left(\Gamma_{k_n}/t_n\right)} \leq x|\Gamma_{k_n} \right\} - P\{Z \leq x\} \right| \xrightarrow{\mathrm{P}} 0,$$

which implies as $n \to \infty$

$$\frac{\widetilde{W}^{(k_n)}(t_n) - t_n\mu_W\left(\Gamma_{k_n}/t_n\right)}{\sqrt{t_n}\sigma_W\left(\Gamma_{k_n}/t_n\right)} \xrightarrow{\mathrm{D}} Z.$$

Remark 6 Theorem 1 of Mason [7] contains the added assumption that $k_n \to \infty$, as $n \to \infty$. An examination of its proof shows that this assumption is unnecessary. Also we note in passing that Theorem 1 implies Theorem 2.

For the convenience of the reader, we state the following results. Corollary 1 is from Mason [7]. The proof of Corollary 2 follows after some obvious changes of notation that of Corollary 1.

Corollary 1 *Assume that $W(t)$, $t \geq 0$, is a subordinator with drift 0, whose Lévy tail function $\overline{\Lambda}_W$ is regularly varying at zero with index $-\alpha$, where $0 < \alpha < 1$. For any sequence of positive integers $\{k_n\}_{n\geq 1}$ converging to infinity and sequence of positive constants $\{t_n\}_{n\geq 1}$ satisfying $k_n/t_n \to \infty$, we have, as $n \to \infty$,*

$$\frac{\widetilde{W}^{(k_n)}(t_n) - t_n \mu_W(k_n/t_n)}{\sqrt{t_n}\sigma_W(k_n/t_n)} \xrightarrow{\text{D}} \sqrt{\frac{2}{\alpha}} Z. \tag{27}$$

Corollary 2 *Assume that $W(t)$, $t \geq 0$, is a subordinator with drift 0, whose Lévy tail function $\overline{\Lambda}_W$ is regularly varying at infinity with index $-\alpha$, where $0 < \alpha < 1$. For any sequence of positive integers $\{k_n\}_{n\geq 1}$ converging to infinity and sequence of positive constants $\{t_n\}_{n\geq 1}$ satisfying $k_n/t_n \to 0$, as $n \to \infty$, we have (27).*

The subordinated subordinator introduced in Special Case 1 above satisfies the conditions of Corollary 1, and the subordinated subordinator in Special Case 2 above fulfills the conditions of Corollary 2. Consider the two cases.

Special Case 1 To see this, notice that in Special Case 1, by (25) necessarily W has Lévy tail function on $(0, \infty)$

$$\overline{\Lambda}_W(y) = \Gamma(1 - \alpha_1\alpha_2)\, y^{-\alpha_1\alpha_2} \mathbf{1}_{\{y>0\}},$$

for $0 < \alpha_1, \alpha_2 < 1$, which is regularly varying at zero with index $-\alpha$, where $0 < \alpha = \alpha_1\alpha_2 < 1$. In this case, from Corollary 1, we get (27) as long as $k_n \to \infty$ and $k_n/t_n \to \infty$, as $n \to \infty$.

Special Case 2 In Special Case 2, observe that $W = V_\alpha(X)$, with $0 < \alpha < 1$, where $V_\alpha = (V_\alpha(t), t \geq 0)$ is an α-stable process with Laplace transform (10), $X = (X(s), s \geq 0)$ is a standard Gamma process, and V_α and X are independent. The process $r^{-1/\alpha} W(r)$ has Laplace transform $(1 + \theta^\alpha/r)^{-r}$, for $\theta \geq 0$, which converges to $\exp(-\theta^\alpha)$ as $r \to \infty$. This implies that for all $t > 0$

$$r^{-1/\alpha} W(rt) \xrightarrow{\text{D}} V_\alpha(t), \text{ as } r \to \infty.$$

By part (ii) of Theorem 15.14 of Kallenberg [5] and (10) for all $x > 0$

$$r\overline{\Lambda}_W\left(r^{1/\alpha}x\right) \to \Gamma(1-\alpha)\, x^{-\alpha}, \text{ as } r \to \infty.$$

This implies that W has a Lévy tail function $\overline{\Lambda}_W(y)$ on $(0, \infty)$, which is regularly varying at infinity with index $-\alpha$, $0 < \alpha < 1$. In this case, by Corollary 2, we can conclude (27) as long as $k_n \to \infty$ and $k_n/t_n \to 0$, as $n \to \infty$.

4 Appendix 1

Recall the notation of Special Case 1. Let V_{α_1}, V_{α_2}, and $(\Gamma_k)_{k\geq 1}$ be independent and $W = V_{\alpha_1}\left(V_{\alpha_2}\right)$. For any $t > 0$, let $m_{V_{\alpha_1}}^{(1)}(V_{\alpha_2}(t)) \geq m_{V_{\alpha_1}}^{(2)}(V_{\alpha_2}(t)) \geq \cdots$ denote the ordered jumps of V_{α_1} on the interval $\left[0, V_{\alpha_2}(t)\right]$. They satisfy

$$\left(m_{\alpha_1}^{(k)}(V_{\alpha_2}(t))\right)_{k\geq 1} \stackrel{\mathrm{D}}{=} \left(c(\alpha_1)\left(\frac{\Gamma_k}{V_{\alpha_2}(t)}\right)^{-1/\alpha_1}\right)_{k\geq 1}.$$

Let $m_W^{(1)}(t) \geq m_W^{(2)}(t) \geq \cdots$ denote the ordered jumps of W on the interval $[0, t]$. In this case, for each $t > 0$

$$\left(m_W^{(k)}(t)\right)_{k\geq 1} \stackrel{\mathrm{D}}{=} \left(c(\alpha_1\alpha_2)\left(\frac{\Gamma_k}{t}\right)^{-1/(\alpha_1\alpha_2)}\right)_{k\geq 1}.$$

Observe that for all $t > 0$

$$W(t) = \sum_{0<s\leq t} \Delta W(s) = \sum_{0<s\leq V_{\alpha_2}(t)} \Delta V_{\alpha_1}(s) = \sum_{k=1}^{\infty} m_{\alpha_1}^{(k)}(V_{\alpha_2}(t)). \tag{28}$$

Note that though (28) holds, $\left(m_{\alpha_1}^{(k)}(V_{\alpha_2}(t))\right)_{k\geq 1}$ is not equal in distribution to $\left(m_W^{(k)}(t)\right)_{k\geq 1}$. To see this, notice that

$$\left(\frac{m_{\alpha_1}^{(k)}(V_{\alpha_2}(t))}{m_{\alpha_1}^{(1)}(V_{\alpha_2}(t))}\right)_{k\geq 1} \stackrel{\mathrm{D}}{=} \left(\left(\frac{\Gamma_k}{\Gamma_1}\right)^{-1/\alpha_1}\right)_{k\geq 1}, \tag{29}$$

whereas

$$\left(\frac{m_W^{(k)}(t)}{m_W^{(1)}(t)}\right)_{k\geq 1} \stackrel{\mathrm{D}}{=} \left(\left(\frac{\Gamma_k}{\Gamma_1}\right)^{-1/(\alpha_1\alpha_2)}\right)_{k\geq 1}. \tag{30}$$

Obviously, the sequences (29) and (30) are not equal in distribution and thus

$$\left(m_{\alpha_1}^{(k)}(V_{\alpha_2}(t))\right)_{k\geq 1} \stackrel{\mathrm{D}}{\neq} \left(m_W^{(k)}(t)\right)_{k\geq 1}.$$

5 Appendix 2

A straightforward modification of the proof of Theorem 1 of Rényi [10] gives the following Anscombe's theorem for Lévy processes.

Theorem A *Let $X = (X(t), t \geq 0)$ be a mean zero Lévy process with $EX^2(t) = t$ for $t \geq 0$, and let $\eta = (\eta(t), t > 0)$ be a random process such that $\eta(t) > 0$ for all $t > 0$ and for some $c > 0$, $\eta(t)/t \xrightarrow{P} c$, as $t \to \infty$, then*

$$X(\eta(t))/\sqrt{\eta(t)} \xrightarrow{D} Z.$$

A version of Anscombe's theorem is given in Gut [2]. See his Theorem 3.1. In our notation, his Theorem 3.1 requires that $\{\eta(t), t \geq 0\}$ be a family of stopping times.

Example A Let $Y_\alpha = (Y_\alpha(y), y \geq 0)$ be the Lévy process with Laplace transform (19) and mean and variance functions (20). We see that

$$X := \left(X(y) = \frac{Y_\alpha(y) - \beta_\alpha y}{\gamma_\alpha}, \ y \geq 0 \right)$$

defines a mean zero Lévy process with $EX^2(y) = y$ for $y \geq 0$. Now let $\eta = (\eta(t), t \geq 0)$ be a standard Gamma process independent of X. Notice that $\eta(t)/t \xrightarrow{P} 1$, as $t \to \infty$. Applying Theorem A, we get as $t \to \infty$,

$$X(\eta(t))/\sqrt{\eta(t)} \xrightarrow{D} Z.$$

In particular, since for each integer $k \geq 1$, $\eta(k) \stackrel{D}{=} \Gamma_k$, this implies that (24) holds for any sequence of positive integers $(k_n)_{n\geq 1}$ converging to infinity as $n \to \infty$, i.e.,

$$\frac{Y_\alpha\left(\Gamma_{k_n}\right) - \beta_\alpha \Gamma_{k_n}}{\gamma_\alpha \sqrt{\Gamma_{k_n}}} \xrightarrow{D} Z.$$

Acknowledgments The author thanks Ross Maller for some useful email discussions that helped him to clarify the two methods of trimming discussed in this chapter. He is also grateful to Péter Kevei and the referee for their careful readings of the manuscript and valuable comments.

References

1. D. Applebaum, *Lévy Processes and Stochastic Calculus*, 2nd edn. Cambridge Studies in Advanced Mathematics, vol. 116 (Cambridge University Press, Cambridge, 2009)
2. A. Gut, Stopped Lévy processes with applications to first passage times. Statist. Probab. Lett. **28**, 345–352 (1996)

3. Ipsen, Y., Maller, R., Resnick, S.: Trimmed Lévy processes and their extremal components. Stochastic Process. Appl. **130**, 2228–2249 (2020)
4. Ipsen, Y., Maller, R., Shemehsavar, S.: Limiting distributions of generalised Poisson-Dirichlet distributions based on negative binomial processes. J. Theoret. Probab. **33** 1974–2000 (2020)
5. Kallenberg, O.: Foundations of modern probability. Second edition. Probability and its Applications (New York). Springer-Verlag, New York, 2002.
6. Kevei, P., Mason, D.M.: The limit distribution of ratios of jumps and sums of jumps of subordinators. ALEA Lat. Am. J. Probab. Math. Stat. **11** 631–642 (2014).
7. Mason, D. M.: Self-standardized central limit theorems for trimmed Lévy processes, J. Theoret. Probab. **34** 2117–2144 (2021).
8. Pillai, R. N.: On Mittag-Leffler functions and related distributions, Ann. Inst. Statist. Math. **42** 157–161 (1990)
9. Rosiński, J.: Series representations of Lévy processes from the perspective of point processes. In Lévy processes, 401–415. Birkhäuser Boston, Boston, MA (2001)
10. Rényi, A.: On the asymptotic distribution of the sum of a random number of independent random variables. Acta Math. Acad. Sci. Hungar. **8** 193–199 (1957)
11. Sato, K.: Lévy Processes and Infinitely Divisible Distributions. Cambridge Univ. Press, Cambridge (2005)
12. Zaĭtsev, A. Yu: On the Gaussian approximation of convolutions under multidimensional analogues of S. N. Bernstein's inequality conditions. Probab. Theory Related Fields **74**, 535–566 (1987)

Functional Central Limit Theorem via Nonstationary Projective Conditions

Florence Merlevède and Magda Peligrad

2020 Mathematics Subject Classification: 60F17; 60G48

1 Introduction and Notations

A time-dependent series, in a discretized form, consists of a triangular array of random variables. Examples of this kind are numerous, and we can cite, for instance, the time-varying regression model. On the other hand, a Markov chain with stationary transition operator is not stationary when it does not start from its equilibrium and it rather starts at a point. Nonstationary type of behavior also appears when we study evolutions in random media. It is also well-known that the blocking procedure, used to weaken the dependence for studying a stationary process or a random field, introduces triangular arrays of variables. Furthermore, many of the results for functions of stationary random fields often incorporate in their proofs complicated inductions, which lead to triangular arrays of random variables.

Historically, the most celebrated limit theorems in nonstationary setting are, among others, the limit theorems involving nonstationary sequences of martingale differences. For more general dependent sequences, one of the basic techniques is to approximate them with martingales. A remarkable early result obtained by using this technique is due to Dobrushin [8], who studied the central limit theorem for nonstationary Markov chains. In order to treat more general dependent structures, McLeish [23, 24] introduced the notion of mixingales, which are martingale-like structures, and imposed conditions to the moments of projections of an individual variable on past sigma fields to derive the functional form of the central limit

F. Merlevède (✉)
Université Gustave Eiffel, LAMA and CNRS UMR 8050, Marne-la-Vallée, France
e-mail: florence.merlevede@univ-eiffel.fr

M. Peligrad
Department of Mathematical Sciences, University of Cincinnati, Cincinnati, OH, USA
e-mail: peligrm@ucmail.uc.edu

R. Adamczak et al. (eds.), *High Dimensional Probability IX*, Progress in Probability 80, https://doi.org/10.1007/978-3-031-26979-0_10

theorem. This method is very fruitful but still involves a large degree of stationarity. In general, the theory of nonstationary martingale approximation has remained much behind the theory of martingale methods for stationary processes. In the stationary setting, the theory of martingale approximations was steadily developed. We mention the well-known results, such as the celebrated results by Gordin [13], Heyde [19], and Maxwell and Woodroofe [22], and the more recent results by Peligrad and Utev [31], Zhao and Woodroofe [44], and Gordin and Peligrad [15], among many others. In the context of random fields, the theory of martingale approximation has been developed in the last decade, with several results by Gordin [14], Volný and Wang [42], Cuny et al. [3], El Machkouri and Giraudo [11], Peligrad and Zhang [33–35], Giraudo [12], and Volný [40, 41]. Due to these results, we know now the necessary and sufficient conditions for various types of martingale approximations, which lead to a variety of maximal inequalities and limit theorems.

The goal of this paper is to survey some results obtained in the recent book [27] and the recent papers [25, 26] concerning the functional form of the central limit theorem (CLT) for non-necessarily stationary dependent structures. These results are obtained by using nonstationary martingale techniques, and, as we shall see, the results are in the spirit of those obtained by McLeish [23, 24]. More precisely, the conditions can be compared to the mixingale conditions imposed in his paper.

Still concerning Gaussian approximation for non-necessarily stationary dependent structures, we would like to mention the paper by Wu and Zhou [43] who show that, under mild conditions, the partial sums of a nonhomogeneous function of an i.i.d. sequence can be approximated, on a richer probability space, by sums of independent Gaussian random variables with nearly optimal errors in probability. As a by-product, a CLT can be derived, provided the underlying random variables have moments of order $2 + \delta$, $\delta > 0$. Their proof combines martingale approximation with m-dependent approximation. The fact that the random variables are functions of an i.i.d. sequence is a crucial assumption in their paper.

We shall point out classes of nonstationary time series, satisfying certain projective criteria (i.e., conditions imposed to conditional expectations), which benefit from a martingale approximation. We shall stress the nonstationary version of the Maxwell-Woodroofe condition, which will be essential for obtaining maximal inequalities and asymptotic results for the following examples: functions of linear processes with nonstationary innovations, quenched version of the functional central limit theorem for a stationary sequence, evolutions in random media such as a process sampled by a shifted Markov chain, and nonstationary $\rho-$mixing and $\alpha-$mixing processes.

The basic setting will be mostly of a sequence of real-valued random variables $(X_k)_{k\geq 1}$ defined on the probability space $(\Omega, \mathcal{K}, P)$, adapted to an increasing filtration $\mathcal{F}_k \subset \mathcal{K}$. Set $S_n = \sum_{i=1}^n X_i$ for $n \geq 1$ and $S_0 = 0$.

We shall also consider real-valued triangular arrays $(X_{k,n})_{1\leq k\leq n}$ adapted to $\mathcal{F}_{k,n} \subset \mathcal{K}$. This means that $X_{k,n}$ is $\mathcal{F}_{k,n}$-measurable and $\mathcal{F}_{k-1,n} \subset \mathcal{F}_{k,n}$ for all $n \geq 1$ and all $1 \leq k \leq n$.

In this case we set $S_k = S_{k,n} = \sum_{i=1}^k X_{i,n}$ $n \geq 1$, and $S_0 = 0$.

We shall be interested in both CLT, i.e.,

$$\frac{S_n - a_n}{b_n} \Rightarrow N(0, \sigma^2),$$

where $\Rightarrow$ denotes the convergence in distribution and N is a normal distributed variable, and also in its functional (FCLT) form, i.e.,

$$\{W_n(t), t \in [0, 1]\} \Rightarrow |\sigma| W \text{ in } (D([0, 1]), \|\cdot\|_\infty),$$

where $W_n(t) = b_n^{-1}(S_{[nt]} - a_{[nt]})$ and W is a standard Brownian motion (here and everywhere in the paper $[x]$ denotes the integer part of x).

We shall consider centered real-valued random variables that are square integrable. The normalizations will be taken $a_n = 0$ and $b_n^2 = n$ or $b_n^2 = \sigma_n^2 = \mathrm{Var}(S_n)$.

In the sequel, we shall often use the notation $\mathbb{E}_i(X) = \mathbb{E}(X|\mathcal{F}_i)$, to replace the conditional expectation. In addition, all along the paper, we shall use the notation $a_n \ll b_n$ to mean that there exists a universal constant C such that, for all n, $a_n \leq C b_n$.

2 Projective Criteria for Nonstationary Time Series

One of the first projection condition, in the nonstationary setting, goes back to McLeish [23]. To simplify the exposition, let us state it in the adapted case, i.e., when $(\mathcal{F}_i)_{i\geq 0}$ is a nondecreasing sequence of σ-algebras, such that X_i is $\mathcal{F}_i$-measurable for any $i \geq 1$.

Theorem 1 *Let $(X_k)_{k\in\mathbb{Z}}$ be a sequence of random variables, centered, with finite second moment and adapted to a nondecreasing sequence $(\mathcal{F}_k)_{k\in\mathbb{Z}}$ of σ-algebras. Assume that $(X_k^2)_{k\in\mathbb{Z}}$ is uniformly integrable and that, for any k and i,*

$$\|\mathbb{E}(X_{i+k}|\mathcal{F}_i)\|_2 \leq C k^{-1/2} (\log k)^{-(1+\varepsilon)}, \tag{1}$$

and there exists a nonnegative constant c^2 such that

$$\frac{\mathbb{E}(S_{[nt]}^2)}{n} \to c^2 t \text{ for any } t \in [0, 1] \text{ and } \frac{\mathbb{E}_{k-m}(S_{k+n} - S_k)^2}{n} \to c^2 \text{ in } \mathbb{L}_1,$$

as $\min(k, m, n) \to \infty$. Then, $\{n^{-1/2} S_{[nt]}, t \in [0, 1]\} \Rightarrow cW$ in $(D([0, 1]), \|\cdot\|_\infty)$, where W is a standard Brownian motion.

However, in the stationary case, a more general projection condition than (1) is known to be sufficient for both CLT and its functional form. Let us describe it briefly.

Let $(X_k)_{k\in\mathbb{Z}}$ be a strictly stationary and ergodic sequence of centered real-valued random variables in $\mathbb{L}^2$, adapted to a strictly stationary filtration $(\mathcal{F}_k)_{k\in\mathbb{Z}}$ and such that

$$\sum_{k\geq 1} \frac{\|\mathbb{E}_0(S_k)\|_2}{k^{3/2}} < \infty. \tag{2}$$

Under condition (2), Maxwell-Woodroofe [22] proved the CLT under the normalization $\sqrt{n}$ and Peligrad-Utev [31] proved its functional form, namely,

$$\{n^{-1/2} S_{[nt]}, t \in [0, 1]\} \Rightarrow cW \text{ in } (D([0, 1]), \| \cdot \|_\infty),$$

where $c^2 = \lim_n n^{-1}\mathbb{E}(S_n^2)$.

It is known that (2) is equivalent to $\sum_{k\geq 0} 2^{-k/2}\|\mathbb{E}_0(S_{2^k})\|_2 < \infty$ and it is implied by

$$\sum_{k>0} k^{-1/2}\|\mathbb{E}_0(X_k)\|_2 < \infty. \tag{3}$$

It should be noted that condition (2) is a sharp condition in the sense that if it is barely violated, then the sequence $(n^{-1/2}S_n)$ fails to be stochastically bounded (see [31]).

The Maxwell-Woodroofe condition is very important for treating the class of $\rho-$mixing sequences whose definition is based on maximum coefficient of correlation. In the stationary case, this is

$$\rho(k) = \sup \operatorname{corr}\big(f(X_{i,}, \ i \leq 0), g(X_j, j \geq k)\big) \to 0,$$

where sup is taken over all functions f, g which are square integrable.

It can be shown that condition (2) is implied by $\sum_{k\geq 0} \rho(2^k) < \infty$ (which is equivalent to $\sum_{k\geq 1} k^{-1}\rho(k) < \infty$). It is therefore well adapted to measurable functions of stationary Gaussian processes. To give another example of a sequence satisfying, (2) let

$$X_k = f\Big(\sum_{i\geq 0} a_i\varepsilon_{k-i}\Big) - \mathbb{E}f\Big(\sum_{i\geq 0} a_i\varepsilon_{k-i}\Big),$$

where (ε_k) are i.i.d. with variance σ^2 and let f be a function such that

$$|f(x) - f(y)| \leq c(|x - y|) \text{ for any } (x, y) \in \mathbb{R}^2,$$

where c is a concave nondecreasing function such that

$$\sum_{k\geq 1} k^{-1/2} c\Big(2\sigma \sum_{i\geq k} |a_i|\Big) < \infty\,.$$

Then, (3) holds (and then (2) also).

The question is, could we have similar results, which extend condition (2) to the nonstationary case and improve on Theorem 1?

2.1 Functional CLT Under the Standard Normalization $\sqrt{n}$

We shall discuss first FCLT in the nonstationary setting under the normalization $\sqrt{n}$. With this aim, we impose the Lindeberg-type condition in the form:

$$\sup_{n\geq 1} \frac{1}{n} \sum_{j=1}^{n} \mathbb{E}(X_j^2) \leq C < \infty \text{ and, for any } \varepsilon > 0,$$

$$\lim_{n\to\infty} \frac{1}{n} \sum_{k=1}^{n} \mathbb{E}\{X_k^2 I(|X_k| > \varepsilon\sqrt{n})\} = 0\,. \tag{4}$$

For any $k \geq 0$, let

$$\delta(k) = \max_{i\geq 0} \|\mathbb{E}(S_{k+i} - S_i|\mathcal{F}_i)\|_2$$

and for any $k, m \geq 0$, let

$$\theta_k^m = m^{-1} \sum_{i=1}^{m-1} \mathbb{E}_k(S_{k+i} - S_k).$$

The following FCLT in the nonstationary setting under the normalization $\sqrt{n}$ was proven by Merlevède et al. [26, 27].

Theorem 2 *Assume that the Lindeberg-type condition (4) holds. Suppose also that*

$$\sum_{k\geq 0} 2^{-k/2} \delta(2^k) < \infty \tag{5}$$

and there exists a constant c^2 such that, for any $t \in [0, 1]$ and any $\varepsilon > 0$,

$$\lim_{m\to\infty} \limsup_{n\to\infty} \mathbb{P}\Big(\Big|\frac{1}{n} \sum_{k=1}^{[nt]} \big(X_k^2 + 2X_k\theta_k^m\big) - tc^2\Big| > \varepsilon\Big) = 0\,. \tag{6}$$

Then, $\{n^{-1/2} S_{[nt]}, t \in [0, 1]\} \Rightarrow cW$ in $(D([0, 1]), \|\cdot\|_\infty)$.

We mention that (5) is equivalent to $\sum_{k>0} k^{-3/2}\delta(k) < \infty$ and it is implied by

$$\sum_{k>0} k^{-1/2} \sup_{i\geq 0} \|\mathbb{E}_i(X_{k+i})\|_2 < \infty. \tag{7}$$

About condition (6) we would like to mention that in the stationary and ergodic case, it is verified under condition (2). Indeed, by the ergodic theorem, for any $k \geq 0$,

$$\lim_{n\to\infty} \mathbb{E}\Big|\frac{1}{n}\sum_{k=1}^{[nt]}(X_k^2 + 2X_k\theta_k^m) - c^2 t\Big| = t\big|\mathbb{E}X_0^2 + 2\mathbb{E}(X_0\theta_0^m) - c^2\big| .$$

Note that, under condition (2), it has been proved in [31] that

$$\frac{1}{m}\mathbb{E}(S_m^2) = \mathbb{E}(X_0^2) + 2\mathbb{E}(X_0\theta_0^m) \to c^2 \text{ as } m \to \infty.$$

Therefore, Theorem 2 is indeed a generalization of the results in Peligrad and Utev [31].

The first application of Theorem 2 is the following:

Example 3 (Application to Stationary Sequences in a Random Time Scenery) We are interested to investigate the limiting behavior of the partial sums associated with the process defined by

$$X_k = \zeta_{k+\phi_k} ,$$

where $\{\zeta_j\}_{j\in\mathbb{Z}}$ is a stationary sequence (observables/random scenery) and $\{\phi_k\}_{k\geq 0}$ is a Markov chain (random time).

The sequence $\{\phi_n\}_{n\geq 0}$ is a "renewal"-type Markov chain defined as follows: $\{\phi_k; k \geq 0\}$ is a discrete Markov chain with the state space $\mathbb{Z}^+$ and transition matrix $P = (p_{i,j})$ given by $p_{k,k-1} = 1$ for $k \geq 1$ and $p_{0,j-1} := p_j = \mathbb{P}(\tau = j)$, $j = 1, 2, \ldots$.

We assume that $\mathrm{e}[\tau] < \infty$, which ensures that $\{\phi_n\}_{n\geq 0}$ has a stationary distribution $\pi = (\pi_i, i \geq 0)$ given by

$$\pi_j = \pi_0 \sum_{i=j+1}^{\infty} p_i \ , \ j = 1, 2 \ldots \text{ where } \pi_0 = 1/\mathrm{e}(\tau) .$$

We also assume that $p_j > 0$ for all $j \geq 0$. Hence, the Markov chain is irreducible.

We are interested by the asymptotic behavior of

$$\Big\{n^{-1/2}\sum_{k=1}^{[nt]} X_k, t \in [0, 1]\Big\}$$

when the Markov chain starts at 0 (so under $\mathbb{P}_{\phi_0=0}$).

Under $\mathbb{P}_{\phi_0=0}$, one can prove that $\mathbb{E}(X_1X_2) \neq \mathbb{E}(X_2X_3)$, and, hence, stationarity is ruled out immediately. Let us assume the following assumption on the random time scenery:

Condition (A_1) $\{\zeta_j\}_{j\geq 0}$ is a strictly stationary sequence of centered random variables in $\mathbb{L}^2$, independent of $(\phi_k)_{k\geq 0}$ and such that

$$\sum_{k \geq 1} \frac{\|\mathbb{E}(\zeta_k|\mathcal{G}_0)\|_2}{\sqrt{k}} < \infty \text{ and } \lim_{n \to \infty} \sup_{j \geq i \geq n} \|\mathbb{E}(\zeta_i \zeta_j|\mathcal{G}_0) - \mathbb{E}(\zeta_i \zeta_j)\|_1 = 0,$$

where $\mathcal{G}_i = \sigma(\zeta_k, k \leq i)$.

Corollary 4 *Assume that* $\mathbb{E}(\tau^2) < \infty$ *and that* $\{\zeta_j\}_{j \geq 0}$ *satisfies condition* (A_1)*. Then, under* $\mathbb{P}_{\phi_0=0}$, $\{n^{-1/2} S_{[nt]}, t \in [0, 1]\}$ *converges in distribution in* $D[0, 1]$ *to a Brownian motion with parameter* c^2 *defined by*

$$c^2 = \mathbb{E}(\zeta_0^2)\Big(1 + 2\sum_{i \geq 1} i\pi_i\Big) + 2\sum_{m \geq 1} \mathbb{E}(\zeta_0 \zeta_m) \sum_{j=1}^{m} (P^j)_{0,m-j} \,,$$

where $(P^j)_{0,b} = \mathbb{P}_{\phi_0=0}(\phi_j = b)$.

The idea of proof is the following. We take $\mathcal{A} = \sigma(\phi_k, k \geq 0)$ and $\mathcal{F}_k = \sigma(\mathcal{A}, X_j, 1 \leq j \leq k)$. One can show that

$$\sup_{k \geq 0} \|\mathbb{E}(X_{k+m}|\mathcal{F}_k)\|_2^2 \leq b^2([m/2]) + b^2(0)\mathbb{P}(\tau > [m/2]),$$

where $b(k) = \|\mathbb{E}(\zeta_k|\mathcal{G}_0)\|_2$. To prove that condition (6) holds, we use, in particular, the ergodic theorem for recurrent Markov chains (together with many tedious computations).

An Additional Comment In the stationary case, other projective criteria can be considered to get the FCLT, such as the so-called Hannan's condition [18]:

$$\mathbb{E}(X_0|\mathcal{F}_{-\infty}) = 0 \text{ a.s. and } \sum_{i \geq 0} \|\mathbf{P}_0(X_i)\|_2 < \infty \,,$$

where $\mathbf{P}_0(\cdot) = \mathbb{E}_0(\cdot) - \mathbb{E}_{-1}(\cdot)$.

Hannan's condition and condition (2) have different areas of applications and are not comparable (see [10]).

If the scenery is a sequence of martingale difference sequence and the process is sampled by the renewal Markov Chain, then under $\mathbb{P}_{\phi_0=0}$, one can prove that

$$\sup_{k \geq 0} \|\mathbf{P}_{k-m}(X_k)\|_2 \sim C\sqrt{\mathbb{P}(\tau > m)}.$$

Hence, in this case, $\sup_{k \geq 0} \|\mathbf{P}_{k-m}(X_k)\|_2$ and $\sup_{k \geq 0} \|\mathbb{E}(X_{k+m}|\mathcal{F}_k)\|_2$ are of the same order of magnitude and

$$\sum_{m \geq 0} \sup_{k \geq 0} \|\mathbf{P}_{k-m}(X_k)\|_2 < \infty \iff \sum_{m \geq 0} \sqrt{\mathbb{P}(\tau > m)} < \infty \,.$$

On the other hand, (7) holds provided $\sum_{k \geq 1} \sqrt{\mathbb{P}(\tau > k)}/\sqrt{k} < \infty$.

2.2 A More General FCLT for Triangular Arrays

Let $\{X_{i,n}, 1 \leq i \leq n\}$ be a triangular array of square integrable ($\mathbb{E}(X_{i,n}^2) < \infty$), centered ($\mathbb{E}(X_{i,n}) = 0$), real-valued random variables adapted to a filtration $(\mathcal{F}_{i,n})_{i\geq 0}$.

We write as before $\mathbb{E}_{j,n}(X) = \mathbb{E}(X|\mathcal{F}_{j,n})$ and set

$$S_{k,n} = \sum_{i=1}^{k} X_{i,n} \text{ and } \theta_{k,n}^{m} = m^{-1} \sum_{i=1}^{m-1} \mathbb{E}_{k,n}(S_{k+i,n} - S_{k,n}) . \tag{8}$$

We assume that the triangular array satisfies the following triangular Lindeberg-type condition:

$$\sup_{n\geq 1} \sum_{j=1}^{n} \mathbb{E}(X_{j,n}^2) \leq C < \infty, \text{ and } \lim_{n\to\infty} \sum_{k=1}^{n} \mathbb{E}\{X_{k,n}^2 I(|X_{k,n}| > \varepsilon)\} = 0,$$
$$\text{for any } \varepsilon > 0. \tag{9}$$

For a nonnegative integer u and positive integers ℓ, m, define the following martingale-type dependence characteristics:

$$A^2(u) = \sup_{n\geq 1} \sum_{k=0}^{n-1} \|\mathbb{E}_{k,n}(S_{k+u,n} - S_{k,n})\|_2^2$$

and

$$B^2(\ell, m) = \sup_{n\geq 1} \sum_{k=0}^{[n/\ell]} \|\bar{S}_{k,n}(\ell, m)\|_2^2 ,$$

where

$$\bar{S}_{k,n}(\ell, m) = \frac{1}{m} \sum_{u=0}^{m-1} \left(\mathbb{E}_{(k-1)\ell+1,n}(S_{(k+1)\ell+u,n} - S_{k\ell+u,n})\right) .$$

We mention that if $X_{k,n} = X_k/\sqrt{n}$,

$$A^2(u) \leq \delta^2(u) \text{ and } B^2(\ell, m) \leq \delta^2(\ell - 1)/\ell.$$

The next theorem was proved by Merlevède et al. [26].

Theorem 5 *Assume that the Lindeberg condition (9) holds and that*

$$\lim_{j\to\infty} 2^{-j/2} A(2^j) = 0 \textit{ and } \liminf_{j\to\infty} \sum_{\ell\geq j} B(2^\ell, 2^j) = 0 . \tag{10}$$

Moreover, assume that there exists a sequence of nondecreasing and right-continuous functions $v_n(\cdot) : [0,1] \to \{0, 1, 2, \ldots, n\}$ *and a nonnegative real* c^2 *such that, for any* $t \in (0, 1]$,

$$\lim_{m\to\infty} \limsup_{n\to\infty} \mathbb{P}\Big(\Big|\sum_{k=1}^{v_n(t)} \big(X_{k,n}^2 + 2X_{k,n}\theta_{k,n}^m\big) - tc^2\Big| > \varepsilon\Big) = 0\,. \tag{11}$$

Then, $\big\{\sum_{k=1}^{v_n(t)} X_{k,n}\,, t \in [0,1]\big\}$ *converges in distribution in* $D([0,1])$ *to* cW*, where* W *is a standard Brownian motion.*

The proof is based on a suitable triangular (nonstationary) martingale approximation. More precisely, for any fixed integer m, we write

$$X_{\ell,n} = D_{\ell,n}^m + \theta_{\ell-1,n}^m - \theta_{\ell,n}^m + Y_{\ell-1,n}^m\,, \tag{12}$$

where $\theta_{\ell,n}^m$ is defined in (8), $Y_{\ell,n}^m = \frac{1}{m}\mathbb{E}_{\ell,n}(S_{\ell+m,n} - S_{\ell,n})$ and, with the notation $\mathbf{P}_{\ell,n}(\cdot) = \mathbb{E}_{\ell,n}(\cdot) - \mathbb{E}_{\ell-1,n}(\cdot)$,

$$D_{\ell,n}^m = \frac{1}{m}\sum_{i=0}^{m-1} \mathbf{P}_{\ell,n}(S_{\ell+i}) = \frac{1}{m}\sum_{i=0}^{m-1} \mathbf{P}_{\ell}(S_{\ell+i} - S_{\ell-1})\,. \tag{13}$$

Then, we show that the FCLT for $\big\{\sum_{k=1}^{v_n(t)} X_{k,n}\,, t \in [0,1]\big\}$ is reduced to prove the FCLT for sums associated to a triangular array of martingale differences, namely, for $\big\{\sum_{k=1}^{v_n(t)} D_{k,n}^{m_n}, t \in [0,1]\big\}$, where (m_n) is a suitable subsequence.

Comment 6 Let us make some comments on the Lindeberg-type condition (9), which is commonly used to prove the CLT when we deal with dependent structures. We refer, for instance, to the papers by Neumann [28] or Rio [37] where this condition is also imposed and examples satisfying such a condition are provided. In addition, in many cases of interest, the considered triangular array takes the following form: $X_{k,n}/\sigma_n$ where $\sigma_n^2 = \mathrm{Var}(S_n)$, and then the first part of (9) reads as the following: there exists a positive constant C such that, for any $n \geq 1$,

$$\sum_{k=1}^{n} \mathbb{E}(X_{k,n}^2) \leq C\mathrm{Var}(S_n)\,, \tag{14}$$

which then imposed a certain growth of the variance of the partial sums. Let us give another example where this condition is satisfied. Assume that $X_i = f_i(Y_i)$ where Y_i is a Markov chain satisfying $\rho_Y(1) < 1$, and then according to [29, Proposition 13], $C \leq (1+\rho_Y(1))(1-\rho_Y(1))^{-1}$. Here, $(\rho_Y(k))_{k\geq 0}$ is the sequence of ρ-mixing coefficients of the Markov chain $(Y_i)_i$. On the other hand, avoiding a condition as (14) is a big challenge and is one of the aims of Hafouta's recent paper [16]. His main new idea is a linearization of the variance of the partial sums, which, to some extent, allows us to reduce the limit theorems to the case when $\mathrm{Var}(S_n)$ grows linearly fast in n. To give more insights, the partial sums are partitioned into blocks, so we write $S_n = \sum_{i=1}^{k_n} Y_{i,n}$, where k_n is of order $\mathrm{Var}(S_n)$ and the summands $Y_{i,n}$ are uniformly bounded in some $\mathbb{L}^p$ (see [16, section 1.4] for more details). Then, the FCLT has to be obtained for the new triangular array $(Y_{i,n}, 1 \le i \le k_n)$.

To verify condition (11), one can use the following proposition proved in [26]:

Proposition 7 *Assume that the Lindeberg-type condition* (9) *holds. Assume also that, for any nonnegative integer* ℓ,

$$\lim_{b\to\infty} \limsup_{n\to\infty} \sum_{k=b+1}^{n} \|\mathbb{E}_{k-b,n}(X_{k,n}X_{k+\ell,n}) - \mathbb{E}_{0,n}(X_{k,n}X_{k+\ell,n})\|_1 = 0$$

and, for any $t \in [0, 1]$,

$$\lim_{m\to\infty} \limsup_{n\to\infty} \mathbb{P}\Big(\Big|\sum_{k=1}^{v_n(t)} \big(\mathbb{E}_{0,n}(X_{k,n}^2) + 2\mathbb{E}_{0,n}(X_{k,n}\theta_{k,n}^m)\big) - tc^2\Big| > \varepsilon\Big) = 0 . \qquad (15)$$

Then, condition (11) is satisfied.

Starting from (12) and summing over ℓ, we get

$$\sum_{\ell=1}^{v_n(t)} (X_{\ell,n}^2 + 2X_{\ell,n}\theta_{\ell,n}^m) = \sum_{\ell=1}^{v_n(t)} (D_{\ell,n}^m)^2 + (\theta_{0,n}^m)^2 - (\theta_{v_n(t),n}^m)^2$$
$$+ \sum_{\ell=1}^{v_n(t)} 2D_{\ell,n}^m(\theta_{\ell-1}^m + Y_{\ell-1,n}^m) + R_n ,$$

where

$$R_n = \sum_{\ell=0}^{v_n(t)-1} (Y_{\ell,n}^m)^2 + 2\sum_{k=0}^{v_n(t)-1} \theta_k^m Y_{k,n}^m .$$

Clearly,

$$\sum_{\ell=1}^{v_n(t)} \mathbb{E}(X_{\ell,n}^2 + 2X_{\ell,n}\theta_{\ell,n}^m) = \sum_{\ell=1}^{v_n(t)} \mathbb{E}(D_{\ell,n}^m)^2 + \mathbb{E}(\theta_{0,n}^m)^2 - \mathbb{E}(\theta_{v_n(t),n}^m)^2 + \mathbb{E}(R_n) .$$

The Lindeberg condition implies that $\mathbb{E}(\theta_{0,n}^m)^2 + \mathbb{E}(\theta_{v_n(t),n}^m)^2$ is tending to zero as $n \to \infty$, whereas

$$\mathbb{E}(R_n) \ll m^{-2}\big(A^2(m) + A(m)\sum_{i=1}^{m} A(i)\big) .$$

Hence, if we assume that $m^{-1}A^2(m) \to 0$ as $m \to \infty$, we derive

$$\lim_{m\to\infty} \limsup_{n\to\infty} \Big| \sum_{\ell=1}^{v_n(t)} \mathbb{E}(X_{\ell,n}^2 + 2X_{\ell,n}\theta_{\ell,n}^m) - \sum_{\ell=1}^{v_n(t)} \mathbb{E}(D_{\ell,n}^m)^2 \Big| = 0 \, .$$

Note also that under the Lindeberg condition and the following reinforced version of condition (10)

$$\lim_{m\to\infty} m^{-1/2} A(m) = 0 \text{ and } \lim_{m\to\infty} \sum_{\ell \geq [\log_2(m)]} B(2^\ell, m) = 0 \, , \tag{16}$$

we have

$$\lim_{m\to\infty} \limsup_{n\to\infty} \Big\| \sum_{\ell=1}^{v_n(t)} X_{\ell,n} - \sum_{\ell=1}^{v_n(t)} D_{\ell,n}^m \Big\|_2 = 0 \, .$$

Lemma 5.4 in [26] can be used to see this (note that, in this lemma, there is a misprint in the statement since, in the last term of the RHS of its inequality, the term $2^{-j/2}$ has to be deleted, as it can be clearly derived from their inequality (5.22)). Therefore, as soon as we consider $\mathcal{F}_{0,n} = \{\emptyset, \Omega\}$ (so $\mathbb{E}_{0,n}(\cdot) = \mathbb{E}(\cdot)$), condition (15) can be verified with the help of the following proposition:

Proposition 8 *Assume that the Lindeberg-type condition* (9) *holds and that* (16) *is satisfied. Assume also that there exists a constant c^2 such that, for any $t \in [0, 1]$,*

$$\mathbb{E}(S_{v_n(t),n}^2) \to c^2 t \, . \tag{17}$$

Then,

$$\lim_{m\to\infty} \limsup_{n\to\infty} \Big| \sum_{k=1}^{v_n(t)} \mathbb{E}\big(X_{k,n}^2 + 2X_{k,n}\theta_{k,n}^m\big) - tc^2 \Big| = 0 \, .$$

3 Applications

3.1 Application to ρ-mixing Triangular Arrays

Theorem 5 gives the following result for ρ-mixing triangular arrays:

Let $\{X_{i,n}, 1 \leq i \leq n\}$ be a triangular array of square integrable centered real-valued random variables. Denote by $\sigma_{k,n}^2 = \mathrm{Var}\big(\sum_{\ell=1}^k X_{\ell,n}\big)$ and $\sigma_n^2 = \sigma_{n,n}^2$. For $0 \leq t \leq 1$, let

$$v_n(t) = \inf\Big\{k;\ 1 \le k \le n\colon \frac{\sigma_{k,n}^2}{\sigma_n^2} \ge t\Big\} \text{ and } W_n(t) = \sigma_n^{-1} \sum_{i=1}^{v_n(t)} X_{i,n}\,. \tag{18}$$

Assume that the triangular array is ρ-mixing in the sense that

$$\rho(k) = \sup_{n\ge 1} \max_{1\le j\le n-k} \rho\big(\sigma(X_{i,n}, 1 \le i \le j), \sigma(X_{i,n}, j+k \le i \le n)\big) \to 0$$

where $\rho(U, V) = \sup\{|corr(X, Y)| : X \in L^2(U), Y \in L^2(V)\}$.

The following is a FCLT for ρ-mixing triangular arrays:

Theorem 9 *Assume that*

$$\sup_{n\ge 1} \sigma_n^{-2} \sum_{j=1}^{n} \mathbb{E}(X_{j,n}^2) \le C < \infty\,,$$

$$\lim_{n\to\infty} \sigma_n^{-2} \sum_{k=1}^{n} \mathbb{E}\{X_{k,n}^2 I(|X_{k,n}| > \varepsilon\sigma_n)\} = 0\,, \ \textit{for any } \varepsilon > 0$$

and

$$\sum_{k\ge 0} \rho(2^k) < \infty\,.$$

Then, $\big\{W_n(t), t \in [0, 1]\big\}$ *converges in distribution in* $D([0, 1])$ *(equipped with the uniform topology) to* W.

This is the functional version of the CLT obtained by Utev [39]. It answers an open question raised by Ibragimov in 1991.

Theorem 9 follows from an application of Theorem 5 to the triangular array $\{\sigma_n^{-1} X_{k,n}, 1 \le k \le n\}_{n\ge 1}$ and the σ-algebras $\mathcal{F}_{k,n} = \sigma(X_{i,n}, 1 \le i \le k)$ for $k \ge 1$ and $\mathcal{F}_{k,n} = \{\emptyset, \Omega\}$ for $k \le 0$.

In what follows, to soothe the notations, we omit the index n involved in the variables and in the σ-algebras.

To check condition (5), we used the fact that, by the definition of the ρ-mixing coefficient, for any $b > a \ge 0$,

$$\|\mathbb{E}_k(S_{k+b} - S_{k+a})\|_2 \le \rho(a)\|S_{k+b} - S_{k+a}\|_2\,,$$

and that, under $\sum_{k\ge 0} \rho(2^k) < \infty$, by the variance inequality of Utev [39], there exists κ such that, for any integers a and b,

$$\|S_b - S_a\|_2^2 \leq \kappa \sum_{i=a+1}^{b} \|X_i\|_2^2 .$$

We then obtain

$$m^{-1}A^2(m) \ll \{\rho^2([\sqrt{m}]) + m^{-1/2}\} \text{ and } B(2^r, m) \ll \rho(2^r - 1) .$$

Since $\rho(n) \to 0$, in order to prove condition (11), we use both Proposition 7 (by recalling that $\mathcal{F}_{0,n}$ is the trivial field $\{\emptyset, \Omega\}$) and Proposition 8. Therefore, the proof of (11) is reduced to show that

$$\sigma_n^{-2}\mathbb{E}(S_{v_n(t)}^2) \to t\,, \quad \text{as } n \to \infty,$$

which holds by the definition of $v_n(t)$ and the Lindeberg condition (9).

For the $\rho-$mixing sequences, we also obtain the following corollary:

Corollary 10 *Let $(X_n)_{n\geq 1}$ be a sequence of centered random variables in $\mathbb{L}^2(\mathbb{P})$. Let $S_n = \sum_{k=1}^{n} X_k$ and $\sigma_n^2 = \mathrm{Var}(S_n)$. Suppose that the Lindeberg condition is satisfied and that $\sum_{k\geq 0} \rho(2^k) < \infty$. In addition, assume that $\sigma_n^2 = nh(n)$, where h is a slowly varying function at infinity. Then, $W_n = \{\sigma_n^{-1}\sum_{k=1}^{[nt]} X_k, t \in (0,1]\}$ converges in distribution in $D([0,1])$ to W, where W is a standard Brownian motion.*

If W_n converges weakly to a standard Brownian motion, then necessarily $\sigma_n^2 = nh(n)$, where $h(n)$ is a slowly varying function. If in Corollary 10 we assume that $\sigma_n^2 = n^{\alpha}h(n)$, where $\alpha > 0$, then one can prove that $W_n \Rightarrow \{G(t), t \in [0,1]\}$ in $D([0,1])$, where $G(t) = \sqrt{\alpha}\int_0^t u^{(\alpha-1)/2}dW(u)$.

In the strictly stationary case, condition $\sum_{k\geq 0}\rho(2^k) < \infty$ implies that $\sigma_n^2/n \to \sigma^2$ and if $\sigma_n^2 \to \infty$, then $\sigma > 0$. Therefore, the functional limit theorem holds under the normalization $\sqrt{n}\sigma$. We then recover the FCLT obtained by Shao [38] (the CLT was first proved by Ibragimov [20]). In this context, condition $\sum_{k\geq 0}\rho(2^k) < \infty$ is minimal as provided by several examples by Bradley, which are discussed in [1, Chap. 34].

Comment 11 In a recent paper, denoting by P_X the law of a random variable X and by G_a the normal distribution $N(0, a)$, Dedecker et al. [7] have proved quantitative estimates for the convergence of P_{S_n/σ_n} to G_1, where S_n is the partial sum associated with either martingale difference sequences or more general dependent sequences, and $\sigma_n^2 = \mathrm{Var}(S_n)$. In particular, they considered the case of ρ-mixing sequences, and, under reinforced conditions compared to those imposed in Theorem 9 or in Corollary 10, they obtained rates in the CLT. Let us describe their result. Let $(X_i)_{i\geq 1}$ be a sequence of centered ($\mathbb{E}(X_i) = 0$ for all i), real-valued bounded random variables, which are ρ-mixing in the sense that

$$\rho(k) = \sup_{j \ge 1} \sup_{v > u \ge j+k} \rho\big(\sigma(X_i, 1 \le i \le j), \sigma(X_u, X_v)\big) \to 0\,, \text{ as } k \to \infty\,,$$

where $\sigma(X_t, t \in A)$ is the σ-field generated by the r.v.'s X_t with indices in A. Let us assume the following set of assumptions:

$$(H) := \begin{cases} 1)\Theta = \sum_{k\ge 1} k\rho(k) < \infty\,. \\ 2)\text{For any } n \ge 1,\, C_n := \max\limits_{1\le \ell \le n} \dfrac{\sum_{i=\ell}^{n} \mathbb{E}(X_i^2)}{\mathbb{E}(S_n - S_{\ell-1})^2} < \infty\,. \end{cases}$$

Denoting by $K_n = \max_{1\le i\le n} \|X_i\|_\infty$, they proved in their Sect. 4.2 that if K_n is uniformly bounded, then, for any positive integer n,

$$\int_{\mathbb{R}} |F_n(t) - \Phi(t)| dt \ll C_n \sigma_n^{-1} \log(2 + C_n \sigma_n^2) \text{ and}$$

$$\|F_n - \Phi\|_\infty \ll \sigma_n^{-1/2} \sqrt{C_n \log(2 + C_n \sigma_n^2)}\,,$$

where F_n is the c.d.f. of S_n/σ_n and Φ is the c.d.f. of a standard Gaussian r.v. We also refer to [16, Section 2.2] for related results concerning rates in the FCLT in terms of the Prokhorov distance.

3.2 *Application to Functions of Linear Processes*

Assume that

$$X_k = f_k\Big(\sum_{i\ge 0} a_i \varepsilon_{k-i}\Big) - \mathbb{E} f_k\Big(\sum_{i\ge 0} a_i \varepsilon_{k-i}\Big),$$

where $(\varepsilon_i)_{i\in\mathbb{Z}}$ are independent random variables such that $(\varepsilon_i^2)_{i\in\mathbb{Z}}$ is a uniformly integrable family and $\sup_{i\in\mathbb{Z}} \|\varepsilon_i\|_2 := \sigma$. The functions f_k are such that, for any k,

$$|f_k(x) - f_k(y)| \le c(|x - y|) \text{ for any } (x, y) \in \mathbb{R}^2\,,$$

where c is concave, nondecreasing, and such that $\lim_{x\to 0} c(x) = 0$ (we shall say that $f_k \in \mathcal{L}(c)$).

Applying Theorem 5 with $X_{k,n} = X_k/\sigma_n$, we derive the following FCLT:

Corollary 12 *Assume that* $\sigma_n^2 = nh(n)$, *where* $h(n)$ *is a slowly varying function at infinity such that* $\liminf_{n\to\infty} h(n) > 0$ *and*

$$\sum_{k\ge 1} k^{-1/2} c\Big(2\sigma \sum_{i\ge k} |a_i|\Big) < \infty\,. \tag{19}$$

Then, $\{\sigma_n^{-1}\sum_{k=1}^{[nt]} X_k, t \in [0,1]\}$ *converges in distribution in* $D([0,1])$ *to a standard Brownian motion.*

The detailed proof can be found in Section 5.6 of [26], but let us briefly describe the arguments allowing to verify conditions (7) and (11) with $v_n(t) = [nt]$ and $X_{k,n} = X_k/\sigma_n$ (recall that (7) implies (5), which in turn implies (10) since $\sigma_n^2 = nh(n)$ with $\liminf_{n\to\infty} h(n) > 0$).

We first consider the following choice of $(\mathcal{F}_i)_{i\geq 0}$: $\mathcal{F}_0 = \{\emptyset, \Omega\}$ and $\mathcal{F}_i = \sigma(X_1, \ldots, X_i)$, for $i \geq 1$. Denote by $\mathbb{E}_\varepsilon$ the expectation with respect to $\varepsilon := (\varepsilon_i)_{i\in\mathbb{Z}}$ and note that since $\mathcal{F}_i \subset \mathcal{F}_{\varepsilon,i}$ where $\mathcal{F}_{\varepsilon,i} = \sigma(\varepsilon_k, k \leq i)$, for any $i \geq 0$, $\|\mathbb{E}(X_{k+i}|\mathcal{F}_i)\|_2 \leq \|\mathbb{E}(X_{k+i}|\mathcal{F}_{\varepsilon,i})\|_2$. Next, for any $i \geq 0$, note that

$$\begin{aligned}\big|\mathbb{E}(X_{k+i}|\mathcal{F}_{\varepsilon,i})\big| = \Big|\mathbb{E}_\varepsilon\Big(f\Big(\sum_{\ell=0}^{k-1} a_\ell \varepsilon'_{k+i-\ell} + \sum_{\ell\geq k} a_\ell \varepsilon_{k+i-\ell}\Big)\Big) \\ - \mathbb{E}_\varepsilon\Big(f\Big(\sum_{\ell=0}^{k-1} a_\ell \varepsilon'_{k+i-\ell} + \sum_{\ell\geq k} a_\ell \varepsilon'_{k+i-\ell}\Big)\Big)\Big|,\end{aligned}$$

where $(\varepsilon'_i)_{i\in\mathbb{Z}}$ is an independent copy of $(\varepsilon_i)_{i\in\mathbb{Z}}$. Therefore, using [6, Lemma 5.1],

$$\|\mathbb{E}(X_{k+i}|\mathcal{F}_i)\|_2 \leq \Big\|c\Big(\sum_{\ell\geq k} |a_\ell||\varepsilon_{k+i-\ell} - \varepsilon'_{k+i-\ell}|\Big)\Big\|_2 \leq c\Big(2\sigma_\varepsilon \sum_{\ell\geq k} |a_\ell|\Big),$$

proving that (7) holds under (19).

On the other hand, to verify condition (11) with $v_n(t) = [nt]$ and $X_{k,n} = X_k/\sigma_n$, Proposition 7 can be used. Hence, because of the Lindeberg condition and the choice of the filtration $(\mathcal{F}_i)_{i\geq 0}$, it is sufficient to prove

$$\lim_{b\to\infty} \limsup_{n\to\infty} \sigma_n^{-2} \sum_{k=b+1}^{n} \|\mathbb{E}_{k-b}(X_k X_{k+\ell}) - \mathbb{E}(X_k X_{k+\ell})\|_1 = 0 \tag{20}$$

and that, for any $t \in [0,1]$,

$$\lim_{m\to\infty} \limsup_{n\to\infty} \frac{1}{\sigma_n^2}\Big|\sum_{k=1}^{[nt]} \Big\{\mathbb{E}(X_k^2) + 2\mathbb{E}(X_k \theta_k^m)\Big\} - t\Big| = 0. \tag{21}$$

Condition (20) can be proved by using similar arguments as those leading to (7). On the other hand, (21) follows from an application of Proposition 8 with $v_n(t) = [nt]$ and $X_{k,n} = X_k/\sigma_n$. Indeed, the Lindeberg condition can be verified, (7) is satisfied, and it is also assumed that $\sigma_n^2 = nh(n)$, where $h(n)$ is a slowly varying function at infinity with $\liminf_{n\to\infty} h(n) > 0$.

3.3 Application to the Quenched FCLT

We should also note that the general FCLT in Theorem 2 also leads as an application to the quenched FCLT under Maxwell-Woodroofe condition (previously proved by Cuny-Merlevède [2], with a completely different proof).

More precisely, the result is the following:

Corollary 13 *Let $(X_k)_{k\in\mathbb{Z}}$ be an ergodic stationary sequence of $\mathbb{L}^2$ centered random variables, adapted to $(\mathcal{F}_k)$ and satisfying*

$$\sum_{k>0} k^{-3/2}\|\mathbb{E}_0(S_k)\|_2 < \infty .$$

Then, $\lim_{n\to\infty} n^{-1/2}\mathbb{E}(S_n^2) = c^2$ and, on a set of probability one, for any continuous and bounded function f from $(D([0,1), \|\cdot\|_\infty)$ to $\mathbb{R}$,

$$\lim_{n\to\infty} \mathbb{E}_0(f(W_n)) = \int f(zc)W(dz)\,,$$

where $W_n = \{n^{-1}\sum_{k=1}^{[nt]} X_k, t \in [0,1]\}$ and W is the distribution of a standard Wiener process.

The idea of proof is to work under $\mathbb{P}_0$ (the conditional probability given $\mathcal{F}_0$) and verify that the conditions of our general FCLT hold with probability one. For instance, we need to verify (6); that is, with probability one, there exists a constant c^2 such that, for any $t \in [0,1]$,

$$\lim_{m\to\infty} \limsup_{n\to\infty} \mathbb{P}_0\Big(\Big|\frac{1}{n}\sum_{k=1}^{[nt]} \big(X_k^2 + \frac{2}{m} X_k \sum_{i=1}^{m-1} \mathbb{E}_k(S_{k+i} - S_k)\big) - tc^2\Big| > \varepsilon\Big) = 0 .$$

But, by the ergodic theorem,

$$\lim_{m\to\infty} \lim_{n\to\infty} \Big|\frac{1}{n}\sum_{k=1}^{[nt]} (X_k^2 + \frac{2}{m} X_k \sum_{i=1}^{m-1} \mathbb{E}_k(S_{k+i} - S_k)) - tc^2\Big| = 0 \ a.s.$$

Hence, by the properties of the conditional expectation, the desired convergence follows.

3.4 Application to Locally Stationary Processes

Let us consider $\{n^{-1/2}\sum_{k=1}^{[nt]} X_{k,n}, t \in [0,1]\}$ when $(X_{k,n}, 1 \le k \le n)$ is a locally stationary process in the sense that $X_{k,n}$ can be locally approximated by a stationary process $\tilde{X}_k(u)$ in some neighborhood of u, i.e., for those k where $|(k/n) - u|$ is small.

Assume that $\mathbb{E}(X_{k,n}) = 0$. For each $u \in [0, 1]$, let $\tilde{X}_k(u)$ be a stationary and ergodic process such that

(S_0) $\max_{1\le j\le n} n^{-1/2}\left|\sum_{k=1}^{j} X_{k,n} - \sum_{k=1}^{j} \tilde{X}_k(k/n)\right| \to^{\mathbb{P}} 0.$

(S_1) $\sup_{u\in[0,1]} \|\tilde{X}_k(u)\|_2 < \infty$ and

$$\lim_{\varepsilon\to 0} \sup_{|u-v|\le\varepsilon} \|\tilde{X}_k(u) - \tilde{X}_k(v)\|_2 = 0.$$

(D) There exists a stationary nondecreasing filtration $(\mathcal{F}_k)_{k\ge 0}$ such that, for each $u \in [0, 1]$, $\tilde{X}_k(u)$ is adapted to $\mathcal{F}_k$ and the following condition holds: $\sum_{k\ge 0} 2^{-k/2}\, \tilde{\delta}(2^k) < \infty$, where $\tilde{\delta}(k) = \sup_{u\in[0,1]} \|\mathbb{E}(\tilde{S}_k(u)|\mathcal{F}_0)\|_2$ and $\tilde{S}_k(u) = \sum_{i=1}^{k} \tilde{X}_i(u)$.

Let us give an example. For any $u \in [0, 1]$, let

$$Y_k(u) = \sum_{i\ge 0} (\alpha(u))^i \varepsilon_{k-i} \text{ and } \tilde{X}_k(u) = f(Y_k(u)) - \mathbb{E} f(Y_k(u))$$

with $f \in \mathcal{L}(c)$ (this space of functions has been defined in Sect. 3.2) and $\alpha(\cdot)$ a Lipschitz continuous function such that $\sup_{u\in[0,1]} |\alpha(u)| = \alpha < 1$.

Define

$$X_{k,n} = \tilde{X}_k(k/n) + n^{-3/2} u_n(\varepsilon_k + \cdots + \varepsilon_{k-n})$$

where $u_n \to 0$.

Condition (S_0) is satisfied and conditions (S_1) and (D) also, provided

$$\int_0^1 \frac{c(t)}{t\sqrt{|\log t|}} dt < \infty .$$

Theorem 14 *Assume the above conditions. Then, there exists a Lebesgue integrable function $\sigma^2(\cdot)$ on $[0, 1]$ such that, for any $u \in [0, 1]$,*

$$\lim_{m\to\infty} \mathbb{E}(\tilde{S}_m(u))^2 = \sigma^2(u)$$

and the sequence of processes $\{n^{-1/2} \sum_{k=1}^{[nt]} X_{k,n},\ t \in [0, 1]\}$ converges in distribution in $D([0, 1])$ to

$$\left\{ \int_0^t \sigma(u) dW(u),\ t \in [0, 1] \right\},$$

where W is a standard Brownian motion.

Compared to the results in Dahlhaus, Richter, and Wu [4], this result has a different range of applications. In addition, we do not need to assume that $\| \sup_{u\in[0,1]} |\tilde{X}_k(u)| \|_2 < \infty$ nor that $\tilde{X}_k(u)$ takes the form $H(u, \eta_k)$ with H a measurable function and $\eta_k = (\varepsilon_j, j \le k)$, where $(\varepsilon_j)_{j\in\mathbb{Z}}$ a sequence of i.i.d. real-valued random variables.

4 The Case of α-Dependent Triangular Arrays

We start this section by defining weak forms of strong-mixing-type coefficients for a triangular array of random variables $(X_{i,n})$. For any integer $i \ge 1$, let $f_{i,n}(t) = \mathbf{1}_{\{X_{i,n}\le t\}} - \mathbb{P}(X_{i,n} \le t)$. For any nonnegative integer k, set

$$\alpha_{1,n}(k) = \sup_{i\ge 0} \max_{i+k\le u} \sup_{t\in\mathbb{R}} \left\| \mathbb{E}\big(f_{u,n}(t)|\mathcal{F}_{i,n}\big) \right\|_1 ,$$

and

$$\alpha_{2,n}(k) = \sup_{i\ge 0} \max_{i+k\le u\le v} \sup_{s,t\in\mathbb{R}} \left\| \mathbb{E}\big(f_{u,n}(t) f_{v,n}(s)|\mathcal{F}_{i,n}\big) - \mathbb{E}\big(f_{u,n}(t) f_{v,n}(s)\big) \right\|_1 ,$$

where, for $i \ge 1$, $\mathcal{F}_{i,n} = \sigma(X_{j,n} 1 \le j \le i\}$ and $\mathcal{F}_{0,n} = \{\emptyset, \Omega\}$. In the definitions above, we extend the triangular arrays by setting $X_{i,n} = 0$ if $i > n$. Assume that

$$\sigma_{n,n}^2 = \mathrm{Var}\Big(\sum_{\ell=1}^{n} X_{\ell,n}\Big) = 1 , \tag{22}$$

and, for $0 \le t \le 1$, define $v_n(t)$ and $W_n(t)$ as in (18).

We shall now introduce two conditions that combine the tail distributions of the variables with their associated α-dependent coefficients:

$$\lim_{m\to\infty} \limsup_{n\to\infty} \sum_{k=1}^{n} \sum_{i=m}^{n} \int_0^{\alpha_{1,n}(i)} Q_{k,n}^2(u)du = 0 \tag{23}$$

and

$$\lim_{m\to\infty} \limsup_{n\to\infty} \sum_{k=1}^{n} \int_0^{\alpha_{2,n}(m)} Q_{k,n}^2(u)du = 0 , \tag{24}$$

where $Q_{k,n}$ is the quantile function of $X_{k,n}$ i.e., the inverse function of $t \mapsto \mathbb{P}(|X_{k,n}| > t)$.

Under the conditions above and using a similar martingale approximation approach as in the proof of Theorem 2, the following result holds (see [25]):

Theorem 15 *Suppose that (9), (22), (23), and (24) hold. Then,* $\{W_n(t), t \in [0, 1]\}$ *converges in distribution in* $D([0, 1])$ *(equipped with the uniform topology) to* W*, where* W *is a standard Brownian motion.*

Under the assumptions of Theorem 15, we then get that $\sum_{k=1}^{n} X_{k,n} \Rightarrow N(0, 1)$. To see this, it suffices to notice that by (22), proving that $\|W_n(1) - \sum_{k=1}^{n} X_{k,n}\|_2 \to 0$ is reduced to prove that $\mathrm{Cov}(\sum_{k=1}^{v_n(1)} X_{k,n}, \sum_{k=1+v_n(1)}^{n} X_{k,n}) \to 0$, which follows from (23) by using Rio's covariance inequality [36] and taking into account the Lindeberg condition.

Very often, for the sake of applications, it is convenient to express the conditions in terms of mixing rates and moments:

Corollary 16 *Assume that conditions (9) and (22) hold. Suppose in addition that, for some* $\delta \in (0, \infty]$,

$$\sup_n \sum_{k=1}^{n} \|X_{k,n}\|_{2+\delta}^2 < \infty \text{ and } \sum_{i \geq 1} i^{2/\delta} \alpha_1(i) < \infty$$

and that

$$\lim_{m \to \infty} \limsup_{n \to \infty} \alpha_{2,n}(m) = 0\,.$$

Then, the conclusion of Theorem 15 holds.

There are numerous counterexamples to the CLT, involving stationary strong mixing sequences, in papers by Davydov [5], Bradley [1], Doukhan et al. [9], and Häggström [17], among others. We know that in the stationary case our conditions reduce to the minimal ones. These examples show that we cannot just assume that only the moments of order 2 are finite. Furthermore, the mixing rate is minimal in some sense (see [9]).

We also would like to mention that a central limit theorem was obtained by Rio [37], which also implies the CLT in Corollary 16.

4.1 Application to Functions of α-Dependent Markov Chains

Let $Y_{i,n} = f_{i,n}(X_i)$, where $\mathbf{X} = (X_i)_{i \in \mathbb{Z}}$ is a stationary Markov process with Kernel operator K and invariant measure ν and, for each i and n, $f_{i,n}$ is such that $\nu(f_{i,n}) = 0$ and $\nu(f_{i,n}^2) < \infty$. Let $\sigma_n^2 = \mathrm{Var}(\sum_{i=1}^{n} Y_{i,n})$ and $X_{i,n} = \sigma_n^{-1} Y_{i,n}$. Note that the weak dependence coefficients $\alpha_1(i)$ of $\mathbf{X}$ can be rewritten as follows: Let BV_1 be the class of bounded variation functions h such that $|h|_v \leq 1$ (where $|h|_v$ is the total variation norm of the measure dh). Then,

$$\alpha_1(i) = \frac{1}{2} \sup_{f \in BV_1} \nu\big(|K^i(f) - \nu(f)|\big) .$$

We mention that $\alpha_2(i)$ will have the same order of magnitude as $\alpha_1(i)$ if the space BV_1 is invariant under the iterates K^n of K, uniformly in n, i.e., there exists a positive constant C such that, for any function f in BV_1 and any $n \geq 1$,

$$|K^n(f)|_v \leq C|f|_v .$$

The Markov chains such that $\alpha_2(n) \to 0$, as $n \to \infty$, are not necessarily mixing in the sense of Rosenblatt.

Let us give an example. In what follows, for $\gamma \in]0, 1[$, we consider the Markov chain $(X_k)_{k\geq 1}$ associated with the transformation T_γ defined from [0, 1] to [0, 1] by

$$T_\gamma(x) = \begin{cases} x(1 + 2^\gamma x^\gamma) & \text{if } x \in [0, 1/2[\\ 2x - 1 & \text{if } x \in [1/2, 1] . \end{cases}$$

This is the so-called LSV [21] map with parameter γ. There exists a unique T_γ-invariant measure ν_γ on [0, 1], which is absolutely continuous with respect to the Lebesgue measure with positive density denoted by h_γ. We denote by K_γ the Perron-Frobenius operator of T_γ with respect to ν_γ (recall that, for any bounded measurable functions f and g, $\nu_\gamma(f \cdot g \circ T_\gamma) = \nu_\gamma(K_\gamma(f)g))$. Then, $(X_i)_{i\geq 0}$ will be the stationary Markov chain with transition Kernel K_γ and invariant measure ν_γ. In addition, we assume that, for any i and n fixed, $f_{i,n}$ is monotonic on some open interval and 0 elsewhere. It follows that the weak dependence coefficients associated with $(X_{i,n})$ are such that $\alpha_{2,n}(k) \leq Ck^{1-1/\gamma}$, where C is a positive constant not depending on n. By applying Corollary 16, we derive that if the triangular array $(X_{i,n})$ satisfies the Lindeberg condition (9) and if

$$\gamma \in (0, 1/2) \text{ and } \sup_{n\geq 1} \frac{1}{\sigma_n^2} \sum_{i=1}^{n} \Big(\int_0^1 f_{i,n}^{2+\delta}(x) x^{-\gamma} dx \Big)^{2/(2+\delta)} < \infty$$

$$\text{for some } \delta > \frac{2\gamma}{1 - 2\gamma} ,$$

then the conclusion of Theorem 15 is satisfied for the triangular array $(X_{i,n})$ defined above.

4.2 *Application to Linear Statistics with α−Dependent Innovations*

We consider statistics of the type

$$S_n = \sum_{j=1}^{n} d_{n,j} X_j , \tag{25}$$

where $d_{n,j}$ are real-valued weights and (X_j) is a strictly stationary sequence of centered real-valued random variables in $\mathbb{L}^2$. This model is also useful to analyze linear processes with dependent innovations and regression models. It was studied in Peligrad and Utev [30] and Rio [37] and also in Peligrad and Utev [32], where a central limit theorem was obtained by using a stronger form of the mixing coefficients.

We assume that the sequence of constants satisfies the following two conditions:

$$\sum_{i=1}^{n} d_{n,i}^2 \to c^2 \text{ and } \sum_{i=1}^{n} (d_{n,j} - d_{n,j-1})^2 \to 0 \text{ as } n \to \infty , \tag{26}$$

where $c^2 > 0$. Also, we impose the conditions

$$\sum_{i \geq 0} \int_0^{\alpha_1(i)} Q^2(u) du < \infty \tag{27}$$

and

$$\alpha_2(m) \to 0, \tag{28}$$

where Q is the quantile function associated with X_0.

Condition (27) implies that $\sum_{k \geq 0} |\mathrm{Cov}(X_0, X_k)| < \infty$ and therefore that the sequence (X_j) has a continuous spectral density $f(x)$. Note also that if the spectral density f is continuous and (26) is satisfied, then

$$\sigma_n^2 = \mathrm{Var}(S_n) \to 2\pi c^2 f(0), \text{ as } n \to \infty .$$

We refer, for instance, to [27, Lemma 1.5] for a proof of this fact. Note also that (26) implies the Lindeberg condition (4). Indeed, condition (26) entails that $\max_{1 \leq \ell \leq n} |d_{n,\ell}| \to 0$, as $n \to \infty$ (see [27, Lemma 12.12]).

By applying Theorem 15, we obtain the following result (see Merlevède-Peligrad [25]):

Theorem 17 *Let $S_n = \sum_{j=1}^{n} d_{n,j} X_j$, where $d_{n,j}$ are real-valued weights and (X_j) is a strictly stationary sequence. Assume that (26), (27), and (28) are satisfied. Then, S_n converges in distribution to $\sqrt{2\pi f(0)}|c|N$, where N is a standard Gaussian random variable. Let $v_{k,n}^2 = \sum_{i=1}^{k} d_{n,i}^2$. Define*

$$v_n(t) = \inf\left\{k;\ 1 \le k \le n:\ v_{k,n}^2 \ge c^2 t\right\} \text{ and } W_n(t) = \sum_{i=1}^{v_n(t)} d_{n,i} X_i\,.$$

Then, $W_n(\cdot)$ converges weakly to $\sqrt{2\pi f(0)}|c|W$, where W is the standard Brownian motion.

Comment 18 To apply Theorem 15, we do not need to impose condition (26) in its full generality. Indeed, this condition can be replaced by the following ones:

$$\sum_{i=1}^{n} d_{n,i}^2 \to c^2 \text{ and } \max_{1\le \ell \le n} |d_{n,\ell}| \to 0 \text{ as } n \to \infty\,, \tag{29}$$

and, for any positive k, there exists a constant c_k such that

$$\lim_{n\to\infty} \frac{\sum_{\ell=1}^{n-k} d_{n,\ell} d_{n,\ell+k}}{\sum_{\ell=1}^{n} d_{n,\ell}^2} \to c_k\,. \tag{30}$$

Indeed, condition (29) implies the Lindeberg condition (4), whereas condition (30) together with $\sum_{k\ge 0} |\mathrm{Cov}(X_0, X_k)| < \infty$ (which is, in particular, implied by (27)) entails that

$$\frac{\sigma_n^2}{\sum_{\ell=1}^{n} d_{n,\ell}^2} \to \sigma^2 = \mathrm{Var}(X_0) + 2\sum_{k\ge 1} c_k \mathrm{Cov}(X_0, X_k)\,, \text{ as } n \to \infty\,. \tag{31}$$

Note that if condition (26) holds, then (30) is satisfied with $c_k = 1$ for all positive integer k and therefore $\sigma^2 = 2\pi f(0)$. Hence, if, in the statement of Theorem 17, condition (26) is replaced by conditions (29) and (30), then its conclusions hold with σ^2 replacing $2\pi f(0)$, where σ^2 is defined in (31). To end this comment, let us give an example where conditions (29) and (30) are satisfied but the second part of (26) fails. With this aim, let x be a real such that $x \notin \pi\mathbb{Z}$ and let $d_{n,k} = \sin(xk)/\sqrt{n}$. For this choice of triangular array, we have $\sum_{i=1}^{n} d_{n,i}^2 \to 1/2$ and, for any positive k, $\sum_{\ell=1}^{n-k} d_{n,\ell} d_{n,\ell+k} \to 2^{-1}\cos(xk)$. Therefore, (30) is satisfied with $c_k = \cos(xk)$ and (26) does not hold.

Remark 19 We refer to Dedecker et al. [7, Section 4] for various results concerning rates of convergence in the central limit theorem for linear statistics of the above type with dependent innovations. In particular, they proved the following result (see their corollary 4.1 and their remark 4.2). Let $p \in (2, 3]$. Assume that

$$\mathbb{P}(|X_0| \ge t) \le C t^{-s} \text{ for some } s > p \text{ and } \sum_{k\ge 1} k(\alpha_2(k))^{2/p-2/s} < \infty\,,$$

and that the spectral density of (X_i) satisfies $\inf_{t\in[-\pi,\pi]} |f(t)| = m > 0$. Then, setting $m_n = \max_{1\le \ell\le n} |d_{n,\ell}|$, the following upper bounds hold: for any positive integer n,

$$\int_{\mathbb{R}} |F_n(t) - \Phi(t)|dt \ll C(n,p) := \begin{cases} \dfrac{m_n^{p-2}}{\sigma_n}\Big(\sum_{\ell=1}^{n} d_{n,\ell}^2\Big)^{(3-p)/2} & \text{if } p \in (2,3) \\ \dfrac{m_n}{\sigma_n}\log\Big(m_n^{-1}\sum_{\ell=1}^{n} d_{n,\ell}^2\Big) & \text{if } p = 3, \end{cases} \tag{32}$$

where we recall that F_n is the c.d.f. of S_n/σ_n and Φ is the c.d.f. of a standard Gaussian r.v. Note that if we replace the condition that the spectral density has to be bounded away from 0 by the weaker one: $f(0) > 0$, and if, as a counterpart, we assume the additional condition $\sum_{k>0} k^2|\mathrm{Cov}(X_0, X_k)| < \infty$, then an additional term appears in (32); namely, we get

$$\int_{\mathbb{R}} |F_n(t) - \Phi(t)|dt \ll C(n,p) + \frac{\Big(\sum_{k=1}^{n+1}(d_{n,k} - d_{n,k-1})^2\Big)^{1/2}}{\sigma_n}.$$

See [7, Corollary 4.2].

In what follows, we apply Theorem 17 to the model of the nonlinear regression with fixed design. Our goal is to estimate the function $\ell(x)$ such that

$$y(x) = \ell(x) + \xi(x),$$

where ℓ is an unknown function and $\xi(x)$ is the noise. If we fix the design points $x_{n,i}$, we get

$$Y_{n,i} = y(x_{n,i}) = \ell(x_{n,i}) + \xi_i(x_{n,i}).$$

According to [32], the nonparametric estimator of $\ell(x)$ is defined to be

$$\hat{\ell}_n(x) = \sum_{i=1}^{n} w_{n,i}(x)Y_{n,i}, \tag{33}$$

where

$$w_{n,i}(x) = K\Big(\frac{x_{n,i} - x}{h_n}\Big) / \sum_{i=1}^{n} K\Big(\frac{x_{n,i} - x}{h_n}\Big).$$

We apply Theorem 17 to find sufficient conditions for the convergence of the estimator $\hat{\ell}_n(x)$. To fix the ideas we shall consider the following setting: The kernel K is a density function, continuous with compact support $[0, 1]$. The design points will be $x_{n,i} = i/n$ and $(\xi_i(x_{n,1}), \ldots, \xi_i(x_{n,i}))$ is distributed as $(X_1, \ldots, X_n)$, where $(X_k)_{k\in\mathbb{Z}}$ is a stationary sequence of centered sequence of random variables satisfying (27) and (28). We then derive the normal asymptotic limit for

$$V_n(x) = \left(\sum_{i=1}^{n} w_{n,i}^2(x)\right)^{-1/2} \left(\hat{\ell}_n(x) - \mathbb{E}(\hat{\ell}_n(x))\right).$$

The following theorem was established in Merlevède-Peligrad [25].

Theorem 20 *Assume for x fixed that $\hat{\ell}_n(x)$ is defined by (33) and the sequence (X_j) is a stationary sequence satisfying (27) and (28). Assume that the kernel K is a density, is square integrable, has compact support, and is continuous. Assume $nh_n \to \infty$ and $h_n \to 0$. Then, $\sqrt{nh_n}(\hat{\ell}_n(x) - \mathbb{E}(\hat{\ell}_n(x)))$ converges in distribution to $\sqrt{2\pi f(0)}|c|N$, where N is a standard Gaussian random variable and c^2 is the second moment of K.*

4.3 Application to Functions of a Triangular Stationary Markov Chain

Let us consider a triangular version of the Markov chain defined in Example 3.

For any positive integer n, $(\xi_{i,n})_{i\ge 0}$ is a homogeneous Markov chain with state space $\mathbb{N}$ and transition probabilities given by

$$\mathbb{P}(\xi_{1,n} = i|\xi_{0,n} = i+1) = 1 \text{ and } \mathbb{P}(\xi_{1,n} = i|\xi_{0,n} = 0) = p_{i+1,n} \text{ for } i \ge 1,$$

where, for $i \ge 2$, $p_{i,n} = c_a/(v_n i^{a+2})$ with $a > 0$, $c_a \sum_{i\ge 2} 1/i^{a+2} = 1/2$, $(v_n)_{n\ge 1}$ a sequence of positive reals and $p_{1,n} = 1 - 1/(2v_n)$. $(\xi_{i,n})_{i\ge 0}$ has a stationary distribution $\pi_n = (\pi_{j,n})_{j\ge 0}$ satisfying

$$\pi_{0,n} = \Big(\sum_{i\ge 1} i p_{i,n}\Big)^{-1} \text{ and } \pi_{j,n} = \pi_{0,n} \sum_{i\ge j+1} p_{i,n} \text{ for } j \ge 1.$$

Let $Y_{i,n} = I_{\xi_{i,n}=0} - \pi_{0,n}$. Let $b_n^2 = \mathrm{Var}\Big(\sum_{k=1}^{n} Y_{k,n}\Big)$ and set $X_{i,n} = Y_{i,n}/b_n$. Provided that $a > 1$ and $v_n/n \to 0$, $(X_{k,n})_{k>0}$ satisfies the functional central limit theorem given in Theorem 15.

Acknowledgments This paper was partially supported by the NSF grant DMS-2054598. The authors would like to thank the referee for carefully reading the manuscript and for the useful suggestions.

References

1. R.C. Bradley, On quantiles and the central limit question for strongly mixing sequences. J. Theor. Probab. **10**, 507–555 (1997)

2. C. Cuny, F. Merlevède, On martingale approximations and the quenched weak invariance principle. Ann. Probab. **42**, 760–793 (2014)
3. C. Cuny, J. Dedecker, D. Volný, A functional central limit theorem for fields of commuting transformations via martingale approximation. Zapiski Nauchnyh Seminarov POMI 441.C. Part 22, 239–263 (2016). J. Math. Sci. **219**, 765–781 (2015)
4. R. Dahlhaus, S. Richter, W.B. Wu, Towards a general theory for nonlinear locally stationary processes. Bernoulli **25**(2), 1013–1044 (2019)
5. Yu.A. Davydov, Mixing conditions for Markov chains. Theory Probab. Appl. **18**, 312–328 (1973)
6. J. Dedecker, Inégalités de Hoeffding et théorème limite central pour les fonctions peu régulières de chaînes de Markov non irréductibles. numéro spécial des Annales de l'ISUP **52**, 39–46 (2008)
7. J. Dedecker, F. Merlevède, E. Rio, Rates of convergence in the central limit theorem for martingales in the non stationary setting. Ann. Inst. H. Poincaré Probab. Statist. **58**(2), 945–966 (2022)
8. R. Dobrushin, Central limit theorems for nonstationary Markov chains I, II. Theory Probab. Appl. **1**, 65–80, 329–383 (1956)
9. P. Doukhan, P. Massart, E. Rio, The functional central limit theorem for strongly mixing processes. Ann. Inst. H. Poincaré Probab. Statist. **30**, 63–82 (1994)
10. O. Durieu, Independence of four projective criteria for the weak invariance principle. ALEA Lat. Am. J. Probab. Math. Stat. **5**, 21–26 (2009)
11. M. El Machkouri, D. Giraudo, Orthomartingale-coboundary decomposition for stationary random fields. Stoch. Dyn. **16**(5), 1650017, 28 pp. (2016)
12. D. Giraudo, Invariance principle via orthomartingale approximation. Stoch. Dyn. **18**(6), 1850043, 29 pp. (2018)
13. M.I. Gordin, The central limit theorem for stationary processes. Soviet. Math. Dokl. **10**, 1174–1176 (1969)
14. M.I. Gordin, Martingale-coboundary representation for a class of random fields. J. Math. Sci. **163**(4), 363–374 (2009)
15. M. Gordin, M. Peligrad, On the functional CLT via martingale approximation. Bernoulli **17**, 424–440 (2011)
16. Y. Hafouta, Convergence rates in the strong, weak and functional invariance principles for nonstationary mixing arrays via variance linearization. arXiv:2107.02234 (2021)
17. O. Häggström, On the central limit theorem for geometrically ergodic Markov chains. Probab. Theory Related Fields **132**, 74–82 (2005)
18. E.J. Hannan, The central limit theorem for time series regression. Stochastic Process. Appl. **9**(3), 281–289 (1979)
19. C.C. Heyde, On the central limit theorem for stationary processes. Z. Wahrsch. verw. Gebiete. **30**, 315–320 (1974)
20. I.A. Ibragimov, A note on the central limit theorem for dependent random variables. Teor. Veroyatnost. i Primenen. **20**(1), 134–140. Theory Probab. Appl. **20**(1), 135–141 (1975)
21. C. Liverani, B. Saussol, S. Vaienti, A probabilistic approach to intermittency. Ergodic Theory Dynam. Syst. **19**(3), 671–685 (1999)
22. M. Maxwell, M. Woodroofe, Central limit theorem for additive functionals of Markov chains. Ann. Probab. **28**, 713–724 (2000)
23. D.L. McLeish, Invariance principles for dependent variables. Z. Wahrscheinlichkeitstheorie und Verw. Gebiete **32**, 165–178 (1975)
24. D.L. McLeish, On the invariance principle for nonstationary mixingales. Ann. Probab. **5**, 616–621 (1977)
25. F. Merlevède, M. Peligrad, Functional CLT for nonstationary strongly mixing processes. Statist. Probab. Lett. **156**, 108581, 10 pp. (2020)
26. F. Merlevède, M. Peligrad, S. Utev, Functional CLT for martingale-like nonstationary dependent structures. Bernoulli **25**(4B), 3203–3233 (2019)

27. F. Merlevède, M. Peligrad, S. Utev, *Functional Gaussian Approximation for Dependent Structures* (Oxford University Press, Oxford, 2019)
28. M.H. Neumann, A central limit theorem for triangular arrays of weakly dependent random variables, with applications in statistics. ESAIM Probab. Stat. **17**, 120–134 (2013)
29. M. Peligrad, Central limit theorem for triangular arrays of nonhomogeneous Markov chains. Probab. Theory Related Fields **154**(3–4), 409–428 (2012)
30. M. Peligrad, S. Utev, Central limit theorem for linear processes. Ann. Probab. **25**, 443–456 (1997)
31. M. Peligrad, S. Utev, A new maximal inequality and invariance principle for stationary sequences. Ann. Probab. **33**, 798–815 (2005)
32. M. Peligrad, S. Utev, Central limit theorem for stationary linear processes. Ann. Probab. **34**, 1608–1622 (2006)
33. M. Peligrad, N.A. Zhang, On the normal approximation for random fields via martingale methods. Stoch. Process. Appl. **128**(4), 1333–1346 (2018)
34. M. Peligrad, N.A. Zhang, Martingale approximations for random fields. Electron. Commun. Probab. **23**, Paper No. 28, 9 pp. (2018)
35. M. Peligrad, N.A. Zhang, Central limit theorem for Fourier transform and periodogram of random fields. Bernoulli **25**(1), 499–520 (2019)
36. E. Rio, Covariance inequalities for strongly mixing processes. Ann. Inst. Henri Poincaré Probab. Stat. **29**, 587–597 (1993)
37. E. Rio, About the Lindeberg method for strongly mixing sequences. ESAIM, Probabilités et Statistiques **1**, 35–61 (1995)
38. Q.M. Shao, On the invariance principle for ρ-mixing sequences of random variables. Chin. Ann. Math. (Ser. B) **10**, 427–433 (1989)
39. S. Utev, Central limit theorem for dependent random variables in *Probability Theory and Mathematical Statistics*, vol. II (Vilnius 1989), "Mokslas", Vilnius (1990), pp. 512–528
40. D. Volný, A central limit theorem for fields of martingale differences. C. R. Math. Acad. Sci. Paris **353**, 1159–1163 (2015)
41. D. Volný, Martingale-coboundary representation for stationary random fields. Stoch. Dyn. **18**(2), 1850011, 18 pp. (2018)
42. D. Volný, Y. Wang, An invariance principle for stationary random fields under Hannan's condition. Stoch. Proc. Appl. **124**, 4012–4029 (2014)
43. W.B. Wu, Z. Zhou, Gaussian approximations for non-stationary multiple time series. Statist. Sinica **21**(3), 1397–1413 (2011)
44. O. Zhao, M. Woodroofe, On Martingale approximations. Ann. Appl. Probab. **18**(5), 1831–1847 (2008)

Part III
Stochastic Processes

Sudakov Minoration for Products of Radial-Type Log-Concave Measures

Witold Bednorz

2020 Mathematics Subject Classification: 60G15; 60G17

1 Introduction

Consider a random vector X, which takes values in $\mathbb{R}^d$. Let T be a subset of $\mathbb{R}^d$, and we can define canonical process as

$$(X_t)_{t\in T}, \quad \text{where } X_t = \sum_{i=1}^{d} t_i X_i = \langle t, X\rangle.$$

One of the basic questions in the analysis of stochastic processes is to characterize $S_X(T) = \mathbf{E}\sup_{t\in T} X_t$, where $(X_t)_{t\in T}$ is a family of random variables. It is well-known that usually $S_X(T) < \infty$ is equivalent to $\mathbf{P}(\sup_{t\in T}|X_t| < \infty) = 1$, i.e., that paths are a.s. bounded. In order to avoid measurability questions formally, $S_X(T)$ should be defined $\sup_{F\subset T}\mathbf{E}\sup_{t\in F} X_t$, where the supremum is over all finite subsets F of T. The main tool invented to study the quantity is the generic chaining. Although the idea works neatly whenever the upper bound is concerned, the lower bound is a much more subtle matter. What one can use toward establishing the lower bound for $\mathbf{E}\sup_{t\in T} X_t$ is the growth condition that can be later used in the partition scheme [18] or construction of a special family of functionals [19]. The core of the approach is the Sudakov minoration. Basically, the minoration means that we can answer the question about understanding $\mathbf{E}\sup_{t\in T} X_t$ in the simplest setting, where points in T are well separated. Before we state the condition, let us make a simple remark.

W. Bednorz (✉)
Institute of Mathematics, University of Warsaw, Warszawa, Poland
e-mail: wbednorz@mimuw.edu.pl

R. Adamczak et al. (eds.), *High Dimensional Probability IX*, Progress in Probability 80, https://doi.org/10.1007/978-3-031-26979-0_11

Remark 1 Let $p \geqslant 1$. Suppose that $|T| \leqslant e^p$, and $\|X_t\|_p \leqslant A$ for each $t \in T$. Then, $\mathbf{E}|\sup_{t\in T} X_t| \leqslant eA$.

Proof Indeed, we have

$$\mathbf{E}|\sup_{t\in T} X_t| \leqslant \mathbf{E}\left[\sum_{t\in T}|X_t|^p\right]^{1/p} \leqslant \left[\mathbf{E}\sum_{t\in T}|X_t|^p\right]^{1/p} \leqslant eA.$$

□

The Sudakov minoration means that something opposite happens. Namely, we require that $|T| \geqslant e^p$, $p \geqslant 1$ and for all $s, t \in T$, $s \neq t$, $\|X_t - X_s\|_p \geqslant A$. These assumptions should imply

$$\mathbf{E}\sup_{t\in T} X_t \geqslant K^{-1}A, \tag{1}$$

where K is an absolute constant. The problem has a long history, which we outline below.

The property was first proved in 1969 for X-Gaussian, i.e., $X = G = (g_i)_{i=1}^d$, g_i are independent standard normal variables. Sudakov [14, 15] provides the result, although some other researchers could be aware of the fact. In 1986 Pajor and Tomczak-Jaegermann [12] observed that the property for Gaussian X can be established in the dual way. Then, Talagrand [16] proved in 1990 that the minoration works for $X = \varepsilon = (\varepsilon_i)_{i=1}^d$, ε_i are independent Rademachers. He invented the result, when studying properties of infinitely divisible processes. Later, in 1994, Talagrand [17] realized that the minoration holds also for exponential-type distributions. The result was improved by Latala [5] in 1997, so that it holds for X, which has independent entries of log-concave tails. The problem to go beyond the class of independent entries occurred to be unexpectedly hard. First result toward this direction was due to Latala [6], 2014 and concerned X of density $\exp(-U(\|x\|_p))$, where U is an increasing convex function and $\|x\|_p = \left(\sum_{i=1}^d |x_i|^p\right)^{1/p}$. Then, in 2015, Latala and Tkocz [7] proved that the minoration in the case of independent entries necessarily requires that entries are α-regular, which basically means that they behave like variables of log-concave distribution. Finally, it has to be mentioned that some new ideas appeared how a proof of the minoration could be established, for example, the dimension reduction—the work of Mendelson, Milman, and Paouris [10]. In this paper they covered some simple cases, though it is known that the program cannot succeed in full generality.

The aim of the paper is to prove that the minoration works for a certain class of log-concave distributions that extends mentioned examples. More precisely, we show that the property holds for X of density

$$\mu_X(dx) = \exp(-\sum_{k=1}^{M} U_k(\|x_k\|_{p_k}^{p_k}))dx, \tag{2}$$

where U_k are increasing convex functions, $p_k > 1$ and $x_k = (x_i)_{i \in J_k}$, and J_k are pairwise disjoint and cover $[d] = \{1, 2, \ldots, d\}$. We use the notation $n_k = |J_k|$ and q_k for p_k^*, i.e., $1/p_k + 1/q_k = 1$. It will be clear from our proof that we can say something about the minoration also if $p_k \geqslant 1$; however, if we want the constant K not to depend on the problem setting, we need some cutoff level for p_k from 1.

The distribution μ_X has some properties that are important for the proof of minoration, namely:

1. Isotropic position
2. One unconditionality
3. Log-concavity
4. Structural condition

Let us comment on them below:

(1) Isotropic position. We say that X is in isotropic position if

$$\mathbf{E}X = 0, \quad \text{i.e.}\mathbf{E}X_i = 0, \quad i \in [d], \quad \mathbf{Cov}X = \mathrm{Id}_d, \quad \text{i.e.} \quad \mathbf{Cov}(X_i X_j) = \delta_{i,j}. \tag{3}$$

Note that by an affine transformation we can impose the property, keeping the structural assumption (2). Note also that it is a very natural modification, which makes the Euclidean distance important, i.e., $\|X_t - X_s\|_2 = d_2(s, t) = \|t - s\|_2$.

(2) We say that X is one-unconditional if

$$(X_1, X_2, \ldots, X_d) \stackrel{d}{=} (\varepsilon_1|X_1|, \varepsilon_2|X_2|, \ldots, \varepsilon_d|X_d|), \tag{4}$$

where $|X_i|$, $i \in [d]$ and independent random signs ε_i, $i \in [d]$ that are independent of all $|X_i|$, $i \in [d]$. Observe that the assumption makes it possible to consider $(X_t)_{t \in T}$ in the following form:

$$X_t = \sum_{i=1}^{d} t_i X_i \stackrel{d}{=} \sum_{i=1}^{d} t_i \varepsilon_i |X_i|.$$

(3) Log-concavity. We assume that X has the distribution

$$\mu(dx) = \exp(-U(x))dx, \tag{5}$$

where U is a convex function. Due to the result of [7], the assumption is almost necessary. More formally, what we need is the comparison of moments. For log-concave and 1-unconditional X, we have that for all $0 < p < q$ and $t \in \mathbb{R}^d$ (see [11] or [1] for the proof)

$$\|X_t\|_q \leqslant \frac{\Gamma(q+1)^{\frac{1}{q}}}{\Gamma(p+1)^{\frac{1}{p}}} \|X_t\|_p \leqslant \frac{q}{p} \|X_t\|_p. \tag{6}$$

The necessary condition for the Sudakov minoration formulated in [7] states that X must be α-regular, which means that there exists $\alpha \geqslant 1$ such that, for all $2 \leqslant p < q$

$$\|X_t - X_s\|_q \leqslant \alpha \frac{q}{p} \|X_t - X_s\|_p. \tag{7}$$

Clearly, for log-concave and 1-unconditional X, the inequality (7) works with $\alpha = 1$.

(4) Structural assumption. As we have explained in (2), we assume that the distribution of X has a certain structure. More precisely, we require that

$$U(x) = \sum_{k=1}^{M} U_k(\|x_k\|_{p_k}^{p_k}), \quad p_k > 1. \tag{8}$$

Therefore, we can treat X as $(X_k)_{k=1}^M$, where vectors X_k with values in $\mathbb{R}^{n_k}$ are independent. The fact, which we use later, is that each X_k has the same distribution as $R_k V_K$, where R_k and V_k are independent and R_k is distributed on $\mathbb{R}_+$ with the density $x^{n-1} \exp(-U_k(x^{p_k}))|\partial B_{p_k}^{n_k}|$ and V_k is distributed on $\partial B_{p_k}^{n_k}$ with respect to the (probabilistic) cone measure ν_k—for details see, e.g., [13]. The most important property of the cone measure is that for any integrable $f : \mathbb{R}^{n_k} \to \mathbb{R}$

$$\int_{\mathbb{R}^{n_k}} f(x)dx = |\partial B_{p_k}^{n_k}| \int_{\mathbb{R}_+} \int_{\partial B_{p_k}^{n_k}} f(r\theta) r^{n_k - 1} \nu_k(d\theta) dr.$$

Note that here $B_{p_k}^{n_k} = \{x \in \mathbb{R}^{n_k} : \|x\|_{p_k} \leqslant 1\}$ and $\partial B_{p_k}^{n_k} = \{x \in \mathbb{R}^{n_k} : \|x\|_{p_k} = 1\}$.

2 Results

Following the previous section, we can formulate the main result of this paper:

Theorem 1 *Suppose that $X \in \mathbb{R}^d$ is a random vector whose density is of the form $\mu_X = \mu_1 \otimes \mu_2 \otimes \ldots \otimes \mu_M$ and μ_k is a log-concave distribution on $\mathbb{R}^{n_k}$ given by*

$$\mu_k(dx_k) = \exp(-U_k(\|x_k\|_{p_k}^{p_k}))dx_k, \quad \text{where } x_k = (x_i)_{i \in J_k},$$

where J_k are disjoint sets such that $\bigcup_{k=1}^M J_k = [d]$ and $p_k \geqslant 1 + \varepsilon$ for some $\varepsilon > 0$. Then, the Sudakov minoration holds, i.e., for every set $T \subset \mathbb{R}^d$ such that $\|X_t - X_s\|_p \geqslant A$, $s \neq t$, $s, t \in T$ for some $p \geqslant 1$, the following inequality holds:

$$\mathbf{E} \sup_{t \in T} X_t \geqslant K^{-1} A,$$

where K depends on ε only.

The proof is quite complicated; that is why it is good to give a sketch of our approach.

1. Simplifications. We show that Theorem 1 has to be established for $A = p$, X in the isotropic position and sets T such that for each $t \in T$, $t = (t_i)_{i=1}^d$, $t_i \in \{0, k_i\}$ for some positive k_i, $i \in [d]$ and $\|X_t - X_s\|_p \geqslant p$, $s \neq t$, $s, t \in T$. Moreover, for each $t \in T$, $\sum_{i=1}^d |t_i|$ and $\sum_{i=1}^d 1_{t_i = k_i}$ are much smaller than p. In other words, minoration should be proved for cube-like sets with thin supports.
2. Moments. The next step is a careful analysis of the condition $\|X_t - X_s\|_p \geqslant p$. In particular, we are going to use the structure assumption $X = (X_1, X_2, \ldots, X_M)$, where entries $X_k \in \mathbb{R}^{n_k}$ are independent. The main trick here is to define a random vector $Y = (Y_1, Y_2, \ldots, Y_M)$ such that Y has all coordinates independent, i.e., not only $Y_k \in \mathbb{R}^{n_k}$ are independent, but also coordinates of each Y_k are independent. Due to our structure assumption, it will be possible to represent $X_k = R_k V_k$, $Y_k = \tilde{R}_k V_k$, where $R_k, \tilde{R}_k \in \mathbb{R}_+$ are some α-regular variables and V_k are distributed with respect to cone measures on $\partial B_{p_k}^{n_k}$, respectively.
3. Split. We observe that we can split each point $t \in T$ into t^* (small part) and $t^\dagger$ (large part). More precisely, we decide whether t_k^* is t_k or 0 and, respectively, $t_k^\dagger$ is 0 or t_k, depending on how large some norm of t_k is. Moreover, we show that either there is a subset $S \subset T$, $|S| \geqslant e^{p/2}$ such that $\|X_{t^*} - X_{s^*}\|_p \gtrsim p$ for any $s \neq t$ $s, t \in S$ or we can find $S \subset T$, $|S| \geqslant e^{p/2}$ such that $\|X_{t^\dagger} - X_{s^\dagger}\|_p \gtrsim p$ for any $s \neq t$, s, t in S.
4. Small part. If the "small part" case holds true, then we show that one may forget about variables $R_k, \tilde{R}_k$ and in this way we may deduce the minoration from such a result for the random vector Y.
5. Large part. If the "large part" case holds true, then we prove that variables V_k are not important; more precisely, we prove, following the approach from the "simplifications" step, that not only $t_i \in \{0, k_i\}$, but we have such a property for $t_k^\dagger = (t_i^\dagger)_{i \in J_k}$, namely, $t_k^\dagger \in \{x_k, 0\}$ for some $x_k \in \mathbb{R}^{n_k}$. Consequently, we may deduce the minoration from such a result for the random vector $(X_{x_k})_{k=1}^M$.

It is a standard argument that having the minoration one can prove the comparison between $\mathbf{E} \sup_{t \in T} X_t$ and $\gamma_X(T)$. However, this approach requires some two additional conditions. First, we need that there exists $\varepsilon > 0$ such that

$$\|X_t\|_{2^{n+1}} \geqslant (1 + \varepsilon) \|X_t\|_{2^n}, \quad n \geqslant 0. \tag{9}$$

The property works if there is a cutoff level for all p_k below ∞. Moreover, there must hold a certain form of measure concentration. What suffices is that there exist constants $K, L \geqslant 1$ such that for any $p \geqslant 2$ and set $T \subset \mathbb{R}^d$

$$\left\| \left(\sup_{t\in T} X_t - K\mathbf{E}\sup_{t\in T} X_t \right)_+ \right\|_p \leqslant L \sup_{s,t\in T} \|X_t - X_s\|_p. \tag{10}$$

The result is known only in few cases. Fortunately, in our setting, it can be derived from the infimum convolution [8]. Note that the problem can be easily reduced to the one where T is a gauge of some norm in $\mathbb{R}^d$. Moreover, the main concentration inequality—$CI(\beta)$ from [8]—holds for radial log-concave densities, and it can be tensorized, namely, if $CI(\beta)$ holds for measures $\mu_1, \mu_2, \ldots, \mu_M$, then $CI(\beta')$ holds for $\mu_1 \otimes \mu_2 \otimes \ldots \otimes \mu_M$ with some dependence between β and β', i.e., β' is some multiplication of β.

Theorem 2 *Under the assumption of Theorem 1 and assuming additionally that* $p_k \leqslant p_\infty < \infty$, *the following holds true*

$$\mathbf{E}\sup_{t\in T} X_t \simeq \gamma_X(T).$$

The result is quite standard and goes through the idea of growth condition—[18].

Before we start the proof of Theorem (1), we need a preliminary result, which explains that the Sudakov minoration has to be established only for sets T that have a cube-like structure.

3 Cube-Like Sets

The first step concerns some basic simplifications of the problem. We first note that the minoration has to be proved only for certain cube-like sets T. We use symbols $\lesssim, \gtrsim, \simeq$ whenever we compare quantities up to a numerical constant comparable to 1.

We have to start from the Bobkov Nazarov [2] inequality or rather from its basic consequence:

$$\|X_t - X_s\|_p \lesssim \|\mathcal{E}_t - \mathcal{E}_s\|_p, \tag{11}$$

where $(\mathcal{E}_i)_{i=1}^d$ are independent symmetric standard exponential variables. The Bobkov Nazarov inequality concerns one-unconditional and log-concave distributions in the isotropic position. However, the inequality (11) may hold in a bit more general setting. For example, it is also true if entries of an isotropic X are independent and α-regular. That is why we may simply refer to (11) as the basic requirement. In the proof of our simplification, the condition is used together with the Kwapien-Gluskin result [3]:

$$\|\mathcal{E}_t\|_p \sim p\|t\|_\infty + \sqrt{p}\|t\|_2. \tag{12}$$

On the other hand, by the result of Hitczenko [4], we also have the lower bound:

$$\|X_t\|_p \gtrsim \|\varepsilon_t\|_p \sim \sum_{1\leqslant i\leqslant p} |t_i^*| + \sqrt{p}\left(\sum_{i>p}^{d} |t_i^*|^2\right)^{1/2}, \tag{13}$$

where $(t_i^*)_{i=1}^d$ is the rearrangement of t such that $|t_1^*| \geqslant |t_2^*| \geqslant \ldots \geqslant |t_d^*|$. The last tool we need is that Sudakov minoration works for the random vector $(\varepsilon_i)_{i=1}^d$. As we have already mentioned, the result was first proved in [16]. We are ready to formulate the main simplification result:

Proposition 1 *Suppose that X is in the isotropic position and fix $p \geqslant 2$. Suppose that:*

1. *X is one-unconditional.*
2. *X is log-concave.*
3. *X satisfies* (11).

Then, to show the Sudakov minoration, it suffices to prove the property for all sets T such that:

1. $\exp(p) \leqslant |T| \leqslant 1 + \exp(p)$ *and* $0 \in T$.
2. *for each $i \in [d] = \{1, 2, \ldots, d\}$*

$$t_i \in \{0, k_i\}, \quad \text{where } k_i \geqslant \rho,$$

   where $\rho \leqslant e^{-1}$ and $\rho/\log\frac{1}{\rho} = 4C\delta$.
3. *for each $t \in T$*

$$\sum_{i\in I(t)} k_i \leqslant 2C\delta p, \quad \text{where } I(t) = \{i \in [d] : t_i = k_i\}. \tag{14}$$

4. *for all $s, t \in T$, $s \neq t$*

$$\|X_t - X_s\|_p \geqslant p,$$

where δ is suitably small and $C \geqslant 1$ a universal constant, and that the following inequality holds true:

$$\mathbf{E}\sup_{t\in T} X_t \geqslant K^{-1}p,$$

where K is a universal constant.

Proof The proof is based on a number of straightforward steps.

Step 1 Obviously, it suffices to show that the Sudakov minoration works for $A = p$. Let $m_p(t,s) = \|\sum_{i=1}^n (t_i - s_i)\varepsilon_i\|_p$. We may assume that $p \geqslant 1$ is suitably large. Moreover, we may consider T such that $0 \in T$, $|T| \geqslant$

$\exp(\frac{p}{2}) + \exp(\frac{3p}{4})$ and $m_p(t, 0) \leqslant \delta p$ for all $t \in T$, where $\delta \leqslant 1$ can be suitably small.

By Talagrand's result [17] (see also [18] for the modern exposition), if $N(T, m_p, u) \geqslant \exp(\frac{p}{4}) - 1$, then $\mathbf{E}\sup_{s,t\in T}\sum_{i=1}^n (t_i - s_i)\varepsilon_i \geqslant L^{-1}u$, for a universal L. Therefore, the result holds true with either $K = 2^{-1}L^{-1}\delta$ or $\mathbf{E}\sup_{s,t\in T}\sum_{i=1}^n (t_i - s_i)\varepsilon_i < 2^{-1}L^{-1}\delta p$ and then

$$N(T, m_p, \frac{1}{2}\delta p) \leqslant \exp(\frac{p}{4}) - 1 \leqslant \frac{\exp(\frac{p}{2})}{1 + \exp(\frac{1}{4}p)}.$$

It implies that there exists $t_0 \in T$ such that

$$\{t \in T : m_p(t, t_0) \leqslant \delta p\}| \geqslant |T|\frac{1 + \exp(\frac{1}{4}p)}{\exp(\frac{p}{2})} \geqslant \exp(\frac{p}{2}) + \exp(\frac{3p}{4}).$$

Therefore, we may consider set $T' = \{t - t_0 : m_p(t, t_0) \leqslant \delta p\}$, which satisfies all the requirements we have promised.

Step 2 Let $\delta \leqslant e^{-1}$ and $\rho/\log(1/\rho) = 4C\delta$. We may assume that $0 \in T$, $|T| \geqslant 1 + \exp(\frac{p}{4})$ and additionally

$$t_i \in (k_i - \rho, k_i + \rho) \cup (-\rho, \rho) \ \text{ for all } t \in T \ \text{ and } \ i \in [d],$$

where k_i are given numbers such that $k_i \geqslant \rho$, where $\rho \leqslant e^{-1}$ and it satisfies

$$\rho \log \frac{1}{\rho} = 4\delta \leqslant e^{-1}, \quad \text{where } C \geqslant 1$$

is a universal constant, which we choose later on.

Indeed, consider measure $\mu = \otimes_{i=1}^d \mu_i$, where $\mu_i(dx) = \frac{1}{2}e^{-|x|}dx$ for all $i \in [d]$. For any $x \in \mathbb{R}^d$, $x = (x_i)_{i=1}^d$ and $t \in T$, $t = (t_i)_{i=1}^d$

$$T_x = \{t \in T : t_i \in (x_i - \rho, x_i + \rho) \cup (-\rho, \rho), \ i \in [d]\}$$

and

$$A_t = \{x \in \mathbb{R}^d : t_i \in (x_i - \rho, x_i + \rho) \cup (-\rho, \rho), \ i \in [d]\}.$$

Now, there are two possibilities, either

$$\mu_i(\{x_i : t_i \in (x_i - \rho, x_i + \rho) \cup (-\rho, \rho)\}) \geqslant \rho e^{-|t_i| - \rho} \tag{15}$$

or

$$\mu_i(\{x_i : t_i \in (x_i - \rho, x_i + \rho) \cup (-\rho, \rho)\}) = 1, \quad \text{if } |t_i| < \rho. \tag{16}$$

Applying (13) we get for some $C \geqslant 1$

$$m_p(t, 0) = \| \sum_{i=1}^{d} t_i \varepsilon_i \|_p \geqslant C^{-1} \left(\sum_{i \leqslant p} |t_i|^* + \sqrt{p} \left(\sum_{i>p} |t_i^*|^2 \right)^{\frac{1}{2}} \right) \tag{17}$$

where $|t_i^*|$ is the nondecreasing rearrangement of $|t_i|$. Also (13) implies that

$$m_p(t, 0) \leqslant C \left(\sum_{i \leqslant p} |t_i|^* + \sqrt{p} \left(\sum_{i>p} |t_i^*|^2 \right)^{\frac{1}{2}} \right), \tag{18}$$

for suitably large $C \geqslant 1$. Therefore, using (17) and $\rho / \log \frac{1}{\rho} = 4C\delta$ we obtain

$$|\{i \in \{1, ..., d\} : |t_i| \geqslant \rho\}| \leqslant \frac{p}{4 \log \frac{1}{\rho}} \leqslant \frac{p}{4}. \tag{19}$$

Again using (17) this shows

$$\sum_{i=1}^{d} |t_i| 1_{|t_i| \geqslant \rho} \leqslant C\delta p. \tag{20}$$

Consequently, by (15,16)

$$\mu(A_t) \geqslant \rho^{\frac{p}{4 \log \frac{1}{\rho}}} \exp(-2C\delta p) \geqslant \exp(-\frac{p}{2}).$$

However, using that $|T| \geqslant \exp(\frac{3p}{4})$, we infer

$$\int \sum_{t \in T} 1_{A_t}(x) \mu(dx) \geqslant |T| \exp(-\frac{p}{2}) \geqslant \exp(\frac{p}{4}).$$

Therefore, we get that there exists at least one point $k \in \mathbb{R}^d$ such that

$$|T_k| \geqslant \exp(\frac{p}{4}).$$

It is obvious that $|k_i|$ may be chosen in a way that $|k_i| \geqslant \rho$. Combining (20) with $|k_i| \geqslant \rho$ and $t_i \in (k_i - \rho, k_i + \rho)$, we obtain

$$\sum_{i=1}^{d}(|k_i|-\rho)\vee\rho 1_{|t_i|\geqslant\rho}\leqslant C\delta p.$$

Clearly, $(|k_i|-\rho)\vee\rho\geqslant\frac{1}{2}|k_i|$, and therefore

$$\frac{1}{2}\sum_{i=1}^{d}|k_i|1_{|t_i|\geqslant\rho}\leqslant C\delta p,$$

which implies (14). Clearly, by the symmetry of each X_i, we may only consider positive $k_i\geqslant\rho$.

Step 3 It suffices to consider set T, which additionally satisfies $t_i\in\{0,k_i\}$ where $k_i\geqslant\rho$. Moreover, $\exp(p/4)\leqslant T\leqslant 1+\exp(p/4)$, $0\in T$ and

$$\|X_t-X_s\|_p\geqslant\frac{p}{2},\quad\text{for all } s,t\in T, s\neq t. \tag{21}$$

Consider the following function:

$$\varphi_i(t_i)=\begin{cases}0 & \text{if } |t_i|<\rho\\ k_i & \text{if } |t_i|\geqslant\rho\end{cases}.$$

Let $\varphi(t)=(\varphi_i(t_i))_{i=1}^{d}$. We show that $\|X_{\varphi(t)}-X_{\varphi(s)}\|_p\geqslant\frac{p}{2}$. It requires some upper bound on $\|X_{t-\varphi(t)}\|_p$. Consider any $s\in T$, then using (12)

$$\|X_s\|_p\leqslant C'(p\|s\|_\infty+\sqrt{p}\|s\|_2).$$

Note that for $s=t-\varphi(t)$ we get by the contraction principle $m_p(s,0)\leqslant p\rho+m_p(t,0)$ and hence using (18)

$$\begin{aligned}\|X_s\|_p&\leqslant C'(p\|s\|_\infty+\sqrt{p}\|s\|_2)\leqslant 2C'\rho p+CC'm_p(s,0)\\&\leqslant C'(2+C)\rho p+CC'm_p(t,0))\leqslant C'((2+C)\rho+C\delta)p\leqslant\frac{p}{4},\end{aligned}$$

for suitably small δ and hence also suitably small ρ. Therefore,

$$\|X_{\varphi(t)}-X_{\varphi(s)}\|_p\geqslant\|X_t-X_s\|_p-\|X_t-X_{\varphi(t)}\|_p-\|X_s-X_{\varphi(s)}\|_p\geqslant\frac{p}{2}.$$

Suppose we can prove the main result for the constructed set T, which satisfies $|T|\geqslant\exp(p/4)$. Formally, we select a subset $S\subset T$ such that $0\in S$ and $\exp(p/4)\leqslant|S|\leqslant 1+\exp(p/4)$. The Sudakov minoration for cube-like sets gives

$$\mathbf{E}\sup_{t\in S} X_{\varphi(t)} \geqslant K^{-1}p,$$

for some universal K. Recall that

$$\|X_{t-\varphi(t)}\|_p \leqslant C'((2+C)\rho + C\delta)p$$

and therefore by Remark 1 and $\exp(p/4) \leqslant |S| \leqslant 1+\exp(p/4)$, $0 \in S$, we get

$$\mathbf{E}\sup_{t\in S} X_{t-\varphi(t)} \leqslant eC'\left((2+C)\rho + C\delta\right)p.$$

Thus,

$$\mathbf{E}\sup_{t\in S} X_t = \mathbf{E}\sup_{t\in S} X_{\varphi(t)} + X_{t-\varphi(t)} \geqslant \mathbf{E}\sup_{t\in S} X_{\varphi(t)} - \mathbf{E}\sup_{t\in S} X_{t-\varphi(t)} \geqslant \frac{1}{2}K^{-1}p,$$

for suitably small δ, i.e., $2eC'\left((2+C)\rho + C\delta\right) \leqslant K^{-1}$. Obviously, the set S is the required simplification in this step.

Step 4 The final step is to replace p by $4p$ so that $\exp(p) \leqslant |T| \leqslant 1+\exp(p)$ and $0 \in T$. Then, obviously, by(7) (with $\alpha = 1$)

$$4\|X_t - X_s\|_p \geqslant \|X_t - X_s\|_{4p} \geqslant 2p.$$

Thus, we may to redefine t_i as $2t_i$. Consequently, the theorem holds true with slightly rearranged constants, namely, we set $\delta' = 2\delta$ instead of δ and ρ' (instead of ρ) that satisfies $\rho'/\log(1/\rho') = 4C\delta'$. Obviously, $\rho' \leqslant e^{-1}$ if δ was suitably small.

□

The above proof is a slightly rearranged version of the argument presented in [5]—we have stated the proof here for the sake of completeness. Note also that without much effort the argument works for α-regular X, i.e., when the inequality (7) is satisfied for all $s, t \in \mathbb{R}^d$.

Remark 2 There are suitably small δ' and δ'' such that for any set T that satisfies properties from Preposition 1, then for all $t \in T$ the following holds true:

$$|I(t)| \leqslant \delta' p, \quad \|t\|_1 = \sum_{i\in I(t)} k_i \leqslant \delta'' p, \tag{22}$$

Proof Indeed by (14) we get $\|t\|_1 \leqslant 2C\delta p$. On the other hand, $\rho|I(t)| \leqslant \|t\|_1 \leqslant 2C\delta p$. Since $4C\delta/\rho = 1/\log(1/\rho)$, this implies

$$|I(t)| \leqslant \frac{1}{2}\frac{1}{\log\frac{1}{\rho}}p.$$

We set $\delta' = 2C\delta$ and $\delta'' = 1/(2\log(1/\rho))$. □

There is another property that we can add to our list of conditions that T has to satisfy. Namely, it suffices to prove minoration only when $\|X_t - X_s\|_p \simeq p$ for all $t \in T$.

Proposition 2 *Suppose that random vector X satisfies* (7), *then it suffices to prove the minoration only for sets T such that $p \leqslant \|X_t - X_s\|_p \leqslant 2\alpha p$.*

Proof The argument is rather standard and can be found, e.g., in [9]. Basically, either there is at least $e^{p/2}$ points that are within the distance $2\alpha p$ from some point in T or one can find at least $e^{p/2}$ points that are $2\alpha p$ separated, i.e., $\|X_t - X_s\|_p \geqslant 2\alpha p$ for all s, t $s \neq t$ in the set. However, then $\|X_t - X_s\|_{p/2} \geqslant \frac{2\alpha}{2\alpha}p = p$, then $\|X_t - X_s\|_{p/2} \geqslant p$. We continue with this set instead of T. It is easy to understand that in this way we have to find a subset T' of T that counts at least $e^{p/2^m}$ elements, where $2^m \leqslant p$, such that

$$p \leqslant \|X_t - X_s\|_{p/2^m} \leqslant 2\alpha p \quad s, t \in T'.$$

In this case, obviously, for $\tilde{t} = t/2^m$ and $\tilde{T} = \{\tilde{t} : t \in T'\}$, we get $\|X_{\tilde{t}} - X_{\tilde{s}}\|_{p/2^m} \simeq p/2^m$. Then, by our assumption that the minoration works for $\tilde{T}$ and $p/2^m$

$$\frac{p}{K2^m} \leqslant \mathbf{E}\sup_{\tilde{t}\in\tilde{T}} X_{\tilde{t}} = \mathbf{E}\sup_{t\in T'} X_{t/2^m} \leqslant \frac{1}{2^m}\mathbf{E}\sup_{t\in T} X_t.$$

Otherwise, if we reach m such that $2^m \leqslant p < 2^{m+1}$, we obtain that there are at least two points $s, t, s \neq t$ such that $\|X_t - X_s\|_2 \geqslant p$. This immediately implies the minoration, since

$$\mathbf{E}\sup_{t\in T} X_t \geqslant \frac{1}{2}\mathbf{E}|X_t - X_s| \geqslant \frac{1}{2\sqrt{2}}\|X_t - X_s\|_2 \gtrsim p.$$

Obviously, by homogeneity

$$\mathbf{E}\sup_{t\in T} X_t \geqslant \mathbf{E}\sup_{t\in\tilde{T}} X_{t/2^m} = \frac{1}{2^m}\mathbf{E}\sup_{t\in T'} X_t \geqslant K^{-1}p,$$

which ends the proof. □

There is one more useful remark. As explained in the previous works on the problem—see Lemma 2.6 in [6] or Section 4 in [10]—the case when $p \geqslant d$ is "easy." More precisely,

Remark 3 If the Sudakov minoration holds for $p = d$, then it holds also for $p \geqslant d$.

4 How to Compute Moments

The second step concerns basic facts on moments of X_t, $t \in T$. Recall that we work with a random vector $X = (X_1, X_2, \ldots, X_d)$, which is 1-unconditional, isotropic, and log-concave. In particular, it implies that for any $t \in \mathbb{R}^d$

$$X_t = \langle X, t\rangle = \sum_{i=1}^{d} t_i X_i \stackrel{d}{=} \sum_{i=1}^{d} t_i \varepsilon_i |X_i|,$$

where $\varepsilon = (\varepsilon_i)_{i=1}^d$ is a vector of independent Rademacher variables, which is independent of X. We start from a series of general facts that are known in this case. We also discuss moments of X with independent, symmetric, and α-regular entries. It will be discussed later that due to our basic simplification—Proposition 1—we have to compute $\|X_t\|_p$ only when $d \leqslant p$, which is a bit simpler than the general case. We start from the characterization proved in [6].

Theorem 3 *Suppose that X has 1-unconditional and log-concave distribution $\mu_X(dx) = \exp(-U(x))dx$. We assume also that X is in the isotropic position. Then, for any $p \geqslant d$*

$$\|X_t\|_p \simeq \sup\left\{\sum_{i=1}^{d} |t_i| a_i : \; U(a) - U(0) \leqslant p\right\}. \tag{23}$$

In [6], there is also an alternative formulation of this result

Theorem 4 *Under the same assumptions as in Theorem 3, we have for any $p \geqslant d$*

$$\|X_t\|_p \simeq \sup\left\{\sum_{i=1}^{d} |t_i| a_i : \; \mathbf{P}\left(\bigcap_{i=1}^{d} \{|X_i| \geqslant a_i\}\right) \geqslant e^{-p}\right\}. \tag{24}$$

Note that a similar characterization works when X has independent and α-regular entries. There is also a version of the above fact formulated in terms of moments, namely,

Theorem 5 *Under the same assumptions as in Theorem 3, we have for any $p \geqslant d$*

$$\|X_t\|_p \simeq \sup\left\{\sum_{i=1}^{d} |t_i| \|X_i\|_{a_i} : \; \sum_{i=1}^{d} a_i \leqslant p\right\}. \tag{25}$$

Proof We prove the result for the sake of completeness. For simplicity we consider only log-concave, one-unconditional, isotropic case. The α-regular case can be proved in the similar way. Consider $a_i \geqslant 0$, $i \in [d]$ such that $\sum_{i=1}^{d} a_i \leqslant p$. Since X is isotropic $a_i < 2$ are unimportant, that is why we use $a_i \vee 2$. It is well-known that moments and quantiles are comparable, namely, for some constant $C \geqslant 1$

$$\mathbf{P}\left(|t_i X_i| \geqslant C^{-1}\|t_i X_i\|_{a_i\vee 2}\right) \geqslant C^{-1}e^{-a_i\vee 2}$$

and hence

$$\prod_{i=1}^{d}\mathbf{P}\left(|t_i X_i| \geqslant C^{-1}\|t_i X_i\|_{a_i\vee 2}\right) \geqslant C^{-d}e^{-\sum_{i=1}^{d} a_i\vee 2}. \tag{26}$$

Consequently,

$$\begin{aligned}\|X_t\|_p &\geqslant C^{-2}\left(\sum_{i=1}^{d}\|t_i X_i\|_{a_i\vee 2}\right)e^{-\sum_{i=1}^{d}\frac{a_i\vee 2}{p}}\\ &\geqslant C^{-2}e^{-3}\sum_{i=1}^{d}\|t_i X_i\|_{a_i\vee 2},\end{aligned}$$

where we have used that

$$\sum_{i=1}^{d} a_i \vee 2 \leqslant 2p + \sum_{i=1}^{d} a_i \leqslant 3p.$$

We turn to prove the converse inequality. Let $|X| = (|X_i|)_{i=1}^d$ and $|X|_t = \sum_{i=1}^d |t_i||X_t|$. Since $p \geqslant d$, $\|X\|_p \leqslant \||X|_t\|_p \leqslant 2\|X_t\|_p$. By the homogeneity of the problem, we can assume that $\||X|_t\|_p = p$ (possibly changing point t by a constant). We have to prove that there exists a_i such that $\sum_{i=1}^d a_i \leqslant p$ whereas $\sum_{i=1}^d \|t_i X_i\|_{a_i} \geqslant C^{-1}p$. It is clear that it suffices to prove the result for p is suitably large, in particular, for $p \geqslant 2$. Let γ be a constant, which we determine later. We define r_i as:

1. $r_i = 2$ if $\|t_i X_i\|_2 \leqslant 2\gamma$.
2. $r_i = p$ if $\|t_i X_i\|_p \geqslant p\gamma$.
3. Otherwise, $r_i = \inf\{r \in [2, p] : \|t_i X_i\|_r = r\gamma\}$.

We first observe that if $r_{i_0} \geqslant p$, then there is nothing to prove since by choosing $a_{i_0} = r_{i_0}$ and other a_i equal 0 we fulfill the requirement $r_{i_0} = \sum_{i=1}^d a_i \leqslant p$, whereas

$$\sum_{i=1}^{d}\|t_i X_i\|_{a_i\vee 2} \geqslant \|t_i X_i\|_{a_{i_0}} \geqslant \gamma a_{i_0} \geqslant \gamma p.$$

Therefore, we may assume that $r_i < p$ for all $i \in [d]$. Now, suppose that we can find a subset $J \subset [d]$ with the property $\sum_{i\in J} r_i \geqslant 3p$ and J does not contain a smaller subset with the property. Therefore, necessarily $\sum_{i\in J} r_i \leqslant 4p$ since $r_i \leqslant p$ for any $i \in [d]$. Obviously,

$$\sum_{i=1}^{d} \|t_i X_i\|_{r_i} \geqslant \sum_{i\in J} \|t_i X_i\|_{r_i} \geqslant \gamma \sum_{i\in J}(r_i - 2) \geqslant \gamma p$$

and hence we can set $a_i = r_i/4$, which implies that $\sum_{i=1}^{d} a_i \leqslant p$ and

$$4\sum_{i=1}^{d} \|t_i X_i\|_{a_i\vee 2} \geqslant \sum_{i=1}^{d} \|t_i X_i\|_{r_i} \geqslant \gamma p.$$

Therefore, to complete the proof, we just need to prove that $\sum_{i=1}^{d} r_i \geqslant 4p$. By the log-concavity for $r_i \geqslant 2$, $u \geqslant 1$

$$\mathbf{P}\left(t_i|X_i| \geqslant eu\|t_i X_i\|_{r_i}\right) \leqslant e^{-r_i u}$$

and therefore

$$\mathbf{P}\left(\frac{t_i|X_i|}{\gamma e} \geqslant r_i u\right) \leqslant e^{-r_i u} = \mathbf{P}(|\mathcal{E}_i| \geqslant r_i u),$$

where we recall that $\mathcal{E}_i$ are symmetric standard exponentials. Consequently,

$$\begin{aligned}\||X|_t\|_p &\leqslant \left(e\gamma\sum_{i=1}^{d} r_i\right) + \\ &+ \left[\mathbf{E}\left(\sum_{i=1}^{d} |t_i||X_i|1_{|t_iX_i|\geqslant \gamma e r_i}\right)^p\right]^{\frac{1}{p}} \leqslant \\ &\leqslant (e\gamma\sum_{i=1}^{d} r_i) + \gamma e\left(\mathbf{E}|\sum_{i=1}^{d} |\mathcal{E}_i|1_{|\mathcal{E}_i|\geqslant r_i}|^p\right)^{\frac{1}{p}} \\ &\leqslant \left(e\gamma\sum_{i=1}^{d} r_i\right) + \gamma e\|Z\|_p,\end{aligned}$$

where Z is of gamma distribution $\Gamma(d, 1)$. Clearly, $\|Z\|_p \leqslant (p + d) \leqslant 2p$ and therefore

$$p = \||X|_t\|_p \leqslant \gamma e\sum_{i=1}^{d} r_i + 4\gamma e p.$$

Therefore, choosing $\gamma = 1/(8e)$

$$\frac{1}{2}\||X|_t\|_p \leqslant \frac{1}{8}\sum_{i=1}^{d} r_i.$$

It proves that $\sum_{i=1}^{d} r_i \geqslant 4p$, which completes the proof. □

We turn to prove some remarks on moments in our case. We are interested in moments of $\|X_t\|_p$ for X that satisfies our structural assumption. In order to explain all the ideas, we need a lot of random variables. It is good to collect their definitions in one place in order to easily find the right reference.

Definition 1 We define the following random variables:

- Let $X \in \mathbb{R}^d$ be a random vector in the isotropic position.
- Let $X = (X_1, X_2, \ldots, X_M)$ where $X_k \in \mathbb{R}^{n_k}$, $k \in [M]$ are independent and X_k has the density

$$\exp(-U_k(\|x_k\|_{p_k}^{p_k})), \quad \text{where } U_k \text{ is convex, increasing, } x_k = (x_i)_{i \in J_k}.$$

- Sets J_k, $1 \leqslant k \leqslant M$ are disjoint and $|J_k| = n_k$.
- We denote $X_k = (X_i)_{i \in J_k}$, and we also use the ordering $X_k = (X_{k1}, X_{k2}, \ldots, X_{kn_k})$.
- Random variables $R_k \in \mathbb{R}_+$, $V_k \in \mathbb{R}^{n_k}$, $k \in [M]$, are such that $X_k = R_k V_k$.
- Random vector $V_k = (V_{k1}, V_{k2}, \ldots, V_{kn_k})$ is distributed with respect to the cone measure on $\partial B_{p_k}^{n_k}$.
- Random variable R_k is distributed on $\mathbb{R}_+$ with respect to the density

$$g_k(s) = s^{n-1}\exp(-U_k(s^{p_k}))|\partial B_{p_k}^{n_k}| 1_{\mathbb{R}_+}(s). \tag{27}$$

- Let $Y = (Y_1, Y_2, \ldots, Y_M)$, where $Y_k \in \mathbb{R}^{n_k}$, $k \in [M]$ are independent and Y_k has the density

$$f_k(x_k) = \prod_{i \in J_k} \frac{b_k^{\frac{1}{p_k}}}{2\Gamma\left(1 + \frac{1}{p_k}\right)} e^{-b_k|x_i|^{p_k}}, \quad \text{where } b_k = \left[\frac{\Gamma(\frac{3}{p_k})}{\Gamma(\frac{1}{p_k})}\right]^{\frac{p_k}{2}}. \tag{28}$$

- We denote $Y_k = (Y_i)_{i \in J_k}$, and we also use the ordering $Y_k = (Y_{k1}, Y_{k2}, \ldots, Y_{kn_k})$.
- Random variables $\tilde{R}_k \in \mathbb{R}_+$, $V_k \in \mathbb{R}^{n_k}$, are such that $Y_k = \tilde{R}_k V_k$; moreover, V_k is already defined and distributed like the cone measure on $\partial B_{p_k}^{n_k}$.
- Random variable $\tilde{R}_k$ is distributed on $\mathbb{R}_+$ with respect to the density

$$\tilde{g}_k(s) = \frac{b_k^{\frac{n_k}{p_k}}}{\frac{1}{p_k}\Gamma(\frac{n_k}{p_k})} s^{n_k-1}\exp(-b_k s^{p_k}) 1_{\mathbb{R}_+}(s), \quad b_k = \left[\frac{\Gamma(\frac{3}{p_k})}{\Gamma(\frac{1}{p_k})}\right]^{\frac{p_k}{2}}. \tag{29}$$

There are some basic consequences of these definitions. Note that two of them, i.e., formulas (27, 29), have been mentioned above.

Remark 4 There is a list of basic properties of variables described in Definition 1.

- Variables R_k, $k \in [M]$ and V_k, $k \in [M]$ are all independent of each other.
- Variables $\tilde{R}_k$, $k \in [M]$ and V_k, $k \in [M]$ are all independent of each other.
- All variables Y_i, $i \in J_k$, $k \in [M]$ are independent and isotropic, in fact Y_i, $i \in J_k$ has the density

$$\frac{b_k^{\frac{1}{p_k}}}{2\Gamma\left(1+\frac{1}{p_k}\right)} e^{-b_k|x|^{p_k}}, \quad x \in \mathbb{R}.$$

- Variables V_k, $k \in [M]$ are 1-unconditional.
- We have $|\partial B_{p_k}^{n_k}| = p_k \left(2\Gamma\left(1 + 1/p_k\right)\right)^{n_k} / \Gamma(n_k/p_k)$.
- Clearly $X_{t_k} = R_k\langle V_k, t_k\rangle$ and hence $\|X_{t_k}\|_r = \|R_k\|_r\|\langle V_k, t_k\rangle\|_r$, $r > 0$. In particular,

$$1 = \mathbf{E}X_{ki}^2 = \mathbf{E}R_k^2\mathbf{E}V_{ki}^2, \quad i = 1, 2, \ldots, n_k. \tag{30}$$

- In the same way $Y_{t_k} = \tilde{R}_k\langle V_k, t_k\rangle$ and hence $\|Y_{t_k}\|_r = \|\tilde{R}_k\|_r\|\langle V_k, t_k\rangle\|_r$, $r > 0$. In particular,

$$1 = \mathbf{E}Y_{ki}^2 = \mathbf{E}\tilde{R}_k^2\mathbf{E}V_{ki}^2, \quad i = 1, 2, \ldots, n_k. \tag{31}$$

Let $I_k(t) = \{i \in J_k : |t_i| > 0\}$. Using Theorem 3 we get for $r \geqslant |I_k(t)|$, where

$$\|\langle Y_k, t_k\rangle\|_r = \|\tilde{R}_k\|_r\|\langle V_k, t_k\rangle\|_r \simeq r^{1/p_k}\|t_k\|_{q_k}. \tag{32}$$

On the other hand,

$$\|\tilde{R}_k\|_r = b_k^{-\frac{1}{p_k}} \frac{\Gamma\left(\frac{n_k+r}{p_k}\right)^{\frac{1}{r}}}{\Gamma\left(\frac{n_k}{p_k}\right)^{\frac{1}{r}}} \simeq (n_k + r)^{\frac{1}{p_k}}.$$

Consequently,

Proposition 3 *The following holds:*

$$\|\langle V_k, t_k\rangle\|_r \simeq \begin{cases} \|t_k\|_{q_k} \frac{r^{\frac{1}{p_k}}}{n_k^{\frac{1}{p_k}}} & \text{if } |I_k(t)| \leqslant r \leqslant n_k \\ \|t_k\|_{q_k} & \text{if } r > n_k \end{cases}$$

It remains to explain the role of $\|R_k\|_r$. For all $k \in [M]$ we define

$$S_k = R_k^{p_k},\quad \tilde{U}_k \text{ satisfies } \exp(-\tilde{U}_k(x))1_{\mathbb{R}_+}(x) = \exp(-U_k(x))\frac{1}{p_k}|\partial B_{p_k}^{n_k}|1_{\mathbb{R}_+}(x). \tag{33}$$

Clearly, S_k has the density

$$h_k(x) = x^{\frac{n_k}{p_k}-1}\exp(-U_k(x))\frac{1}{p_k}|\partial B_{p_k}^{n_k}|1_{\mathbb{R}_+}(x) = x^{\frac{n_k}{p_k}-1}\exp(-\tilde{U}_k(x))1_{\mathbb{R}_+}(x).$$

Note that $\|R_k\|_r = \|S_k\|_{r/p_k}^{\frac{1}{p_k}}$. We use the result of Ball [1]—Lemma 4.

Lemma 1 *For any $0 < p < q$, the following holds:*

$$\begin{aligned} &e^{-\tilde{U}(0)q}\Gamma(p+1)^{q+1}\left[\int_{\mathbb{R}_+} x^q e^{-\tilde{U}(x)}dx\right]^{p+1} \\ &\leqslant e^{-\tilde{U}(0)p}\Gamma(q+1)^{p+1}\left[\int_{\mathbb{R}_+} x^p e^{-\tilde{U}(x)}dx\right]^{q+1}, \end{aligned} \tag{34}$$

where $\tilde{U}(x)$ is a convex, increasing function on $\mathbb{R}_+$.

Note that we use the above result for $\tilde{U}_k$ defined in (33). Consequently, we get

Corollary 1 *For any $r \geqslant 1$, we have*

$$\|R_k\|_r = \|S_k\|_{\frac{r}{p_k}}^{\frac{1}{p_k}} \leqslant e^{\frac{1}{n_k}\tilde{U}_k(0)}\frac{\Gamma\left(\frac{n_k+r}{p_k}\right)^{\frac{1}{r}}}{\Gamma\left(\frac{n_k}{p_k}\right)^{\frac{1}{r}+\frac{1}{n_k}}}.$$

We have to bound $\exp(-\tilde{U}_k(0)/n_k)$. First, using the isotropic position of Y_k, we get

$$1 = \mathbf{E}V_{ki}^2\mathbf{E}\tilde{R}_k^2 = \frac{1}{b_k^{\frac{2}{p_k}}}\frac{\Gamma\left(\frac{n_k+2}{p_k}\right)}{\Gamma\left(\frac{n_k}{p_k}\right)}\mathbf{E}V_{ki}^2,$$

hence

$$\mathbf{E}V_{ki}^2 = \frac{b_k^{\frac{2}{p_k}}\Gamma\left(\frac{n_k}{p_k}\right)}{\Gamma\left(\frac{n_k+2}{p_k}\right)}.$$

Therefore, by Lemma 1, we get

$$\frac{1}{\mathbf{E}(V_{ki})^2} = \mathbf{E}R_k^2 \leqslant e^{\frac{2}{n_k}\tilde{U}_k(0)} \frac{\Gamma\left(\frac{n_k+2}{p_k}\right)}{\Gamma\left(\frac{n_k}{p_k}\right)^{1+\frac{2}{n_k}}}.$$

It proves the following inequality:

$$e^{\frac{1}{n_k}\tilde{U}_k(0)} \geqslant b_k^{-\frac{1}{p_k}} \Gamma\left(\frac{n_k}{p_k}\right)^{\frac{1}{n_k}}. \tag{35}$$

In order to prove the upper bound, we need a Hensley-type inequality. First, observe that

$$\mathbf{P}(R_k \leqslant t) = \int_0^t x^{n_k-1} e^{-\tilde{U}_k(x^{p_k})} dx \leqslant \int_0^t x^{n_k-1} e^{-\tilde{U}_k(0)} dx = \frac{1}{n_k} t^{n_k} e^{-\tilde{U}_k(0)}.$$

Let $F_k(t) = \frac{1}{n_k} t^{n_k} e^{-\tilde{U}_k(0)}$ for $0 \leqslant t \leqslant t_*$, where $t_* = \left(n_k e^{\tilde{U}_k(0)}\right)^{1/n_k}$. We can calculate

$$\begin{aligned}
\frac{1}{\mathbf{E}(V_k)_i^2} &= \mathbf{E}R_k^2 = 2\int_0^\infty t\mathbf{P}(R_k > t)dt \\
&\geqslant 2\int_0^{t_*} t(1 - F_k(t))dt \geqslant 2\int_0^{t_*} t(1 - t^{n_k})dt \\
&= t_*^2 - 2\frac{1}{n_k+2} t_*^{n_k+2} e^{-\tilde{U}_k(0)} \frac{1}{n_k} \\
&= t_*^2 - \frac{2}{n_k+2} t_*^2 = \frac{n_k}{n_k+2} t_*^2.
\end{aligned}$$

Therefore,

$$\left(n_k e^{\tilde{U}_k(0)}\right)^{\frac{2}{n_k}} \leqslant \frac{n_k+2}{n_k} \frac{\Gamma\left(\frac{n_k+2}{p_k}\right)}{b_k^{\frac{n_k}{p_k}} \Gamma\left(\frac{n_k}{p_k}\right)}.$$

This means

$$e^{\frac{1}{n_k}\tilde{U}_k(0)} \leqslant \frac{1}{n_k^{\frac{1}{n_k}}} \frac{(n_k+2)^{\frac{1}{2}}}{n_k^{\frac{1}{2}}} \frac{\Gamma\left(\frac{n_k+2}{p_k}\right)^{\frac{1}{2}}}{b_k^{\frac{1}{p_k}} \Gamma\left(\frac{n_k}{p_k}\right)^{\frac{1}{2}}}. \tag{36}$$

It finally proves

$$\frac{\Gamma\left(\frac{n_k}{p_k}\right)^{\frac{1}{n_k}}}{b_k^{\frac{1}{p_k}}} \leqslant e^{\frac{1}{n_k}\tilde{U}_k(0)} \leqslant \frac{1}{n_k^{\frac{1}{n_k}}} \frac{(n_k+2)^{\frac{1}{2}}}{n_k^{\frac{1}{2}}} \frac{\Gamma\left(\frac{n_k+2}{p_k}\right)^{\frac{1}{2}}}{b_k^{\frac{1}{p_k}}\Gamma\left(\frac{n_k}{p_k}\right)^{\frac{1}{2}}}$$

and hence

$$e^{\frac{1}{n_k}\tilde{U}_k(0)} \simeq \frac{1}{\|V_{ki}\|_2} \simeq n_k^{\frac{1}{p_k}}.$$

Together with Corollary 1, it shows that

$$\|R_k\|_r \lesssim n_k^{\frac{1}{p_k}} \frac{\Gamma\left(\frac{n_k+r}{p_k}\right)^{\frac{1}{r}}}{\Gamma\left(\frac{n_k}{p_k}\right)^{\frac{1}{n_k}+\frac{1}{r}}}. \tag{37}$$

Note that

$$\frac{\Gamma\left(\frac{n_k+r}{p_k}\right)^{\frac{1}{r}}}{\Gamma\left(\frac{n_k}{p_k}\right)^{\frac{1}{n_k}+\frac{1}{r}}} \simeq \left(1+\frac{r}{n_k}\right)^{\frac{n_k+r}{p_k r}}.$$

That is why

$$\|R_k\|_r \lesssim (n_k+r)^{1/p_k}. \tag{38}$$

In particular, $\|R_k\|_r \simeq n_k^{1/p_k}$ for $1 \leqslant r \leqslant n_k$. We introduce variables $P_k = R_k/\|R_k\|_2$, $Q_k = \tilde{R}_k/\|\tilde{R}_k\|_2$, $W_k = \|R_k\|_2 V_k$. This helps since now vectors $(\varepsilon_k P_k)_{k=1}^M$, $(\varepsilon_k Q_k)_{k=1}^M$ are in the isotropic position, where ε_k, $k = 1, 2, \ldots, M$ are independent Rademacher variables, independent of all P_k and Q_k. Obviously, also W_k, $k = 1, 2, \ldots, M$ are isotropic, independent and independent of all P_k. Thus, in particular, $\|P_k\|_r, \|Q_k\|_r \simeq 1$, for $1 \leqslant r \leqslant n_k$. Once again due to Lemma 1

$$e^{-\tilde{U}_k(0)(\frac{n_k+q}{p_k}-1)}\Gamma\left(\frac{n_k+p}{p_k}\right)^{\frac{n_k+q}{p_k}}\left[\int_{\mathbb{R}_+} x^{\frac{n_k+q}{p_k}-1}e^{-\tilde{U}_k(x)}dx\right]^{\frac{n_k+p}{p_k}}$$
$$\leqslant e^{-\tilde{U}_k(0)(\frac{n_k+p}{p_k}-1)}\Gamma\left(\frac{n_k+q}{p_k}\right)^{\frac{n_k+p}{p_k}}\left[\int_{\mathbb{R}_+} x^{\frac{n_k+p}{p_k}-1}e^{-\tilde{U}_k(x)}dx\right]^{\frac{n_k+q}{p_k}}.$$

Therefore,

$$\|R_k\|_q^{\frac{q(n_k+p)}{p_k}} \leqslant e^{\tilde{U}_k(0)\frac{(q-p)}{p_k}} \frac{\Gamma\left(\frac{n_k+q}{p_k}\right)^{\frac{n_k+p}{p_k}}}{\Gamma\left(\frac{n_k+p}{p_k}\right)^{\frac{n_k+q}{p_k}}} \|R_k\|_p^{\frac{p(n_k+q)}{p_k}}.$$

which is

$$\|R_k\|_q \leqslant e^{\tilde{U}_k(0)\frac{(q-p)}{q(n_k+p)}} \frac{\Gamma\left(\frac{n_k+q}{p_k}\right)^{\frac{1}{q}}}{\Gamma\left(\frac{n_k+p}{p_k}\right)^{\frac{n_k+q}{q(n_k+p)}}} \|R_k\|_p^{\frac{p(n_k+q)}{q(n_k+p)}}. \tag{39}$$

Note that

$$\frac{\Gamma\left(\frac{n_k+q}{p_k}\right)^{\frac{1}{q}}}{\Gamma\left(\frac{n_k+p}{p_k}\right)^{\frac{n_k+q}{q(n_k+p)}}} \simeq \frac{(n_k+q)^{\frac{n_k+q}{p_k q}}}{(n_k+p)^{\frac{n_k+q}{p_k q}}} \simeq \left(\frac{n_k+q}{n_k+p}\right)^{\frac{n_k+q}{qp_k}}.$$

Moreover, for $p \geqslant 1$

$$e^{\tilde{U}_k(0)\frac{(q-p)}{q(n_k+p)}} \lesssim (n_k^{\frac{1}{p_k}})^{\frac{(q-p)n_k}{q(n_k+p)}} \lesssim \|R_k\|_p^{\frac{(q-p)n_k}{q(n_k+p)}}$$

and since

$$\frac{(q-p)n_k}{q(n_k+p)} + \frac{p(n_k+q)}{q(n_k+p)} = 1$$

it proves that

$$\|R_k\|_q \lesssim \left(\frac{n_k+q}{n_k+p}\right)^{\frac{1}{p_k}} \|R_k\|_p, \tag{40}$$

which means that R_k is α-regular. In particular, $\varepsilon_k R_k$, $k = 1, 2, \ldots, M$ are independent symmetric α-regular variables. It also implies that $\tilde{R}_k$ have the fastest growth among all the possible distributions of R_k.

We end this section with the characterization of $\|X_t - X_s\|_p$. Recall our notation $P_k = R_k/\|R_k\|_2$ and $W_k = \|R_k\|_2 V_k$. We show

Proposition 4 *Under the assumptions as in Theorem 3 and assuming that all the properties mentioned in Preposition 1 are satisfied, the following result holds:*

$$\|X_t - X_s\|_p \simeq \sup\left\{\sum_{k=1}^{M} \|P_k\|_{r_k} \|\langle W_k, t_k - s_k\rangle\|_{r_k} 1_{r_k \geqslant |I_k(t)\triangle I_k(s)||} : \sum_{k=1}^{M} r_k = p\right\} \tag{41}$$

$$\|Y_t - Y_s\|_p \simeq \sup\left\{\sum_{k=1}^{M} \|Q_k\|_{r_k} \|\langle W_k, t_k - s_k\rangle\|_{r_k} 1_{r_k \geqslant |I_k(t)\triangle I_k(s)||} : \sum_{k=1}^{M} r_k = p\right\}. \tag{42}$$

Moreover, if $r_k \geqslant |I_k(t)\triangle I_k(s)||$, *then*

$$\begin{aligned}\|Q_k\|_{r_k} \|\langle W_k, t_k - s_k\rangle\|_{r_k} = \|Y_{t_k}\|_{r_k} \sim r_k^{1/p_k} \|t_k\|_{q_k} \lesssim \|P_k\|_{r_k} |\langle W_k, t_k - s_k\rangle\|_{r_k} \\ = \|X_{t_k}\|_{r_k}.\end{aligned}$$

Proof We use that $X_t - X_s = \sum_{k=1}^{M} \langle X_k, t_k - s_k\rangle$ and X_k are independent, log-concave vectors. Observe that

$$X_{t_k - s_k} = \langle X_k, t_k - s_k\rangle = P_k\langle W_k, t_k - s_k\rangle.$$

Let us denote also $I_k(t) = J_k \cap I(t)$. By Proposition 1—see (22)—the number of $k \in M$ for which $t_k - s_k$ is nonzero is much smaller than p. We use Theorem 5 treating $X_t - X_s$ as $\sum_{k=1}^{M} X_{t_k - s_k}$, and it yields

$$\|X_t - X_s\|_p \simeq \sup\left\{\sum_{k=1}^{M} \|X_{t_k}\|_{r_k} : \sum_{k=1}^{M} r_k = p\right\}.$$

Since $\|X_{t_k - s_k}\|_{r_k} = \|P_k\|_{r_k} \|\langle W_k, t_k - s_k\rangle\|_{r_k}$, we get

$$\|X_t - X_s\|_p \simeq \sup\left\{\sum_{k=1}^{M} \|P_k\|_{r_k} \|\langle W_k, t_k - s_k\rangle\|_{r_k} : \sum_{k=1}^{M} r_k = p\right\}. \tag{43}$$

Moreover,

$$\|P_k\|_r \simeq 1, \quad \text{for } r \leqslant n_k, \quad \|P_k\|_r \lesssim \frac{r_k^{\frac{1}{p_k}}}{n_k^{\frac{1}{p_k}}}, \quad r > n_k.$$

and

$$\|\langle W_k, t_k - s_k\rangle\|_r \simeq \begin{cases} r^{1/p_k} \|t_k - s_k\|_{q_k} \text{ for } |I_k(t)\triangle I_k(s)| \leqslant r \leqslant n_k \\ n_k^{1/p_k} \|t_k - s_k\|_{q_k} \text{ for } r > n_k \end{cases}.$$

We stress that we should not care about $r_k < |I_k(t)\triangle I_k(s)|$ since, by (22), $\sum_{k=1}^{M} |I_k(t)\triangle I_k(s)|$ is much smaller than p. Thus, $r_k < |I_k(t)\triangle I_k(s)|$ can be ignored when estimating $\|X_t - X_s\|_p$. Thus, finally,

$$\|X_t - X_s\|_p \simeq \left\{\sum_{k=1}^{M} \|P_k\|_{r_k} \|\langle W_k, t_k - s_k\rangle\|_{r_k} 1_{r_k \geqslant |I_k(t)\triangle I_k(s)||} : \sum_{k=1}^{M} r_k = p\right\}$$

as required. The proof for $\|Y_t - Y_s\|_p$ follows the same scheme: we have to use (32). □

We will use the above result many times in the subsequent proofs, even without mentioning it.

5 Positive Process

We are going to show that it suffices to prove that for all $p \geqslant 1$

$$\mathbf{E} \sup_{t\in T} \sum_{k=1}^{M} |X_{t_k}| \geqslant \frac{1}{K} p.$$

For simplicity we define $|X|_t = \sum_{k=1}^{M} |X_{t_k}|$. We show

Lemma 2 *Suppose that T satisfies simplifications from Proposition 1. Then, $\mathbf{E}\sup_{t\in T} |X|_t \geqslant \frac{1}{K} p$, which implies also that $\mathbf{E}\sup_{t\in T} X_t \geqslant \frac{1}{4K} p$.*

Proof Before we start, let introduce notation: writing $\mathbf{E}_I$ for $I \subset [M]$ we mean the integration over vectors $(X_k)_{k\in I}$. Note, in particular, that $\mathbf{E} = \mathbf{E}_{I,I^c}$.
Let us observe that

$$\mathbf{E} \sup_{t\in T} X_t = \mathbf{E} \sup_{t\in T} \sum_{k=1}^{M} X_{t_k}$$

$$= \mathbf{E} \sup_{t\in T} \sum_{k=1}^{M} \varepsilon_k X_{t_k} \geqslant \mathbf{E} \sup_{t\in T} \sum_{k=1}^{M} \varepsilon_k |X_{t_k}|,$$

where $(\varepsilon_k)_{k=1}^{M}$ are independent random signs. The last inequality is due to Bernoulli comparison and the inequality

$$||X_{t_k}| - |X_{s_k}|| \leqslant |X_{t_k} - X_{s_k}|.$$

Now, the standard argument works, namely,

$$\mathbf{E}\sup_{t\in T}\sum_{k=1}^{M}\varepsilon_k|X_{t_k}| = \sum_{I\subset[M]}\frac{1}{2^M}\mathbf{E}_{I,I^c}\sup_{t\in T}\sum_{k\in I}|X_{t_k}| - \sum_{k\in I^c}|X_{t_k}|$$

$$\geqslant \sum_{I\subset[M]}\frac{1}{2^M}\mathbf{E}_I\sup_{t\in T}\left(\sum_{k\in I}|X_{t_k}| - \sum_{k\in I^c}\mathbf{E}_{I^c}|X_{t_k}|\right)$$

$$\geqslant \sum_{I\subset[M]}\frac{1}{2^M}\mathbf{E}\sum_{k\in I}|X_{t_k}| - \delta"p \geqslant \frac{1}{2}\mathbf{E}\sup_{t\in T}\sum_{k=1}^{M}|X_{t_k}| - \delta"p,$$

where we have used Jensen's inequality and (22) which implies

$$\sum_{k\in I^c}\mathbf{E}|X_{t_k}| \leqslant \sum_{i\in I(t)}k_i\mathbf{E}|X_i| \leqslant \sum_{i\in I(t)}k_i \leqslant \delta"p.$$

It suffices that $\delta" \leqslant 1/(4K)$. □

The above result enables us to split points t_k into small and large part. More precisely, let us split $t_k = t_k^* + t_k^\dagger$, where

$$\textbf{small part}: \; t_k^* = t_k \text{ if } D^{q_k}\|t_k\|_{q_k}^{q_k} \leqslant A^{p_k}n_k \text{ and } 0 \text{ otherwise,} \quad (44)$$

$$\textbf{large part}: \; t_k^\dagger = t_k \text{ if } D^{q_k}\|t_k\|_{q_k}^{q_k} > A^{p_k}n_k \text{ and } 0 \text{ otherwise.} \quad (45)$$

We will need that $A, D \geqslant 1$ and A is such that $A^{p_k} \geqslant D^{q_k}$ for each $k \in [M]$. This is the moment when we take advantage of the cutoff level $p_k \geqslant 1+\varepsilon$, since then we can find such A that does depend on ε only. In fact, we could do better since our split is not important when $q_k > n_k$. Indeed, then $n_k^{1/q_k} \simeq 1$, $\|t_k\|_{q_k} \geqslant \|t_k\|_\infty$, and it is well-known that we can require in the proof of minoration that each $|t_i|$, $i \in I(t)$ is large enough. That means, for p_k too close to 1, the first case simply cannot happen. In general, we do not want to bound n_k from above, but, fortunately, it is possible to reduce the dimension of our problem by a simple trick. Note that T has about e^p points whose supports are thin—much smaller than p. Thus, there are only pe^p important coordinates on which we can condition our basic vector X. Moreover, due to the Prekopa-Leindler theorem, such a reduced problem is still of log-concave distribution type. Since it does not affect the computation of $\|X_t - X_s\|_p$ moments s, t, we end up in the question, where $d \leqslant pe^p$, which together with Remark 3 shows that we should care only for $p \leqslant d \leqslant pe^p$. The trouble with this approach is that it introduces dependence of ε on p, which we do not like in our considerations.

We have two possibilities:

- There exists $S \subset T$, $|S| \geqslant e^{p/2}$ such that for any $s, t \in S$, $s \neq t$ and

$$\sum_{k=1}^{M} D^{q_k} \|t_k^* - s_k^*\|_{q_k}^{q_k} = \sum_{k=1}^{M} \sum_{i \in I_k(t^*) \triangle I_k(s^*)} D^{q_k} k_i^{q_k} \geqslant p. \tag{46}$$

- There is a point $t_0 \in T$ and a subset $S \subset T$, $|S| \geqslant e^{p/2}$ such that for any $t \in S$ the following inequality holds true

$$\sum_{k=1}^{M} D^{q_k} \|t_k^* - t_{0k}^*\|_{q_k}^{q_k} = \sum_{k=1}^{M} \sum_{i \in I_k(t^*) \triangle I_k(t_0^*)} D^{q_k} k_i^{q_k} < p. \tag{47}$$

Let $\mathbf{E}_P, \mathbf{E}_W$ denote the integration with respect to $(P_k)_{k=1}^M$ and $(W_k)_{k=1}^M$. Note that we have

$$\begin{aligned} \mathbf{E} \sup_{t \in T} |X|_t &\geqslant \mathbf{E} \sup_{t \in T} \sum_{k=1}^{M} |P_k| |\langle W_k, t_k \rangle| \\ &\geqslant \mathbf{E}_W \sup_{t \in T} \sum_{k=1}^{M} \mathbf{E}_P |P_k| |\langle W_k, t_k \rangle| \simeq \mathbf{E} \sup_{t \in T} \sum_{k=1}^{M} |\langle W_k, t_k \rangle|. \end{aligned} \tag{48}$$

For simplicity, we denote

$$|\tilde{X}|_t = \sum_{k=1}^{M} |\langle W_k, t_k \rangle|.$$

Obviously, $|\tilde{X}|_t \geqslant |\tilde{X}|_{t^*}$. We are going to prove that, in the case of small coefficients (cf. (44)), necessarily $\mathbf{E} \sup_{t \in T} |\tilde{X}|_{t^*} \geqslant K^{-1} p$.

6 Small Coefficients

Recall that we work in the cube-like setting introduced in Proposition 1. We are going to prove the Sudakov minoration, in the setting where there are a lot of well-separated points in T, in the sense of (46). Toward this goal, we need the process $(Y_t)_{t \in T}$—see (28). We slightly modify the process, namely, let

$$Z_t = \sum_{k=1}^{M} \sum_{i \in I_k(t)} \varepsilon_i \, (k_i |Y_i|) \wedge D^{q_k} k_i^{q_k}, \quad |Z|_t = \sum_{k=1}^{M} \sum_{i \in I_k(t)} (k_i |Y_i|) \wedge D^{q_k} k_i^{q_k},$$

where as usual ε_i, $i \in [d]$ are independent random signs. Since now the entrances are independent, in order to prove that $\mathbf{E} \sup_{t \in S} Z_{t^*} \geqslant K^{-1} p$, it suffices to show that $\|Z_{t^*} - Z_{s^*}\|_p \gtrsim p/D$. We prove that this works under the condition (46).

Lemma 3 *Suppose that* (46) *holds true. Then,* $\|Z_{t^*} - Z_{s^*}\|_p \gtrsim p/D$.

Proof Note that for some $\sum_{k=1}^M \sum_{i\in I_k} r_{ki} = p$

$$\|Z_{t^*} - Z_{s^*}\|_p \simeq \sum_{k=1}^{M} \sum_{i\in I_k(t^*)\triangle I_k(s^*)} \|(k_i|Y_i|) \wedge D^{q_k} k_i^{q_k}\|_{r_{ki}}.$$

However, it is clear that $\|(k_i|Y_i|) \wedge D^{q_k} k_i^{q_k}\|_{r_{ki}} \simeq \|k_i Y_i\|_{r_{ki}} \wedge D^{q_k} k_i^{q_k}$. Since $\|k_i Y_i\|_{r_{ki}} \simeq k_i r_{ki}^{\frac{1}{p_k}}$, we get

$$\|Z_{t^*} - Z_{s^*}\|_p \simeq \sum_{k=1}^{M} \sum_{i\in I_k(t^*)\triangle I_k(s^*)} \left(k_i r_{ki}^{\frac{1}{p_k}}\right) \wedge D^{q_k} k_i^{q_k}.$$

Our assumption is that

$$\sum_{k=1}^{M} \sum_{i\in I_k(t^*)\triangle I_k(s^*)} D^{q_k} k_i^{q_k} \geqslant p,$$

which means that we can select sets $I_k \subset I_k(t^*)\triangle I_k(s^*)$ such that $r_{ki} = D^{q_k} k_i^{q_k}$ for each $i \in I_k$ and $r_{ki} = 0$ for $i \in J_k \backslash I_k$, which satisfy $\sum_{k=1}^M \sum_{i\in I_k} D^{q_k} k_i^{q_k} \simeq p$. Then,

$$\|Z_{t^*} - Z_{s^*}\|_p \gtrsim \sum_{k=1}^{M} \sum_{i\in I_k} D^{q_k - 1} k_i^{q_k} \gtrsim p/D.$$

This proves the result. □

We have proved that $\mathbf{E}\sup_{t\in S} Z_{t^*} \geqslant K^{-1} p$. This implies that

$$K^{-1} p \leqslant \mathbf{E}\sup_{t\in S} \sum_{k=1}^{M} \left(\sum_{i\in I_k(t^*)} k_i \varepsilon_i |Y_i| \right) \wedge D^{q_k} \|t_k\|_{q_k}^{q_k}.$$

The crucial thing is to establish a similar inequality replacing $\left|\sum_{i\in I_k(t)} k_i \varepsilon_i |Y_i|\right|$ with $\sum_{i\in I_k(t)} k_i |Y_i|$. This can be done following the approach we have used in the proof of the positive process lemma—Lemma 2. Namely, we have

Lemma 4 *The following inequality holds:*

$$\mathbf{E}\sup_{t\in S} \sum_{k=1}^{M} \left(\sum_{i\in I_k(t^*)} k_i \varepsilon_i |Y_i| \right) \wedge D^{q_k} \|t_k\|_{q_k}^{q_k} \geqslant K^{-1} p. \tag{49}$$

Proof Let us observe that

$$\mathbf{E}\sup_{t\in S}\sum_{k=1}^{M}\Big|\sum_{i\in I_k(t^*)} k_i Y_i\Big| \wedge \|t_k\|_{q_k}^{q_k}$$

$$=\frac{1}{2^d}\sum_{I\subset[d]}\mathbf{E}_{I,I^c}\sup_{t\in S}\left(\sum_{k=1}^{M}\Big|\sum_{i\in I_k(t^*)\cap I} k_i|Y_i| - \sum_{i\in I_k(t^*)\cap I^c} k_i|Y_i|\Big| \wedge D^{q_k}\|t_k\|_{q_k}^{q_k}\right)$$

$$\geqslant \frac{1}{2^d}\sum_{I\subset[d]}\mathbf{E}_I\sup_{t\in S}\left(\sum_{k=1}^{M}\left(\sum_{i\in I_k(t^*)\cap I} k_i|Y_i|\right)\wedge D^{q_k}\|t_k\|_{q_k}^{q_k} - \mathbf{E}_{I^c}\sum_{i\in I_k(t^*)\cap I^c} k_i|Y_i|\right)$$

$$\geqslant \frac{1}{2^d}\sum_{I\subset[d]}\mathbf{E}\sup_{t\in S}\sum_{k=1}^{M}\left(\sum_{i\in I_k(t^*)\cap I} k_i|Y_i|\right)\wedge D^{k_q}\|t_k\|_{q_k}^{q_k} - \delta'' p$$

$$\geqslant \frac{1}{2}\mathbf{E}\sup_{t\in S}\sum_{k=1}^{M}\left(\sum_{i\in I_k(t^*} k_i|Y_i|\right)\wedge D^{q_k}\|t_k\|_{q_k}^{q_k} - \delta'' p,$$

where we have used $\mathbf{E}|Y_i| \leqslant 1$ and $\sum_{i\in I(t)} k_i \leqslant \delta'' p$—Proposition 1, see (22). We have also used that for positive a, b, c, inequalities $|a-b|\wedge c \geqslant a\wedge c - b$ and $a\wedge c + b\wedge c \geqslant (a+b)\wedge c$ hold true. For suitably small δ'', this implies

$$(4K)^{-1}p \leqslant \mathbf{E}\sup_{t\in S}\sum_{k=1}^{M}\Big|\sum_{i\in I_k(t^*)} k_i\varepsilon_i|Y_i|\Big| \wedge D^{q_k}\|t_k\|_{q_k}^{q_k}.$$

□

Let us recall our notation $Q_k = \tilde{R}_k/\|\tilde{R}_k\|_2$, $W_k = V_k\|\tilde{R}_k\|_2$, $Q_k\langle W_k, t_k\rangle = Y_{t_k}$. Consequently, the above result can be rewritten in the following form:

$$(4K)^{-1}p \leqslant \mathbf{E}\sup_{t\in S}\sum_{k=1}^{M}\big(Q_k|\langle W_k, t_k^*\rangle|\big)\wedge D^{q_k}\|t_k\|_{q_k}^{q_k}. \tag{50}$$

and finally

$$\mathbf{E}\sup_{t\in S}\sum_{k=1}^{M}\big(Q_k|\langle W_k, t_k^*\rangle|\big)\wedge D^{q_k}\|t_k\|_{q_k}^{q_k}$$

$$\leqslant \mathbf{E}\sup_{t\in S}\sum_{k=1}^{M}\big(C|\langle W_k, t_k^*\rangle|\big)\wedge D^{q_k}\|t_k\|_{q_k}^{q_k}$$

$$+\,\mathbf{E}\sup_{t\in S}\sum_{k=1}^{M}\big((Q_k - C)_+|\langle W_k, t_k^*\rangle|\big)\wedge D^{q_k}\|t_k\|_{q_k}^{q_k}. \tag{51}$$

We prove that the latter term is small comparable to p. Recall that $|S| \leqslant |T| \leqslant 1 + e^p$, we are done if we show that

$$\left\| \sum_{k=1}^{M} \left((Q_k - C)_+ |\langle W_k, t_k^* \rangle|\right) \wedge D^{q_k} \|t_k\|_{q_k}^{q_k} \right\|_p \leqslant cp \tag{52}$$

for c is suitably small. We prove this result in the next theorem. This is the main estimate in this section.

Theorem 6 *For suitably large C, (52) holds with c, which can be suitably small.*

Proof Clearly, for $u > \tilde{C}$

$$\mathbf{P}(Q_k > u + \tilde{C}) = \frac{b_k^{\frac{n_k}{p_k}}}{\frac{1}{p_k}\Gamma(\frac{n_k}{p_k})} \int_{\|\tilde{R}_k\|_2(\tilde{C}+u)}^{\infty} x^{n_k-1} e^{-b_k x^{p_k}} dx$$

$$= \frac{b_k^{\frac{n_k}{p_k}}}{\frac{1}{p_k}\Gamma(\frac{n_k}{p_k})} \int_{\|\tilde{R}_k\|_2 u}^{\infty} (\|\tilde{R}_k\|_2 \tilde{C} + x)^{n_k-1} e^{-b_k(\|\tilde{R}_k\|_2\tilde{C}+x)^{p_k}} dx$$

$$\leqslant 2^{n_k-1} \exp(-b_k \|\tilde{R}_k\|_2^{p_k} C^{p_k}) \frac{b_k^{\frac{n_k}{p_k}}}{\frac{1}{p_k}\Gamma(\frac{n_k}{p_k})} \int_{\|\tilde{R}_k\|_2 u}^{\infty} x^{n_k-1} e^{-b_k x^{p_k}} dx$$

$$= 2^{n_k-1} \exp(-b_k \|\tilde{R}_k\|_2^{p_k} \tilde{C}^{p_k}) \mathbf{P}(Q_k > u).$$

Therefore, choosing $\tilde{C} = C/2$, we have $(Q_k - C)_+$, which can be replaced by $\delta_k Q_k$, where δ_k is independent of Q_k; moreover, $\delta_k \in \{0, 1\}$ and

$$\mathbf{P}(\delta_k = 1) = 2^{n_k-1} \exp(-b_k \|\tilde{R}_k\|_2^{p_k} \tilde{C}^{p_k}).$$

Note that

$$b_k \|\tilde{R}\|_2^{p_k} = \frac{\Gamma\left(\frac{n_k+2}{p_k}\right)^{\frac{p_k}{2}}}{\Gamma\left(\frac{n_k}{p_k}\right)^{\frac{p_k}{2}}} \simeq \frac{n_k+2}{p_k},$$

and therefore, for suitably large C,

$$\mathbf{P}(\delta_k = 1) \leqslant \exp(-n_k (C/4)^{p_k}).$$

We denote by $\mathbf{E}_\delta, \mathbf{E}_{Q,W}$ integration with respect to variables $(\delta_k)_{k=1}^M$ and $(Q_k, W_k)_{k=1}^M$. We have

$$\begin{aligned}
&\|\sum_{k=1}^M \left((Q_k - C)_+|\langle W_k, t_k^*\rangle|\right) \wedge D^{q_k}\|t_k\|_{q_k}^{q_k}\|_p \\
&\quad\leqslant \|\sum_{k=1}^M \delta_k \left(Q_k|\langle W_k, t_k^*\rangle|\right) \wedge D^{q_k}\|t_k\|_{q_k}^{q_k}\|_p \\
&= \left[\mathbf{E}_\delta \sum_{K\subset[M]} \prod_{k\in K} 1_{\delta_k=1} \prod_{l\in K^c} 1_{\delta_l=0} \mathbf{E}_{Q,W}\left[\sum_{k\in K} (Q_k|\langle W_k, t_k\rangle|) \wedge D^{q_k}\|t_k\|_{q_k}^{q_k}\right]^p\right]^{\frac{1}{p}}.
\end{aligned} \tag{53}$$

We use Proposition 4—Eq. (42) to get

$$\begin{aligned}
&\left(\mathbf{E}_{Q,W}\left[\sum_{k\in K} \left(Q_k|\langle W_k, t_k^*\rangle|\right) \wedge D^{q_k}\|t_k\|_{q_k}^{q_k}\right]^p\right)^{\frac{1}{p}} \\
&\leqslant \|\sum_{k=1}^M Q_k|\langle W_k, t_k^*\rangle| \wedge D^{q_k}\|t_k\|_{q_k}^{q_k}\|_p \\
&\lesssim \sum_{k=1}^M \left(r_k^{\frac{1}{p_k}}\|t_k^*\|_{q_k} \wedge D^{q_k}\|t_k\|_{q_k}^{q_k}\right) 1_{r_k>|I_k(t)|}
\end{aligned}$$

for some $\sum_{k=1}^M r_k = p$. Let us denote by K_0 the subset of $[M]$ that consists of k such that $r_k^{1/p_k}\|t_k^*\|_{q_k} \leqslant Dr_k$. Clearly,

$$\sum_{k\in K_0} r_k^{\frac{1}{p_k}}\|t_k^*\|_{q_k} \wedge D^{q_k}\|t_k\|_{q_k}^{q_k} \leqslant D\sum_{k\in K_0} r_k \lesssim Dp.$$

On the other hand, if for some $k \in [M]$, $r_k^{1/p_k}\|t_k^*\|_{q_k} > Dr_k$, then, obviously, $t_k = t_k^*$ and $r_k \leqslant D^{q_k}\|t_k\|_{q_k}^{q_k}$. However, in the case of small coefficients (44), we have $D^{q_k}\|t_k\|_{q_k}^{q_k} \leqslant A^{p_k}n_k$ and hence $r_k \leqslant A^{p_k}n_k$ for all $k \in K_0^c$.. Once again by Proposition 4-Eq. (41)

$$\sum_{k\in K_0^c} r_k^{1/p_k}\|t_k^*\|_{q_k} 1_{r_k>|I_k(t)|} \lesssim A\|\sum_{k=1}^M |P_k|\langle W_k, t_k^*\rangle|\|_p \lesssim A\|X_{t^*}\|_p \lesssim Ap,$$

where we have used the simplification from Proposition 2—i.e., $\|X_t\|_p \leqslant 2p$ and the Bernoulli comparison, which gives $\|X_{t^*}\|_p \leqslant \|X_t\|_p$. In this way we get

$$\left(\mathbf{E}_{Q,W}\left[\sum_{k\in K}(Q_k|\langle W_k,t_k^*\rangle|)\wedge D^{q_k}\|t_k\|_{q_k}^{q_k}\right]^p\right)^{\frac{1}{p}} \lesssim \max\{A,D\}p.$$

Let us return to our bound (53). Note that we should only care for $K\subset M$ such that

$$\|\sum_{k\in K}(Q_k|\langle W_k,t_k^*\rangle|)\wedge D^{q_k}\|t_k\|_{q_k}^{q_k}\|_p \geqslant pc.$$

We use now that $D^{q_k}\|t_k^*\|_{q_k}^{q_k}\leqslant A^{p_k}n_k$, and hence $\sum_{k\in K}A^{p_k}n_k\geqslant pc$. But then

$$\mathbf{E}_\delta\prod_{k\in K}1_{\delta_k=1}=\exp(-\sum_{k\in K}(C/4)^{p_k}n_k)\leqslant \exp(-(C/(4A))cp)\leqslant \exp(-p/c),$$

whenever $C\geqslant 4A/c^2$. Finally, we should observe that due to (22) we have $|I(t)|\leqslant \delta' p$. Therefore, there are at most $2^{p\delta'}$ sets K for which we have to use the second method. With respect to (53) the above bounds imply

$$\|\sum_{k=1}^{M}((Q_k-C)_+|\langle W_k,t_k^*\rangle|)\wedge D^{q_k}\|t_k\|_{q_k}^{q_k}\|_p \lesssim cp+2^{\delta'}e^{-\frac{1}{c}}\max\{A,D\}p\lesssim cp$$

if c is suitably small. It proves the result. □

Consequently, by (1)

$$\mathbf{E}\sup_{t\in S}\sum_{k=1}^{M}((Q_k-C)_+|\langle W_k,t_k^*\rangle|)\wedge D^{q_k}\|t_k\|_{q_k}^{q_k}\leqslant ecp.$$

which together with (50) and (51) gives

$$\mathbf{E}\sup_{t\in S}\sum_{k=1}^{M}|\langle W_k,t_k^*\rangle|\wedge D^{q_k}\|t_k\|_{q_k}^{q_k}\geqslant C^{-1}(8K)^{-1}p,$$

for c is suitably small, i.e., $ec\leqslant(8K)^{-1}$. Hence, $\mathbf{E}\sup_{t\in S}|\bar{X}|_t\geqslant(8CK)^{-1}p$. As stated in (48) we have solved the case of small coefficients (cf. (44)).

7 Large Coefficients

We work in the cube-like setting formulated in Proposition 1 accompanied by the simplification from Proposition 2. Our goal is to prove the minoration for large coefficients—see (45). Recall now that we can work under the condition (47). Therefore, we may assume that there is a point $t_0 \in T$ and a subset $S \subset T$ such that $|S| \geqslant e^{p/2}$ and

$$\sum_{k=1}^{M} \sum_{i \in I_k(t^*) \triangle I_k(t_0^*)} D^{q_k} k_i^{q_k} < p.$$

We are going to show that in this case $\|X_{t^\dagger} - X_{s^\dagger}\|_p \gtrsim p$ for all $s \neq t$, $s, t \in S$. It suffices to prove that $\|X_{t^*} - X_{t_0^*}\|_p$ and $\|X_{s^*} - X_{t_0^*}\|_p$ are a bit smaller than p. More precisely,

Lemma 5 *If D is suitably large, then $\|X_{t^*} - X_{t_0^*}\|_p \leqslant p/4$ for any $t \in S$.*

Proof Consider $\sum_{k=1}^{M} r_k = p$ and

$$\|X_{t^*} - X_{t_0^*}\|_p \simeq \sum_{k=1}^{M} \|P_k\|_{r_k} \|\langle W_k, t_k^* - t_{0k}^*\rangle\|_{r_k}.$$

Not that we have the inequality

$$\|P_k\|_{r_k} |\langle W_k, t_k^* - t_{0k}^*\rangle\|_{r_k} \lesssim r_k^{\frac{1}{p_k}} \|t_k^* - t_{0k}^*\|_{q_k}.$$

Using that $a^{1/p_k} b^{1/q_k} \leqslant a/p_k + b/q_k$, we get

$$\|P_k\|_{r_k} |\langle W_k, t_k^* - t_{0k}^*\rangle\|_{r_k} \leqslant p_k^{-1} \frac{r_k}{D} + q_k^{-1} D^{\frac{q_k}{p_k}} \|t_k^* - t_{0k}^*\|_{q_k}^{q_k}.$$

Clearly, $q_k/p_k = 1/(p_k - 1)$. Now, we can benefit from the condition (47) namely, we have

$$\sum_{k=1}^{M} \sum_{i \in I_k(t^*) \triangle I_k(t_0^*)} D^{\frac{1}{p_k - 1}} k_i^{q_k} < p/D.$$

For D is suitably small, it proves that $\|X_{t^*} - X_{t_0^*}\|_p \leqslant p/4$. □

Consequently, since $\|X_t - X_s\|_p \geqslant p$ for $s \neq t$, $s, t \in S$, we obtain

$$\|X_{t^\dagger} - X_{s^\dagger}\|_p \geqslant \|X_t - X_s\|_p - \|X_{t^*} - X_{t_0^*}\|_p - \|X_{s^*} - X_{t_0^*}\|_p \geqslant p/2. \tag{54}$$

The last result we need stems from the fact that $\|X_{t^\dagger}\|_p \leqslant \|X_t\|_p \lesssim p$, which is due to Bernoulli comparison and Proposition 2. Namely, we have

Lemma 6 *For any $t \in T$, the following inequality holds true:*

$$\sum_{k=1}^{M} n_k^{\frac{1}{p_k}} \|t_k^\dagger\|_{q_k} \lesssim p.$$

Moreover, $\sum_{k=1}^{M} n_k 1_{t_k^\dagger \neq 0} \leqslant pc$, where c can be suitably small.

Proof Obviously, $\|X_{t^\dagger}\|_p = \|\sum_{k=1}^{M} P_k\langle W_k, t_k^\dagger\rangle\|_p$, so since $\|X_{t^\dagger}\|_p \lesssim p$ there must exist r_k, $1 \leqslant k \leqslant M$ such that $\sum_{k=1}^{M} r_k = p$ and

$$\sum_{k=1}^{M} \|P_k\|_{r_k} \|\langle W_k, t_k^\dagger\rangle\|_{r_k} \lesssim p.$$

If $r_k \geqslant n_k$, we can use

$$\|P_k\|_{r_k} \|\langle W_k, t_k^\dagger\rangle\|_{r_k} \gtrsim n_k^{\frac{1}{p_k}} \|t_k^\dagger\|_{q_k}.$$

Our aim is to show that we can use $r_k \geqslant n_k$ for all k. This is possible if

$$\sum_{k=1}^{M} n_k 1_{t^\dagger k \neq 0} \leqslant p.$$

Suppose conversely that $\sum_{k \in J} n_k 1_{t_k^\dagger \neq 0} \simeq p$, for some $J \subset [M]$, then the inequality $\|X_{t^\dagger}\|_p \lesssim p$ gives

$$\sum_{k \in J} n_k^{\frac{1}{p_k}} \|t_k^\dagger\|_{q_k} \lesssim p.$$

However, if $t_k^\dagger \neq 0$, then $D\|t_k^\dagger\|_{q_k} > A^{p_k-1} n_k^{\frac{1}{q_k}}$ by (45). This is exactly the moment, where we need our technical assumption that p_k are not too close to 1. Namely, if $A \gtrsim \max\{D^{\frac{1}{p_k-1}}\}$, we get $\sum_{k \in J} \sum_{k=1}^{M} n_k 1_{t_k^\dagger \neq 0} > p$, which is a contradiction. Consequently, we can select $r_k \geqslant n_k$ for any k such that $t_k^\dagger \neq 0$. Thus, finally,

$$\sum_{k=1}^{M} n_k^{\frac{1}{p_k}} \|t_k^\dagger\|_{q_k} \lesssim p.$$

Moreover,

$$\sum_{k=1}^{M} n_k \frac{A^{p_k-1}}{D} \lesssim p.$$

Since A can be much larger than $\max D^{\frac{1}{p_k-1}}$, it completes the proof. □

We can start the main proof in the case of large coefficients—cf. (45). We are going to use a similar trick to the approach presented in the "simplification lemma"—Proposition 1. The point is that on each $\mathbb{R}^{n_k} = \mathbb{R}^{|J_k|}$ we impose a radial-type distribution. Namely, let μ_k be the probability distribution on $\mathbb{R}^{n_k}$ with density

$$\mu_k(dx) = \exp\left(-\frac{1}{B} n_k^{\frac{1}{p_k}} \|x\|_{q_k}\right) \frac{n_k^{\frac{n_k}{p_k q_k}}}{B^{\frac{n_k}{q_k}} \Gamma\left(\frac{n_k}{q_k}\right) |\partial B_{q_k}^{n_k}|} dx.$$

The fundamental property of μ_k is that

$$\mu_k(n^{-\frac{1}{p_k}} B u B_{q_k}^{n_k}) = \frac{1}{\Gamma\left(\frac{n_k}{q_k}\right)} \int_0^u s^{\frac{n_k}{q_k}-1} e^{-s} ds.$$

The median value of u_0 the distribution $\Gamma\left(\frac{n_k}{q_k}, 1\right)$ is comparable to $\frac{n_k}{q_k}$ and hence for $u = \rho \frac{n_k}{q_k}$, where ρ is smaller than 1 we have

$$\frac{1}{\Gamma\left(\frac{n_k}{q_k}\right)} \int_0^u s^{\frac{n_k}{q_k}-1} e^{-s} ds \gtrsim \rho^{\frac{n_k}{q_k}} \geqslant \frac{1}{2} \left(\frac{u}{u_0}\right)^{\frac{n_k}{q_k}}.$$

Since $u_0 \leqslant n_k / q_k$, it gives

$$\mu_k(\rho B n_k^{\frac{1}{q_k}} q_k^{-1}) \geqslant \rho^{\frac{n_k}{q_k}},$$

for any ρ which is a bit smaller than 1. Furthermore, by the construction

$$\begin{aligned}\mu_k\left(t_k^\dagger + \rho B n_k^{\frac{1}{q_k}} q_k^{-1} B_{q_k}^{n_k}\right) &\gtrsim \rho^{\frac{n_k}{q_k}} e^{-B^{-1} n_k^{\frac{1}{p_k}} \|t_k^\dagger\|_{q_k}} \\ &= \exp\left(-\frac{n_k}{q_k} \log \frac{1}{\rho} - B^{-1} n_k^{\frac{1}{p_k}} \|t_k^\dagger\|_{q_k}\right).\end{aligned}$$

Due to Lemma 6 we may find constant $B \geqslant 1$ in such a way that for any $t \in T$,

$$B^{-1} \sum_{k=1}^{M} n_k^{\frac{1}{p_k}} \|t_k^{\dagger}\|_{q_k} \leqslant p/8. \tag{55}$$

Moreover, by the same result, we may require that

$$\log\left(\frac{1}{\rho}\right) \cdot \sum_{k=1}^{M} n_k 1_{t_k^{\dagger} \neq 0} \leqslant p/8 \tag{56}$$

for some suitably small ρ.

Let μ be a measure defined on $\mathbb{R}^d$ by $\mu = \mu_1 \otimes \mu_2 \otimes \ldots \otimes \mu_M$. We define also

$$A_t = \Big(x \in \mathbb{R}^d : \ \|t_k^{\dagger} - x_k\|_{q_k} \leqslant \rho B n_k^{\frac{1}{q_k}} q_k^{-1}, \ \text{ or } \ \|t_k^{\dagger}\|_{q_k} \leqslant \rho B n_k^{\frac{1}{q_k}} q_k^{-1}, \\ \text{for all } \ k \in [M] \Big).$$

Using (55) and (56), we get

$$\mu(A_t) \geqslant \exp\left(- \sum_{k=1}^{M} \left(\frac{n_k}{q_k} \log \frac{1}{\rho} 1_{t_k^{\dagger} \neq 0} + n_k^{\frac{1}{p_k}} \|t_k^{\dagger}\|_{q_k} \right) \right) \geqslant e^{-p/4}.$$

However, there are at least $e^{p/2}$ points in S and hence

$$\sum_{t \in S} \mu(A_t) \geqslant e^{p/4}.$$

We define

$$A_x = \Big(t \in T : \ \|t_k^{\dagger} - x_k\|_{q_k} \leqslant \rho B n_k^{\frac{1}{q_k}} q_k^{-1}, \ \text{ or } \ \|t_k^{\dagger}\|_{q_k} \leqslant \rho B n_k^{\frac{1}{q_k}} q_k^{-1}, \\ \text{for all } \ k \in [M] \Big)$$

and observe that

$$\int |A_x| \mu(dx) = \sum_{t \in T} \int 1_{t \in A_x} \mu(dx) = \sum_{t \in T} \int 1_{x \in A_t} \mu(dx) \\ = \sum_{t \in T} \mu(A_t) \geqslant e^{p/4}.$$

Consequently, there must exists $x \in \mathbb{R}^d$ such that $|A_x|$ counts at least $e^{p/4}$ points. Note that we can require that $\|x_k\|_{q_k} > \rho B n_k^{\frac{1}{q_k}} q_k^{-1}$. Let us denote the improved set S by $\bar{S}$, in particular, $|\bar{S}| > e^{p/4}$.

The final step is as follows. We define the function:

$$\varphi_k(t) = \begin{cases} x_k \in \mathbb{R}^{n_k} \text{ if } \|t_k\|_{q_k} > \rho B n^{\frac{1}{q_k}} q_k^{-1}, \\ 0 \in \mathbb{R}^{n_k} \quad \text{if } \|t_k\|_{q_k} \leqslant \rho B n_k^{\frac{1}{q_k}} q_k^{-1} \end{cases}$$

Let $\varphi = (\varphi_k)_{k=1}^M$. Note that if $\varphi_k(t) = 0$, then $t_k^\dagger = 0$ by (45). Therefore, we are sure that $\|t_k^\dagger - \varphi_k(t)\|_{q_k} \leqslant \rho B n_k^{\frac{1}{q_k}} q_k^{-1}$.

Lemma 7 *The following inequality holds:*

$$\|X_{t^\dagger} - X_{\varphi(t)}\|_p \lesssim cp,$$

where c is suitably small.

Proof Indeed it suffices to check for $\sum_{k=1}^M r_k = p$ that

$$\sum_{k=1}^M r_k^{\frac{1}{p_k}} \|t_k^\dagger - \varphi_k(t)\|_{q_k} \leqslant cp.$$

We use the inequality $a^{1/p_k} b^{1/q_k} \leqslant a/p_k + b/q_k$ and $\|t_k - \varphi_k(t)\|_{q_k} \leqslant \rho B n_k^{\frac{1}{q_k}} q_k^{-1}$ to get

$$r_k^{\frac{1}{p_k}} \|t_k^\dagger - \varphi_k(t)\|_{q_k} \leqslant \frac{1}{p_k} c r_k + \frac{1}{q_k} c^{-\frac{1}{p_k - 1}} \rho^{q_k} B^{q_k} q_k^{-q_k} n_k 1_{t_k^\dagger \neq 0}.$$

It remains to notice that if only $\rho B q_k^{-1} \leqslant c$

$$\sum_{k=1}^M \frac{1}{q_k} c^{-\frac{1}{p_k - 1}} \rho^{q_k} B^{q_k} q_k^{-q_k} n_k 1_{t_k^\dagger \neq 0} \leqslant c \sum_{k=1}^M n_k 1_{t_k^\dagger \neq 0} \leqslant pc.$$

However, ρ can be suitably small; thus, the result follows: □

Now, if c is suitably small, we get that

$$\|X_{\varphi(t)} - X_{\varphi(s)}\|_p \geqslant p/4.$$

Indeed, this is the consequence of (54) and $\|X_{\varphi(t)} - X_{t^\dagger}\|_p \leqslant p/8$ and $\|X_{\varphi(s)} - X_{\varphi(t_0)}\|_p \leqslant p/8$, namely,

$$\|X_{\varphi(t)} - X_{\varphi(s)}\|_p \geqslant \|X_{t^\dagger} - X_{s^\dagger}\|_p - \|X_{\varphi(t)} - X_{t^\dagger}\|_p - \|X_{\varphi(s)} - X_{\varphi(t_0)}\|_p.$$

Moreover, by Remark 1 and $\bar{S} \subset S \subset T$, $0 \in T$ and $|T| \leqslant 1 + e^p$

$$\mathbf{E} \sup_{t \in \bar{S}} (X_{\varphi(t)} - X_{t^\dagger}) \leqslant ecp.$$

It proves that

$$\mathbf{E} \sup_{t \in \bar{S}} X_{\varphi(t)} \leqslant \mathbf{E} \sup_{t \in \bar{S}} X_{t^\dagger} + \mathbf{E} \sup_{t \in \bar{S}} (X_{\varphi(t)} - X_{t^\dagger}) \leqslant \mathbf{E} \sup_{t \in \bar{S}} X_{t^\dagger} + ecp.$$

It remains to observe that $X_{\varphi_k(t)}$ is either 0 or X_{x_k}. But vector $(X_{x_k})_{k=1}^M$ is one-unconditional and log-concave with independent entries, so we may use the standard Sudakov minoration—the main result of [5]. Therefore,

$$\mathbf{E} \sup_{t \in \bar{S}} X_{\varphi(t)} \geqslant \frac{1}{K} p.$$

It completes the proof of Sudakov minoration in the case of large coefficients (cf. (45)). In this way we have completed the program described in Sect. 2.

8 The Partition Scheme

Having established the Sudakov minoration, one can prove that partition scheme works and in this way establish the characterization of $S_X(T) = \mathbf{E} \sup_{t \in T} X_t$. We need here additionally that (9) holds.

Let us recall the general approach to obtain/get the lower bound on $S_X(T)$. We define a family of distances $d_n(s,t) = \|X_t - X_s\|_{2^n}$. Moreover, let $B_n(t, \varepsilon)$ be the ball centered at t with radius ε in d_n distance and in similarly $\Delta_n(A)$ be the diameter of $A \subset T$ in d_n distance. By (6) we know that $d_{n+1} \leqslant 2d_n$, and moreover, the condition (9) reads as $(1+\varepsilon)d_n \leqslant d_{n+1}$. Let us recall that this property holds true if all p_k are smaller than some $p_\infty < \infty$. Note that the assumption is not easily removable, since for Bernoulli canonical processes the theory which we describe below does not work.

We follow the generic chaining approach for families of distances described in [18]. Let $N_n = 2^{2^n}$, $n \geqslant 1$, $N_0 = 1$. The natural candidate for the family $F_{n,j}$ is $F_{n,j} = F$, where $F(A) = K \mathbf{E} \sup_{t \in A} X_t$, $A \subset T$ for suitably large constant K. We have to prove the growth condition, namely, that for some $r = 2^{\kappa - 2} \geqslant 4$ and fixed $n_0 \geqslant 0$ for any given $n \geqslant n_0$ and $j \in \mathbb{Z}$ if we can find points $t_1, t_2, \ldots t_{N_n} \in B_n(t, 2^n r^{-j})$, which are $2^{n+1} r^{-j-1}$ separated in d_{n+1} distance, then for any sets $H_i \subset B_{n+\kappa}(t_i, 2^{n+\kappa} r^{-j-2})$ the following inequality holds true:

$$F\left(\bigcup_{i=1}^{N_n} H_i\right) \geqslant 2^n r^{-j-1} + \min_{1\leqslant i\leqslant N_n} F(H_i). \tag{57}$$

Note that in our setting

$$B_n(t, 2^n r^{j-2}) \subset B_{n+1}(t, 2^{n+1} r^{j-2}) \subset \ldots \subset B_{n+\kappa}(t, 2^{n+\kappa} r^{-j-2}) \tag{58}$$

and also

$$B_{n+\kappa}(t, 2^{n+\kappa} r^{-j-2}) \subset B_n\left(t, \frac{2^{n+\kappa} r^{-j-2}}{(1+\varepsilon)^\kappa}\right) \subset B_n\left(t, \frac{2^{n+2} r^{-j-1}}{(1+\varepsilon)^\kappa}\right). \tag{59}$$

In particular, we get from (59) that

$$H_i \subset B_{n+\kappa}(t_i, 2^{n+\kappa} r^{-j-2}) \subset B_n\left(t_i, \frac{2^{n+2} r^{-j-2}}{(1+\varepsilon)^\kappa}\right) \tag{60}$$

and hence we have that H_i are small when compared with the separation level r^{-j-1}. Clearly, if $d_{n+1}(t_i, t_j) \geqslant 2^{n+1} r^{-j-1}$, then

$$d_n(t_i, t_j) \geqslant 2^{-1} d_{n+1}(t_i, t_j) \geqslant 2^n r^{-j-1}.$$

The last property we need is a special form of concentration, i.e.,

$$\|(\mathbf{E} \sup_{t\in A} X_t - \sup_{t\in A} X_t)_+\|_p \leqslant L \sup_{t\in A} \|X_t\|_p. \tag{61}$$

Let us recall that due to the result of [8] and some straightforward observations fortunately the inequality holds in our setting. We turn to prove the main result of this section.

Proposition 5 *Suppose that X_t satisfies assumptions on the regularity of distances as well as the concentration inequality* (61), *then $F_{n,j} = F$, $F(A) = K\mathbf{E}\sup_{t\in A} X_t$ satisfies the growth condition* (57) *for some r and n_0.*

Proof Let us define $A = \bigcup_{i=1}^{N_n} H_i$ and $H_i \subset B_{n+1}(t_i, 2^{n+1} r^{-j-1})$. We have

$$\begin{aligned}
F(A) &\geqslant K\mathbf{E} \sup_{1\leqslant i\leqslant N_n} X_{t_i} + \sup_{t\in H_i}(X_t - X_{t_i}) \\
&\geqslant K\mathbf{E} \sup_{1\leqslant i\leqslant N_n} X_{t_i} + \min_{1\leqslant i\leqslant N_n} F(H_i - t_i) \\
&\quad - K(\mathbf{E} \sup_{1\leqslant i\leqslant N_n} (\mathbf{E}\sup_{t\in H_i}(X_t - X_{t_i}) - \sup_{t\in H_i}(X_t - X_{t_i})) \\
&\geqslant 2^{n+1} r^{-j-1} + \min_{1\leqslant i\leqslant N_n} F(H_i) - 2KL \sup_{1\leqslant i\leqslant N_n} \|(\mathbf{E}\sup_{t\in H_i}(X_t - X_{t_i})
\end{aligned}$$

$$- \sup_{t\in H_i}(X_t - X_{t_i}))_+\|_{2^n}$$

$$\geqslant 2^{n+1}r^{-j-1} + \min_{1\leqslant i\leqslant N_n} F(H_i) - 2KL \max_{1\leqslant i\leqslant N} \Delta_n(H_i),$$

where we have used here the Sudakov minoration. Using (60), we get

$$\Delta_n(H_i) \leqslant \frac{2^{n+3}r^{j-1}}{(1+\varepsilon)^\kappa}.$$

Therefore, choosing sufficiently large κ, we can guarantee that

$$F(A) \geqslant 2^n r^{-j-1}$$

as required. □

The basic result of [18] is that, having the growth condition for families of distances, it is true that $S_X(T) = \mathbf{E}\sup_{t\in T} X_t$ is comparable with $\gamma_X(T)$, where

$$\gamma_X(T) = \inf_{\mathcal{A}} \sup_{t\in T} \sum_{n=0}^{\infty} \Delta_n(A_n(t)).$$

It completes the proof of Theorem 2. Namely, for some absolute constant K,

$$K^{-1}\gamma_X(T) \leqslant \mathbf{E}\sup_{t\in T} X_t \geqslant K\gamma_X(T).$$

In this way it establishes a geometric characterization of $S_X(T)$, assuming our list of conditions.

Acknowledgments Research partially supported by NCN Grant UMO-2019/35/B/ST1/04292.

References

1. K. Ball, Logarithmically convex functions and sections of convex sets in $\mathbb{R}^n$. Studia Math. **88**, 69–84 (1988)
2. S.G. Bobkov, F.L. Nazarov, On convex bodies and log-concave probability measures with unconditional basis, in *Geometric Aspects of Functional Analysis, Lecture Notes in Mathematics*, vol. 1807 (Springer, Berlin , 2013)
3. E.D. Gluskin, S. Kwapien, Tails and moment estimates for sums of independent random variables with logarithmically concave tails. Studia Math. **114**(3), 303–309 (1995)
4. P. Hitczenko, Domination inequality for martingale transforms of Rademacher sequence. Israel J. Math. **84**, 161–178 (1993)
5. R. Latala, Sudakov minoration principle and supremum of some processes. Geom. Funct. Anal. **7**, 936–953 (1997)

6. R. Latala, Sudakov-type minoration for log-concave vectors. Studia Math. **223**, 251–274 (2014)
7. R. Latala, T. Tkocz, A note on suprema of canonical processes based on random variables with regular moments. Electron. J. Probab. **20**(36), 1–17 (2015)
8. R. Latala, J. Wojtaszczyk, On the infimum convolution inequality. Studia Math. **189**, 147–187 (2008)
9. M. Ledoux, M. Talagrand, Probability in Banach spaces. Isoperimetry and processes, in *Results in Math. and Rel. Areas (3).*, vol. 23 (Springer, Berlin, 1991), xii+480 pp.
10. S. Mendelson, E. Milman, G. Paouris, Generalized dual Sudakov minoration via dimension reduction – a program. Arxiv (2016)
11. P. Nayar, K. Oleszkiewicz, Khinchine type inequalities with optimal constants via ultra log-concavity. Positivity **16**, 359–371 (2012)
12. A. Pajor, N. Tomczak-Jaegermann, Subspaces of small codimension of finite dimensional Banach spaces. Proc. Amer. Math. Soc. **97**, 637–642 (1986)
13. M. Pilipczuk, J.O. Wojtaszczyk, The negative association property for the absolute values of random variables equidistributed on a generalized Orlicz ball. Positivity **12**, 421–474 (2008)
14. V.N. Sudakov, Gaussian measures, Cauchy measures and ε-entropy. Soviet. Math. Dokl. **12**, 412–415 (1969)
15. V.N. Sudakov, A remark on the criterion of continuity of Gaussian sample functions, in *Proceedings of the Second Japan-USSR Symposium on Probability Theory.* Lecture Notes in Mathematics, vol. 330 (Springer, Berlin, 1973), pp. 444–454
16. M. Talagrand, Characterization of almost surely continuous 1-stable random Fourier series and strongly stationary processes. Ann. Probab. **18**, 85–91 (1990)
17. M. Talagrand, The supremum of certain canonical processes. Am. J. Math. **116**, 283–325 (1994)
18. M. Talagrand, Upper and lower bounds for stochastic processes, in *A Series of Modern Surveys in Mathematics*, vol. 60 (Springer, Berlin, 2014)
19. R. van Hadel, Chaining, interpolation and convexity II: The contraction principle. Ann. Probab. **46**, 1764–1805 (2018)

Lévy Measures of Infinitely Divisible Positive Processes: Examples and Distributional Identities

Nathalie Eisenbaum and Jan Rosiński

2020 Mathematics Subject Classification: 60E05; 60E07; 60G51; 60G57; 60G55

1 Introduction

A random process is infinitely divisible if all its finite dimensional marginals are infinitely divisible. Let $\psi = (\psi(x), x \in E)$ be a nonnegative infinitely divisible process with no drift. The infinite divisibility of ψ is characterized by the existence of a unique measure ν on $\mathbb{R}_+^E$, the space of all functions from E into $\mathbb{R}_+$, such that for every $n > 0$, every $\alpha_1, .., \alpha_n$ in $\mathbb{R}_+$ and every $x_1, .., x_n$ in E:

$$\mathbb{E}[\exp\{-\sum_{i=1}^{n} \alpha_i \psi(x_i)\}] = \exp\{-\int_{\mathbb{R}_+^E} (1 - e^{-\sum_{i=1}^{n} \alpha_i y(x_i)})\nu(dy)\}. \tag{1.1}$$

The measure ν is called the Lévy measure of ψ. The existence and uniqueness of such measures was established in complete generality in [16]. In Sect. 2, we recall some definitions and facts about Lévy measures.

It might be difficult to obtain an expression for the Lévy measure ν directly from (1.1). In [3], a general expression for ν has been established. Its proof is based on several identities involving ψ. Among them:
For every $a \in E$ with $0 < \mathbb{E}[\psi(a)] < \infty$, there exists a nonnegative process $(r^{(a)}(x), x \in E)$ independent of ψ such that

N. Eisenbaum
MAP5, CNRS - Université de Paris, Paris, France
e-mail: nathalie.eisenbaum@parisdescartes.fr

J. Rosiński (✉)
Department of Mathematics, University of Tennessee, Knoxville, TN, USA
e-mail: jrosinsk@utk.edu

R. Adamczak et al. (eds.), *High Dimensional Probability IX*, Progress in Probability 80, https://doi.org/10.1007/978-3-031-26979-0_12

$$\psi + r^{(a)} \text{ has the law of } \psi \text{ under } \mathbb{E}\Big[\frac{\psi(a)}{\mathbb{E}[\psi(a)]}, \cdot\Big] \tag{1.2}$$

Actually, the existence of $(r^{(a)}, a \in E)$ characterizes the infinite divisibility of ψ. This characterization has been established in [2] (see also [16, Proposition 4.7]).

Under an assumption of stochastic continuity for ψ, the general expression for ν obtained in [3] is the following:

$$\nu(F) = \int_E \mathbb{E}\Big[\frac{F(r^{(a)})}{\int_E r^{(a)}(x)m(dx)}\Big]\mathbb{E}[\psi(a)]m(da), \tag{1.3}$$

for any measurable functional F on $\mathbb{R}_+^E$, where m is any σ-finite measure with support equal to E such that $\int_E \mathbb{E}[\psi(x)]m(dx) < \infty$.

Moreover, the law of $r^{(a)}$ is connected to ν as follows (see [3, 16]):

$$\mathbb{E}[F(r^{(a)})] = \frac{1}{\mathbb{E}[\psi(a)]} \int_{\mathbb{R}_+^E} y(a)F(y)\,\nu(dy). \tag{1.4}$$

The problem of determining ν is hence equivalent to the one of the law of $r^{(a)}$ for every a in E. But knowing ν, one can not only write (1.2) but many other identities of the same type. In each one, the process $r^{(a)}$ is replaced by a process with an absolutely continuous law with respect to ν (see [16, Theorem 4.3(a)]).

Some conditionings on ψ lead to a splitting of ν. This allows to obtain decompositions of ψ into independent infinitely divisible components (see [3, Theorems 1.1, 1.2 and 1.3]). As an example:
For every $a \in E$, there exists a nonnegative infinitely divisible process $(\mathcal{L}^{(a)}(x), x \in E)$ independent of an infinitely divisible process $((\psi(x), x \in E)|\psi(a) = 0)$ such that

$$\psi \overset{(\text{law})}{=} (\psi \mid \psi(a) = 0) \; + \; \mathcal{L}^{(a)}. \tag{1.5}$$

By Eisenbaum [3, Theorem 1.2], the processes $(\psi \mid \psi(a) = 0)$ and $\mathcal{L}^{(a)}$ have the respective Lévy measures ν_a and $\tilde{\nu}_a$, where

$$\nu_a(dy) = \mathbb{1}_{\{y(a)=0\}}\nu(dy) \text{ and } \quad \tilde{\nu}_a(dy) = \mathbb{1}_{\{y(a)>0\}}\nu(dy)\,. \tag{1.6}$$

In Sect. 3, to illustrate the relations and identities (1.1)–(1.5), we choose to consider simple examples of nonnegative infinitely divisible processes. In each case the Lévy measure is directly computable from (1.1) or from the stochastic integral representation of ψ (see [12]). Thanks to (1.2) and its extensions and (1.5), we present remarkable identities satisfied by the considered nonnegative infinitely divisible processes. Moreover, the general expression (1.3) provides alternative formulas for the Lévy measure, which are also remarkable. We treat the cases of Poisson processes, Sato processes, stochastic convolutions, and tempered stable subordinators. We also point out a connection with infinitely divisible random measures. We end Sect. 3 by reminding the case of infinitely divisible permanental processes, which is the first case for which identities in law of the same type as (1.2) have been established. In this case, such identities in law are called "isomorphism theorems" in reference to the very first one established by Dynkin [1], the so-called "Dynkin isomorphism theorem."

When ψ is an infinitely divisible permanental process, the two processes $r^{(a)}$ and $\mathcal{L}^{(a)}$ have the same law. If, moreover, ψ is a squared Gaussian process, Marcus and Rosen [9] have established correspondences between path properties of ψ and the ones of $\mathcal{L}^{(a)}$. The extension of these correspondences to general infinitely divisible permanental processes has been undertaken by several authors (see [3, 4, 10] or [11]). Similarly, in Sect. 4, we consider a general infinitely divisible nonnegative process ψ and state some trajectory correspondences between ψ and $\mathcal{L}^{(a)}$, resulting from an iteration of (1.5) (see also [16]).

Finally, observing that, given an infinitely divisible positive process ψ, $r^{(a)}$ is not a priori "naturally" connected to ψ, we present, in Sect. 5, $r^{(a)}$ as the limit of a sequence of processes naturally connected to ψ.

2 Preliminaries on Lévy Measures

In this section we recall some definitions and facts about general Lévy measures given in [16, Section 2]. Some additional material can be found in [15]. Let $(\xi(x), x \in E)$ be a real-valued infinitely divisible process, where E is an arbitrary nonempty set. A measure ν defined on the cylindrical σ-algebra $\mathcal{R}^E$ of $\mathbb{R}^E$ is called the Lévy measure of ξ if the following two conditions hold:

(i) For every $x_1, \ldots, x_n \in E$, the Lévy measure of the random vector $(\xi(x_1), \ldots, \xi(x_n))$ coincides with the projection of ν onto $\mathbb{R}^{\{x_1,\ldots,x_n\}}$, modulo the mass at the origin.
(ii) $\nu(A) = \nu_*(A \setminus 0_E)$ for all $A \in \mathcal{R}^E$, where ν_* denotes the inner measure and 0_E is the origin of $\mathbb{R}^E$.

The Lévy measure of an infinitely divisible process always exists and (ii) guarantees its uniqueness. Condition (i) implies that $\int_{\mathbb{R}^E}(f(x)^2 \wedge 1)\,\nu(df) < \infty$ for every $x \in E$.

A Lévy measure ν is σ-finite if and only if then there exists a countable set $E_0 \subset E$ such that

$$\nu\{f \in \mathbb{R}^E : f_{|E_0} = 0\} = 0. \tag{2.1}$$

Actually, if (i) and (2.1) hold, then does so (ii) and ν is a σ-finite Lévy measure.

Condition (2.1) is usually easy to verify. For instance, if an infinitely divisible process $(\xi(x), x \in E)$ is separable in probability, then its Lévy measure satisfies (2.1), so is σ-finite. The separability in probability is a weak assumption. It says that there is a countable set $E_1 \subset E$ such that for every $x \in E$ there is a sequence $(x_n) \subset E_1$ such that $\xi(x_n) \to \xi(x)$ in $\mathbb{P}$. Infinitely divisible processes whose Lévy measures do not satisfy (2.1) include such pathological cases as an uncountable family of independent Poisson random variables with mean 1.

If the process ξ has paths in some "nice" subspace of $\mathbb{R}^E$, then, due to the transfer of regularity [16, Theorem 3.4], its Lévy measure ν is carried by the same subspace of $\mathbb{R}^E$. Thus, one can investigate the canonical process on $(\mathbb{R}^E, \mathcal{R}^E)$ under the law of ξ and also under the measure ν and relate their properties. This approach was successful in the study of distributional properties of subadditive functionals of paths of infinitely divisible processes [17] and the decomposition and classification of stationary stable processes [13], among others.

If an infinitely divisible process ξ without Gaussian component has the Lévy measure ν, then it can be represented as

$$(\xi(x), x \in E) \overset{\text{(law)}}{=} \Big(\int_{\mathbb{R}^E} f(x)\,[N(df) - \chi(f(t))\nu(df)] + b(x),\ x \in E \Big) \tag{2.2}$$

where N is a Poisson random measure on $(\mathbb{R}^E, \mathcal{R}^E)$ with intensity measure ν, $\chi(u) = \mathbb{1}_{[-1,1]}(u)$, and $b \in \mathbb{R}^E$ is deterministic. Relation (2.2) can be strengthen to the equality almost surely under some minimal regularity conditions on the process ξ, provided the probability space is rich enough (see [16, Theorem 3.2]). This is an extension to general infinitely divisible processes of the celebrated Lévy-Itô representation.

Obviously, all the above apply to processes presented in the introduction but in a more transparent form. Namely, if $(\psi(x), x \in E)$ is an infinitely divisible nonnegative process, then its Lévy measure ν is concentrated on $\mathbb{R}_+^E$ and (2.2) becomes

$$(\psi(x), x \in E) \overset{\text{(law)}}{=} \Big(f_0(x) + \int_{\mathbb{R}_+^E} f(x)\, N(df),\ x \in E \Big), \tag{2.3}$$

where N is a Poisson random measure on $\mathbb{R}_+^E$ with intensity measure ν such that $\int_{\mathbb{R}_+^E} (f(x) \wedge 1)\, \nu(df) < \infty$ for every $x \in E$. Moreover, $\mathbb{E}[\psi(x)] < \infty$ if and only if $\int_{\mathbb{R}_+^E} f(x)\, \nu(df) < \infty$ and $f_0 \geq 0$ is a deterministic drift.

Since N can be seen as a countable random subset of $\mathbb{R}_+^E$, one can write (2.3) as

$$(\psi(x), x \in E) \overset{\text{(law)}}{=} \Big(f_0(x) + \sum_{f \in N} f(x),\ x \in E \Big). \tag{2.4}$$

We end this section with a necessary and sufficient condition for a measure ν to be the Lévy measure of a nonnegative infinitely divisible process. It is a direct consequence of [16] section 2. From now on we will assume that ψ has no drift, in which case $f_0 = 0$ in (2.3)–(2.4).

Let ν be a measure on $(\mathbb{R}_+^E, \mathcal{B}^E)$, where $\mathcal{B}^E$ denotes the cylindrical σ-algebra associated to $\mathbb{R}_+^E$ the space of all functions from E into $\mathbb{R}_+$. There exists an infinitely divisible nonnegative process $(\psi(x), x \in E)$ such that for every $n > 0$, every $x_1, .., x_n$ in E:

$$\mathbb{E}[\exp\{-\sum_{i=1}^{n} \alpha_i \psi(x_i)\}] = \exp\{-\int_{\mathbb{R}_+^E} (1 - e^{-\sum_{i=1}^n \alpha_i y(x_i)}) \nu(dy)\},$$

if and only if ν satisfies the two following conditions:

(L1) For every $x \in E$ $\nu(y(x) \wedge 1)) < \infty$.
(L2) For every $A \in \mathcal{B}^E$, $\nu(A) = \nu_*(A \setminus 0_E)$, where ν_* is the inner measure.

3 Illustrations

By a standard uniform random variable we mean a random variable with the uniform law on [0, 1]. A random variable with exponential law and mean 1 will be called standard exponential.

3.1 *Poisson Process*

A Poisson process $(N_t, t \geq 0)$ with intensity λm, where $\lambda > 0$ and m is the Lebesgue measure on $\mathbb{R}_+$, is the simplest Lévy process, but its Lévy measure ν is even simpler. It is a σ-finite measure given by

$$\nu(F) = \lambda \int_0^\infty F\big(\mathbb{1}_{[s,\infty)}\big)\, ds, \tag{3.1}$$

for every measurable functional $F : \mathbb{R}_+^{[0,\infty)} \mapsto \mathbb{R}_+$. Thus, (3.1) says that ν is the image of λm by the mapping $s \mapsto \mathbb{1}_{[s,\infty)}$ from $\mathbb{R}_+$ into $\mathbb{R}_+^{[0,\infty)}$.

Formula (3.1) is a special case of [16, Example 2.23]. We will derive it here for the sake of illustration and completeness.

Let $(N_t, t \geq 0)$ be a Poisson process as above. By a routine computation of the Laplace transform, we obtain that for every $0 \leq t_1 < \cdots < t_n$ the Lévy measure $\nu_{t_1,\ldots,t_n}$ of $(N_{t_1}, \ldots, N_{t_n})$ is of the form

$$\nu_{t_1,\ldots,t_n} = \sum_{i=1}^{n} \lambda \Delta t_i \, \delta_{\mathbf{u}_i},$$

where $\Delta t_i = t_i - t_{i-1}$, $t_0 = 0$, and $\mathbf{u}_i = (\underbrace{0, \ldots, 0}_{i-1 \text{ times}}, 1, \ldots, 1) \in \mathbb{R}^n$, $i = 1, \ldots, n$.

To verify that (3.1) satisfies (i) of Sect. 1, consider a finite dimensional functional F, that is, $F(f) = F_0(f(t_1), \ldots, f(t_n))$, where $F_0 : \mathbb{R}_+^n \mapsto \mathbb{R}_+$ is a Borel function with $F_0(0, \ldots, 0) = 0$ and $0 \le t_1 < \cdots < t_n$. From (3.1) we have

$$\begin{aligned}\nu(F) &= \lambda \int_0^\infty F\big(\mathbb{1}_{[s,\infty)}\big)\, ds = \lambda \int_0^\infty F_0(\mathbb{1}_{[s,\infty)}(t_1), \ldots, \mathbb{1}_{[s,\infty)}(t_n))\, ds \\ &= \lambda \sum_{i=1}^n \int_{t_{i-1}}^{t_i} F_0(\mathbf{u}_i)\, ds = \int_{\mathbb{R}_+^n} F_0(x)\, \nu_{t_1,\ldots,t_n}(dx)\end{aligned}$$

which proves (i). Condition (2.1) holds for any unbounded set, for instance, $E_0 = \mathbb{N}$. Indeed,

$$\nu\{f \in \mathbb{R}_+^{[0,\infty)} : f_{|\mathbb{N}} = 0\} = \lambda \int_0^\infty \mathbb{1}\{s : \mathbb{1}_{[s,\infty)}(n) = 0\ \forall n \in \mathbb{N}\})\, ds = 0,$$

so that ν is the Lévy measure of $(N_t, t \ge 0)$.

The next proposition exemplifies remarkable identities resulting from (1.5) and (1.2). It also gives an alternative "probabilistic" form of the Lévy measure ν.

Proposition 3.1 *Let $N = (N_t, t \ge 0)$ be a Poisson process with intensity λm, where m is the Lebesgue measure on $\mathbb{R}_+$ and $\lambda > 0$.*

(a1) *Given $a > 0$, let $r^{(a)}$ be the process defined by $r^{(a)}(t) := \mathbb{1}_{[aU,\infty)}(t)$, $t \ge 0$, where U is a standard uniform random variable independent of $(N_t, t \ge 0)$. Then, $(r^a(t), t \ge 0)$ satisfies (1.2), that is,*

$$(N_t + \mathbb{1}_{[aU,\infty)}(t),\ t \ge 0) \overset{\text{(law)}}{=} (N_t,\ t \ge 0) \text{ under } \mathbb{E}\Big[\frac{N_a}{\lambda a};\ \cdot\ \Big].$$

(b1) *For any nonnegative random variable Y whose support equals $\mathbb{R}_+$ and $\mathbb{E}Y < \infty$, the Lévy measure ν of $(N_t, t \ge 0)$ can be represented as*

$$\nu(F) = \lambda\, \mathbb{E}\big[Y h(UY)\, F\big(\mathbb{1}_{[UY,\infty)}\big)\big]$$

for every measurable functional $F : \mathbb{R}_+^{[0,\infty)} \mapsto \mathbb{R}_+$, where U is a standard uniform random variable independent of Y and $h(x) = 1/\mathbb{P}[Y \ge x]$.

In particular, if Y is a standard exponential random variable independent of U, then

$$\nu(F) = \lambda\, \mathbb{E}\big[Y e^{UY}\, F\big(\mathbb{1}_{[UY,\infty)}\big)\big].$$

(c1) *The components of the decomposition (1.5): $N \overset{\text{(law)}}{=} (N \mid N_a = 0) + \mathcal{L}^{(a)}$, can be identified as*

$$(N_t, t \geq 0 \mid N_a = 0) \overset{\text{(law)}}{=} (N_{t\vee a} - N_a,\ t \geq 0).$$

and

$$(\mathcal{L}_t^{(a)},\ t \geq 0) \overset{\text{(law)}}{=} (N_{t\wedge a},\ t \geq 0).$$

The Lévy measures ν_a and $\tilde{\nu}_a$ of $(N_t, t \geq 0 \mid N_a = 0)$ and of $(\mathcal{L}_t^{(a)}, t \geq 0)$, respectively, are given by

$$\nu_a(F) = \lambda \int_a^\infty F\big(\mathbb{1}_{[s,\infty)}\big)\, ds,$$

and

$$\tilde{\nu}_a(F) = \lambda \int_0^a F\big(\mathbb{1}_{[s,\infty)}\big)\, ds,$$

for every measurable functional $F : \mathbb{R}_+^{[0,\infty)} \mapsto \mathbb{R}_+$.

Proof

(a1) By (1.4) we have for any measurable functional $F : \mathbb{R}^{[0,\infty)} \mapsto \mathbb{R}_+$

$$\begin{aligned}
\mathbb{E}F(r_t^{(a)}, t \geq 0) &= \frac{1}{\mathbb{E}N_a} \int F(y)\, y(a)\, \nu(dy) \\
&= \frac{1}{a} \int_0^\infty F(\mathbb{1}_{[s,\infty)})\, \mathbb{1}_{[s,\infty)}(a)\, ds \\
&= \frac{1}{a} \int_0^a F(\mathbb{1}_{[s,\infty)})\, ds = \mathbb{E}F(\mathbb{1}_{[aU,\infty)}).
\end{aligned}$$

Thus, $(r_t^{(a)}, t \geq 0) \overset{\text{(law)}}{=} (\mathbb{1}_{[aU,\infty)}(t), t \geq 0)$. Choosing U independent of N, we have (1.2) for $r_t^{(a)} = \mathbb{1}_{[aU,\infty)}(t)$, $t \geq 0$, which completes the proof of (a1).

(b1) This point is an illustration of the invariance property in m of (1.3). Indeed, since the process $(N_t, t \geq 0)$ is stochastically continuous, we have for every σ-finite measure $\tilde{m}$ whose support is $[0, \infty)$ and $\int_0^\infty t\, \tilde{m}(dt) < \infty$

$$\begin{aligned}
\nu(F) &= \int_0^\infty \mathbb{E}\left[\frac{F(r^{(a)})}{\int_0^\infty r_s^{(a)}\, \tilde{m}(ds)}\right] \mathbb{E}[N_a]\, \tilde{m}(da) \\
&= \lambda \int_0^\infty \mathbb{E}\left[\frac{F(\mathbb{1}_{[aU,\infty)})}{\tilde{m}([aU,\infty))}\right] a\, \tilde{m}(da).
\end{aligned}$$

If $\tilde{m}$ is the law of a nonnegative random variable Y, then

$$\begin{aligned}\nu(F) &= \lambda \int_0^\infty \mathbb{E}\left[a\, h(aU) F(\mathbb{1}_{[aU,\infty)})\right] \tilde{m}(da) \\ &= \lambda \mathbb{E}\left[Y h(UY) F(\mathbb{1}_{[UY,\infty)})\right],\end{aligned}$$

which is the formula in (b1).

(c1) Since $(N_t, t \geq 0 \,|\, N_a = 0)$ has the Lévy measure $\nu_a(dy) = \mathbb{1}_{\{y(a)=0\}}\nu(dy)$ (see [3]), by (3.1) we get

$$\begin{aligned}\nu_a(F) &= \int F(y)\, \mathbb{1}_{\{y(a)=0\}}\nu(dy) \\ &= \lambda \int_0^\infty F(\mathbb{1}_{[s,\infty)})\mathbb{1}_{\{\mathbb{1}_{[s,\infty)}(a)=0\}}\, ds \\ &= \lambda \int_a^\infty F(\mathbb{1}_{[s,\infty)})\, ds\,.\end{aligned}$$

Since $\tilde{\nu}_a = \nu - \nu_a$, by (3.1) we have

$$\tilde{\nu}_a(F) = \lambda \int_0^a F(\mathbb{1}_{[s,\infty)})\, ds.$$

Let $0 = t_0 < t_1 < \cdots < t_n$ be such that $t_m = a$ for some $m \leq n$. For $\alpha_i > 0$ we obtain

$$\begin{aligned}\mathbb{E}\exp\Big\{-\sum_{i=1}^n \alpha_i(\mathcal{L}_{t_i}^{(a)} - \mathcal{L}_{t_{i-1}}^{(a)})\Big\} &= \exp\{-\tilde{\nu}_a(1 - e^{-\sum_{i=1}^n \alpha_i(y(t_i)-y(t_{i-1}))})\} \\ &= \exp\{-\lambda \int_0^a (1 - e^{-\sum_{i=1}^n \alpha_i(\mathbb{1}_{[s,\infty)}(t_i)-\mathbb{1}_{[s,\infty)}(t_{i-1}))})\, ds\} \\ &= \exp\{-\lambda \sum_{i=1}^m \int_{t_{i-1}}^{t_i} (1 - e^{-\sum_{i=1}^n \alpha_i(\mathbb{1}_{[s,\infty)}(t_i)-\mathbb{1}_{[s,\infty)}(t_{i-1}))})\, ds\} \\ &= \exp\{-\lambda \sum_{i=1}^m (t_i - t_{t-1})(1 - e^{-\alpha_i})\} \\ &= \mathbb{E}\exp\Big\{-\sum_{i=1}^n \alpha_i(N_{t_i\wedge a} - N_{t_{i-1}\wedge a})\Big\}\end{aligned}$$

which shows that $(\mathcal{L}_t^{(a)},\ t \geq 0) \overset{\text{(law)}}{=} (N_{t\wedge a},\ t \geq 0)$.

Since $(N_{t\wedge a},\ t \geq 0)$ and $(N_{t\vee a} - N_a,\ t \geq 0)$ are independent and they add to $(N_t, t \geq 0)$, $(N_t, t \geq 0 \,|\, N_a = 0) \overset{\text{(law)}}{=} (N_{t\vee a} - N_a,\ t \geq 0)$. □

Remarks 3.2

(1) By Proposition 3.1(b1) the Lévy measure ν of N can be viewed as the law of the stochastic process

$$(\mathbb{1}_{[UY,\infty)}(t),\ t \geq 0)$$

under the infinite measure $\lambda Y h(UY)\, d\mathbb{P}$. This point of view provides some intuition about the support of a Lévy measure and better understanding how its mass is distributed on the path space.

(2) The process $(r_t^{(a)}, t \geq 0)$ of Proposition 3.1(a1) is not infinitely divisible. Indeed, for each $t > 0$, $r_t^{(a)}$ is a Bernoulli random variable.

(3) While the decomposition (1.5) is quite intuitive in case (c1), it is not so for general infinitely divisible random fields (cf. [3]).

3.2 *Sato Processes*

Recall that a process $X = (X_t, t \geq 0)$ is H-self-similar, $H > 0$, if for every $c > 0$

$$(X_{ct}, t \geq 0) \stackrel{\text{(law)}}{=} (c^H X_t, t \geq 0)\,.$$

It is well-known that a Lévy process is H-self-similar if and only if it is strictly α-stable with $\alpha = 1/H \in (0, 2]$ (see [19, Proposition 13.5]). In short, there are only obvious examples of self-similar Lévy processes.

Sato [18] showed that, within a larger class of additive processes, there is a rich family of self-similar processes, which is generated by self-decomposable laws. These processes are known as *Sato processes* and will be precisely defined below.

Recall that the law of a random variable S is said to be self-decomposable if for every $b > 1$ there exists an independent of S random variable R_b such that

$$S \stackrel{\text{(law)}}{=} b^{-1} S + R_b.$$

Wolfe [21] and Jurek and Vervaat [7] showed that a random variable S is self-decomposable if and only if

$$S \stackrel{\text{(law)}}{=} \int_0^\infty e^{-s}\, dY_s \tag{3.2}$$

for some Lévy process $Y = (Y_s,\ s \geq 0)$ with $\mathbb{E}(\ln^+ |Y_1|) < \infty$. Moreover, there is a 1–1 correspondence between the laws of S and Y_1. The process Y is called the background driving Lévy process (BDLP) of S.

Sato [18] proved that a random variable S has the self-decomposable law if and only if for each $H > 0$ there exists a unique additive H-self-similar process $(X_t, t \geq 0)$ such that $X_1 \overset{\text{(law)}}{=} S$. An additive self-similar process, whose law at time 1 is self-decomposable, will be called a Sato process.

Jeanblanc et al. [6, Theorem 1] gave the following representation of Sato processes. Let Y be the BDLP specified in (3.2) and let $\hat{Y} = (\hat{Y}_s, s \geq 0)$ be an independent copy of Y. Then, for each $H > 0$, the process

$$X_r := \begin{cases} \int_{\ln(r^{-1})}^{\infty} e^{-Ht}\, d_t(Y_{Ht}) & \text{if } 0 \leq r \leq 1 \\ X_1 + \int_0^{\ln r} e^{Ht}\, d_t(\hat{Y}_{Ht}) & \text{if } r \geq 1. \end{cases} \tag{3.3}$$

is the Sato process with self-similarity exponent H. Stochastic integrals in (3.2) and (3.3) can be evaluated pathwise by parts due to the smoothness of the integrants. We will give another form of this representation that is easier to use for our purposes.

Theorem 3.3 *Let $\bar{Y} = (\bar{Y}_s, s \in \mathbb{R})$ be a double-sided Lévy process such that $\bar{Y}_0 = 0$ and $\mathbb{E}(\ln^+ |\bar{Y}_1|) < \infty$. Then, for each $H > 0$, the process*

$$X_t := \int_{\ln(t^{-H})}^{\infty} e^{-s}\, d\bar{Y}_s\,, \quad t \geq 0, \tag{3.4}$$

is a Sato process with self-similarity exponent H. Conversely, any Sato process with self-similarity exponent H has a version given by (3.4).

Proof By definition, a double-sided Lévy process $\bar{Y}$ is indexed by $\mathbb{R}$, has stationary and independent increments, càdlàg paths, and $\bar{Y}_0 = 0$ a.s. Since (3.4) coincides with (3.2) when $t = 1$, the improper integral $X_1 = \int_0^{\infty} e^{-s}\, d\bar{Y}_s$ converges a.s. and it has a self-decomposable distribution. Moreover,

$$X_{0+} = \lim_{t \downarrow 0} \int_{\ln(t^{-H})}^{\infty} e^{-s}\, d\bar{Y}_s = 0 \quad a.s.$$

For every $0 < t_1 < \cdots < t_n$ and $u_k = \ln(t_k^{-H})$ the increments

$$X_{t_k} - X_{t_{k-1}} = \int_{u_k}^{\infty} e^{-s}\, d\bar{Y}_s - \int_{u_{k-1}}^{\infty} e^{-s}\, d\bar{Y}_s = \int_{u_k}^{u_{k-1}} e^{-s}\, d\bar{Y}_s, \quad k = 2, \ldots, n$$

are independent as $\bar{Y}$ has independent increments. Thus, X is an additive process.

To prove the H-self-similarity of X, notice that since X is an additive process, it is enough to show that for every $c > 0$ and $0 < t < u$

$$X_{cu} - X_{ct} \overset{\text{(law)}}{=} c^H (X_u - X_t). \tag{3.5}$$

Since $\bar{Y}$ has stationary increments, we get

$$X_{cu} - X_{ct} = \int_{\ln((cu)^{-H})}^{\ln((ct)^{-H})} e^{-s}\, d\bar{Y}_s = \int_{\ln(u^{-H})+\ln(c^{-H})}^{\ln(t^{-H})+\ln(c^{-H})} e^{-s}\, d\bar{Y}_s$$

$$\overset{\text{(law)}}{=} \int_{\ln(u^{-H})}^{\ln(t^{-H})} e^{-s-\ln(c^{-H})}\, d\bar{Y}_s = c^H (X_u - X_t)\,,$$

which proves (3.5).

Conversely, let $X = (X_t : t \geq 0)$ be a H-self-similar Sato process. By (3.2) there exists a unique in law Lévy process $Y = (Y_t : t \geq 0)$ such that $\mathbb{E}(\ln^+ |Y_1|) < \infty$ and

$$X_1 \overset{\text{(law)}}{=} \int_0^\infty e^{-s}\, dY_s\,.$$

Let $Y^{(1)}$ and $Y^{(2)}$ be independent copies of the Lévy process Y. Define $\bar{Y}_s = Y_s^{(1)}$ for $s \geq 0$ and $\bar{Y}_s = Y^{(2)}_{(-s)-}$ for $s < 0$. Then $\bar{Y}$ is a double-sided Lévy process with $\bar{Y}_1 \overset{\text{(law)}}{=} Y_1$. Then

$$\tilde{X}_t := \int_{\ln(t^{-H})}^\infty e^{-s}\, d\bar{Y}_s\,, \quad t \geq 0,$$

is a version of X. □

Corollary 3.4 *Let $X = (X_t : t \geq 0)$ be a H-self-similar Sato process given by* (3.4). *Let ρ be the Lévy measure of $\bar{Y}_1$. Then the Lévy measure ν of X is given by*

$$\nu(F) = \int_{\mathbb{R}} \int_{\mathbb{R}} F(xe^{-s} 1\!\!1_{[e^{-s/H},\infty)})\, \rho(dx) ds. \tag{3.6}$$

Proof We can write (3.4) as $X_t = \int_{\mathbb{R}} f_t(s)\, d\bar{Y}_s$, where $f_t(s) = e^{-s} 1\!\!1_{[e^{-s/H},\infty)}(t)$. It follows from [12, Theorem 2.7(iv)] that the Lévy measure ν of X is the image of $m \otimes \rho$ by the map $(s, x) \mapsto x f_{(\cdot)}(s)$ from $\mathbb{R}^2$ into $\mathbb{R}^{[0,\infty)}$. □

From now on we will consider a *H-self-similar nonnegative Sato process with finite mean and no drift*. By Theorem 3.3 we have

$$\psi(t) = \int_{\ln(t^{-H})}^\infty e^{-s}\, d\bar{Y}_s\,, \quad t \geq 0, \tag{3.7}$$

where $\bar{Y} = (\bar{Y}_t, t \in \mathbb{R})$ is a double-sided subordinator without drift such that $\bar{Y}_0 = 0$ and $\mathbb{E}\bar{Y}_1 < \infty$. Consequently, $\mathbb{E}\psi(t) = \kappa t^H, t \geq 0$, where $\kappa := \mathbb{E}\psi(1) = \mathbb{E}\bar{Y}_1$.

Proposition 3.5 *Let $(\psi(t), t \geq 0)$ be a nonnegative H-self-similar Sato process given by* (3.7)*. Therefore, the Lévy measure ρ of $\bar{Y}_1$ is concentrated on $\mathbb{R}_+$.*

(a2) *Given $a > 0$, let $(r^{(a)}(t), t \geq 0)$ be the process defined by:*

$$r^{(a)}(t) := a^H UV \mathbb{1}_{[aU^{1/H},\infty)}(t), t \geq 0,$$

where U is a standard uniform random variable and V has the distribution $\kappa^{-1} x\rho(dx)$, with U, V and $(\psi(t), t \geq 0)$ independent. Then $r^{(a)}$ satisfies (1.2)*, that is,*

$$\{\psi(t) + a^H UV \mathbb{1}_{[aU^{1/H},\infty)}(t),\ t \geq 0\} \overset{\text{(law)}}{=} \{\psi(t),\ t \geq 0\} \text{ under } \mathbb{E}\Big[\frac{\psi(a)}{\kappa a^H};\ \cdot\Big].$$

(b2) *Let G be a standard exponential random variable, U and V be as above, and assume that G, U, and V are independent. Then the Lévy measure ν of the process $(\psi(t), t \geq 0)$ can be represented as*

$$\nu(F) = \kappa \mathbb{E}\Big[(UV)^{-1} e^{GU^{1/H}} F(G^H UV \mathbb{1}_{[GU^{1/H},\infty)})\Big]$$

for every measurable functional $F : \mathbb{R}_+^{[0,\infty)} \mapsto \mathbb{R}_+$. Therefore, ν is the law of the process $(G^H UV \mathbb{1}_{[GU^{1/H},\infty)}(t),\ t \geq 0)$ under the measure $\kappa(UV)^{-1} e^{GU^{1/H}}\, d\mathbb{P}$.

(c2) *The components of the decomposition* (1.5)*: $\psi \overset{\text{(law)}}{=} (\psi \mid \psi(a) = 0) + \mathcal{L}^{(a)}$, can be identified as*

$$(\psi(t), t \geq 0 \mid \psi(a) = 0) \overset{\text{(law)}}{=} (\psi(t \vee a) - \psi(a),\ t \geq 0).$$

and

$$(\mathcal{L}_t^{(a)},\ t \geq 0) \overset{\text{(law)}}{=} (\psi(t \wedge a),\ t \geq 0).$$

The Lévy measures ν_a and $\tilde{\nu}_a$ of $(\psi(t), t \geq 0 \mid \psi(a) = 0)$ and of $(\mathcal{L}_t^{(a)}, t \geq 0)$, respectively, are given by

$$\nu_a(F) = \int_{-\infty}^{\ln(a^{-H})} \int_0^\infty F(xe^{-s} \mathbb{1}_{[e^{-s/H},\infty)})\, \rho(dx) ds$$

and

$$\tilde{\nu}_a(F) = \int_{\ln(a^{-H})}^\infty \int_0^\infty F(xe^{-s} \mathbb{1}_{[e^{-s/H},\infty)})\, \rho(dx) ds,$$

for every measurable functional $F : \mathbb{R}_+^{[0,\infty)} \mapsto \mathbb{R}_+$.

Proof

(a2) By (1.4) we have for any measurable functional $F : \mathbb{R}^{[0,\infty)} \mapsto \mathbb{R}_+$

$$\begin{aligned}\mathbb{E}F(r_t^a,\ t \geq 0) &= \frac{1}{\mathbb{E}\psi(a)} \int_{\mathbb{R}_+^E} F(y)y(a)\,\nu(dy) \\ &= \frac{1}{a^H \mathbb{E}\psi(1)} \int_{\mathbb{R}} \int_{\mathbb{R}_+} F(xe^{-s}\mathbb{1}_{[e^{-s/H},\infty)}) \\ &\quad \times xe^{-s}\mathbb{1}_{[e^{-s/H},\infty)}(a)\,\rho(dx)ds \\ &= \frac{a^{-H}}{\mathbb{E}\psi(1)} \int_{\ln(a^{-H})}^{\infty} \int_{\mathbb{R}_+} F(xe^{-s}\mathbb{1}_{[e^{-s/H},\infty)})\ x\,\rho(dx)\,e^{-s}ds \\ &= a^{-H} \int_{\ln(a^{-H})}^{\infty} \mathbb{E}F(Ve^{-s}\mathbb{1}_{[e^{-s/H},\infty)})\,e^{-s}ds \\ &= \mathbb{E}\Big[F(a^H UV\mathbb{1}_{[aU^{1/H},\infty)})\Big].\end{aligned}$$

Thus, $(r_t^a,\ t \geq 0) \overset{\text{(law)}}{=} (a^H UV\mathbb{1}_{[aU^{1/H},\infty)}(t),\ t \geq 0)$. Since U, V, and ψ are independent, (1.2) completes the proof of (a2).

(b2) Since the process $(\psi(t), t \geq 0)$ is stochastically continuous, we have for every σ-finite measure $\tilde{m}$ whose support is $[0,\infty)$ and $\int_0^\infty t^H\,\tilde{m}(dt) < \infty$

$$\begin{aligned}\nu(F) &= \int_0^\infty \mathbb{E}\left[\frac{F(r^{(a)})}{\int_0^\infty r_s^{(a)}\,\tilde{m}(ds)}\right]\mathbb{E}[\psi(a)]\,\tilde{m}(da) \\ &= \mathbb{E}[\psi(1)] \int_0^\infty \mathbb{E}\left[\frac{F(a^H UV\mathbb{1}_{[aU^{1/H},\infty)})}{UV\,\tilde{m}([aU^{1/H},\infty))}\right]\tilde{m}(da).\end{aligned}$$

If $\tilde{m}$ is the law of a nonnegative random variable W, then

$$\begin{aligned}\nu(F) &= \mathbb{E}[\psi(1)] \int_0^\infty \mathbb{E}\left[\frac{h(aU^{1/H})}{UV}\,F(a^H UV\mathbb{1}_{[aU^{1/H},\infty)})\right]\tilde{m}(da) \\ &= \mathbb{E}[\psi(1)]\mathbb{E}\left[\frac{h(U^{1/H}W)}{UV}\,F(UVW^H\mathbb{1}_{[U^{1/H}W,\infty)})\right]\end{aligned}$$

which is the formula in (b2).

(c2) Since the conditional process $(\psi(t), t \geq 0 \,|\, \psi(a) = 0)$ has the Lévy measure $\nu_a(dy) = \mathbb{1}_{\{y(a)=0\}}\nu(dy)$ (see [3]), by (3.6) we obtain for any measurable functional $F : \mathbb{R}_+^{[0,\infty)} \mapsto \mathbb{R}_+$ and $a > 0$

$$\begin{aligned}\nu_a(F) &= \int F(y)\,\mathbb{1}_{\{y(a)=0\}}\nu(dy)\\ &= \int_{\mathbb{R}}\int_{\mathbb{R}_+} F(xe^{-s}\mathbb{1}_{[e^{-s/H},\infty)})\mathbb{1}_{\{xe^{-s}\mathbb{1}_{[e^{-s/H},\infty)}(a)=0\}}\,\rho(dx)ds\\ &= \int_{-\infty}^{\ln(a^{-H})}\int_0^\infty F(xe^{-s}\mathbb{1}_{[e^{-s/H},\infty)})\,\rho(dx)ds\,.\end{aligned}$$

Since $\tilde{\nu}_a = \nu - \nu_a$,

$$\tilde{\nu}_a(F) = \int_{\ln(a^{-H})}^{\infty}\int_0^\infty F(xe^{-s}\mathbb{1}_{[e^{-s/H},\infty)})\,\rho(dx)ds$$

Let $0 = t_0 < t_1 < \cdots < t_n$ be such that $t_m = a$ for some $m \le n$. For $\alpha_i > 0$ we obtain

$$\begin{aligned}&\mathbb{E}\exp\Big\{-\sum_{i=1}^n \alpha_i(\mathcal{L}_{t_i}^{(a)} - \mathcal{L}_{t_{i-1}}^{(a)})\Big\} = \exp\{-\tilde{\nu}_a(1-e^{-\sum_{i=1}^n \alpha_i(y(t_i)-y(t_{i-1}))})\}\\ &= \exp\Big\{-\int_{\ln(a^{-H})}^{\infty}\int_0^\infty (1-e^{-\sum_{i=1}^n \alpha_i xe^{-s}(\mathbb{1}_{[e^{-s/H},\infty)}(t_i)-\mathbb{1}_{[e^{-s/H},\infty)}(t_{i-1}))})\\ &\quad\times \rho(dx)ds\Big\}\\ &= \exp\Big\{-\int_{\ln(a^{-H})}^{\infty}\int_0^\infty (1-e^{-\sum_{i=1}^n \alpha_i xe^{-s}\mathbb{1}_{(t_{i-1},t_i]}(e^{-s/H})})\,\rho(dx)ds\Big\}\\ &= \exp\Big\{-\sum_{i=1}^m\int_{\ln(t_i^{-H})}^{\ln(t_{i-1}^{-H})}\int_0^\infty (1-e^{-\alpha_i xe^{-s}})\,\rho(dx)ds\Big\}\\ &= \prod_{i=1}^m \mathbb{E}\exp\{-\alpha_i(\psi(t_i)-\psi(t_{i-1}))\}\\ &= \mathbb{E}\exp\{-\sum_{i=1}^n \alpha_i(\psi(t_i\wedge a)-\psi(t_{i-1}\wedge a))\}\,,\end{aligned}$$

which shows that $(\mathcal{L}_t^{(a)},\ t \ge 0) \overset{\text{(law)}}{=} (\psi(t\wedge a),\ t\ge 0)$.

Since $(\psi(t\wedge a),\ t\ge 0)$ and $(\psi(t\vee a)-\psi(a),\ t\ge 0)$ are independent and they add to $(\psi(t),\ t\ge 0)$, we get $(\psi(t), t\ge 0 \mid \psi(a)=0) \overset{\text{(law)}}{=} (\psi(t\vee a)-\psi(a),\ t\ge 0)$. □

3.3 Stochastic Convolution

Let $Z = (Z_t, t \geq 0)$ be a subordinator with no drift. For a fixed function $f : \mathbb{R}_+ \mapsto \mathbb{R}_+$ and $t \geq 0$, the stochastic convolution $f * Z$ is given by

$$(f * Z)(t) = \int_0^t f(t-s)\, dZ_s \,.$$

Assume that $\kappa := \mathbb{E}Z_1 \in (0, \infty)$ and $\int_0^t f(s)\, ds < \infty$ for every $t > 0$. Therefore, $\mathbb{E}[(f * Z)(t)] = \kappa \int_0^t f(s)\, ds < \infty$. Set $f(u) = 0$ when $u < 0$.

We will consider the stochastic convolution process

$$\psi(t) := \int_0^t f(t-s)\, dZ_s \,, \quad t \geq 0. \tag{3.8}$$

Clearly, $(\psi(t),\ t \geq 0)$ is an infinitely divisible process. To determine, its Lévy measure we write $\psi(t) = \int_0^\infty f_t(s)\, dZ_s$, where $f_t(s) = f(t-s)$. It follows from [12, Theorem 2.7(iv)] that the Lévy measure ν of the process ψ is the image of $m \otimes \rho$ by the map $(s, x) \mapsto x f_{(\cdot)}(s)$ acting from $\mathbb{R}_+^2$ into $\mathbb{R}^{[0,\infty)}$. That is,

$$\nu(F) = \int_0^\infty \int_0^\infty F(xf(t-s),\ t \geq 0)\, \rho(dx) ds \tag{3.9}$$

for every measurable functional $F : \mathbb{R}_+^{[0,\infty)} \mapsto \mathbb{R}_+$.

Proposition 3.6 *Let $(\psi(t),\ t \geq 0)$ be a stochastic convolution process as in* (3.8). *Let ρ be the Lévy measure of Z_1 and $I(a) := \int_0^a f(s)\, ds$.*

(a3) *Given $a > 0$ such that $I(a) > 0$, let $r^{(a)}$ be the process defined by:*

$$r^{(a)}(t) := Vf(t - U_a), \quad t \geq 0$$

where the random variable U_a has density $\dfrac{f(a-s)}{I(a)}$ on $[0, a]$, V has the law $\kappa^{-1} x \rho(dx)$ on $\mathbb{R}_+$, and U_a, V, and $(\psi(t) : t \geq 0)$ are independent. Then $r^{(a)}$ satisfies (1.2), *that is,*

$$\big(\psi(t) + Vf(t - U_a),\ t \geq 0\big) \overset{(\text{law})}{=} \big(\psi(t),\ t \geq 0\big) \text{ under } \mathbb{E}\Big[\frac{\psi(a)}{\kappa I(a)};\ \cdot\,\Big]$$

(b3) *Suppose that $\int_0^\infty e^{-\theta s} f(s)\, ds < \infty$ for some $\theta > 0$. Let Y be a random variable with the exponential law of mean θ^{-1} and independent of V specified in (a3). Then the Lévy measure ν of $(\psi(t), t \geq 0)$ can be represented as*

$$\nu(F) = \frac{\kappa}{\theta}\mathbb{E}\left[V^{-1}e^{\theta Y} F(Vf(t-Y),\ t \geq 0)\right].$$

for every measurable functional $F : \mathbb{R}_+^{[0,\infty)} \mapsto \mathbb{R}_+$. *Therefore,* ν *is the law of the process* $(Vf(t-Y),\ t \geq 0)$ *under the measure* $\kappa\theta^{-1}V^{-1}e^{\theta Y}\,d\mathbb{P}$.

(c3) *The components of the decomposition* (1.5)*:* $\psi \overset{(\text{law})}{=} (\psi \mid \psi(a) = 0) + \mathcal{L}^{(a)}$*, can be identified as*

$$(\psi(t), t \geq 0 \mid \psi(a) = 0) \overset{(\text{law})}{=} \left(\int_0^t f(t-s)\mathbb{1}_{D_a}(s)\,dZ_s,\ t \geq 0\right)$$

and

$$(\mathcal{L}_t^{(a)},\ t \geq 0) \overset{(\text{law})}{=} \left(\int_0^t f(t-s)\mathbb{1}_{D_a^c}(s)\,dZ_s,\ t \geq 0\right)$$

where $D_a = \{s \geq 0 : f(a-s) = 0\}$ *and* $D_a^c = \mathbb{R}_+ \setminus D_a$.

The Lévy measures ν_a *and* $\tilde{\nu}_a$ *of* $(\psi(t), t \geq 0 \mid \psi(a) = 0)$ *and of* $(\mathcal{L}_t^{(a)}, t \geq 0)$*, respectively, are given by*

$$\nu_a(F) = \int_{D_a}\int_0^\infty F(xf(t-s),\ t \geq 0)\,\rho(dx)ds$$

and

$$\tilde{\nu}_a(F) = \int_{D_a^c}\int_0^\infty F(xf(t-s),\ t \geq 0)\,\rho(dx)ds,$$

for every measurable functional $F : \mathbb{R}_+^{[0,\infty)} \mapsto \mathbb{R}_+$.

Proof

(a3) From (1.4) and (3.9), we get

$$\begin{aligned}
\mathbb{E}F(r_t^{(a)},\ t \geq 0) &= \frac{1}{\mathbb{E}\psi(a)}\int F(y)\,y(a)\,\nu(dy) \\
&= \frac{1}{\kappa I(a)}\int_0^\infty\int_0^\infty F\big(xf(t-s),\ t \geq 0\big)\,xf(a-s)\,\rho(dx)ds \\
&= \int_0^a\int_0^\infty F(xf(t-s), t \geq 0)\,\frac{x\rho(dx)}{\kappa}\,\frac{f(a-s)ds}{I(a)} \\
&= \mathbb{E}\big[F(Vf(t-U_a),\ t \geq 0)\big].
\end{aligned}$$

(b3) Since ψ is stochastically continuous, using (1.3) and (a3), we have for every σ-finite measure $\tilde{m}$ whose support is $[0,\infty)$ and $\int_0^\infty I(a)\,\tilde{m}(da) < \infty$

$$\nu(F) = \int_0^\infty \mathbb{E}\left[\frac{F(r^{(a)})}{\int_0^\infty r_s^{(a)}\,\tilde{m}(ds)}\right]\mathbb{E}[\psi(a)]\,\tilde{m}(da)$$
$$= \kappa \int_0^\infty \mathbb{E}\left[\frac{F(Vf(t-U_a),\ t \geq 0)}{V\int_0^\infty f(s-U_a)\,\tilde{m}(ds)}\right] I(a)\,\tilde{m}(da).$$

Since $\tilde{m}$ is the law of Y in our case, it is easy to check that $\beta := \int_0^\infty I(a)\,\tilde{m}(da) < \infty$. Also,

$$\int_0^\infty f(s-U_a)\,\tilde{m}(ds) = \beta\theta e^{-\theta U_a}.$$

Then we get

$$\nu(F) = \frac{\kappa}{\beta\theta}\int_0^\infty \mathbb{E}\Big[V^{-1}e^{\theta U_a}F(Vf(t-U_a),\ t \geq 0)\Big]\, I(a)\theta e^{-\theta a}\,da$$
$$= \frac{\kappa}{\beta\theta}\int_0^\infty\int_0^a \mathbb{E}\Big[V^{-1}e^{\theta s}F(Vf(t-s),\ t \geq 0)\Big]\, f(a-s)\,ds\,\theta e^{-\theta a}\,da$$
$$= \frac{\kappa}{\theta}\int_0^\infty \mathbb{E}\Big[V^{-1}e^{\theta s}F(Vf(t-s),\ t \geq 0)\Big]\,\theta e^{-\theta s}\,ds$$
$$= \frac{\kappa}{\theta}\mathbb{E}\Big[V^{-1}e^{\theta Y}F(Vf(t-Y),\ t \geq 0)\Big].$$

(c3) Since the conditional process $(\psi(t), t \geq 0 \mid \psi(a) = 0)$ has the Lévy measure $\nu_a(dy) = \mathbb{1}_{\{y(a)=0\}}\nu(dy)$ (see [3]), by (3.6) we obtain for any measurable functional $F : \mathbb{R}_+^{[0,\infty)} \mapsto \mathbb{R}_+$ and $a > 0$

$$\nu_a(F) = \int F(y)\,\mathbb{1}_{\{y(a)=0\}}\nu(dy)$$
$$= \int_0^\infty\int_0^\infty F(xf(t-s),\ t \geq 0)\,\mathbb{1}_{\{(x,s):\,xf(a-s))=0\}}\,\rho(dx)ds$$
$$= \int_0^\infty\int_0^\infty F(xf(t-s),\ t \geq 0)\,\mathbb{1}_{D_a}(s)\,\rho(dx)ds\,.$$

Using again [12, Theorem 2.7(iv)] we see that ν_a is the Lévy measure of the process

$$\Big(\int_0^t f(t-s)\mathbb{1}_{D_a}(s)\,dZ_s,\ t \geq 0\Big)$$

which is a nonnegative infinitely divisible process without drift. Since the law of such process is completely characterized by its Lévy measure, we infer that

$$(\psi(t), t \geq 0 \mid \psi(a) = 0) \overset{\text{(law)}}{=} \left(\int_0^t f(t-s) \mathbb{1}_{D_a}(s)\, dZ_s,\ t \geq 0 \right).$$

Since $\tilde{\nu}_a = \nu - \nu_a$ and $\psi \overset{\text{(law)}}{=} (\psi \mid \psi(a) = 0) + \mathcal{L}^{(a)}$, we can apply the same argument as above to get

$$(\mathcal{L}_t^{(a)},\ t \geq 0) \overset{\text{(law)}}{=} \left(\int_0^t f(t-s) \mathbb{1}_{D_a}(s)\, dZ_s,\ t \geq 0 \right).$$

□

3.4 Tempered Stable Subordinator

Tempered α-stable subordinators behave at short time like α-stable subordinators and may have all moments finite, while the latter have the first moment infinite. Therefore, we can make use of tempered stable subordinators to illustrate identities (1.2)–(1.5). For concreteness, consider a tempered α-stable subordinator $(\psi(t), t \geq 0)$ determined by the Laplace transform

$$\mathbb{E} e^{-u\psi(1)} = \exp\{1 - (1+u)^\alpha\} \tag{3.10}$$

where $\alpha \in (0, 1)$. When $\alpha = 1/2$, ψ is also known as the inverse Gaussian subordinator. A systematic treatment of tempered α-stable laws and processes can be found in [14]. In particular, the Lévy measure of $\psi(1)$ is given by

$$\rho(dx) = \frac{1}{|\Gamma(-\alpha)|} x^{-\alpha-1} e^{-x}\, dx, \quad x > 0,$$

Rosiński [14, Theorems 2.3 and 2.9(2.17)]. Therefore, the Lévy measure ν of the process ψ is given by

$$\begin{aligned} \nu(F) &= \int_0^\infty \int_0^\infty F(x\mathbb{1}_{[s,\infty)})\, \rho(dx) ds \\ &= \frac{1}{|\Gamma(-\alpha)|} \int_0^\infty \int_0^\infty F(x\mathbb{1}_{[s,\infty)})\, x^{-\alpha-1} e^{-x}\, dx ds, \end{aligned} \tag{3.11}$$

for every measurable functional $F : \mathbb{R}_+^{[0,\infty)} \mapsto \mathbb{R}_+$.

Proposition 3.7 *Let $(\psi(t), t \geq 0)$ be a tempered α-stable subordinator as above.*

(a4) *Given $a > 0$, let $r^{(a)}$ be the process defined by:*

$$r^{(a)}(t) := G\mathbb{1}_{[aU,\infty)}(t), \quad t \geq 0$$

where G has a Gamma$(1-\alpha, 1)$ *law and U is a standard uniform random variable independent of G. Then* $r^{(a)}$ *satisfies* (1.2), *that is,*

$$(\psi(t) + G\mathbb{1}_{[aU,\infty)}(t),\ t \geq 0) \overset{\text{(law)}}{=} (\psi(t),\ t \geq 0) \text{ under } \mathbb{E}\Big[\frac{\psi(a)}{\alpha a};\ .\Big]$$

(b4) *The Lévy measure* ν *of* $(\psi(t), t \geq 0)$ *can be represented as*

$$\nu(F) = \alpha^{-1}\mathbb{E}\big[G^{-1}Ye^{UY}\, F(G\mathbb{1}_{[UY,\infty)})\big]$$

for every measurable functional $F : \mathbb{R}_+^{[0,\infty)} \mapsto \mathbb{R}_+$. *Here, G, U are as in* (a4), *Y is a standard exponential variable, and G, U, and Y are independent. Consequently,* ν *is the law of the process* $(G\mathbb{1}_{[UY,\infty)},\ t \geq 0)$ *under the measure* $\alpha^{-1}G^{-1}Ye^{UY}\, d\mathbb{P}$.

(c4) *The components of the decomposition* (1.5)*:* $\psi \overset{\text{(law)}}{=} (\psi \mid \psi(a) = 0) + \mathcal{L}^{(a)}$, *can be identified as*

$$(\psi(t), t \geq 0 \,|\, \psi(a) = 0) \overset{\text{(law)}}{=} (\psi(t \vee a) - \psi(a),\ t \geq 0).$$

and

$$(\mathcal{L}_t^{(a)},\ t \geq 0) \overset{\text{(law)}}{=} (\psi(t \wedge a),\ t \geq 0).$$

The Lévy measures ν_a *and* $\tilde{\nu}_a$ *of* $(\psi(t), t \geq 0 \,|\, \psi(a) = 0)$ *and of* $(\mathcal{L}_t^{(a)}, t \geq 0)$*, respectively, are given by*

$$\nu_a(F) = \frac{1}{|\Gamma(-\alpha)|}\int_a^\infty \int_0^\infty F\big(x\mathbb{1}_{[s,\infty)}\big)\, x^{-\alpha-1}e^{-x}\, dxds$$

and

$$\tilde{\nu}_a(F) = \frac{1}{|\Gamma(-\alpha)|}\int_0^a \int_0^\infty F\big(x\mathbb{1}_{[s,\infty)}\big)\, x^{-\alpha-1}e^{-x}\, dxds,$$

for every measurable functional $F : \mathbb{R}_+^{[0,\infty)} \mapsto \mathbb{R}_+$.

Proof

(a4) From (3.10) we get $\mathbb{E}\psi(a) = \alpha a$. Using (3.11). and (1.4), we get

$$\begin{aligned}
\mathbb{E}F(r_t^{(a)},\ t \geq 0) &= \frac{1}{\mathbb{E}\psi(a)}\int F(y)\, y(a)\, \nu(dy) \\
&= \frac{1}{\alpha a}\int_0^\infty \int_0^\infty F(x\mathbb{1}_{[s,\infty)})\, x\mathbb{1}_{[s,\infty)}(a)\, \rho(dx)ds
\end{aligned}$$

$$= \frac{1}{\Gamma(1-\alpha)a} \int_0^a \int_0^\infty F(x\mathbb{1}_{[s,\infty)})\, x^{-\alpha} e^{-x}\, dxds$$
$$= \mathbb{E}\big[F(G\mathbb{1}_{[aU,\infty)})\big].$$

(b4) We apply [3, Theorem 1.2] to $(\psi(t), t \geq 0)$ and $(r_t^{(a)}, t \geq 0)$ specified in (a4). Proceeding analogously to the previous examples, we get for any σ-finite measure $\tilde{m}$ whose support equals $\mathbb{R}_+$ and $\int_{\mathbb{R}_+} a\,\tilde{m}(da) < \infty$

$$\nu(F) = \int_0^\infty \mathbb{E}\left[\frac{F(r^{(a)})}{\int_0^\infty r_s^{(a)}\, \tilde{m}(ds)}\right] \mathbb{E}[\psi(a)]\, \tilde{m}(da)$$
$$= \frac{1}{\alpha} \int_0^\infty \mathbb{E}\left[\frac{F(G\mathbb{1}_{[U_a,\infty)})}{G\tilde{m}([aU,\infty))}\right] a\,\tilde{m}(da).$$

When $\tilde{m}$ is the law of a standard exponential random variable, we obtain

$$\nu(F) = \frac{1}{\alpha} \int_0^\infty \mathbb{E}\left[e^{aU} G^{-1} F(G\mathbb{1}_{[aU,\infty)})\right] ae^{-a} da$$
$$= \alpha^{-1}\mathbb{E}\left[G^{-1} Y e^{UY} F(G\mathbb{1}_{[UY,\infty)})\right].$$

(c4) We will omit this proof as it is similar to the proof of (c1) in the Poisson case. □

3.5 *Connection with Infinitely Divisible Random Measures*

Let $\mathcal{M}(S)$ denote the space of finite measures on a Borel space $(S, \mathcal{S})$. $\mathcal{M}(S)$ is a Borel space under the topology of weak convergence of finite measures. A measurable map $\xi : \Omega \mapsto \mathcal{M}(S)$ is called a random measure on S. Any random measure ξ can also be viewed as a stochastic process indexed by $\mathcal{S}$ and having paths in $\mathcal{M}(S) \subset \mathbb{R}^{\mathcal{S}}$, $\xi = \{\xi(A),\ A \in \mathcal{S}\}$. A random measure is called infinitely divisible if the corresponding stochastic process is infinitely divisible.

3.5.1 Cluster Representation

The key result on infinitely divisible random measures is the cluster representation. It says that any infinitely divisible random measure ξ on $(S, \mathcal{S})$ is of the form

$$\xi = m + \int_{\mathcal{M}(S)} \mu\, \Lambda(d\mu) \quad a.s. \tag{3.12}$$

where Λ is a Poisson random measure on $\mathcal{M}(S)$ with intensity λ satisfying

$$\int_{\mathcal{M}(S)} (\mu(A) \wedge 1)\, \lambda(d\mu) < \infty, \quad A \in \mathcal{S} \tag{3.13}$$

and $m \in \mathcal{M}(S)$ is nonrandom (see [8, Theorem 3.20]). Notice that this result follows from (2.3) of Sect. 2 when $E = \mathcal{S}$ and $\psi = \xi$. We will sketch a proof to this claim. Indeed, since (2.3) in this case states that

$$\big(\xi(A), A \in \mathcal{S}\big) \overset{\text{(law)}}{=} \Big(f_0(A) + \int_{\mathbb{R}_+^{\mathcal{S}}} f(A)\, N(df),\ \ A \in \mathcal{S}\Big),$$

pathwise additivity of ξ implies that ν, the Lévy measure of ξ, is concentrated on finite additive functions $f : \mathcal{S} \mapsto \mathbb{R}_+$. Since the σ-algebra $\mathcal{S}$ is countably generated and ξ is pathwise σ-additive, ν is a σ-finite measure concentrated on $\mathcal{M}(S)$ with $\nu(\{0\}) = 0$. It follows that $f_0 \in \mathcal{M}(S)$. Hence

$$\big(\xi(A), A \in \mathcal{S}\big) \overset{\text{(law)}}{=} \Big(m(A) + \int_{\mathcal{M}(S)} \mu(A)\, N(d\mu),\ \ A \in \mathcal{S}\Big).$$

This equality can be strengthen to the almost sure equality by the usual argument. Hence (3.12)–(3.13) hold with $\lambda = \nu$, $\Lambda = N$, and $m = f_0$.

3.5.2 A Characterization of Infinitely Divisible Random Measures

One can make use of (1.2) for nonnegative processes indexed by $\mathcal{S}$ to obtain the following characterization of infinitely divisible random measures on S.

A random measure ξ on S is infinitely divisible if and only if, for every A in $\mathcal{S}$ such that $0 < \mathbb{E}[\xi(A)] < \infty$, there exists a random measure $r^{(A)}$ on S, independent of ξ such that:

$$\xi + r^{(A)} \overset{\text{(law)}}{=} \xi \ \textit{under}\ \mathbb{E}\Big[\frac{\xi(A)}{\mathbb{E}[\xi(A)]}, \cdot\Big] \tag{3.14}$$

The characterization (3.14) can be connected to another characterization given in [8, Theorem 6.17]. Namely, assume that ξ has a σ-finite intensity n, then ξ is infinitely divisible if and only if for every a in S there exists a random measure $R^{(a)}$ on S, independent of ξ such that

$$\xi + R^{(a)} \overset{\text{(law)}}{=} \xi^a, \tag{3.15}$$

where ξ^a is the Palm measure of ξ at point a.

By definition, the Palm measures $\{\xi^a, a \in S\}$ of ξ satisfy for every A in $\mathcal{S}$ and every measurable subset L of $\mathcal{M}(S)$

$$\mathbb{E}[\xi(A); \xi \in L] = \int_A n(da)\mathbb{P}[\xi^a \in L],$$

which leads to the following relation for A such that $0 < n(A) < \infty$

$$\mathbb{P}[\,\xi + r^{(A)} \in L] = \frac{1}{n(A)} \int_A n(da)\mathbb{P}[\,\xi + R^{(a)} \in L].$$

By computing the Laplace transforms, one finally has:

$$r^{(A)} \overset{(\text{law})}{=} \frac{1}{n(A)} \int_A n(da) R^{(a)}. \tag{3.16}$$

In the special case of a point a of S such that $\mathbb{P}[\xi(\{a\}) > 0] > 0$ (e.g., S is discrete), one obtains $R^{(a)} \overset{(\text{law})}{=} r^{(\{a\})}$.

3.5.3 A Decomposition Formula

Given an infinitely divisible random measure ξ on S, one can take advantage of (1.6) to obtain, for every A such that $0 < \mathbb{E}[\xi(A)] < \infty$, the existence of an infinitely divisible random measure $\mathcal{L}^{(A)}$ on S such that:

$$\xi \overset{(\text{law})}{=} (\xi \mid \xi(A) = 0) \;+\; \mathcal{L}^{(A)}, \tag{3.17}$$

with the two measures on the right-hand side independent.

3.5.4 Some Remarks

In this section we take $S = \mathbb{R}_+^E$. Let χ be a finite infinitely divisible random measure on $\mathbb{R}_+^E$ with no drift and Lévy measure λ. Assume now that for every a in E:

$$\int_{\mathbb{R}_+^E} f(a) \int_{\mathcal{M}(\mathbb{R}_+^E)} \mu(df)\, \lambda(d\mu) < \infty.$$

Consider then the nonnegative process ψ on E defined by $\psi(x) = \int_{\mathbb{R}_+^E} f(x)\chi(df)$. The process ψ is infinitely divisible and nonnegative. The following proposition gives its Lévy measure.

Proposition 3.8 *The infinitely divisible nonnegative process* $(\int_{\mathbb{R}_+^E} f(x)\chi(df), x \in E)$ *admits for the Lévy measure* ν *given by:*

$$\nu = \int_{\mathcal{M}(\mathbb{R}_+^E)} \mu \, \lambda(d\mu).$$

Proof From (2.3), we know that there exists a Poisson point process $\tilde{N}$ on $\mathbb{R}_+^E$ with intensity the Lévy measure of ψ satisfying $(\psi(x), x \in E) = (\int_{\mathbb{R}_+^E} f(x)\tilde{N}(df), x \in E)$. Besides, χ admits the following expression: $\chi = \int_{\mathcal{M}(\mathbb{R}_+^E)} \mu \; N(d\mu)$, with N Poisson point process on $\mathcal{M}(\mathbb{R}_+^E)$ with intensity λ. One obtains:

$$(\psi(x), x \in E) = \Big(\int_{\mathbb{R}_+^E} f(x) \int_{\mathcal{M}(\mathbb{R}_+^E)} \mu(df) \, N(d\mu), \quad x \in E \Big).$$

Using then Campbell formula for every measurable subset A of $\mathbb{R}_+^E$, one computes the intensity of the Poisson point process $\int_{\mathcal{M}(\mathbb{R}_+^E)} \mu(df) \; N(d\mu)$

$$\begin{aligned} \mathbb{E}[\int_{\mathbb{R}_+^E} 1_A(f) \int_{\mathcal{M}(\mathbb{R}_+^E)} \mu(df) \, N(d\mu)] &= \int_{\mathbb{R}_+^E} 1_A(f) \int_{\mathcal{M}(\mathbb{R}_+^E)} \mu(df) \, \lambda(d\mu) \\ &= \nu(A)\,. \end{aligned}$$

□

Proposition 3.8 allows to write every Lévy measure ν on $\mathbb{R}_+^E$ in terms of a Lévy measure on $\mathcal{M}(\mathbb{R}_+^E)$. Indeed, given a Lévy measure ν on $\mathbb{R}_+^E$, denote by $(\psi(x), x \in E)$ the corresponding infinitely divisible nonnegative process without drift. From (2.3), we know that ψ admits the representation $(\int_{\mathbb{R}_+^E} f(x)\chi(df), x \in E)$ with χ Poisson random measure on $\mathcal{M}(\mathbb{R}_+^E)$. The random measure χ is hence infinitely divisible. Proposition 3.8 gives us:

$$\nu = \int_{\mathcal{M}(\mathbb{R}_+^E)} \mu \, \lambda(d\mu) \tag{3.18}$$

where λ is the Lévy measure of χ. Proposition 3.8 allows to see that, given ν, the Lévy measure λ satisfying (3.18) is not unique.

Proposition 3.9 *The intensity ν of a Poisson random measure on $\mathbb{R}_+^E$ with Lévy measure λ satisfies:*

$$\nu = \int_{\mathcal{M}(\mathbb{R}_+^E)} \mu \, \lambda(d\mu).$$

3.6 Infinitely Divisible Permanental Processes

A permanental process $(\psi(x), x \in E)$ with index $\beta > 0$ and kernel $k = (k(x, y), (x, y) \in E \times E)$ is a nonnegative process with finite dimensional Laplace transforms satisfying, for every $\alpha_1, .., \alpha_n \geq 0$ and every $x_1, x_2,..,x_n$ in E:

$$\mathbb{E}[\exp\{-\frac{1}{2}\sum_{i=1}^{n} \alpha_i \psi(x_i)\}] = \det(I + \alpha K)^{-\beta} \tag{3.19}$$

where α is the diagonal matrix with diagonal entries $(\alpha_i)_{1\leq i\leq n}$, I is the $n\times n$-identity matrix and K is the matrix $(k(x_i, x_j))_{1\leq i,j\leq n}$.

Note that the kernel of a permanental process is not unique.

In case $\beta = 1/2$ and k can be chosen symmetric positive semi-definite, $(\psi(x), x \in E)$ equals in law $(\eta_x^2, x \in E)$ where $(\eta_x, x \in E)$ is a centered Gaussian process with covariance k. The permanental processes hence represent an extension of the definition of squared Gaussian processes.

A necessary and sufficient condition on (β, k) for the existence of a permanental process $(\psi(x), x \in E)$ satisfying (3.19) has been established by Vere-Jones [20]. Since we are interested by the subclass of infinitely divisible permanental processes, we will only remind a necessary and sufficient condition for a permanental process to be infinitely divisible. Remark that if $(\psi(x), x \in E)$ is infinitely divisible, then, for every measurable nonnegative d, $(d(x)\psi(x), x \in E)$ is also infinitely divisible. Up to the product by a deterministic function, $(\psi(x), x \in E)$ is infinitely divisible if and only if it admits for kernel the 0-potential densities (the Green function) of a transient Markov process on E (see [4] and [5]).

Consider an infinitely divisible permanental process $(\psi(x), x \in E)$ admitting for kernel the Green function $(g(x, y), (x, y) \in E \times E)$ of a transient Markov process $(X_t, t \geq 0)$ on E. For simplicity assume that ψ has index $\beta = 1$. For $a \in E$ such that $g(a, a) > 0$, denote by $(L_\infty^{(a)}(x), x \in E)$ the total accumulated local time process of X conditioned to start at a and killed at its last visit to a. In [3], (1.5) has been explicitly written for ψ:

$$\psi \overset{\text{(law)}}{=} (\psi | \psi(a) = 0) + \mathcal{L}^{(a)}$$

with $\mathcal{L}^{(a)}$ independent process of $(\psi|\psi(a) = 0)$, such that $\mathcal{L}^{(a)} \overset{\text{(law)}}{=} (2L_\infty^{(a)}(x), x \in E)$. Moreover, $(\psi|\psi(a) = 0)$ is a permanental process with index 1 and with kernel the Green function of X killed at its first visit to a.

One can also explicitly write (1.2) for ψ with $(r^{(a)}(x), x \in E) \overset{\text{(law)}}{=} (2L_\infty^{(a)}(x), x \in E)$. Hence the case of infinitely divisible permanental processes is a special case since $r^{(a)}$ is infinitely divisible and $r^{(a)} \overset{\text{(law)}}{=} \mathcal{L}^{(a)}$.

The easiest way to obtain the Lévy measure ν of ψ is to use (1.3) with m σ-measure with support equal to E such that: $\int_E g(x, x)m(dx) < \infty$, to obtain

$$\nu(F) = \int_E \mathbb{E}[\frac{F(2L_\infty^{(a)})}{\int_E L_\infty^{(a)}(x)m(dx)}]g(a,a)m(da),$$

for any measurable functional F on $\mathbb{R}_+^E$.

If, moreover, the 0-potential densities $(g(x, y), (x, y) \in E \times E)$ were taken with respect to m, then, for every a, $\int_E L_\infty^{(a)}(x)m(dx)$ represents the time of the last visit to a by X starting from a.

4 Transfer of Continuity Properties

Using (1.6), a nonnegative infinitely divisible process $\psi = (\psi(x), x \in E)$ with Lévy measure ν and no drift is hence connected to a family of nonnegative infinitely divisible processes $\{\mathcal{L}^{(a)}, a \in E\}$. In the case when ψ is an infinitely divisible squared Gaussian process, Marcus and Rosen [9] have established correspondences between path properties of ψ and the ones of $\mathcal{L}^{(a)}$, $a \in E$. To initiate a similar study for a general ψ, we assume that (E, d) is a separable metric space with a dense set $D = \{a_k, k \in \mathbb{N}^*\}$.

One immediately notes that if ψ is a.s. continuous with respect to d, then, for every a in E, $\mathcal{L}^{(a)}$ is a.s. continuous with respect to d and the measure ν is supported by the continuous functions from E into $\mathbb{R}_+$, i.e., $r^{(a)}$ is continuous with respect to d, for every a in E.

Conversely, if $\mathcal{L}^{(a)}$ is continuous with respect to d for every a in E, what can be said about the continuity of ψ ?

As noticed in [16] (Proposition 4.7) the measure ν admits the following decomposition:

$$\nu = \sum_{k=1}^{\infty} 1_{A_k} \nu_k, \tag{4.1}$$

where $A_1 = \{y \in \mathbb{R}_+^E : y(a_1) > 0\}$ and for $k > 1$,

$$A_k = \{y \in \mathbb{R}_+^E : y(a_i) = 0, \forall i < k \text{ and } y(a_k) > 0\}$$

and ν_k is defined by

$$\nu_k(F) = \mathbb{E}[\frac{\mathbb{E}(\psi(a_k)}{r_{a_k}^{(a_k)}} 1_{A_k}(r^{(a_k)})F(r^{(a_k)})]$$

for every measurable functional $F : \mathbb{R}_+^E \mapsto \mathbb{R}_+$.

For every k the measure ν_k is a Lévy measure. Since the supports of these measures are disjoint, they correspond to independent nonnegative infinitely divisible processes that we denote by $L(k)$, $k \geq 1$. As a consequence of (4.1), ψ admits the following decomposition:

$$\psi \overset{\text{(law)}}{=} \sum_{k=1}^{\infty} L(k). \tag{4.2}$$

Note that

$$L(1) \overset{\text{(law)}}{=} \mathcal{L}^{(a_1)}$$

and similarly for every $k > 1$:

$$L(k) \overset{\text{(law)}}{=} (\mathcal{L}^{(a_k)} | \mathcal{L}^{(a_k)}_{|\{a_1,\dots,a_{k-1}\}} = 0).$$

Consequently, for every $k \geq 1$, $L(k)$ is continuous with respect to d.

From (4.2), one obtains all kind of $0 - 1$ laws for ψ. For example:

- $\mathbb{P}[\psi$ is continuous on $E] = 0$ or 1.
- ψ has a deterministic oscillation function w, such that for every a in E:

$$\liminf_{x \to a} \psi(x) = \psi(a) \text{ and } \limsup_{x \to a} \psi(x) = \psi(a) + w(a).$$

Exactly as in [3], one shows the following propositions:

Proposition 4.1 *If for every a in E, $\mathcal{L}^{(a)}$ is a.s. continuous, then there exists a dense subset Δ of E such that a.s. ψ is continuous at each point of Δ and $\psi_{|\Delta}$ is continuous.*

Proposition 4.2 *Assume that ψ is stationary. If, for every a in E, $\mathcal{L}^{(a)}$ is a.s. continuous, then ψ is continuous.*

5 A Limit Theorem

Given a nonnegative infinitely divisible without drift process $(\psi_x, x \in E)$, the following result gives an intrinsic way to obtain $r^{(a)}$ for every a in E:

Theorem 5.1 *For a nonnegative infinitely divisible process $(\psi_x, x \in E)$ with Lévy measure ν, denote by $\psi^{(\delta)}$ an infinitely divisible process with Lévy measure $\delta\nu$. Then, for any a in E such that $\mathbb{E}[\psi_a] > 0$, $r^{(a)}$ is the limit in law of the processes $\psi^{(\delta)}$ under $\mathbb{E}\Big[\frac{\psi_a^{(\delta)}}{\mathbb{E}[\psi_a^{(\delta)}]}; \cdot\Big]$, as $\delta \to 0$.*

Proof We remind (1.4): $\mathbb{P}[r^{(a)} \in dy] = \frac{y(a)}{\mathbb{E}[\psi_a]} \nu(dy)$. Since $\mathbb{E}[\psi_a^{(\delta)}] = \delta \mathbb{E}[\psi_a]$, one obtains immediately $\mathbb{P}[r^{(a)} \in dy] = \frac{y(a)}{\mathbb{E}[\psi_a^{(\delta)}]} \delta \nu(dy)$. Consequently, $r^{(a)}$ satisfies:

$$\psi^{(\delta)} + r^{(a)} \overset{(\text{law})}{=} \psi^{(\delta)} \text{ under } \mathbb{E}[\frac{\psi_a^{(\delta)}}{\mathbb{E}[\psi_a^{(\delta)}]}; \ . \ ].$$

As $\delta \to 0$, $\psi^{(\delta)}$ converges to the 0-process in law, so $\psi^{(\delta)}$ under $\mathbb{E}[\frac{\psi_a^{(\delta)}}{\mathbb{E}[\psi_a^{(\delta)}]}; \cdot]$ must converge in law to $r^{(a)}$. □

From (1.2) and (1.5), one obtains in particular:

$$\mathcal{L}^{(a)} + r^{(a)} \overset{(\text{law})}{=} \mathcal{L}^{(a)} \text{ under } \mathbb{E}[\frac{\mathcal{L}_a^{(a)}}{\mathbb{E}[\mathcal{L}_a^{(a)}]}; \ .] \tag{5.1}$$

We know from [3] that the Lévy measure of $\mathcal{L}^{(a)}$ is $\nu(dy) 1_{y(a)>0}$. Denote by $\ell^{(a,\delta)}$ a nonnegative process with Lévy measure $\delta \nu(dy) 1_{y(a)>0}$. Using Theorem 5.1, one obtains that $r^{(a)}$ is also the limit in law of $\ell^{(a,\delta)}$ under $\mathbb{E}[\frac{\ell_a^{(a,\delta)}}{\mathbb{E}[\ell_a^{(a,\delta)}]}; \ . \ ]$.

Acknowledgement Rosinski's research was partially supported by the Simons Foundation grant 281440.

References

1. E.B. Dynkin, Local times and quantum fields, in *Seminar on Stochastic Processes* (Birkhauser, Basel, 1983), pp. 64–84
2. N. Eisenbaum, A Cox process involved in the Bose-Einstein condensation. Ann. Henri Poincaré **9**(6), 1123–1140 (2008)
3. N. Eisenbaum, Decompositions of infinitely divisible nonnegative processes. Electron. J. Probab. **24**(109), 1–25 (2019)
4. N. Eisenbaum, H. Kaspi,On permanental processes. Stochastic. Process. Appl. **119**(5), 1401–1415 (2009)
5. N. Eisenbaum, F. Maunoury, Existence conditions of permanental and multivariate negative binomial distributions. Ann. Probab. **45**(6B), 4786–4820 (2017)
6. M. Jeanblanc, J. Pitman, M. Yor, Self-similar processes with independent increments associated with Lévy and Bessel processes. Stochastic. Process. Appl. **100**, 223–231 (2002)
7. Z.J. Jurek, W. Vervaat, An integral representation for self-decomposable Banach space valued random variables. Z. Wahrsch. Verw. Gebiete. **62**(2), 247–262 (1983)
8. O. Kallenberg, *Random Measures, Theory and Applications* (Springer, Berlin, 2017)
9. M. Marcus, J. Rosen, *Markov Processes, Gaussian Processes and Local Times* (Cambridge University Press, Cambridge, 2006)
10. M. Marcus, J. Rosen, A sufficient condition for the continuity of permanental processes with applications to local times of Markov processes. Ann. Probab. **41**(2), 671–698 (2013)

11. M. Marcus, J. Rosen, Sample path properties of permanental processes. Electron. J. Probab. **23**(58), 47 (2018)
12. B.S. Rajput, J. Rosiński, Spectral representation of infinitely divisible processes. Probab. Th. Rel. Fields **82**, 451–487 (1989)
13. J. Rosiński, On the structure of stationary stable processes. Ann. Probab. **23**(3), 1163–1187 (1995)
14. J. Rosiński, Tempering stable processes. Stochastic. Proc. Appl. **117**, 677–707 (2007)
15. J. Rosiński, Lévy and related jump-type infinitely divisible processes, in *Lecture Notes, Cornell University, Ithaca, NY* (2007)
16. J. Rosiński, Representations and isomorphism identities for infinitely divisible processes. Ann. Probab. **46**(6), 3229–3274 (2018)
17. J. Rosiński, G. Samorodnitsky, Distributions of Subadditive functionals of sample paths of infinitely divisible processes. Ann. Probab. **21**(2), 996–1014 (1993)
18. K. Sato, Self-similar processes with independent increments. Probab. Th. Rel. Fields **89**(3), 285–300 (1991)
19. K. Sato, *Lévy processes and Infinitely Divisible Distributions* (Cambridge University Press, Cambridge, 1999)
20. D. Vere-Jones, Alpha-permanents and their applications to multivariate gamma, negative binomial and ordinary binomial distributions. New Zealand J. Math. **26**(1), 125–149 (1997)
21. S.J. Wolfe, On a continuous analogue of the stochastic difference equation $X_n = \rho X_{n-1+B_n}$. Stochastic Process. Appl. **12**(3), 301–312 (2002)

Bounding Suprema of Canonical Processes via Convex Hull

Rafał Latała

2020 Mathematics Subject Classification: 60E15; 60G17; 60G50

1 Formulation of the Problem

Let $X = (X_1, \ldots, X_n)$ be a centered random vector with independent coordinates. To simplify the notation, we will write

$$X_t = \langle t, X \rangle = \sum_i t_i X_i \quad \text{for } t = (t_1, \ldots, t_n) \in \mathbb{R}^n.$$

Our aim is to estimate the expected value of the supremum of the process $(X_t)_{t \in T}$, i.e., the quantity

$$b_X(T) := \mathbb{E} \sup_{t \in T} X_t, \quad T \subset \mathbb{R}^n \text{ nonempty bounded.}$$

There is a long line of research devoted to bounding $b_X(T)$ via the chaining method (cf. the monograph [11]). However, chaining methods do not work well for heavy-tailed random variables. In this paper, we will investigate another approach based on the convex hull method.

First, let us discuss an easy upper bound. Suppose that there exists $t_0, t_1, \ldots \in \mathbb{R}^n$ such that

$$T - t_0 \subset \overline{\text{conv}}\{\pm t_i : \ i \geq 1\} \tag{1}$$

R. Latała (✉)
Institute of Mathematics, University of Warsaw, Warszawa, Poland
e-mail: rlatala@mimuw.edu.pl

R. Adamczak et al. (eds.), *High Dimensional Probability IX*, Progress in Probability 80, https://doi.org/10.1007/978-3-031-26979-0_13

Then, for any $u > 0$,

$$\mathbb{E}\sup_{t\in T} X_t = \mathbb{E}\sup_{t\in T} X_{t-t_0} \le \mathbb{E}\sup_{i\ge 1}|X_{t_i}| \le u + \sum_{i\ge 1}\mathbb{E}|X_{t_i}|I_{\{|X_{t_i}|\ge u\}}.$$

Indeed the equality above follows since $X_{t-t_0} = X_t - X_{t_0}$ and $\mathbb{E}X_{t_0} = 0$ and all inequalities are pretty obvious. To make the notation more compact, let us define for nonempty countable sets $S \subset \mathbb{R}^n$

$$M_X(S) = \inf_{u>0}\Big[u + \sum_{t\in S}\mathbb{E}|X_t|I_{\{|X_t|\ge u\}}\Big],$$

$$\widetilde{M}_X(S) = \inf\Big\{m > 0\colon \sum_{t\in S}\mathbb{E}|X_t|I_{\{|X_t|\ge m\}} \le m\Big\}.$$

It is easy to observe that

$$\widetilde{M}_X(S) \le M_X(S) \le 2\widetilde{M}_X(S). \tag{2}$$

To see the lower bound, let us fix $u > 0$ and set $m = u + \sum_{t\in S}\mathbb{E}|X_t|I_{\{|X_t|\ge u\}}$ then

$$\sum_{t\in S}\mathbb{E}|X_t|I_{\{|X_t|\ge m\}} \le \sum_{t\in S}\mathbb{E}|X_t|I_{\{|X_t|\ge u\}} \le m,$$

so $\widetilde{M}_X(s) \le m$. For the upper bound, it is enough to observe that for $u > \widetilde{M}_X(S)$, we have $\sum_{t\in S}\mathbb{E}|X_t|I_{\{|X_t|\ge u\}} \le u$.

Thus, we have shown that

$$b_X(T) \le M_X(S) \le 2\widetilde{M}_X(S) \quad \text{if} \quad T - t_0 \subset \overline{\mathrm{conv}}(S \cup -S). \tag{3}$$

Remark 1 The presented proof of (3) did not use independence of coordinates of X, the only required property is mean zero.

Main Question *When can we reverse bound* (3)*—what should be assumed about variables X_i (and the set T) in order that*

$$T - t_0 \subset \overline{\mathrm{conv}}(S \cup -S) \quad \text{and} \quad M_X(S) \lesssim \mathbb{E}\sup_{t\in T} X_t \tag{4}$$

for some $t_0 \in \mathbb{R}^n$ and nonempty countable set $S \subset \mathbb{R}^n$?

Remark 2 It is not hard to show (see Sect. 3 below) that $M_X(S) \sim \mathbb{E}\max_i |X_{t_i}| = b_X(S \cup -S)$ if $S = \{t_1, \ldots, t_k\}$ and variables $(X_{t_i})_i$ are independent. Thus, our main question asks whether the parameter $b_X(T)$ may be explained by enclosing a translation of T into the convex hull of points $\pm t_i$ for which variables X_{t_i} behave as though they are independent.

Remark 3 The main question is related to Talagrand conjectures about suprema of positive selector processes, c.f. [11, Section 13.1], i.e., the case when $T \subset \mathbb{R}^n_+$ and $\mathbb{P}(X_i \in \{0, 1\}) = 1$. Talagrand investigates the possibility of enclosing T into a solid convex hull, which is bigger than the convex hull. On the other hand, we think that in our question, some regularity conditions on variables X_i are needed (such as $4 + \delta$ moment condition (10), which is clearly not satisfied for nontrivial classes of selector processes).

Remark 4

(i) In the one-dimensional case if $a = \inf T$, $b = \sup T$, then $T \subset [a, b] = \frac{a+b}{2} + \operatorname{conv}\{\frac{a-b}{2}, \frac{b-a}{2}\}$. Hence

$$b_{X_1}(T) = \mathbb{E}\max\{aX_1, bX_1\} = \frac{a+b}{2}\mathbb{E}X_1 + \mathbb{E}\Big|\frac{b-a}{2}X_1\Big|$$
$$= \frac{b-a}{2}\mathbb{E}|X_1| \geq \widetilde{M}_{X_1}\Big(\Big\{\frac{b-a}{2}\Big\}\Big),$$

so this case is trivial. Thus, in the sequel, it is enough to consider $n \geq 2$.

(ii) The set $V := \overline{\operatorname{conv}}(S \cup -S)$ is convex and origin-symmetric. Hence, if $T = -T$ and $T - t_0 \subset V$, then $T + t_0 = -(-T - t_0) = -(T - t_0) \subset V$ and $T \subset \operatorname{conv}((T - t_0) \cup (T + t_0)) \subset V$. Thus, for symmetric sets, it is enough to consider only $t_0 = 0$.

(iii) Observe that $b_X(\operatorname{conv}(T)) = b_X(T)$ and $T - t_0$ is a subset of a convex set if and only if $\operatorname{conv}(T) - t_0$ is a subset of this set. Moreover, if $T - T \subset V$, then $T - t_0 \subset V$ for any $t_0 \in V$ and $b_X(T - T) = b_X(T) + b_X(-T) = b_X(T) + b_{-X}(T)$. So if X is symmetric, it is enough to consider symmetric convex sets T.

Notation Letters c, C will denote absolute constants which value may differ at each occurence. For two nonnegative functions f and g, we write $f \gtrsim g$ (or $g \lesssim f$) if $g \leq Cf$. Notation $f \sim g$ means that $f \gtrsim g$ and $g \gtrsim f$. We write $c(\alpha)$, $C(\alpha)$ for constants depending only on a parameter α and define accordingly relations $\gtrsim_\alpha$, $\lesssim_\alpha$, $\sim_\alpha$.

Organization of the paper In Sect. 2, we present another quantity $m_X(S)$, defined via L_p-norms of $(X_t)_{t\in S}$ and show that for regular variables X_i, it is equivalent to $M_X(S)$. We also discuss there the relation of the convex hull method to the chaining functionals. In Sect. 3, we show that for $T = B_1^n$, the bound (3) may be reversed for arbitrary independent $X_1, \ldots, X_n$ and $S = \{e_1, \ldots, e_n\}$. Section 4 is devoted to the study of ellipsoids. First, we show that for $T = B_2^n$ and symmetric p-stable random Variables, $1 < p < 2$, one cannot reverse (3). Then, we prove that under $4 + \delta$ moment condition, our main question has the affirmative answer for $T = B_2^n$ and more general case of ellipsoids. We extend this result to the case of linear images of B_q^n-balls, $q \geq 2$ in Sect. 5. We conclude by discussing some open questions in the last section.

2 Regular Growth of Moments

In this section, we consider variables with regularly growing moments in a sense that

$$\|X_i\|_{2p} \le \alpha \|X_i\|_p < \infty \quad \text{for } p \ge 1, \tag{5}$$

where $\|X\|_p = (\mathbb{E}|X|^p)^{1/p}$.

For such variables, we will prove that there is alternate quantity equivalent to $M_X(S)$, namely,

$$m_X(S) := \inf \sup_i \|X_{t_i}\|_{\log(e+i)}.$$

where the infimum runs over all numerations of $S = \{t_i :\ 1 \le i \le N\}$, $N \le \infty$.

It is not hard to check (cf. Lemma 4.1 in [6]) that (5) yields

$$\|X_t\|_{2p} \le C_0(\alpha)\|X_t\|_p \quad \text{for } p \ge 1 \tag{6}$$

and as a consequence, we have for $p > 0$,

$$\mathbb{P}(|X_t| \ge e\|X_t\|_p) \le e^{-p}, \quad \mathbb{P}(|X_t| \ge c_1(\alpha)\|X_t\|_p) \ge \min\{c_2(\alpha), e^{-p}\}, \tag{7}$$

where the first bound follows by Chebyshev's inequality and the second one by the Paley-Zygmund inequality.

Proposition 5 *Suppose that X_i are independent r.v's satisfying condition* (5). *Then,* $M_X(S) \sim_\alpha m_X(S)$.

Proof Let $S = \{t_i :\ 1 \le i \le N\}$ and $m := \sup_i \|X_{t_i}\|_{\log(e+i)}$. Then, for $u > 1$,

$$\sum_{s\in S} \mathbb{P}(|X_s| \ge um) \le \sum_{i=1}^N \mathbb{P}(|X_{t_i}| \ge u\|X_{t_i}\|_{\log(e+i)}) \le \sum_{i=1}^N u^{-\log(e+i)}.$$

Therefore,

$$\begin{aligned}\sum_{s\in S} \mathbb{E}|X_s| I_{\{|X_s|\ge e^2 m\}} &= \sum_{s\in S}\Big(e^2 m\mathbb{P}(|X_s| \ge e^2 m) + m\int_{e^2}^\infty \mathbb{P}(|X_s| \ge um)du\Big) \\ &\le m\sum_{i=1}^N \Big(e^{2-2\log(e+i)} + \int_{e^2}^\infty u^{-\log(e+i)}du\Big) \\ &\le m\sum_{i=1}^N \Big((e+i)^{-2}\Big(e^2 + \frac{1}{\log(e+i)-1}\Big)\Big) \le 100m,\end{aligned}$$

which shows that $M_X(S) \leq 100 m_X(S)$ (this bound does not use neither regularity neither independence of X_i).

To establish the reverse inequality, let us take any $m > 2M_X(S) \geq \widetilde{M}_X(S)$ and enumerate elements of S as $t_1, t_2, \ldots$ in such a way that that $i \to \mathbb{P}(|X_{t_i}| \geq m)$ is nonincreasing. By the definition of $\widetilde{M}_X(S)$, we have

$$\sum_{i=1}^{N} \mathbb{P}(|X_{t_i}| \geq m) \leq \frac{1}{m} \sum_{i=1}^{N} \mathbb{E}|X_{t_i}| I_{\{|X_{t_i}| \geq m\}} \leq 1.$$

In particular, it means that $\mathbb{P}(|X_{t_i}| \geq m) \leq 1/i$. By (7) this yields that for $i > 1/c_2(\alpha)$ $\|X_{t_i}\|_{\log(i)} \leq m/c_1(\alpha)$. Since $\log(e+i)/\log(i) \leq 2$ for $i \geq 3$, we have $\|X_{t_i}\|_{\log(e+i)} \leq C(\alpha) m$ for large i. For $i \leq \max\{3, 1/c_2(\alpha)\}$, it is enough to observe that $\log(e+i) \leq 2^{k(\alpha)}$, so

$$\|X_{t_i}\|_{\log(e+i)} \leq C_0(\alpha)^{k(\alpha)} \mathbb{E}|X_{t_i}| \leq C_0(\alpha)^{k(\alpha)} M_X(S).$$

This shows that $\|X_{t_i}\|_{\log(e+i)} \lesssim_\alpha m$ for all i, and therefore, $m_X(S) \lesssim_\alpha M_X(S)$.

□

2.1 γ_X-Functional

The famous Fernique-Talagrand theorem [3, 10] states that suprema of Gaussian processes may be estimated in geometrical terms by γ_2-functional. This result was extended in several directions. One of them is based on the so-called γ_X functional.

For a nonempty subset $T \subset \mathbb{R}^n$, we define

$$\gamma_X(T) := \inf \sup_{t \in T} \sum_{n=0}^{\infty} \Delta_{n,X}(A_n(t)),$$

where the infimum runs over all increasing sequences of partitions $(\mathcal{A}_n)_{n \geq 0}$ of T such that $\mathcal{A}_0 = \{T\}$ and $|\mathcal{A}_n| \leq N_n := 2^{2^n}$ for $n \geq 1$, $A_n(t)$ is the unique element of $\mathcal{A}_n$ which contains t, and $\Delta_{n,X}(A)$ denotes the diameter of A with respect to the distance $d_n(s,t) := \|X_s - X_t\|_{2^n}$.

It is not hard to check that $b_X(T) \lesssim \gamma_X(T)$. The reverse bound was discussed in [7], where it was shown that it holds (with constants depending on β and λ) if

$$\|X_i\|_p \leq \beta \frac{p}{q} \|X_i\|_q \text{ and } \|X_i\|_{\lambda p} \geq 2\|X_i\|_p \text{ for all } i \text{ and } p \geq q \geq 2. \tag{8}$$

Moreover, the condition $\|X_i\|_p \leq \beta \frac{p}{q} \|X_i\|_q$ is necessary in the i.i.d. case if the estimate $\gamma_X(T) \leq C b_X(T)$ holds with a constant independent on n and $T \subset \mathbb{R}^n$.

The next result may be easily deduced from the proof of [7, Corollary 2.7], but we provide its proof for the sake of completeness.

Proposition 6 *Let X_i be independent and satisfy condition* (5) *and let T be a nonempty subset of $\mathbb{R}^n$ such that $\gamma_X(T) < \infty$. Then, there exists set $S \subset \mathbb{R}^n$ such that for any $t_0 \in T$, $T - t_0 \subset T - T \subset \overline{\mathrm{conv}}(S \cup -S)$ and $M_X(S) \lesssim m_X(S) \lesssim_\alpha \gamma_X(T)$.*

Proof Wlog (since it is only a matter of rescaling) we may assume that $\mathbb{E}X_i^2 = 1$.

By the definition of $\gamma_X(T)$, we may find an increasing sequence of partitions $(\mathcal{A}_n)$ such that $\mathcal{A}_0 = \{T\}$, $|\mathcal{A}_j| \le N_j$ for $j \ge 1$ and

$$\sup_{t\in T} \sum_{n=0}^{\infty} \Delta_{n,X}(A_n(t)) \le 2\gamma_X(T). \tag{9}$$

For any $A \in \mathcal{A}_n$, let us choose a point $\pi_n(A) \in A$ and set $\pi_n(t) := \pi_n(A_n(t))$.

Let $M_n := \sum_{j=0}^{n} N_j$ for $n = 0, 1, \ldots$ (we put $N_0 := 1$). Then, $\log(M_n + 2) \le 2^{n+1}$. Notice that there are $|\mathcal{A}_n| \le N_n$ points of the form $\pi_n(t) - \pi_{n-1}(t)$, $t \in T$. So we may define s_k, $M_{n-1} \le k < M_n$, $n = 1, 2, \ldots$ as some rearrangement (with repetition if $|\mathcal{A}_n| < N_n$) of points of the form $(\pi_n(t) - \pi_{n-1}(t))/\|X_{\pi_n(t)} - X_{\pi_{n-1}(t)}\|_{2^{n+1}}$, $t \in T$. Then, $\|X_{s_k}\|_{\log(k+e)} \le 1$ for all $k \ge 1$.

Observe that

$$\|t-\pi_n(t)\|_2 = \|X_t - X_{\pi_n(t)}\|_2 \le \Delta_{2,X}(A_n(t)) \le \Delta_{n,X}(A_n(t)) \to 0 \quad \text{for } n \to \infty.$$

For any $s, t \in T$, we have $\pi_0(s) = \pi_0(t)$ and thus

$$\begin{aligned} s - t &= \lim_{n\to\infty} (\pi_n(s) - \pi_n(t)) \\ &= \lim_{n\to\infty} \left(\sum_{k=1}^{n} (\pi_k(s) - \pi_{k-1}(s)) - \sum_{k=1}^{n} (\pi_k(t) - \pi_{k-1}(t)) \right). \end{aligned}$$

This shows that

$$T - T \subset R\, \overline{\mathrm{conv}}\{\pm s_k : \ k \ge 1\},$$

where

$$\begin{aligned} R := 2\sup_{t\in T} \sum_{n=1}^{\infty} d_{n+1}(\pi_n(t), \pi_{n-1}(t)) &\le 2\sup_{t\in T} \sum_{n=1}^{\infty} \Delta_{n+1,X}(A_{n-1}(t)) \\ \le C(\alpha) \sup_{t\in T} \sum_{n=1}^{\infty} \Delta_{n-1,X}(A_{n-1}(t)) &\le 2C(\alpha)\gamma_X(T), \end{aligned}$$

where the second inequality follows by (6). Thus, it is enough to define $S := \{Rs_k\colon\ k \geq 1\}$. □

Remark 7 Proposition 6 together with the equivalence $b_X(T) \sim_{\alpha,\lambda} \gamma_X(T)$ shows that the main question has the affirmative answer for any bounded nonempty set T if symmetric random variables X_i satisfy moment bounds (5). We strongly believe that the condition $\|X_i\|_{\lambda p} \geq 2\|X_i\|_p$ is not necessary—equivalence of $b_X(T)$ and the convex hull bound was established in the case of symmetric Bernoulli r."s ($\mathbb{P}(X_i = \pm 1) = 1/2$) in [1, Corollary 1.2]. However, to treat the general case of r.v's satisfying only the condition $\|X_i\|_p \leq \beta\frac{p}{q}\|X_i\|_q$, one should most likely combine γ_X functional with a suitable decomposition of the process $(X_t)_{t\in T}$, as was done for Bernoulli processes.

3 Toy Case: ℓ_1-Ball

Let us now consider a simple case of $T = B_1^n = \{t \in \mathbb{R}^n\colon\ \|t\|_1 \leq 1\}$. Let

$$u_0 := \inf\left\{u > 0\colon\ \mathbb{P}\big(\max_i |X_i| \geq u\big) \leq \frac{1}{2}\right\}.$$

Since

$$\mathbb{P}\big(\max_i |X_i| \geq u\big) \geq \frac{1}{2}\min\left\{1, \sum_i \mathbb{P}(|X_i| \geq u)\right\}$$

we get

$$\begin{aligned}
\mathbb{E}\sup_{t\in B_1^n} X_t &= \mathbb{E}\max_{1\leq i\leq n}|X_i| = \int_0^\infty \mathbb{P}\big(\max_{1\leq i\leq n}|X_i| \geq u\big)\mathrm{d}u \\
&\geq \frac{1}{2}u_0 + \int_{u_0}^\infty \frac{1}{2}\sum_{i=1}^n \mathbb{P}(|X_i| \geq u)\mathrm{d}u \\
&= \frac{1}{2}u_0 + \frac{1}{2}\sum_{i=1}^n \int_{u_0}^\infty \mathbb{P}(|X_i| \geq u)\mathrm{d}u = \frac{1}{2}u_0 + \frac{1}{2}\sum_{i=1}^n \mathbb{E}(|X_i| - u_0)_+.
\end{aligned}$$

Therefore,

$$2u_0 + \sum_{i=1}^n \mathbb{E}|X_i| I_{\{|X_i|\geq 2u_0\}} \leq 2u_0 + 2\sum_{i=1}^n \mathbb{E}(|X_i| - u_0)_+ \leq 4\mathbb{E}\sup_{t\in B_1^n} X_t,$$

so that $M_X(\{e_i:\ i\le n\})\le 4\mathbb{E}\sup_{t\in B_1^n}X_t$, where $(e_i)_{i\le n}$ is the canonical basis of $\mathbb{R}^n$. Since $B_1^n\subset \mathrm{conv}\{\pm e_1,\dots,\pm e_n\}$, we get the affirmative answer to the main question for $T=B_1^n$.

Proposition 8 *If $T = B_1^n$, then estimate* (4) *holds for arbitrary independent integrable r.v's $X_1,\dots,X_n$ with $S=\{e_1,\dots,e_n\}$ and $t_0=0$.*

4 Case II. Euclidean Balls

Now, we move to the case $T=B_2^n$. Then, $\sup_{t\in T}\langle t,x\rangle=|x|$, where $|x|=\|x\|_2$ is the Euclidean norm of $x\in\mathbb{R}^n$.

4.1 *Counterexample*

In this subsection, $X=(X_1,X_2,\dots,X_n)$, where X_k have symmetric p-stable distribution with characteristic function $\varphi_{X_k}(t)=\exp(-|t|^p)$ and $p\in(1,2)$. We will assume for convenience that n is even. Let G be a canonical n-dimensional Gaussian vector, independent of X. Then,

$$\mathbb{E}|X|=\mathbb{E}_X\mathbb{E}_G\sqrt{\frac{\pi}{2}}|\langle X,G\rangle|=\sqrt{\frac{\pi}{2}}\mathbb{E}_G\mathbb{E}_X|\langle X,G\rangle|=\sqrt{\frac{\pi}{2}}\mathbb{E}_G\|G\|_p\mathbb{E}|X_1|$$
$$\sim_p \mathbb{E}\|G\|_p\sim(\mathbb{E}\|G\|_p^p)^{1/p}\sim n^{1/p}.$$

Observe also that for $u>0$, $\mathbb{P}(|X_1|\ge u)\sim_p\min\{1,u^{-p}\}$, so

$$\mathbb{E}|X_1|I_{\{|X_1|\ge u\}}\sim_p u\min\{1,u^{-p}\}+\int_u^\infty \min\{1,v^{-p}\}dv\sim_p\min\{1,u^{1-p}\},\qquad u>0$$

and

$$\mathbb{E}|X_t|I_{\{|X_t|\ge u\}}=\|t\|_p\mathbb{E}|X_1|I_{\{|X_1|\ge u/\|t\|_p\}}\sim_p\min\{\|t\|_p,u^{1-p}\|t\|_p^p\},\quad u>0,\ t\in\mathbb{R}^n.$$

Hence,

$$\sum_{t\in S}\|t\|_p^p\lesssim_p u^p\qquad\text{for } u>\widetilde{M}_X(S).$$

Suppose that $B_2^n\subset\overline{\mathrm{conv}}(S\cup -S)$ and $M_X(S)\sim\widetilde{M}_X(S)<\infty$. We may then enumerate elements of S as $(t_k)_{k=1}^N$, $N\le\infty$ in such a way that $(\|t_k\|_p)_{k=1}^N$ is nonincreasing. Obviously $N\ge n$ (otherwise $\mathrm{conv}(S\cup -S)$ would have empty interior). Take $u > \widetilde{M}_X(S)$ and set $E := \operatorname{span}(\{t_k\colon\ k \le n/2\})$. Then, $\|t_k\|_p^p \le C_p u^p/n$ for $k > n/2$. Thus,

$$B_2^n \subset \overline{\operatorname{conv}}(S \cup -S) \subset E + \overline{\operatorname{conv}}(\{\pm t_k\colon\ k > n/2\}) \subset E + \Big(\frac{C_p}{n}\Big)^{1/p} u B_p^n.$$

Let $F = E^{\perp}$ and P_F denotes the ortogonal projection of $\mathbb{R}^n$ onto the space F. Then, $\dim F = \dim E = n/2$ and

$$B_2^n \cap F = P_F(B_2^n) \subset \Big(\frac{C_p}{n}\Big)^{1/p} u P_F(B_p^n).$$

In particular,

$$n^{-1/2} \sim \operatorname{vol}_{n/2}^{2/n}(B_2^n \cap F) \le \Big(\frac{C_p}{n}\Big)^{1/p} u \operatorname{vol}_{n/2}^{2/n}(P_F(B_p^n)).$$

By the Rogers-Shephard inequality [8] and inclusion $B_2^n \subset n^{1/p-1/2} B_p^n$, we have

$$\begin{aligned}\operatorname{vol}_{n/2}(P_F(B_p^n)) &\le \binom{n}{n/2} \frac{\operatorname{vol}_n(B_p^n)}{\operatorname{vol}_{n/2}(B_p^n \cap E)} \le 2^n \frac{\operatorname{vol}_n(B_p^n)}{\operatorname{vol}_{n/2}(n^{1/2-1/p} B_2^n \cap E)}\\ &\le (Cn^{-1/p})^{n/2}.\end{aligned}$$

This shows that $u \gtrsim_p n^{2/p-1/2}$. Thus, $M_X(S) \gtrsim_p n^{2/p-1/2} \gg n^{1/p} \sim_p b_X(B_2^n)$, and our question has a negative answer in this case.

4.2 $4 + \delta$ *Moment Condition*

In this part we establish positive answer to the main question in the case $T = B_2^n$ under the following $4 + \delta$ moment condition:

$$\exists_{r \in (4,8], \lambda < \infty}\ (\mathbb{E} X_i^r)^{1/r} \le \lambda (\mathbb{E} X_i^2)^{1/2} < \infty \quad i = 1, \ldots, n. \tag{10}$$

The restriction $r \le 8$ is just for convenience. The following easy consequence of (10) will be helpful in the sequel.

Lemma 9 *Suppose that $X_1, \ldots, X_n$ are independent mean zero r.'s satisfying condition* (10). *Then, for any $1 \le p \le r$,*

$$\Big\|\sum_{i=1}^n u_i X_i\Big\|_p \sim_\lambda \Big\|\sum_{i=1}^n u_i X_i\Big\|_2 = \Big(\sum_{i=1}^n u_i^2 \mathbb{E} X_i^2\Big)^{1/2} \tag{11}$$

and

$$\Big\| \sum_{1\le i<j\le n} u_{ij}X_iX_j \Big\|_p \sim_\lambda \Big\| \sum_{1\le i<j\le n} u_{ij}X_iX_j \Big\|_2 = \Big(\sum_{1\le i<j\le n} u_{ij}^2 \mathbb{E}X_i^2 \mathbb{E}X_j^2 \Big)^{1/2}. \tag{12}$$

Proof Since it is only a matter of scaling wlog, we may and will assume that $\mathbb{E}X_i^2 = 1$ for all i.

Rosenthal's inequality [9] gives for $2 \le p \le r$ (recall that $r \in (4, 8]$, so constants below do not depend on r)

$$\begin{aligned} \Big\| \sum_{i=1}^n u_i X_i \Big\|_p &\sim \Big(\sum_i \mathbb{E}|u_iX_i|^2 \Big)^{1/2} + \Big(\sum_i \mathbb{E}|u_iX_i|^p \Big)^{1/p} \\ &\sim_\lambda \Big(\sum_i u_i^2 \Big)^{1/2} + \Big(\sum_i |u_i|^p \Big)^{1/p} \\ &\sim \Big(\sum_i u_i^2 \Big)^{1/2}. \end{aligned}$$

To estimate $\|S\|_p$ for $1 \le p \le 2$ and $S = \sum_{i=1}^n u_iX_i$, it is enough to note that $\|S\|_1 \le \|S\|_p \le \|S\|_2$ and $\|S\|_2 \le \|S\|_4^{1/3} \|S\|_1^{2/3} \sim_\lambda \|S\|_2^{1/3}\|S\|_1^{2/3}$, so $\|S\|_p \sim_\lambda \|S\|_2$.

To prove the last part of the assertion we will use the hypercontractive method. Observe that for a real number u, there exists $\theta \in [0, 1]$ such that

$$\begin{aligned} (1+u)^r &\le \Big(1 + ru + \frac{r(r-1)}{2}(1+\theta u)^{r-2}u^2\Big) I_{\{|u|<1\}} + (2|u|)^r I_{\{|u|\ge 1\}} \\ &\le 1 + ru + r^2 2^{r-3} u^2 + 2^r |u|^r. \end{aligned}$$

Hence (note that $\lambda \ge 1$, $\mathbb{E}X_i = 0$, $\mathbb{E}X_i^2 = 1$ and $\mathbb{E}|X_i|^r \le \lambda^r$)

$$\begin{aligned} \mathbb{E}\Big(1 + \frac{1}{32\lambda}uX_i\Big)^r &\le 1 + r^2 2^{r-3}\frac{u^2}{1024} + 2^{-4r}|u|^r \le 1 + \frac{ru^2}{4} + \frac{|u|^r}{2} \\ &\le 1 + \max\Big\{\frac{r}{2}u^2, |u|^r\Big\}. \end{aligned}$$

Since

$$(\mathbb{E}(1+uX_i)^2)^{r/2} = (1+u^2)^{r/2} \ge 1 + \max\Big\{\frac{r}{2}u^2, |u|^r\Big\}$$

we get $\|1 + \frac{1}{32\lambda}uX_i\|_r \le \|1 + uX_i\|_2$ for any $u \in \mathbb{R}$ and the hypercontractivity method (cf. [5, Theorem 6.5.2]) yields (12) for $p = r$. The case $1 \le p \le r$ may be obtained in the same way as in the proof of (11). □

Observe that (10) implies that $\mathrm{Var}(X_i^2) \le (\lambda^4 - 1)(\mathbb{E}X_i^2)^2$, so $\mathrm{Var}(|X|^2) \le \sum_i (\lambda^4 - 1)(\mathbb{E}X_i^2)^2 \le (\lambda^4 - 1)(\mathbb{E}|X|^2)^2$. This yields that $\mathbb{E}|X|^4 \le \lambda^4(\mathbb{E}|X|^2)^2$ and $(\mathbb{E}|X|^2)^{1/2} \le \lambda^2\mathbb{E}|X|$.

The next fact is pretty standard; we prove it for completeness.

Lemma 10 *For any k there exists $T \subset B_2^k$ with $|T| \le 5^k$ such that $B_2^k \subset 2\mathrm{conv}(T)$.*

Proof Let T be the maximal $\frac{1}{2}$-separated set in B_2^k, the standard volumetric argument shows that $|T| \le 5^k$. We have $B_2^k \subset T + \frac{1}{2}B_2^k \subset \mathrm{conv}(T) + \frac{1}{2}B_2^k$, so $B_2^k \subset 2\mathrm{conv}(T)$. □

The next lemma comes from [4].

Lemma 11 *For any $1 \le k \le n$, there exists $T \subset B_2^n$ with $|T| \le \frac{2n}{k}5^k$ such that $B_2^n \subset 2\sqrt{\frac{2n}{k}}\mathrm{conv}(T)$.*

Proof Let $l = \lceil n/k \rceil \le 2n/k$ and $\mathbb{R}^n = F_1 \oplus \cdots \oplus F_l$ be an orthogonal decomposition of $\mathbb{R}^n$ into spaces of dimension at most k. By Lemma 10 we can find $T_i \subset B_2(F_i) := B_2^n \cap F_i$ such that $B_2(F_i) \subset 2\mathrm{conv}(T_i)$ and $|T_i| \le 5^k$. Let $T := \bigcup_{i \le l} T_i$. Then, $T \subset B_2^n$ and $|T| \le l5^k \le \frac{2n}{k}5^k$.

Fix now $x \in B_2^n$ and x_i denotes its orthogonal projection on F_i. Observe that

$$\sum_{i \le l} \|x_i\| \le \sqrt{l}\Big(\sum_{i \le l} \|x_i\|^2\Big)^{1/2} \le \sqrt{l}.$$

Therefore,

$$x \subset \sqrt{l}\mathrm{conv}\Big\{0, \frac{x_1}{\|x_1\|}, \ldots, \frac{x_l}{\|x_l\|}\Big\} \subset \sqrt{l}\mathrm{conv}\Big(\bigcup_{i \le l} B_2(F_i)\Big) \subset 2\sqrt{l}\mathrm{conv}(T).$$

□

Lemma 12 *Let Y be a vector uniformly distributed over S^{n-1}. Then,*

$$\mathbb{E}|\langle Y, t\rangle| I_{\{|\langle Y,t\rangle| \ge u\}} \le \min\Big\{\frac{|t|}{\sqrt{n}}, \frac{2(|t|^2 + nu^2)}{nu} e^{-nu^2/(2|t|^2)}\Big\} \quad t \in \mathbb{R}^n,\ u > 0.$$

Proof Observe that $\langle Y, t\rangle$ is distributed as $|t|Y_1$. Hence,

$$\mathbb{E}|\langle Y, t\rangle| I_{\{|\langle Y,t\rangle| \ge u\}} = |t|\mathbb{E}|Y_1| I_{\{|Y_1| \ge u/|t|\}}.$$

We have $\mathbb{E}|Y_1| \le (\mathbb{E}|Y_1|^2)^{1/2} = n^{-1/2}$. Moreover, $\mathbb{P}(Y_1 \ge v) \le \exp(-nv^2/2)$ for $v \ge 0$ (cf. [12]). Therefore

$$\mathbb{E}|Y_1|I_{\{|Y_1|\geq u\}} \leq u\mathbb{P}(|Y_1|\geq u) + \int_u^\infty \mathbb{P}(|Y_1|\geq v)\mathrm{d}v$$
$$\leq 2ue^{-nu^2/2} + 2\int_u^\infty e^{-nv^2/2}\mathrm{d}v$$
$$\leq 2ue^{-nu^2/2} + 2\int_u^\infty \frac{nv}{nu}e^{-nv^2/2}\mathrm{d}v = \frac{2(1+nu^2)}{nu}e^{-nu^2/2}.$$

□

Now, we are able to show that (4) holds for $T = B_2^n$ under $4+\delta$ moment condition.

Proposition 13 *Let $X_1,\ldots,X_n$ be independent centered r.v's with variance 1 satisfying condition* (10)*. Then, there exists $S\subset\mathbb{R}^n$ such that $|S|\leq 10n^2$, $B_2^n\subset \mathrm{conv}(S)$ and*

$$M_X(S) \lesssim_{r,\lambda} \sqrt{n} \sim_\lambda \mathbb{E}|X| = b_X(B_2^n).$$

Proof By the Rosenthal inequality [9], we have (recall that $r\in(4,8]$),

$$\big\||X|^2-n\big\|_{r/2} = \Big\|\sum_{i=1}^n (X_i^2-1)\Big\|_{r/2} \lesssim \Big(\sum_{i=1}^n \mathrm{Var}(X_i^2)\Big)^{1/2} + \Big(\sum_{i=1}^n \mathbb{E}|X_i^2-1|^{r/2}\Big)^{2/r}$$
$$\lesssim_\lambda n^{1/2} + n^{2/r} \leq 2n^{1/2}.$$

Therefore,

$$\mathbb{E}|X|I_{\{|X|\geq\sqrt{2n}\}} \leq \mathbb{E}\sqrt{2(|X|^2-n)}I_{\{|X|\geq\sqrt{2n}\}} \leq \sqrt{2}n^{1/2-r/2}\mathbb{E}(|X|^2-n)^{r/2}$$
$$\leq C(\lambda)n^{1/2-r/4}. \tag{13}$$

By Lemma 11 (applied with $k = c(r)\log n$), there exists $t_1,\ldots,t_N$ such that $B_2^n\subset\mathrm{conv}\{t_1,\ldots,t_N\}$, $N\leq 10n^{1/2+r/8}$ and $|t_i|\leq C(r)\sqrt{n/\log n}$, $1\leq i\leq N$. Let U be the random rotation (uniformly distributed on $O(n)$) then Ut_i is distributed as $|t_i|Y$, where Y has uniform distribution on S^{n-1}. Thus, by Lemma 12,

$$\mathbb{E}_U\mathbb{E}_X|\langle X,Ut_i\rangle|I_{\{|\langle X,Ut_i\rangle|\geq u\}}$$
$$= \mathbb{E}_X\mathbb{E}_Y|\langle Y,|t_i|X\rangle|I_{\{|\langle Y,|t_i|X\rangle|\geq u\}}$$
$$\leq \mathbb{E}\min\left\{\frac{|t_i||X|}{\sqrt{n}}, \frac{2(|t_i|^2|X|^2+nu^2)}{nu}e^{-nu^2/(2|t_i|^2|X|^2)}\right\}$$
$$\leq \frac{|t_i|}{\sqrt{n}}\mathbb{E}|X|I_{\{|X|\geq\sqrt{2n}\}} + \frac{4|t_i|^2+2u^2}{u}e^{-u^2/(4|t_i|^2)}.$$

Recall that $|t_i| \lesssim_r \sqrt{n/\log n}$ so for sufficiently large $C(r)$, we get by (13),

$$\mathbb{E}_U \mathbb{E}_X |\langle X, Ut_i\rangle| I_{\{|\langle X, Ut_i\rangle| \geq C(r)\sqrt{n}\}} \leq C(\lambda) n^{-r/4} |t_i| + n^{-2} \leq C(r,\lambda) n^{1/2 - r/4}.$$

As a consequence, there exists $U \in O(n)$ such that

$$\sum_{i=1}^{N} \mathbb{E}_X |\langle X, Ut_i\rangle| I_{\{|\langle X, Ut_i\rangle| \geq C(r)\sqrt{n}\}} \leq N C(r,\lambda) n^{1/2 - r/4} \leq 10 C(r,\lambda) n^{1-r/8}. \tag{14}$$

Thus, if we put $S := \{Ut_1, \ldots, Ut_N\}$ we will have $\mathrm{conv}(S) = U\mathrm{conv}\{t_1, \ldots, t_N\} \supset B_2^n$ and $M_X(S) \leq C'(r,\lambda)\sqrt{n}$. □

4.3 Ellipsoids

We now extend the bounds from the previous subsection to the case of ellipsoids, i.e., sets of the form

$$\mathcal{E} := \Big\{ t \in \mathbb{R}^n : \sum_{i=1}^{n} \frac{\langle t, u_i\rangle^2}{a_i^2} \leq 1 \Big\}, \tag{15}$$

where $u_1, \ldots, u_n$ is an orthonormal system in $\mathbb{R}^n$ and $a_1, \ldots, a_n > 0$.

Observe that

$$\sup_{t \in \mathcal{E}} \langle t, x\rangle = \sqrt{\sum_{i=1}^{n} a_i^2 \langle x, u_i\rangle^2}.$$

To treat this case, we will need the following Lemma:

Lemma 14 *Let $X = (X_1, \ldots, X_n)$, where X_i are independent mean zero and variance one r.v's satisfying $4+\delta$ condition* (10).

(i) For any $a_1, \ldots, a_n \geq 0$ and any o.n. vectors $u_1, \ldots, u_n$,

$$\mathbb{E}\Big(\sum_{k=1}^{n} a_k^2 \langle X, u_k\rangle^2\Big)^{1/2} \sim_\lambda \Big(\mathbb{E}\sum_{k=1}^{n} a_k^2 \langle X, u_k\rangle^2\Big)^{1/2} = \Big(\sum_{k=1}^{n} a_k^2\Big)^{1/2}.$$

(ii) For any $n \times n$ matrix B,

$$\Big(\mathbb{E}(|BX|^2 - \|B\|_{\mathrm{HS}}^2)^{r/2}\Big)^{2/r} \leq C(\lambda) \|B^T B\|_{\mathrm{HS}}^{1/2}.$$

In particular, for any linear supspace $E \subset \mathbb{R}^n$ od dimension $k \in \{1, \ldots, n\}$,

$$\left(\mathbb{E}(|P_E X|^2 - k)^{r/2}\right)^{2/r} \le C(\lambda)k^{1/2}.$$

Proof Part (i) follows from Lemma 9.

To show part (ii) let $B = (b_{ij})_{i,j=1}^n$, $e_1, e_2, \ldots, e_n$ be the canonical basis of $\mathbb{R}^n$ and let

$$\sigma_{i,j} := \sum_{l=1}^n b_{l,i} b_{l,j} = \langle Be_i, Be_j \rangle, \quad 1 \le i, j \le n.$$

Then,

$$\begin{aligned}\big\| |BX|^2 - \|B\|_{\mathrm{HS}}^2 \big\|_{r/2} &= \Big\| \sum_{i=1}^n (X_i^2 - 1)\sigma_{i,i} + \sum_{1 \le i \ne j \le n} X_i X_j \sigma_{i,j} \Big\|_{r/2} \\ &\le \Big\| \sum_{i=1}^n (X_i^2 - 1)\sigma_{i,i} \Big\|_{r/2} + \Big\| \sum_{1 \le i \ne j \le n} X_i X_j \sigma_{i,j} \Big\|_{r/2}.\end{aligned}$$

Applying Rosenthal's inequality, we get

$$\begin{aligned}\Big\| \sum_{i=1}^n (X_i^2 - 1)\sigma_{i,i} \Big\|_{r/2} &\lesssim \Big(\sum_{i=1}^n \mathrm{Var}(X_i^2)\sigma_{i,i}^2\Big)^{1/2} + \Big(\sum_{i=1}^n \mathbb{E}(X_i^2 - 1)^{r/2}\sigma_{i,i}^{r/2}\Big)^{2/r} \\ &\lesssim_\lambda \Big(\sum_{i=1}^n \sigma_{i,i}^2\Big)^{1/2} + \Big(\sum_{i=1}^n \sigma_{i,i}^{r/2}\Big)^{2/r} \le 2\Big(\sum_{i=1}^n \sigma_{i,i}^2\Big)^{1/2}.\end{aligned}$$

Hypercontractive method (as in the proof of Lemma 9) yields

$$\Big\| \sum_{i \ne j} X_i X_j \sigma_{i,j} \Big\|_{r/2} \lesssim_\lambda \Big\| \sum_{i \ne j} X_i X_j \sigma_{i,j} \Big\|_2 = \Big(\sum_{i \ne j} \sigma_{i,j}^2\Big)^{1/2}.$$

Finally,

$$\Big(\sum_{i=1}^n \sigma_{i,i}^2\Big)^{1/2} + \Big(\sum_{i \ne j} \sigma_{i,j}^2\Big)^{1/2} \le 2\Big(\sum_{i,j} \sigma_{i,j}^2\Big)^{1/2} = 2\|B^T B\|_{\mathrm{HS}}.$$

□

Now, we state and prove the main result of this section.

Theorem 15 *Let $X_1, \ldots, X_n$ be independent centered r.v's satisfying the condition* (10) *and let T be an ellipsoid in $\mathbb{R}^n$. Then, there exists $S \subset \mathbb{R}^n$ such that $|S| \le 10n^2$, $T \subset \mathrm{conv}(S)$ and*

$$M_X(S) \lesssim_{r,\lambda} b_X(T).$$

Proof Since it is only a matter of scaling, we may and will assume that $\mathbb{E}X_i^2 = 1$ for all i. Let $T = \mathcal{E}$ be an ellipsoid of the form (15). Then, the first part of Lemma 14 yields

$$\mathbb{E}\sup_{t\in\mathcal{E}} X_t = \mathbb{E}\Big(\sum_{k=1}^{n} a_k^2\langle X, u_k\rangle^2\Big)^{1/2} \sim_\lambda \sqrt{\sum_{k=1}^{n} a_k^2}.$$

By homogenity, we may assume that $\sum_{k=1}^{n} a_k^2 = 1$.

Define

$$I_k := \{i\colon\ 2^{-k-1} < a_i \le 2^{-k}\},\quad n_k := |I_k|,\quad J := \{k \in \mathbb{Z}\colon\ I_k \neq \emptyset\},$$
$$E_k := \operatorname{span}\{u_i : i \in I_k\}.$$

Then,

$$1 \le \sum_{k\in J} n_k 2^{-2k} < 4. \tag{16}$$

In particular, J is a subset of nonnegative integers.

We claim that for any positive sequence $(c_k)_{k\in J}$ such that $\sum_k c_k^{-2} \le 1$,

$$\mathcal{E} \subset \operatorname{conv}\Big(\bigcup_{k\in J} c_k 2^{-k} B_2^{I_k}\Big), \quad \text{where} \quad B_2^{I_k} := B_2^n \cap E_k.$$

Indeed, let $P_k x := \sum_{i\in I_k}\langle x, u_i\rangle u_i$ be the projection of x onto E_k, then

$$x = \sum_{k\in J} c_k^{-1} 2^k |P_k x| c_k 2^{-k}\frac{P_k x}{|P_k x|}$$

and for $x \in \mathcal{E}$,

$$\sum_{k\in J} c_k^{-1} 2^k |P_k x| \le \sqrt{\sum_{k\in J} c_k^{-2}}\sqrt{\sum_{k\in J} 2^{2k}|P_k x|^2} \le \sqrt{\sum_{k\in J}\sum_{i\in I_k}\frac{\langle x, u_i\rangle^2}{a_i^2}} \le 1.$$

Let us for a moment fix $k \in J$. By Lemma 11 (applied with $k = c(r)\log n_k$), there exists $t_1, \ldots, t_{N_k} \in E_k$ such that $B_2^{I_k} \subset \operatorname{conv}\{t_1, \ldots, t_{N_k}\}$, $N_k \le 10 n_k^{1/2+r/8}$ and $|t_i| \le C(r)\sqrt{n_k/\log(n_k)}$. Let U be the random rotation of E_k (uniformly distributed on $O(E_k)$) then Ut_i is distributed as $|t_i|Y$, where Y has uniform distribution on $S^{I_k} := S^{n-1} \cap E_k$. Thus, by Lemma 12,

$$\begin{aligned}
&\mathbb{E}_U\mathbb{E}_X|\langle X, Ut_i\rangle|I_{\{|\langle X,Ut_i\rangle|\geq u\}}\\
&\quad= \mathbb{E}_X\mathbb{E}_Y|\langle Y, |t_i|P_{E_k}X\rangle|I_{\{|\langle Y,|t_i|P_{E_k}X\rangle|\geq u\}}\\
&\quad\leq \mathbb{E}\min\left\{\frac{|t_i||P_{E_k}X|}{\sqrt{n_k}}, \frac{2(|t_i|^2|P_{E_k}X|^2+n_ku^2)}{n_ku}e^{-n_ku^2/(2|t_i|^2|P_{E_k}X|^2)}\right\}\\
&\quad\leq \frac{|t_i|}{\sqrt{n_k}}\mathbb{E}|P_{E_k}X|I_{\{|P_{E_k}X|\geq\sqrt{2n_k}\}} + \frac{4|t_i|^2+2u^2}{u}e^{-u^2/(4|t_i|^2)}.
\end{aligned}$$

We have

$$\begin{aligned}
\mathbb{E}|P_{E_k}X|I_{\{|P_{E_k}X|\geq\sqrt{2n_k}\}} &\leq \sqrt{2}\mathbb{E}(|P_{E_k}X|^2-n_k)^{1/2}I_{\{|P_{E_k}X|\geq\sqrt{2n_k}\}}\\
&\leq \sqrt{2}n_k^{1/2-r/2}\mathbb{E}(|P_{E_k}X|^2-n_k)^{r/2}\leq C(\lambda)n_k^{1/2-r/4},
\end{aligned}$$

where the last bound follows by Lemma 14. Recall that $|t_i|\lesssim_r\sqrt{n_k/\log n_k}$; thus, for sufficiently large $C(r)$, we get

$$\mathbb{E}_U\mathbb{E}_X|\langle X, Ut_i\rangle|I_{\{|\langle X,Ut_i\rangle|\geq C(r)\sqrt{n}\}}\leq C(\lambda)n_k^{-r/4}|t_i|+n_k^{-2}\leq C(r,\lambda)n_k^{1/2-r/4}.$$

As a consequence, there exists $U\in O(E_k)$ such that

$$\sum_{i=1}^{N_k}\mathbb{E}_X|\langle X, Ut_i\rangle|I_{\{|\langle X,Ut_i\rangle|\geq C(r)\sqrt{n_k}\}}\leq N_kC(r,\lambda)n_k^{1/2-r/4}\leq 10C(r,\lambda)n_k^{1-r/8}.$$

Define $S_k = \{t_{k,1},\ldots,t_{k,N_k}\} := \{Ut_1,\ldots,Ut_{N_k}\}$. Then, $\operatorname{conv}(S_k) = U\operatorname{conv}\{t_1,\ldots,t_{N_k}\}\supset B_2^{I_k}$, $N_k\leq 10n_k^{1/2+r/8}\leq 10n_k^2$ and

$$\sum_{i=1}^{N_k}\mathbb{E}_X|\langle X, t_{k,i}\rangle|I_{\{|\langle X,t_{k,i}\rangle|\geq C(r)\sqrt{n_k}\}}\leq C(r,\lambda)n_k^{1-r/8}.$$

Set $c_k := 2^{k+2}(2^k+n_k)^{-1/2}$. By (16) we get $\sum_{k\in J}c_k^{-2}\leq 1$, so

$$\mathcal{E}\subset\operatorname{conv}\Big(\bigcup_{k\in J}c_k2^{-k}B_2^{I_k}\Big)\subset\operatorname{conv}(\{c_k2^{-k}t_{k,i}\colon\ k\in J, i\leq N_k\}) := \operatorname{conv}(S).$$

We have

$$|S| = \sum_{k\in J}N_k\leq\sum_{k\in J}10n_k^2\leq 10\Big(\sum_{k\in J}n_k\Big)^2 = 10n^2.$$

Moreover,

$$\begin{aligned}
&\sum_{s\in S}\mathbb{E}|\langle s,X\rangle|I_{\{|\langle s,X\rangle|\geq 4C(r)\}}\\
&\quad=\sum_{k\in J}2^{-k}c_k\sum_{i=1}^{N_k}\mathbb{E}|\langle t_{k,i},X\rangle|I_{\{2^{-k}c_k|\langle t_{k,i},X\rangle|\geq 4C(r)\}}\\
&\quad\leq\sum_{k\in J}4(2^k+n_k)^{-1/2}\sum_{i=1}^{N_k}\mathbb{E}|\langle t_{k,i},X\rangle|I_{\{|\langle t_{k,i},X\rangle|\geq C(r)\sqrt{n_k}\}}\\
&\quad\leq\sum_{k\in J}4(2^k+n_k)^{-1/2}C(r,\lambda)n_k^{1-r/8}\\
&\quad\leq 4C(r,\lambda)\sum_{k\in J}(2^k+n_k)^{1/2-r/8}\\
&\quad\leq 4C(r,\lambda)\sum_{k\geq 0}2^{k(1/2-r/8)}\\
&\quad\leq C'(r,\lambda),
\end{aligned}$$

which shows that $M_X(S)\sim\widetilde{M}_X(S)\lesssim_{\lambda,r}1\sim b_X(\mathcal{E})$.

□

5 Case III. ℓ_q^n-Balls, $2<q\leq\infty$

It turns out that results of the previous sections may be easily applied to get estimates in the case when $T=B_q^n$ is the unit ball in ℓ_q^n and $q\in(2,\infty]$. In the whole section by q', we will denote the Hölder dual of q, i.e., $q'=\frac{q}{q-1}$, $2\leq q<\infty$ and $q'=1$ for $q=\infty$.

Proposition 16 *Let $X_1,\ldots,X_n$ be independent centered r.v's with variance 1 satisfying condition* (10). *Then, there exists $S\subset\mathbb{R}^n$ such that $|S|\leq 10n^2$, $B_q^n\subset\mathrm{conv}(S)$ and*

$$M_X(S)\lesssim_{r,\lambda}n^{1/q'}\sim_\lambda b_X(B_q^n).$$

Proof Since $q'\in(1,2]$, condition (10) yields $\|X_i\|_{q'}\sim_\lambda\|X_i\|_{2q'}\sim_\lambda\|X_i\|_2=1$ and hence $(\mathbb{E}\|X\|_{q'}^{2q'})^{1/(2q')}\sim_\lambda(\mathbb{E}\|X\|_{q'}^{q'})^{1/q'}$. Therefore,

$$b_X(B_q^n)=\mathbb{E}\sup_{t\in B_q^n}\langle t,X\rangle=\mathbb{E}\|X\|_{q'}\sim_\lambda\left(\mathbb{E}\|X\|_{q'}^{q'}\right)^{1/q'}\sim_\lambda n^{1/q'}.$$

Hölder's inequality implies $B_q^n \subset n^{1/2-1/q} B_2^n = n^{1/q'-1/2} B_2^n$ and the assertion easily follows from Proposition 13. □

Now, let us consider the case of linear transformation of ℓ_q^n-ball, i.e. $T = AB_q^n$. Next simple lemma shows how to estimate $b_X(T)$.

Lemma 17 *Let $X = (X_1, \ldots, X_n)$, where X_i are independent mean zero and variance one r.v's satisfying $4 + \delta$ condition* (10)*. Then, for any $n \times n$ matrix A and $2 \le q \le \infty$, we have*

$$b_X(AB_q^n) = b_{A^T X}(B_q^n) \sim_\lambda \Big(\sum_{i=1}^n |Ae_i|^{q'}\Big)^{1/q'}.$$

Proof Observe that

$$\sup_{t \in AB_q^n} \langle X, t\rangle = \sup_{t \in B_q^n} \langle A^T X, t\rangle = \Big(\sum_{i=1}^n |\langle A^T X, e_i\rangle|^{q'}\Big)^{1/q'} = \Big(\sum_{i=1}^n |\langle X, Ae_i\rangle|^{q'}\Big)^{1/q'}.$$

Condition (10) (see Lemma 9) implies that

$$\|\langle X, Ae_i\rangle\|_{2q'} \sim_\lambda \|\langle X, Ae_i\rangle\|_{q'} \sim_\lambda \|\langle X, Ae_i\rangle\|_2 = |Ae_i|.$$

Hence, $\|\sup_{t\in AB_q^n}\langle X, t\rangle\|_{2q'} \sim_\lambda \|\sup_{t\in AB_q^n}\langle X, t\rangle\|_{q'}$ and

$$b_X(AB_q^n) = \Big\|\sup_{t \in AB_q^n} \langle X, t\rangle\Big\|_1 \sim_\lambda \Big\|\sup_{t \in AB_q^n} \langle X, t\rangle\Big\|_{q'} \sim_\lambda \Big(\sum_{i=1}^n |Ae_i|^{q'}\Big)^{1/q'}.$$

□

As in the proof of Proposition 16, we may include linear image of B_q^n into ellipsoid with the comparable b_X-bound and deduce from Theorem 15 the following more general result.

Theorem 18 *Let $X_1, \ldots, X_n$ be independent centered r.v's satisfying condition* (10) *and let $T = AB_q^n$ for some $2 \le q \le \infty$ and an $n \times n$ matrix A. Then, there exists $S \subset \mathbb{R}^n$ such that $|S| \le 10n^2$, $T \subset \mathrm{conv}(S)$ and*

$$M_X(S) \lesssim_{r,\lambda} b_X(T).$$

Proof Since it is only a matter of scaling, we may and will assume that $\mathbb{E}X_i^2 = 1$ for all i. By Lemma 17 it is enough to show that

$$M_X(S) \lesssim_{r,\lambda} \Big(\sum_{i=1}^n |Ae_i|^{q'}\Big)^{1/q'}.$$

By homogenity, we may assume that $\sum_{i=1}^{n} |Ae_i|^{q'} = 1$. Case $q = 2$ was treated in Theorem 15, so we may assume that $q > 2$, i.e. $q' < 2$. Moreover, we may assume that $Ae_i \neq 0$ for all i.

Let $\lambda_i := |Ae_i|^{1-q'/2}$. Observe that if $t \in B_q^n$, then by Hölder's inequality

$$\sum_{i=1}^{n} |\lambda_i t_i|^2 \leq \Big(\sum_{i=1}^{n} |t_i|^q\Big)^{2/q} \Big(\sum_{i=1}^{n} |\lambda_i|^{2q/(q-2)}\Big)^{(q-2)/q}$$
$$= \Big(\sum_{i=1}^{n} |t_i|^q\Big)^{2/q} \Big(\sum_{i=1}^{n} |Ae_i|^{q'}\Big)^{(q-2)/q} \leq 1.$$

This shows that $D^{-1}B_q^n \subset B_2^n$, where $D := \mathrm{diag}(d_1, \ldots, d_n)$ and $d_i := |Ae_i|^{q'/2-1}$. Hence, $AB_q^n \subset ADB_2^n$ and

$$b_X(ADB_2^n) \sim_\lambda \Big(\sum_{i=1}^{n} |ADe_i|^2\Big)^{1/2} = \Big(\sum_{i=1}^{n} |Ae_i|^{q'}\Big)^{1/2} = 1.$$

We get the assertion applying Theorem 15 for the ellipsoid ADB_2^n. □

6 Concluding Remarks and Open Questions

We have shown that the main question has the affirmative answer in the case T is an ellipsoid (or more general linear image of ℓ_q^n-ball, $2 \leq q \leq n$) if X_i are independent mean zero r.v's satisfying the $4+\delta$ moment condition (10). The following questions are up to our best knowledge open.

- Does (4) holds for $T = B_q^n$, $1 < q < 2$ and X_i satisfying $4+\delta$ moment condition?
- John's theorem states that for any convex symmetric set T in $\mathbb{R}^n$, there exists an ellipsoid $\mathcal{E}$ such that $\mathcal{E} \subset T \subset \sqrt{n}\mathcal{E}$. Hence, Theorem 15 implies that under $4 + \delta$ condition (10) one may find finite set S such that $T \subset \mathrm{conv}(S \cup -S)$ and $M_X(S) \leq C(r, \lambda)\sqrt{n} b_X(T)$. We do not whether one may improve upon $\sqrt{n}$ factor for general sets T.
- Are there heavy-tailed random variables X_i such that (4) holds for arbitrary set T (for heavy-tailed r.v's approach via chaining functionals described in Sect. 2.1 fails to work)?
- Let X_i be heavy-tailed symmetric Weibull r.v's (i.e. symmetric variables with tails $\exp(-t^r)$, $0 < r < 1$). Bogucki [2] was able to obtain two-sided bounds for $b_X(T)$ with the use of random permutations (which may be eliminated if T is permutationally invariant). We do not know if the convex hull method works in this case.

Acknowledgement Supported by the National Science Centre, Poland grant 2015/18/A/ST1/00553.

References

1. W. Bednorz, R. Latała, On the boundedness of Bernoulli processes. Ann. Math.(2) **180**, 1167–1203 (2014)
2. R. Bogucki, Suprema of canonical Weibull processes. Statist. Probab. Lett. **107**, 253–263 (2015)
3. X. Fernique, Regularité des trajectoires des fonctions aléatoires gaussiennes, in *École d'Été de Probabilités de Saint-Flour, IV-1974*. Lecture Notes in Mathematics, vol. 480 (Springer, Berlin, 1975), pp. 1–96
4. M. Kochol, Constructive approximation of a ball by polytopes. Math. Slovaca **44**, 99–105 (1994)
5. S. Kwapień, W.A. Woyczyński, *Random Series and Stochastic Integrals: Single and Multiple* (Birkhauser, Boston, 1992)
6. R. Latała, M. Strzelecka, Comparison of weak and strong moments for vectors with independent coordinates. Mathematika **64**, 211–229 (2018)
7. R. Latała, T. Tkocz, A note on suprema of canonical processes based on random variables with regular moments. Electron. J. Probab. **20**(36), 1–17 (2015)
8. C.A. Rogers, G.C. Shephard, Convex bodies associated with a given convex body. J. London Math. Soc. **33**, 270–281 (1958)
9. H.P. Rosenthal, On the subspaces of L^p $(p > 2)$ spanned by sequences of independent random variables. Israel J. Math. **8**, 273–303 (1970)
10. M. Talagrand, Regularity of Gaussian processes. Acta Math. **159**, 99–149 (1987)
11. M. Talagrand, *Upper and Lower Bounds for Stochastic Processes*, 2nd edn. (Springer, Cham, 2021)
12. T. Tkocz, An upper bound for spherical caps. Amer. Math. Monthly **119**, 606–607 (2012)

Part IV
High-Dimensional Statistics

Random Geometric Graph: Some Recent Developments and Perspectives

Quentin Duchemin and Yohann De Castro

2020 Mathematics Subject Classification: 05C80; 62G05; 60E15

1 Introduction

1.1 Random Graph Models

Graphs are nowadays widely used in applications to model real-world complex systems. Since they are high-dimensional objects, one needs to assume some structure on the data of interest to be able to efficiently extract information on the studied system. To this purpose, a large number of models of random graphs have been already introduced. The most simple one is the Erdös-Renyi model $G(n, p)$ in which each edge between pairs of n nodes is present in the graph with some probability $p \in (0, 1)$. One can also mention the scale-free network model of Barabasi and Albert [11] or the small-world networks of Watts and Strogatz [93]. We refer to [27] for an introduction to the most famous random graph models. On real-world problems, it appears that there often exist some relevant variables accounting for the heterogeneity of the observations. Most of the time, these explanatory variables are unknown and carry a precious information on the system studied. To deal with such cases, latent space models for network data emerged (see [88]). One of the most studied latent models are the *community -random graphs* where each node is assumed to belong to one (or multiple) community, while the connection probabilities between two nodes in the graph depend on their respective membership. The well-known stochastic block model has received

Q. Duchemin (✉)
LAMA, Universite Gustave Eiffel, Universite Paris-Est Creteil, CNRS, Marne-la-Vallée, France
e-mail: quentin.duchemin@univ-eiffel.fr

Y. De Castro
Institut Camille Jordan - École Centrale de Lyon, Lyon, France
e-mail: yohann.de-castro@ec-lyon.fr

R. Adamczak et al. (eds.), *High Dimensional Probability IX*, Progress in Probability 80, https://doi.org/10.1007/978-3-031-26979-0_14

increasing attention in the recent years, and we refer to [1] for a nice introduction to this model and the statistical and algorithmic questions at stake. In the previous mentioned latent space models, the intrinsic geometry of the problem is not taken into account. However, it is known that the underlying spatial structure of network is an important property since geometry affects drastically the topology of networks (see [12] and [88]). To deal with embedded complex systems, spatial random graph models have been studied such as the random geometric graph (RGG). This paper surveys the recent developments in the theoretical analysis of RGGs through the prism of modern statistics and applications.

The theoretical analysis of random graph models is interesting by itself since it often involves elegant and important information theoretic, combinatorial, or probabilistic tools. In the following, we adopt this mindset trying to provide a faithful picture of the state-of-the-art results on RGGs focusing mainly on high-dimensional settings and nonparametric inference while underlining the main technical tools used in the proofs. We want to illustrate how the theory can impact real-data applications. To this end, we will essentially be focused on the following questions:

- **Detecting geometry in RGGs.** Nowadays, real-world problems often involve high-dimensional feature spaces. The first natural work is to identify the regimes where the geometry is lost in the dimension (see Eq. (1) for a formal definition). Several recent papers have made significant progress toward the resolution of this question that can be formalized as follows. Given a graph of n nodes, a latent geometry of dimension $d = d(n)$ and edge density $p = p(n)$, for what triples (n, d, p) is the model indistinguishable from $G(n, p)$?
- **Nonparametric estimation in RGGs.** By considering other rules for connecting latent points, the RGG model can be naturally extended to cover a larger class of networks. In such a framework, we will wonder what can be learned in an adaptive way from graphs with an underlying spatial structure. We will address nonparametric estimation in RGGs and its extension to growth model.
- **Connections between RGGs and community-based latent models.** Until recently, community and geometric-based random graph models have been mainly studied separately. Recent works try to investigate graph models that account for both cluster and spatial structures. We present some of them, and we sketch interesting research directions for future works.

1.2 Brief Historical Overview of RGGs

The RGG model was first introduced by Gilbert [51] to model the communications between radio stations. Gilbert's original model was defined as follows: pick points in $\mathbb{R}^2$ according to a Poisson point process of intensity one and join two if their distance is less than some parameter $r > 0$. The Gilbert model has been intensively studied, and we refer to [91] for a nice survey of its properties including connectivity, giant component, coverage, or chromatic number. The most closely related model is the random geometric graph where n nodes are independently and identically distributed on the space. A lot of results are actually transferable from one model to the other as presented in [80, Section 1.7]. In this paper, we will focus on the n points i.i.d. model which is formally defined in the next subsection (see Definition 1). The random geometric graph model was extended to other latent spaces such as the hypercube $[0, 1]^d$, the Euclidean sphere, or the compact Lie group [74]. A large body of literature has been devoted to studying the properties of low-dimensional random geometric graphs [17, 31, 80]. RGGs have found applications in a very large span of fields. One can mention wireless networks [55, 72], gossip algorithms [92], consensus [46], spread of a virus [84], protein-protein interactions [56], and citation networks [97]. One can also cite an application to motion planning in [89], a problem which consists in finding a collision-free path for a robot in a workspace cluttered with static obstacles. The ubiquity of this random graph model to faithfully represent real-world networks has motivated a great interest for its theoretical study.

1.3 Outline

In Sect. 2, we formally define the RGG and several variant models that will be useful for this article. In Sects. 3, 4 and 5, we describe recent results related to high-dimensional statistic, nonparametric estimation, and temporal prediction. Note that in these three sections, we will be working with the d-dimensional sphere $\mathbb{S}^{d-1}$ as latent space. $\mathbb{S}^{d-1}$ will be endowed with the Euclidean metric $\|\cdot\|$ which is the norm induced by the inner product $\langle \cdot, \cdot \rangle : (x, y) \in \left(\mathbb{S}^{d-1}\right)^2 \mapsto \sum_{i=1}^d x_i y_i$. The choice of this latent space is motivated by both recent theoretical developments in this framework [3, 26, 32, 58] and by applications [82, 83]. We further discuss in Sect. 6 recent works that investigate the connections between community-based random graph models and RGGs. Contrary to the previous sections, our goal is not to provide an exhaustive review of the literature in Sect. 6 but rather to shed light on some pioneering papers. Table 1 summarizes the organization of the paper.

Table 1 Outline of the paper. Models are defined in Sect. 2

Section	Questions tackled	Model
3	Geometry detection	RGG on $\mathbb{S}^{d-1}$
4	Nonparametric estimation	TIRGG on $\mathbb{S}^{d-1}$
5	Nonparametric estimation and temporal prediction	MRGG on $\mathbb{S}^{d-1}$
6	Connections between community-based models and RGGs	

2 The Random Geometric Graph Model and Its Variants

The questions that we tackle here can require some additional structure on the model. In this section, we define the variants of the RGG that will be useful for our purpose. Figure 1 shows the connections between these different models.

2.1 (Soft-) Random Geometric Graphs

Definition 1 (Random Geometric Graph: RGG) Let $(\mathcal{X}, \rho)$ be a metric space and m be a Borel probability measure on $\mathcal{X}$. Given a positive real number $r > 0$, the random geometric graph with $n \in \mathbb{N}\backslash\{0\}$ points and level $r > 0$ is the random graph G such that

- The n vertices $X_1, \dots, X_n$ of G are chosen randomly in $\mathcal{X}$ according to the probability measure $m^{\otimes n}$ on $\mathcal{X}^n$.
- For any $i, j \in [n]$ with $i \neq j$, an edge between X_i and X_j is present in G if and only if $\rho(X_i, X_j) \leq r$.

We denote $\mathrm{RGG}(n, m, (\mathcal{X}, \rho))$ the distribution of such random graphs.

Motivated by wireless *ad hoc* networks, Soft-RGGs have been more recently introduced (see [81]). In such models, we are given some function $H : \mathbb{R}_+ \to [0, 1]$ and two nodes at distance ρ in the graph are connected with probability $H(\rho)$.

Definition 2 (Soft Random Geometric Graph: Soft-RGG) Let $(\mathcal{X}, \rho)$ be a metric space, m be a Borel probability measure on $\mathcal{X}$ and consider some function $H : \mathbb{R}_+ \to [0, 1]$. The Soft (or probabilistic) Random Geometric Graph with $n \in \mathbb{N}\backslash\{0\}$ points with connection function H is the random graph G such that

- the n vertices $X_1, \dots, X_n$ of G are chosen randomly in $\mathcal{X}$ according to the probability measure $m^{\otimes n}$ on $\mathcal{X}^n$.
- for any $i, j \in [n]$ with $i \neq j$, we draw an edge between nodes X_i and X_j with probability $H\big(\rho(X_i, X_j)\big)$.

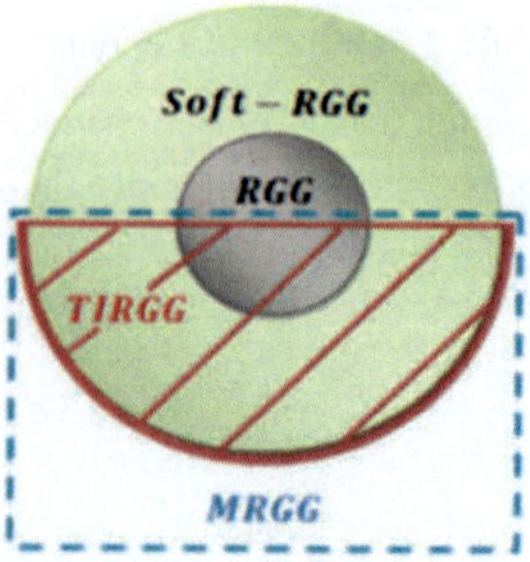

Fig. 1 Venn diagram of the different random graph models

We denote Soft-RGG$(n, m, (\mathscr{X}, \rho))$ the distribution of such random graphs.

Note that the RGG model with level $r > 0$ is a particular case of the Soft-RGG model where the connection function H is chosen as $\rho \mapsto \mathbb{1}_{\rho \leq r}$. The obvious next special case to consider of Soft-RGG is the so-called percolated RGG introduced in [75] which is obtained by retaining each edge of a RGG of level $r > 0$ with probability $p \in (0, 1)$ (and discarding it with probability $1 - p$). This reduces to consider the connection function $H : \rho \mapsto p \times \mathbb{1}_{\rho \leq r}$. Particular common choices of connection function are the *Rayleigh fading* activation functions which take the form

$$H^{Rayleigh}(\rho) = \exp\left[-\zeta \left(\frac{\rho}{r}\right)^{\eta}\right], \quad \zeta > 0, \eta > 0.$$

We refer to [35] and references therein for a nice overview of Soft-RGGs in particular the most classical connection functions and the question of connectivity in the resulting graphs.

2.2 Translation Invariant Random Geometric Graphs

One possible nonparametric generalization of the (Soft)-RGG model is given by the W random graph model (see, for example, [38]) based on the notion of graphon. In this model, given latent points $x_1, \ldots, x_n$ uniformly and independently sampled in $[0, 1]$, the probability to draw an edge between i and j is $\Theta_{i,j} := W(x_i, x_j)$ where W is a symmetric function from $[0, 1]^2$ onto $[0, 1]$, referred to as a graphon. Hence, the adjacency matrix A of this graph satisfies

$$\forall i, j \in [n], \quad A_{i,j} \sim \text{Ber}(\Theta_{i,j}),$$

where for any $p \in [0, 1]$, $\text{Ber}(p)$ is the Bernoulli distribution with parameter p.

Remark Let us point out that graphon models can also be defined by replacing the latent space $[0, 1]$ by the Euclidean sphere $\mathbb{S}^{d-1} := \{x \in \mathbb{R}^d \mid \|x\|_2 = 1\}$ in which case latent points are sampled independently and uniformly on $\mathbb{S}^{d-1}$.

This model has been widely studied in the literature (see [70]), and it is now well-known that, by construction, graphons are defined on an equivalent class *up to a measure preserving homomorphism*. More precisely, two graphons U and W define the same probability distribution if and only if there exist measure preserving maps $\varphi, \psi : [0, 1] \to [0, 1]$ such that $U(\varphi(x), \varphi(y)) = W(\psi(x), \psi(y))$ almost everywhere. Hence, it can be challenging to have a simple description from an observation given by sampled graph—since one has to deal with all possible composition of a bivariate function by any measure preserving homomorphism. Such difficulty arises in [94] or in [62] that use, respectively, maximum likelihood and least square estimators to approximate the graphon W from the adjacency matrix A. In those works, the error measures are based on the so-called *cut-distance* that is defined as an infimum over all measure-preserving transformations. This statistical issue motivates the introduction of (Soft)-RGGs with latent metric spaces for which the distance is invariant by translation (or conjugation) of pairs of points. This natural assumption leads to consider that the latent space has some group structure, namely, it is a compact Lie group or some compact symmetric space.

Definition 3 (Translation Invariant Random Geometric Graph: TIRGG) Let (S, γ) be a compact Lie group with an invariant Riemannian metric γ normalized so that the range of γ equals $[0, \pi]$. Let m be the uniform probability measure on S and let us consider some map $\mathbf{p} : [-1, 1] \to [0, 1]$, called the envelope function. The Translation Invariant Random Geometric Graph with $n \in \mathbb{N}\backslash\{0\}$ points is the random graph G such that

- The n vertices $X_1, \dots, X_n$ of G are chosen randomly in S according to the probability measure $m^{\otimes n}$ on S^n.
- For any $i, j \in [n]$ with $i \neq j$, we draw an edge between nodes X_i and X_j with probability $\mathbf{p}\left(\cos \gamma(X_i, X_j)\right)$.

In Sect. 4, we present recent results regarding nonparametric estimation in the TIRGG model with $S := \mathbb{S}^{d-1}$ the Euclidean sphere of dimension d from the observation of the adjacency matrix. A related question was addressed in [62] where the authors derived sharp rates of convergence for the L^2 loss for the stochastic block model (which belongs to the class of graphon models). Let us point out that a general approach to control the L^2 loss between the probability matrix and a eigenvalue-tresholded version of the adjacency matrix is the USVT method introduced by Chatterjee [28], which was further investigated by Xu [98]. In Sect. 4, another line of work is presented to estimate the envelope function $\mathbf{p}$ where the difference between the adjacency matrix and the *matrix of probabilities* Θ is controlled in operator norm. The cornerstone of the proof is the convergence of the spectrum of the matrix of probabilities toward the spectrum of some integral operator associated with the envelope function $\mathbf{p}$. Based on the analysis of [63], the proof of this convergence includes in particular matrix Bernstein inequality from [90] and concentration inequality for order 2 U-statistics with bounded kernels that was first studied by Arcones and Gine [6] and remains an active field of research (see [52, 57], or [60]).

2.3 *Markov Random Geometric Graphs*

In the following, we will refer to *growth models* to denote random graph models in which a node is added at each new time step in the network and is connected to other vertices in the graph according to some probabilistic rule that needs to be specified. In the last decade, growth models for random graphs with a spatial structure have gained an increased interest. One can mention [61, 78], and [99] where geometric variants of the preferential attachment model are introduced with one new node entering the graph at each time step. More recently, [96] and [95] studied a growing variant of the RGG model. Note that in the latter works, the birth time of each node is used in the connection function while nodes are still sampled independently in $\mathbb{R}^2$. Still motivated by nonparametric estimation, the TIRGG model can be extended to a growth model by considering a Markovian sampling scheme of the latent positions. Considering a Markovian latent dynamic can be relevant to model customer behavior for item recommendation or to study bird migrations where animals have regular seasonal movement between breeding and wintering grounds [cf. 39].

Definition 4 (Markov Random Geometric Graph: MRGG) Let (S, γ) be a compact Lie group with an invariant Riemannian metric γ normalized so that the range of γ equals $[0, \pi]$. Let us consider some map $\mathbf{p} : [-1, 1] \to [0, 1]$, called the envelope function. The Markov Random Geometric Graph with $n \in \mathbb{N}\backslash\{0\}$ points is the random graph G such that

- The sequence of n vertices $(X_1, \ldots, X_n)$ of G is a Markov chain on S.
- For any $i, j \in [n]$ with $i \neq j$, we draw an edge between nodes X_i and X_j with probability $\mathbf{p}\left(\cos \gamma(X_i, X_j)\right)$.

In Sect. 5, we shed light on a recent work from [40] that achieves nonparametric estimation in MRGGs on the Euclidean sphere of dimension d. The theoretical study of such graphs becomes more challenging because of the dependence induced by the latent Markovian dynamic. Proving the consistency of the nonparametric estimator of the envelope function $\mathbf{p}$ proposed in Sect. 5 requires in particular a new concentration inequality for U-statistics of order 2 of uniformly ergodic Markov chains. By solving link prediction problems, [40] also reveal that MRGGs are convenient tools to extract temporal information on growing graphs with an underlying spatial structure.

2.4 *Other Model Variants*

Choice of the Metric Space The Euclidean sphere and the unit square in $\mathbb{R}^d$ are the most studied latent spaces in the literature for RGGs. By the way, [3] offers an interesting comparison of the different topological properties of RGGs working on one or the other of these two spaces. Nevertheless, one can find variants such as in [4] where Euclidean balls are considered. More recently, some researchers left the Euclidean case to consider negatively curved—i.e., hyperbolic—latent spaces. Random graphs with a hyperbolic latent space seem promising to faithfully model real-world networks. Actually, [65] showed that the RGG built on the hyperbolic geometry is a scale-free network, that is the proportion of node of degree k is of order $k^{-\gamma}$ where γ is between 2 and 3. The scale-free property is found in the most part of real networks as highlighted by Xie et al. [96].

Different Degree Distributions It is now well-known that the average degree of nodes in random graph models is a key property for their statistical analysis. Let us highlight some important regimes in the random graph community that will be useful in this paper. The dense regime corresponds to the case where the expected normalized degree of the nodes (i.e., degree divided by n) is independent of the number of nodes in the graph. The other two important regimes are the relatively sparse and the sparse regimes where the average degree of nodes scales, respectively, as $\log(n)/n$ and $1/n$ with the number of nodes n. A direct and important consequence of these definitions is that in the (relatively) sparse regime, the envelope function $\mathbf{p}$ from Definitions 3 and 4 depends on n, while it remains independent of n in the dense regime. Similarly, in the (relatively) sparse regime, the radius threshold r from Definition 1 (resp. the connection function H from Definition 2) depends on n contrary to the dense regime.

3 Detecting Geometry in RGGs

To quote [17], *"One of the main aims of the theory of random graphs is to determine when a given property is likely to appear."* In this direction, several works tried to identify structure in networks through testing procedure; see, for example, [23, 50], or [49]. Regarding RGGs, most of the results have been established in the low-dimensional regime $d \leq 3$ [12, 77, 80, 81]. Goel et al. [53] proved in particular that all monotone graph properties (i.e., property preserved when adding edges to the graph) have a sharp threshold for RGGs that can be distinguished from the one of Erdös-Rényi random graphs in low dimensions. However, applications of RGGs to cluster analysis and the interest in the statistics of high-dimensional data sets have motivated the community to investigate the properties of RGGs in the case where $d \to \infty$. If the ambitious problem of recognizing if a graph can be realized as a geometric graph is known to be NP-hard [24], one can take a step back and wonder if a given RGG still carries some spatial information as $d \to \infty$ or if geometry is lost in high dimensions (see Eq. (1) for a formal definition), a problem known as geometry detection. In the following, we present some recent results related to geometry detection in RGGs with latent space the Euclidean sphere $\mathbb{S}^{d-1}$, and we highlight several interesting directions for future research.

Notations Given two sequences $(a_n)_{n\in\mathbb{N}}$ and $(b_n)_{n\in\mathbb{N}}$ of positive numbers, we write $a_n = \mathcal{O}_n(b_n)$ or $b_n = \Omega_n(a_n)$ if the sequence $(a_n/b_n)_{n\geq 0}$ is bounded, and we write $a_n = o_n(b_n)$ or $b_n = \omega_n(a_n)$ if $a_n/b_n \underset{n\to+\infty}{\to} 0$. We further denote $a_n = \Theta(b_n)$ if $a_n = \mathcal{O}_n(b_n)$ and $b_n = \mathcal{O}_n(a_n)$. In the following, we will denote $G(n, p, d)$ the distribution of random graphs of size n where nodes $X_1, \dots, X_n$ are sampled uniformly on $\mathbb{S}^{d-1}$ and where distinct vertices $i \in [n]$ and $j \in [n]$ are connected by an edge if and only if $\langle X_i, X_j\rangle \geq t_{p,d}$. The threshold value $t_{p,d} \in [-1, 1]$ is such that $\mathbb{P}\left(\langle X_1, X_2\rangle \geq t_{p,d}\right) = p$. Note that $G(n, p, d)$ is the distribution of RGGs on $(\mathbb{S}^{d-1}, \|\cdot\|)$ sampling nodes uniformly with connection function $H : t \mapsto$

$\mathbb{1}_{t \leq \sqrt{2-2t_{p,d}}}$. In the following, we will also use the notation $G(n, d, p)$ to denote a graph sampled from this distribution. We also introduce some definitions of standard information theoretic tools with Definition 5.

Definition 5 Let us consider two probability measures P and Q defined on some measurable space $(E, \mathscr{E})$. The total variation distance between P and Q is given by

$$\mathrm{TV}(P, Q) := \sup_{A \in \mathscr{E}} |P(A) - Q(A)|.$$

Assuming further that $P \ll Q$ and denoting dP/dQ the density of P with respect to Q,

- The χ^2-divergence between P and Q is defined by

$$\chi^2(P, Q) := \int_E \left(\frac{dP}{dQ} - 1 \right)^2 dQ.$$

- The Kullback-Leibler divergence between P and Q is defined by

$$\mathrm{KL}(P, Q) := \int_E \log \left(\frac{dP}{dQ} \right) dP.$$

Considering that both p and d depend on n, we will say in this paper that *geometry is lost* if the distributions $G(n, p)$ and $G(n, p, d)$ are indistinguishable, namely, if

$$\mathrm{TV}(G(n, p), G(n, p, d)) \to 0 \quad \text{as } n \to \infty. \tag{1}$$

3.1 Detecting Geometry in the Dense Regime

Devroye et al. [37] is the first to consider the case where $d \to \infty$ in RGGs. In this paper, the authors proved that the number of cliques in the dense regime in $G(n, p, d)$ is close to the one of $G(n, p)$ provided $d \gg \log n$ in the asymptotic $d \to \infty$. This work allowed them to show the convergence of the total variation (see Definition 5) between RGGs and Erdös-Renyi graphs as $d \to \infty$ for fixed p and n. Bubeck et al. [26] closed the question of geometry detection in RGGs in the dense regime showing that a phase transition occurs when d scales as n^3 as stated by Theorem 1.

Theorem 1 ([26, Theorem 1])

(i) Let $p \in (0, 1)$ be fixed, and assume that $d/n^3 \to 0$. Then

$$\mathrm{TV}(G(n, p), G(n, p, d)) \to 1 \quad \text{as } n \to \infty.$$

(ii) Furthermore, if $d/n^3 \to \infty$, then

$$\sup_{p\in(0,1)} \mathrm{TV}(G(n,p), G(n,p,d)) \to 0 \quad \textit{as } n \to \infty.$$

The proof of Theorem 1.(i) relies on a count of *signed* triangles in RGGs. Denoting A the adjacency matrix of the RGG, the number of triangles in A is $\mathrm{Tr}(A^3)$, while the total number of signed triangles is defined as

$$\begin{aligned}\tau(G(n,p,d)) &:= \mathrm{Tr}((A-p(J-I))^3)\\ &= \sum_{\{i,j,k\}\in\binom{[n]}{3}} \left(A_{i,j}-p\right)\left(A_{i,k}-p\right)\left(A_{j,k}-p\right),\end{aligned}$$

where I is the identity matrix and $J \in \mathbb{R}^{n\times n}$ is the matrix with every entry equals to 1. The analogous quantity in Erdös Renyi graphs $\tau(G(n,p))$ is defined similarly. Bubeck et al. [26] showed that the variance of $\tau(G(n,p,d))$ is of order n^3, while the one of the number of triangles is of order n^4. This smaller variance for signed triangles is due to the cancellations introduced by the centering of the adjacency matrix. Lemma 1 provides the precise bounds obtained on the expectation and the variance of the statistic of signed triangles. Theorem 1.(i) follows from the lower bounds (resp. the upper bounds) on the expectations (resp. the variances) of $\tau(G(n,p))$ and $\tau(G(n,p,d))$ presented in Lemma 1.

Lemma 1 ([26, Section 3.4]) *For any $p \in (0,1)$ and any $n, d \in \mathbb{N}\backslash\{0\}$ it holds*

$$\mathbb{E}[\tau(G(n,p))] = 0, \quad \mathbb{E}[\tau(G(n,p,d))] \geq \binom{n}{3}\frac{C_p}{\sqrt{d}}$$

$$\textit{and} \quad \max\left\{\mathbb{V}ar\left[\tau(G(n,p))\right], \mathbb{V}ar\left[\tau(G(n,p,d))\right]\right\} \leq n^3 + \frac{3n^4}{d},$$

where $C_p > 0$ is a constant depending only on p.

Let us now give an overview of the proof of the indistinguishable part of Theorem 1. Bubeck et al. [26] proved that in the dense regime, the phase transition for geometry detection occurs at the regime at which Wishart matrices becomes indistinguishable from GOEs (Gaussian Orthogonal Ensemble). In the following, we draw explicitly this link in the case $p = 1/2$.

An $n \times n$ Wishart matrix with d degrees of freedom is a matrix of inner products of n d-dimensional Gaussian vectors denoted by $W(n,d)$, while an $n \times n$ GOE random matrix is a symmetric matrix with i.i.d. Gaussian entries on and above the diagonal denoted by $M(n)$. Let $\mathbb{X}$ be an $n \times d$ matrix where the entries are i.i.d. standard normal random variables, and let $W = W(n,d) = \mathbb{X}\mathbb{X}^\top$ be the corresponding $n \times n$ Wishart matrix. Then recalling that for $X_1 \sim \mathcal{N}(0, I_d)$ a standard gaussian vector of dimension d, $X_1/\|X_1\|_2$ is uniformly distributed on the sphere $\mathbb{S}^{d-1}$, we get that the $n \times n$ matrix A defined by

$$\forall i, j \in [n], \quad A_{i,j} = \begin{cases} 1 \text{ if } W_{i,j} \geq 0 \text{ and } i \neq j \\ 0 \text{ otherwise.} \end{cases}$$

has the same distribution as the adjacency matrix of a graph sampled from $G(n, 1/2, d)$. We denote H the map that takes W to A. Analogously, one can prove that $G(n, 1/2)$ can be seen as a function of an $n \times n$ GOE matrix. Let $M(n)$ be a symmetric $n \times n$ random matrix where the diagonal entries are i.i.d. normal random variables with mean zero and variance 2, and the entries above the diagonal are i.i.d.standard normal random variables, with the entries on and above the diagonal all independent. Then $B = H(M(n))$ is distributed as the adjacency matrix of $G(n, 1/2)$. We then get

$$\begin{aligned} \mathrm{TV}\left(G(n, 1/2, d), G(n, 1/2)\right) &= \mathrm{TV}\left(H(W(n, d)), H(M(n))\right) \\ &\leq \mathrm{TV}\left(W(n, d)), M(n)\right). \end{aligned} \tag{2}$$

If a simple application of the multivariate central limit theorem proves that the right hand side of (2) goes to zero as $d \to \infty$ for fixed n, more work is necessary to address the case where $d = d(n) = \omega_n(n^3)$ and $n \to \infty$. The distributions of $W(n, d)$ and $M(n)$ are known and allow explicit computations leading to Theorem 2. This proof can be adapted for any $p \in (0, 1)$ leading to Theorem 1.(ii) from (2).

Theorem 2 ([26, Theorem 7]) *Define the random matrix ensembles $W(n, d)$ and $M(n)$ as above. If $d/n^3 \to \infty$, then*

$$\mathrm{TV}(W(n, d), M(n)) \to 0.$$

Extensions Considering $\mathbb{R}^d$ as latent space endowed with the Euclidean metric, [25] extended Theorem 2 and proved an information theoretic phase transition. To give an overview of their result, let us consider the $n \times n$ Wigner matrix $\mathscr{M}_n$ with zeros on the diagonal and i.i.d. standard Gaussians above the diagonal. For some $n \times d$ matrix $\mathbb{X}$ with i.i.d. entries from a distribution μ on $\mathbb{R}^d$ that has mean zero and variance 1, we also consider the following rescaled Wishart matrix associated with $\mathbb{X}$

$$\mathscr{W}_{n,d} := \frac{1}{\sqrt{d}} \left(\mathbb{X}\mathbb{X}^\top - \mathrm{diag}(\mathbb{X}\mathbb{X}^\top) \right),$$

where the diagonal was removed. Using a high-dimensional entropic central limit theorem, [25] proved Theorem 3 which implies that geometry is lost in $RGG(n, \mu, (\mathbb{R}^d, \|\cdot\|))$ as soon as $d \gg n^3 \log^2(d)$ provided that the measure μ is sufficiently smooth (namely log-concave) and the rate is tight up to logarithmic factors. We refer to [85] for a friendly presentation of this result. Note that the comparison between Wishart and GOE matrices also naturally arise when dealing with covariance matrices. For example, Theorem 2 was used in [19] to study the informational-computational tradeoff of sparse principal component analysis.

Theorem 3 ([25, Theorem 1]) *If the distribution* μ *is log-concave and* $\frac{d}{n^3 \log^2(d)} \to \infty$*, then* $\mathrm{TV}(\mathscr{W}_{n,d}, \mathscr{M}_n) \to 0$.

On the other hand, if μ *has a finite fourth moment and* $\frac{d}{n^3} \to 0$*, then* $\mathrm{TV}(\mathscr{W}_{n,d}, \mathscr{M}_n) \to 1$.

3.2 Failure to Extend the Proof Techniques to the Sparse Regime

Bubeck et al. [26] provided a result in the sparse regime where $p = \frac{c}{n}$ showing that one can distinguish between $G(n, \frac{c}{n})$ and $G(n, \frac{c}{n}, d)$ as long as $d \ll \log^3 n$. The authors conjectured that this rate is tight for the sparse regime (see Conjecture 1).

Conjecture 1 ([26, Conjecture 1]) Let $c > 0$ be fixed, and assume that $d/\log^3(n) \to \infty$. Then,

$$\mathrm{TV}\left(G\left(n, \frac{c}{n}\right), G\left(n, \frac{c}{n}, d\right)\right) \to 0 \quad \text{as } n \to \infty.$$

The testing procedure from [26] to prove the distinguishability result in the sparse regime was based on a simple counting of triangles. Indeed, when p scales as $\frac{1}{n}$, the signed triangle statistic τ does not give significantly more power than the triangle statistic which simply counts the number of triangles in the graph. Recently, [8] provided interesting results that give credit to Conjecture 1. First, they proved that in the sparse regime, the clique number of $G(n, p, d)$ is almost surely at most 3 under the condition $d \gg \log^{1+\varepsilon} n$ for any $\varepsilon > 0$. This means that in the sparse regime, $G(n, p, d)$ does not contain any complete subgraph larger than a triangle like Erdös-Renyi graphs. Hence, it is hopeless to prove that Conjecture 1 is false considering the number of k-cliques for $k \geq 4$. Nevertheless, one could believe that improving the work of [26] by deriving sharper bounds on the number of 3-cliques (i.e., the number of triangles), it could possible to statistically distinguish between $G(n, p, d)$ and $G(n, p)$ in the sparse regime even for some $d \gg \log^3 n$. In a regime that can be made arbitrarily close to the sparse one, [8] proved that this is impossible as stated by Theorem 4.

Theorem 4 ([8, Theorem 5]) *Let us suppose that* $d \gg \log^3 n$ *and* $p = \theta(n)/n$ *with* $n^m \leq \theta(n) \ll n$ *for some* $m > 0$*. Then, the expected number of triangles—denoted* $\mathbb{E}[T(n, p, d)]$*—in RGGs sampled from* $G(n, p, d)$ *is of order* $\binom{n}{3} p^3$*, meaning that there exist two universal constants* $c, C > 0$ *such that for n large enough it holds*

$$c\binom{n}{3} p^3 \leq \mathbb{E}[T(n, p, d)] \leq C\binom{n}{3} p^3.$$

In a nutshell, the work from [8] suggests that a negative result regarding Conjecture 1 cannot be obtained using statistics based on clique numbers. This discussion naturally gives rise to the following more general question:

Given a random graph model with n nodes, latent geometry in dimension $d = d(n)$ and edge density $p = p(n)$, for what triples (n, d, p) is the model $G(n, p, d)$ indistinguishable from $G(n, p)$? $(\mathscr{Q})$

3.3 Toward the Resolution of Geometry Detection

3.3.1 A First Improvement When $d > n$

A recent work from [21] tackled the general problem $(\mathscr{Q})$ and proved Theorem 5.

Theorem 5 ([21, Theorem 2.4]) *Suppose $p = p(n) \in (0, 1/2]$ satisfies that $n^{-2}\log n = \mathcal{O}_n(p)$ and*

$$d \gg \min\left\{pn^3 \log p^{-1}\ ,\ p^2 n^{7/2}(\log n)^3\sqrt{\log p^{-1}}\right\},$$

where d also satisfies that $d \gg n\log^4 n$. Then,

$$\mathrm{TV}(G(n, p), G(n, p, d)) \to 0 \quad \text{as } n \to \infty.$$

Remarks In the dense regime, Theorem 5 recovers the optimal guarantee from Theorem 1. In the sparse regime, Theorem 5 states that if $d \gg n^{3/2}(\log n)^{7/2}$, then geometry is lost in $G(n, \frac{c}{n}, d)$ (where $c > 0$). This result improves the work from [26]. Nevertheless, regarding Conjecture 1, it remains a large gap between the rates $\log^3 n$ and $n^{3/2}(\log n)^{7/2}$ where nothing is known up to date. Let us sketch the main elements of the proof of Theorem 5. In the following, we denote $G = G(n, p, d)$ with set of edges $E(G)$, and for any $i, j \in [n]$, $i \neq j$, we denote $G_{\sim\{i,j\}}$ the set of edges other than $\{i, j\}$ in G. One first important step of their approach is the following tenzorization Lemma for the Kullback-Leibler divergence.

Lemma 2 ([64, Lemma 3.4]) *Let us consider $(X, \mathscr{B})$ a measurable space with X a Polish space and $\mathscr{B}$ its Borel σ-field. Consider some probability measure μ on the product space X^k with $\mu = \mu_1 \otimes \mu_2 \otimes \cdots \otimes \mu_k$. Then, for any other probability measure ν on X^k, it holds*

$$\mathrm{KL}(\nu||\mu) \le \sum_{i=1}^{k} \mathbb{E}_{x\sim\nu}\left[\mathrm{KL}\left(\nu_i(\cdot|x_{\sim i})||\mu_i\right)\right],$$

where ν_i *is the probability distribution corresponding to the* i*-th marginal of* ν *and where* $x_{\sim i} := (x_1, \ldots, x_{i-1}, x_{i+1}, \ldots, x_k)$.

$$\begin{aligned}
&2\mathrm{TV}(G(n,p,d), G(n,p))^2 \\
&\quad \le \mathrm{KL}(G(n,p,d)||G(n,p)) \qquad \text{from Pinsker's inequality} \\
&\quad \le \sum_{1\le i<j\le n} \mathbb{E}\left[\mathrm{KL}\left(\mathscr{L}(\mathbb{1}_{\{i,j\}\in E(G)}|\sigma(G_{\sim\{i,j\}}))||\mathrm{Bern}(p)\right)\right] \qquad \text{from Lemma 2} \\
&\quad \le \binom{n}{2} \times \mathbb{E}\left[\chi^2\left(\mathscr{L}(\mathbb{1}_{e_0\in E(G)}|\sigma(G_{\sim e_0})), \mathrm{Bern}(p)\right)\right] \\
&\quad = \binom{n}{2} \times \mathbb{E}\left[\frac{(Q-p)^2}{p(1-p)}\right],
\end{aligned}$$

where $Q := \mathbb{P}\left(e_0 \in E(G)|G_{\sim e_0}\right)$ is a $\sigma(G_{\sim e_0})$-measurable random variable corresponding to the probability that a specific edge is included in the graph given the rest of the graph. The proof then consists in showing that with high probability, Q concentrates near p. To do so, they use a coupling argument that gives an alternative way to generate X_1 that provides a direct description of $\mathbb{1}_{e_0\in E(G)}$ in terms of the random variables introduced in the coupling. If this step may seem computationally involved, it is not conceptually difficult since it turns out to be a simple re-parametrization of the problem. An integration of this concentration result for Q implies that the convergence of Theorem 5 holds when $d \gg pn^3 \log p^{-1}$. To get the convergence result in the regime where $d \gg p^2 n^{7/2}(\log n)^3\sqrt{\log p^{-1}}$ – which gives the improvement over [26] in the sparse case – one additional step of coupling is required. More precisely, they decompose $\mathbb{E}[(Q-p)^2]$ as $\mathbb{E}[(Q-p)\times(Q-p)]$. The previous coupling argument gives a concentration inequality allowing to bound with high probability the first term $|Q-p|$. It remains then to upper bound $\mathbb{E}[|Q-p|]$ which relies on a simple observation given by the following proposition.

Proposition 1 ([21, Proposition 5.3]) *Let* $\nu_{\sim e_0}$ *denote the marginal distribution of* G *restricted to all edges that are not* e_0*, and let* $\nu^+_{\sim e_0}$ *denote the distribution of* G *conditioned on the event* $e_0 \in E(G)$*. It holds*

$$\mathbb{E}[|Q-p|] = 2p \times \mathrm{TV}\left(\nu^+_{\sim e_0}, \nu_{\sim e_0}\right). \tag{3}$$

The proof is then concluded by using another coupling argument between $\nu^+_{\sim e_0}$ and $\nu_{\sim e_0}$ to upper bound the total variation distance involved in Eq. (3), and we give a sketch of proof in the following. Given latent positions $X_1, \ldots, X_n$ uniformly and independently sampled on $\mathbb{S}^{d-1}$, we can consider without loss of generality that

$X_1 = (1, 0, \dots 0)$. Denoting $X_2 = (X_{2,j})_{j\in[d]}$ and φ_d the density of $X_{2,1}$,[1] one can define $\gamma = \sqrt{\frac{1-\tau^2}{1-X_{2,1}^2}}$ and $X_2^+ := (\tau, \gamma X_{2,2}, \dots, \gamma X_{2,d})$ where τ is a random variable in $[-1, 1]$ with density $\varphi_{d,p}^+(x) = p^{-1}\mathbb{1}_{x\geq t_{p,d}}\varphi_d(x)$. Denoting further $G_{\sim e_0}$ (resp. $G_{\sim e_0}^+$) the RGG with threshold $t_{p,d}$ induced by the latent points $(X_i)_{i\in[n]}$ (resp. $(X_1, X_2^+, X_3, \dots, X_n)$) without the edge $e_0 = \{1, 2\}$, $G_{\sim e_0}$ (resp.$G_{\sim e_0}^+$) is distributed as $\nu_{\sim e_0}$ (resp. $\nu_{\sim e_0}^+$). Hence, it holds

$$\begin{aligned}\mathbb{E}[|Q - p|] &\leq 2p \times \mathrm{TV}\left(\nu_{\sim e_0}^+, \nu_{\sim e_0}\right) \leq 2p \times \mathbb{P}(G_{\sim e_0} \neq G_{\sim e_0}^+)\\ &\leq 2p \sum_{i=3}^{n} \mathbb{P}\left(\mathbb{1}_{\langle X_2, X_i\rangle \geq t_{p,d}} \neq \mathbb{1}_{\langle X_2^+, X_i\rangle\geq t_{p,d}}\right).\end{aligned}$$

The proof is concluded using standard concentration arguments.

3.3.2 Reaching the Polylogarithmic Regime

Very recently, [69] came with novels ideas and improved upon the previous bounds for geometry detection by polynomial factors in the sparse regime. This significant breakthrough presented in Theorem 6 almost solves Conjecture 1.

Theorem 6 ([69, Theorem 1.2]) *For any fixed constant $c \geq 1$, if $d \gg \log^{36} n$, then*

$$\mathrm{TV}\left(G\left(n, \frac{c}{n}\right), G\left(n, \frac{c}{n}, d\right)\right) \to 0 \quad \text{as } n \to \infty.$$

The authors do not limit their analysis to the sparse regime but also provide results holding for any regime interpolating between the sparse and the dense cases as shown with Theorem 7.

Theorem 7 ([69, Theorem 1.1 and Lemma A.1])

- *For any fixed constant $c > 0$, if $\frac{c}{n} < p < \frac{1}{2}$ and $d \gg p^2 n^3$, then*

$$\mathrm{TV}\left(G\left(n, p\right), G\left(n, p, d\right)\right) \to 0 \quad \text{as } n \to \infty.$$

- *If $\frac{1}{n^2} \ll p \leq 1 - \delta$ for any fixed constant $\delta > 0$, then as long as $d \ll (nH(p))^3$,*

$$\mathrm{TV}\left(G\left(n, \frac{c}{n}\right), G\left(n, \frac{c}{n}, d\right)\right) \to 1 \quad \text{as } n \to \infty,$$

[1] I.e., φ_d is the density of a one-dimensional marginal of a uniform random point on $\mathbb{S}^{d-1}$.

where $H(p) = p \log \frac{1}{p} + (1-p) \log \frac{1}{1-p}$ is the binary entropy function. This result can be achieved using the signed triangle statistic following an approach strictly analogous to [26].

Liu et al. [69] extend the work from [26] and prove that the signed statistic distinguishes between $G(n, p)$ and $G(n, p, d)$ not only in the sparse and dense cases but also for most p, as long as $d \ll (nH(p))^3$. We provide in the Appendix "Appendix: Outline of the Proofs of Theorems 6 and 7" a synthetic description of the proofs of Theorems 6 and 7. Let us mention that the proofs rely on a new concentration result for the area of the intersection of a random sphere cap with an arbitrary subset of $\mathbb{S}^{d-1}$, which is established using optimal transport maps and entropy-transport inequalities on the unit sphere. Liu et al. [69] make use of this set-cap intersection concentration lemma for the theoretical analysis of the belief propagation algorithm.

3.4 Open Problems and Perspectives

The main results we have presented so far look as follows:

Task	Current state of knowledge	Ref.
Recognizing if a graph can be realized as a RGG	NP-hard	1998
Testing between $G(n,p,d)$ and $G(n,p)$ in high-dimension for $p \in (0,1)$ fixed	0 — Polynomial time test — n^3 — Undistinguishable — d	2016
Testing between $G(n,\frac{c}{n},d)$ and $G(n,\frac{c}{n})$ in high-dimension for $c > 0$	0 — Polynomial time test — $\log^3 n$ — ? — $\log^{36} n$ — Undistinguishable — d	2016 & 2021

With new proof techniques based on combinatorial arguments, direct couplings, and applications of information inequalities, [21] were the first to make a progress toward Conjecture 1. Nevertheless, their proof was heavily relying on a coupling step involving a De Finetti-type result that requires the dimension d to be larger than the number of points n. Liu et al. [69] improved upon the previous bounds by polynomial factors with innovative proof arguments. In particular, their analysis makes use of the belief propagation algorithm and the cavity method and relies on a new sharp estimate for the area of the intersection of a random sphere cap with an arbitrary subset of $\mathbb{S}^{d-1}$. The proof of this new concentration result is an application of optimal transport maps and entropy-transport inequalities. Despite this recent progress, a large span of research directions remain open, and we discuss some of them in the following:

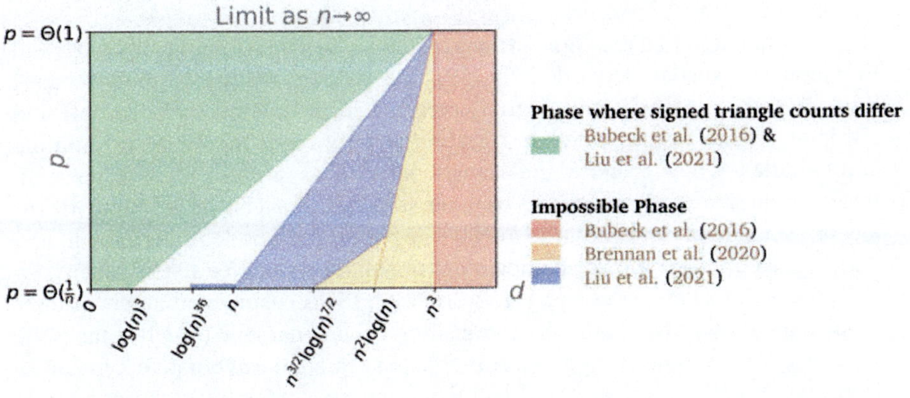

Fig. 2 Phase diagram of the (d, p) regions where geometry detection (on the Euclidean sphere) is known to be information theoretically impossible or possible (in polynomial time). Note that the figure only presents a simplified illustration of the current state of knowledge for the problem of geometry detection on $\mathbb{S}^{d-1}$ since the true scales are not respected

1. *Closing the gaps for geometry detection on the Euclidean sphere* $\mathbb{S}^{d-1}$.
   Figure 2 shows that there are still important research directions to investigate to close the question of geometry detection regarding RGGs on $\mathbb{S}^{d-1}$. First, in the sparse regime, it would be desirable to finally know if Conjecture 1 is true, meaning that the phase transition occurs when the latent dimension is of the order of $\log^3 n$. It could be fruitful to see if some steps in the approach from [69] could be sharpened in order to get down to the threshold $\log^3 n$. A question that seems even more challenging is to understand what happens in the regimes where $p = p(n) \in (\frac{1}{n}, 1)$ and $d = d(n) \in ([H(p)n]^3, p^2n^3)$ (corresponding to the white region on Fig. 2). To tackle this question, one could try to extend the methods used in the sparse case by Liu et al. [69] to denser cases. Another possible approach to close this gap would be to dig deeper into the connections between the Wishart and GOE ensembles. One research direction to possibly improve the existing impossibility results regarding geometry detection would be to avoid the use of the data-processing inequality in Eq. (2) which makes us lose the fact that we do not observe the matrices $W(n, d)$ and $M(n)$ themselves. To some extent, we would like to take into account that some information is lost by observing only the adjacency matrices. In a recent work, [22] made the first step in this direction. They study the total variation distance between the Wishart and GOE ensembles when some given mask is applied beforehand. They proved that the combinatorial structure of the revealed entries, viewed as the adjacency matrix of a graph G, drives the distance between the two distributions of interest. More precisely, they provide regimes for the latent dimension d based exclusively on the number of various small subgraphs in G, for which the total variation distance goes to either 0 or 1 as $n \to \infty$.

2. *How specific is the signed triangle statistic to RGGs?*

   Let us mention that the signed triangle statistic has found applications beyond the scope of spatial networks. In [59], the authors study community-based random graphs (namely, the degree corrected mixed membership model) and are interested in testing whether a graph has only one community or multiple communities. They propose the signed polygon as a class of new tests. In that way, they extend the signed triangle statistic to m-gon in the network for any $m \geq 3$. Contrary to [26], the average degree of each node is not known, and the degree corrected mixed membership model allows degree heterogeneity. In [59], the authors define the signal-to-noise ratio (SNR) using parameters of their model, and they prove that a phase transition occurs, namely, (i) when the SNR goes to $+\infty$, the signed polygon test is able to separate the alternative hypothesis from the null asymptotically and (ii) when the SNR goes to 0 (and additional mild conditions), then the alternative hypothesis is inseparable from the null.
3. *How the phase transition phenomenon in geometry detection evolves when other latent spaces are considered?*

   This question is related to the robustness of the previous results with respect to the latent space. Inspired by Bubeck et al. [26] and Eldan and Mikulincer [43] provided a generalization of Theorem 1 considering an ellipsoid rather than the sphere $\mathbb{S}^{d-1}$ as latent space. They proved that the phase transition also occurs at n^3 provided that we consider the appropriate notion of dimension which takes into account the anisotropy of the latent structure.

   In [31], the clustering coefficient of RGGs with nodes uniformly distributed on the hypercube shows systematic deviations from the Erdos-Rényi prediction.
4. *What is inherent to the connection function?*

   Considering a fixed number of nodes, [45] use a multivariate version of the central limit theorem to show that the joint probability of rescaled distances between nodes is normal-distributed as $d \to \infty$. They provide a way to compute the correlation matrix. This work allows them to evaluate the average number of M-cliques, i.e., of fully connected subgraphs with M vertices, in high-dimensional RGGs and Soft-RGGs. They can prove that the infinite dimensional limit of the average number of M-cliques in Erdös-Rényi graphs is the same of the one obtained from for Soft-RGGs with a continuous activation function. On the contrary, they show that for classical RGGs, the average number of cliques does not converge to the Erdös-Rényi prediction. This paper leads to think that the behavior of local observables in Soft-RGGs can heavily depend on the connection function considered. The work from [45] is one of the first to address the emerging questions concerning the high-dimensional fluctuations of some statistics in RGGs. If they focused on the number of M-cliques, one can also mention the recent work from [54] that provide a central limit theorem for the edge counting statistic as the space dimension d tends to infinity. Their work shows that the Malliavin–Stein approach for Poisson functionals that was first introduced in stochastic geometry can also be used to deal with spatial random models in high dimensions.

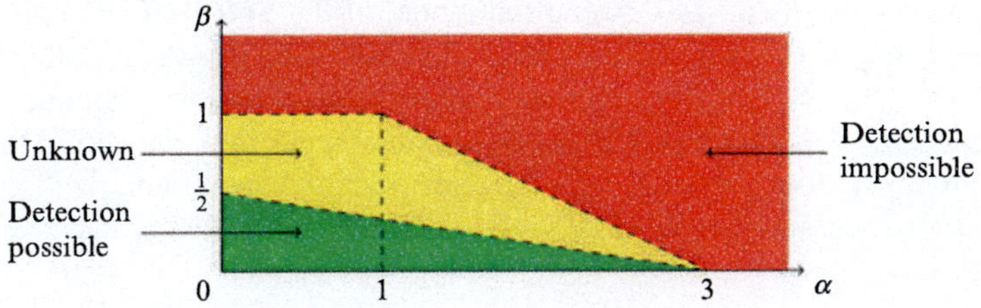

Fig. 3 Phase diagram for detecting geometry in the soft random geometric graph $G(n, p, d, q)$. Here, $d = n^\alpha$ and $q = n^{-\beta}$ for some $\alpha, \beta > 0$

In a recent work, [67] are interested in extending the previous mentioned results on geometry detection in RGGs to Soft-RGGs with some specific connection functions. The authors consider the dense case where the average degree of each node scales with the size of the graph n and study geometry detection with graphs sampled from Soft-RGGs that interpolate between the standard RGG on the sphere $\mathbb{S}^{d-1}$ and the Erdös-Rényi random graph. Hence, the null hypothesis remains that the observed graph G is a sample from $G(n, p)$, while the alternative becomes that the graph is the Soft-RGG where we draw an edge between nodes i and j with probability

$$(1-q)p + q\mathbb{1}_{t_{p,d} \le \langle X_i, X_j \rangle},$$

where $(X_i)_{i \ge 1}$ are randomly and independently sampled on $\mathbb{S}^{d-1}$ and where $q \in [0, 1]$ can be interpreted as the geometric strength of the model. Denoting the random graph model $G(n, p, d, q)$, one can easily notice that $G(n, p, d, 1)$ is the standard RGG on the Euclidean sphere $\mathbb{S}^{d-1}$ while $G(n, p, d, 0)$ reduces to the Erdös-Rényi random graph. Hence, by taking $q = 1$ in Theorem 8, we recover Theorem 1 from [26]. One can further notice that Theorem 8 depicts a polynomial dependency on q for geometry detection but when $q < 1$ there is a gap between the upper and lower bounds as illustrated by Fig. 3 taken from [67]. As stated in [67], *[...] a natural direction of future research is to consider [geometry detection] for other connection functions or underlying latent spaces, in order to understand how the dimension threshold for losing geometry depends on them.*

Theorem 8 ([67, Theorem 1.1]) *Let* $p \in (0, 1)$ *be fixed.*

(i) If $n^3q^6/d \to \infty$, *then*

$$\mathrm{TV}(G(n, p), G(n, p, d, q)) \to 1 \quad \text{as } n \to \infty.$$

(ii) If $nq \to 0$ *or* $n^3q^2/d \to 0$, *then*

$$\mathrm{TV}(G(n, p), G(n, p, d, q)) \to 0 \quad \text{as } n \to \infty.$$

The same authors in [68] extend the model of the Soft-RGG by considering the latent space $\mathbb{R}^d$ where the latent positions $(X_i)_{i\in[n]}$ are i.i.d. sampled with $X_1 \sim \mathcal{N}(0, I_d)$. Two different nodes $i, j \in [n]$ are connected with probability $\mathbf{p}(\langle X_i, X_j\rangle)$ where $\mathbf{p}$ is a monotone increasing connection function. More precisely, they consider a connection function $\mathbf{p}$ parametrized by (i) a cumulative distribution function $F : \mathbb{R} \to [0, 1]$ and (ii) a scalar $r > 0$ and given by

$$\mathbf{p} : t \mapsto F\left(\frac{t - \mu_{p,d,r}}{r\sqrt{d}}\right),$$

where $\mu_{p,d,r}$ is determined by setting the edge density in the graph to be equal to p, namely, $\mathbb{E}\left[\mathbf{p}(\langle X_1, X_2\rangle)\right] = p$. They work in the dense regime by considering that $p \in (0, 1)$ is independent of n. The parameter r encodes the flatness of the connection function and is typically a function of n. The authors prove phase transitions of detecting geometry in this framework in terms of the dimension of the underlying geometric space d and the variance parameter r. The larger the r, the smaller the dimension d at which the phase transition occurs. When $r \underset{n\to\infty}{\to} 0$, the connection function becomes an indicator function and the transition appears at $d \asymp n^3$ (recovering the result from Theorem 1 established for RGGs on the Euclidean sphere).

5. *Suppose that we know that the latent variables are embedded in a Euclidean sphere, can we estimate the dimension d from the observation of the graph?*
   When $p = 1/2$, [26] obtained a bound on the difference of the expected number of signed triangles between consecutive dimensions leading to Theorem 9.

Theorem 9 ([26, Theorem 5]) *There exists a universal constant $C > 0$, such that for all integers n and $d_1 < d_2$, one has*

$$\mathrm{TV}(G(n, 1/2, d_1), G(n, 1/2, d_2)) \geq 1 - C\left(\frac{d_1}{n}\right)^2.$$

The bound provided by Theorem 9 is tight in the sense that when $d \gg n$, $G(n, 1/2, d)$ and $G(n, 1/2, d+1)$ are indistiguishable as proved in [42]. More recently, [5] proposed a method to infer the latent dimension of a Soft-RGG on the Euclidean sphere in the low-dimensional setting. Their approach is proved to correctly recover the dimension d in the relatively sparse regime as soon as the connection function belongs to some Sobolev class and satisfies a spectral gap condition.

6. *Extension to hypergraphs and information-theoretic/computational gaps.*
   Let us recall that a hypergraph is a generalization of a graph in which an edge can join any number of vertices. Extensions of RGGs to hypergraphs have already been proposed in the literature (see, for example, [71]). A nice research

direction would consist in investigating the problem of geometry detection in these geometric representations of random hypergraphs. As already discussed, it has been conjectured that the problem of geometry detection in RGGs on $\mathbb{S}^{d-1}$ does not present a statistical-to-algorithmic gap meaning that whenever it is information theoretically possible to differ $G(n, p, d)$ from $G(n, p)$, we can do it with a computational complexity polynomial in n (using the signed triangle statistic). Dealing with hypergraphs, one can legitimately think that statistical-to-algorithmic gaps could emerge. This intuition is based on the fact that most of the time, going from a matrix problem to a tensor problem brings extra challenges. One can take the example of principal component analysis of Gaussian k-tensors with a planted rank-one spike (cf. [13]). In this problem, we assume that we observe for any $l \in [n]$,

$$\mathbf{Y}^l = \lambda u^{\otimes k} + \mathbf{W}^l,$$

where $u \in \mathbb{S}^{d-1}$ is deterministic, $\lambda \geq 0$ is the signal-to-noise ratio, and where $(\mathbf{W}^l)_{l \in [n]}$ are independent Gaussian k-tensor (we refer to [13] for further details). The goal is to infer the "planted signal" or "spike," u. In the matrix case (i.e., when $k = 2$), whenever the problem is information theoretically solvable, we can also recover the spike with a polynomial time algorithm (using, for example, a spectral method). If we look at the tensor version of this problem where $k \geq 3$, there is a regime of signal-to-noise ratios for which it is information theoretically possible to recover the signal but for which there is no known algorithm to approximate it in polynomial time in n. This is a statistical-to-algorithmic gap, and we refer to [20, Section 3.8] and references therein for more details.

7. *Can we describe the properties of high-dimensional RGGs in the regimes where* $\mathrm{TV}(G(n, p), G(n, p, d)) \to 1$ *as* $n \to \infty$?

   In the low-dimensional case, RGGs have been extensively studied: their spectral or topological properties, chromatic number or clustering number, are now well known (see, e.g., [80, 91]). One of the first work studying the properties of high-dimensional RGGs is [8] where the authors are focused on the clique structure. These questions are essential to understand how good high-dimensional RGGs are as models for the theory of network science.

8. *How to find a relevant latent space given a graph with an underlying geometric structure?*

   As stated in [85], *Perhaps the ultimate goal is to find good representations of network data, and hence to faithfully embed the graph of interest into an appropriate metric space.* This task is known as *manifold learning* in the machine learning community. Recently, [88] proved empirically that the eigenstructure of the Laplacian of the graph provides information on the curvature of the latent space. This is an interesting research direction to propose model selection procedure and infer a relevant latent space for a graph.

4 Nonparametric Inference in RGGs

In this section, we are interested in nonparametric inference in TIRGGs (see Definition 3) on the Euclidean sphere $\mathbb{S}^{d-1}$. The methods presented rely mainly on spectral properties of such random graphs. Note that spectral aspects in (Soft-)RGGs have been investigated for a long time (see, for example, [86]), and it is now well known that the spectra of RGGs are very different from the one of other random graph models since the appearance of particular subgraphs give rise to multiple repeated eigenvalues (see [76] and [16]). Recent works took advantage of the information captured by the spectrum of RGGs to study topological properties such as [2]. In this section, we will see that random matrix theory is a powerful and convenient tool to study the spectral properties of RGGs as already highlighted by Dettmann et al. [36].

4.1 Description of the Model and Notations

We consider a Soft-RGG on the Euclidean sphere $\mathbb{S}^{d-1}$ endowed with the geodesic distance ρ. We consider that the connection function H is of the form $H : t \mapsto \mathbf{p}(\cos(t))$ where $\mathbf{p} : [-1, 1] \to [0, 1]$ is an unknown function that we want to estimate. This Soft-RGG belongs to the class of TIRGG has defined in Sect. 2 and corresponds to a graphon model where the graphon W is given by

$$\forall x, y \in \mathbb{S}^{d-1}, \quad W(x, y) := \mathbf{p}(\langle x, y\rangle).$$

W, viewed as an integral operator on square-integrable functions, is a compact convolution (on the left) operator

$$\mathbb{T}_W : f \in L^2(\mathbb{S}^{d-1}) \mapsto \int_{\mathbb{S}^{d-1}} W(x, \cdot) f(x) \sigma(dx) \in L^2(\mathbb{S}^{d-1}), \tag{4}$$

where σ is the Haar measure on $\mathbb{S}^{d-1}$. The operator $\mathbb{T}_W$ is Hilbert-Schmidt, and it has a countable number of bounded real eigenvalues λ_k^* with zero as the only accumulation point. The eigenfunctions of $\mathbb{T}_W$ have the remarkable property that they do not depend on p (see [30, Lemma 1.2.3]): they are given by the real Spherical Harmonics. We denote $\mathcal{H}_l$ the space of real Spherical Harmonics of degree l with dimension d_l and with orthonormal basis $(Y_{l,j})_{j \in [d_l]}$. We end up with the following spectral decomposition of the *envelope* function $\mathbf{p}$

$$\forall x, y \in \mathbb{S}^{d-1}, \quad \mathbf{p}(\langle x, y\rangle) = \sum_{l \geq 0} p_l^* \sum_{j=1}^{d_l} Y_{l,j}(x) Y_{l,j}(y) = \sum_{l \geq 0} p_l^* c_l G_l^{\beta}(\langle x, y\rangle), \tag{5}$$

where $\lambda^* := (p_0^*, p_1^*, \dots, p_1^*, \dots, p_l^*, \dots, p_l^*, \dots)$ meaning that each eigenvalue p_l^* has multiplicity d_l and G_l^β is the Gegenbauer polynomial of degree l with parameter $\beta := \frac{d-2}{2}$ and $c_l := \frac{2l+d-2}{d-2}$. $\mathbf{p}$ is assumed bounded and as a consequence $\mathbf{p} \in L^2((-1,1), w_\beta)$ where the weight function w_β is defined by $w_\beta(t) := (1-t^2)^{\beta-1/2}$. Note that the decomposition (5) shows that it is enough to estimate the eigenvalues $(p_l^*)_l$ to recover the envelope function $\mathbf{p}$.

4.2 *Estimating the Matrix of Probabilities*

Let us denote A the adjacency matrix of the Soft-RGG G given by entries $A_{i,j} \in \{0,1\}$ where $A_{i,j} = 1$ if the nodes i and j are connected and $A_{i,j} = 0$ otherwise. We denote by Θ the $n \times n$ symmetric matrix with entries $\Theta_{i,j} = \mathbf{p}\left(\langle X_i, X_j \rangle\right)$ for $1 \le i < j \le n$ and zero diagonal entries. We consider the scaled version of the matrices A and Θ given by

$$\widehat{T_n} = \frac{1}{n}A \quad \text{and} \quad T_n = \frac{1}{n}\Theta.$$

Bandeira and van Handel [10] proved a near-optimal error bound for the operator norm of $\widehat{T_n} - T_n$. Coupling this result with the Weyl's perturbation theorem gives a control on the difference between the eigenvalues of the matrices $\widehat{T_n}$ and T_n, namely, with probability greater than $1 - \exp(-n)$, it holds,

$$\forall k \in [n], \quad |\lambda_k(\widehat{T_n}) - \lambda_k(T_n)| \le \|\widehat{T_n} - T_n\| = O(1/\sqrt{n}), \tag{6}$$

where $\lambda_k(M)$ is the k-th largest eigenvalue of any symmetric matrix M. This result shows that the spectrum of the scaled adjacency matrix $\widehat{T_n}$ is a good approximation of the one of the scaled matrix of probabilities T_n.

4.3 *Spectrum Consistency of the Matrix of Probabilities*

For any $R \ge 0$, we denote

$$\tilde{R} := \sum_{l=0}^{R} d_l, \tag{7}$$

which corresponds to the dimension of the space of Spherical Harmonics with degree at most R. Proposition 2 states that the spectrum of T_n converges toward the one of the integral operator $\mathbb{T}_W$ in the δ_2 metric which is defined as follows:

Definition 6 Given two sequences x, y of reals—completing finite sequences by zeros—such that $\sum_i x_i^2 + y_i^2 < \infty$, we define the ℓ_2 rearrangement distance $\delta_2(x, y)$ as

$$\delta_2^2(x, y) := \inf_{\sigma \in \mathfrak{S}_n} \sum_i (x_i - y_{\sigma(i)})^2 ,$$

where $\mathfrak{S}_n$ is the set of permutations with finite support. This distance is useful to compare two spectra.

Proposition 2 ([32, Proposition 4]) *There exists a universal constant $C > 0$ such that for all $\alpha \in (0, 1/3)$ and for all $n^3 \geq \tilde{R} \log(2\tilde{R}/\alpha)$, it holds*

$$\delta_2(\lambda(T_n), \lambda^*) \leq 2 \left[\sum_{l > R} d_l \left(p_l^*\right)^2 \right]^{1/2} + C\sqrt{\tilde{R}\left(1 + \log(\tilde{R}/\alpha)\right)/n}, \tag{8}$$

with probability at least $1 - 3\alpha$.

Proposition 2 shows that the ℓ_2 rearrangement distance between λ^* and $\lambda(T_n)$ decomposes as the sum of a bias term and a variance term. The second term on the right-hand side of (8) corresponds to the variance. The proof leading to this variance bound relies on the Hoffman-Wielandt inequality and borrows ideas from [63]. It makes use of recent developments in random matrix concentration by applying a Bernstein-type concentration inequality (see [90], for example) to control the operator norm of the sum of independent centered symmetric matrices given by

$$\sum_{i=1}^{n} \left(\mathbf{Y}(X_i)\mathbf{Y}(X_i)^\top - \mathbb{E}\left[\mathbf{Y}(X_i)\mathbf{Y}(X_i)^\top \right] \right), \tag{9}$$

with $\mathbf{Y}(x) = \big(Y_{0,0}(x), Y_{1,1}(x), \ldots, Y_{1,d_1}(x), Y_{2,1}(x), \ldots, Y_{2,d_2}(x), \ldots, Y_{R,1}(x), \ldots, Y_{R,d_R}(x)\big)^\top \in \mathbb{R}^{\tilde{R}}$ for all $x \in \mathbb{S}^{d-1}$. The proof of Proposition 2 also exploits concentration inequality for U-statistic dealing with a bounded, symmetric, and σ-canonical kernel (see [33, Definition 3.5.1]). The first term on the right-hand side of (8) is the bias arising from choosing a resolution level equal to R. Its behavior as a function of R can be analyzed by considering some regularity condition on the envelope $\mathbf{p}$. Assuming that $\mathbf{p}$ belongs to the Sobolev class $Z^s_{w_\beta}((-1, 1))$ (with regularity encoded by some parameter $s > 0$) defined by

$$\left\{ g = \sum_{k \geq 0} g_k^* c_k G_k^\beta \in L^2((-1, 1), w_\beta) \;\middle|\; \|g\|^*_{Z^s_{w_\beta}((-1,1))} \right.$$

$$\left. := \left[\sum_{l=0}^{\infty} d_l |g_l^*|^2 \left(1 + (l(l + 2\beta))^s\right) \right]^{1/2} < \infty \right\},$$

and choosing the resolution level $R_{opt} = \lceil (n/\log n)^{\frac{1}{2s+d-1}} \rceil$ to balance the bias/variance tradeoff appearing on the right-hand side of (8), we get that

$$\mathbb{E}\left[\delta_2^2\left(\lambda(T_n), \lambda^*\right)\right] \lesssim \left[\frac{n}{\log n}\right]^{-\frac{2s}{2s+(d-1)}}.$$

Thus, we recover a classical nonparametric rate of convergence for estimating a function with smoothness s in a space of dimension $d-1$. This is also the rate toward the probability matrix obtained by Xu [98]. Note that the choice of R_{opt} requires the knowledge of the regularity parameter s. To overcome this issue, [32] proposed an adaptive procedure using the Goldenshluger-Lepski method.

4.4 Estimation of the Envelope Function

Let us denote $\lambda := \lambda(\widehat{T}_n)$. For a prescribed model size $R \in \mathbb{N}\backslash\{0\}$, [32] define the estimator $\widehat{\lambda}^R$ of the truncated spectrum $\lambda^{*R} := (p_0^*, p_1^*, \ldots, p_1^*, \ldots, p_R^*, \ldots, p_R^*)$ of λ^* as

$$\widehat{\lambda}^R := (p_0^R(\hat{\sigma}), p_1^R(\hat{\sigma}), \ldots, p_1^R(\hat{\sigma}), \ldots, p_1^R(\hat{\sigma}), \ldots, p_R^R(\hat{\sigma}), \ldots, p_R^R(\hat{\sigma})),$$

with

$$\hat{\sigma} \in \arg\min_{\sigma \in \mathfrak{S}_n} \sum_{l=0}^{R} \sum_{k=\widetilde{l-1}}^{\tilde{l}} \left(p_l^R(\sigma) - \lambda_{\sigma(k)}\right)^2 + \sum_{k=\tilde{R}+1}^{n} \lambda_{\sigma(k)}^2 \quad \text{and}$$

$$p_l^R(\sigma) = \frac{1}{d_l} \sum_{k=\widetilde{l-1}}^{\tilde{l}} \lambda_{\sigma(k)},$$

where $\mathfrak{S}_n$ is the set of permutations of $[n]$ and where we used the notation (7) with the convention $\widetilde{-1} = 1$. Using the results of the two previous subsections, namely, (6) and Proposition 2, we obtain [32, Theorem.6] which states that

$$\mathbb{E}\left[\delta_2^2\left(\widehat{\lambda}^{R_{opt}}, \lambda^*\right)\right] \lesssim \left[\frac{n}{\log n}\right]^{-\frac{2s}{2s+(d-1)}}.$$

The envelope function $\mathbf{p}$ can then be approximated by the plug-in estimator $\widehat{\mathbf{p}} \equiv \sum_{l=0}^{R_{opt}} p_l^{R_{opt}}(\hat{\sigma}) c_l G_l^{\beta}$ based on the decomposition (5). One drawback of this approach is the exponential complexity in R of the computation of $\widehat{\lambda}^R$. In the next section, we will describe an approach based on a Hierarchical Agglomerative Clustering algorithm to estimate the envelope function $\mathbf{p}$ efficiently.

4.5 Open Problems and Perspectives

The minimax rate of estimating a s-regular function on a space of (Riemannian) dimension $d-1$ such as $\mathbb{S}^{d-1}$ from n observations is known to be of order $n^{-\frac{s}{2s+d-1}}$. In the framework of this section, even if the domain of the envelope function $\mathbf{p}$ is $[-1, 1]$, inputs of $\mathbf{p}$ are the pairwise distances given by inner products of points embedded in $\mathbb{S}^{d-1}$. Hence, it is still an open question to know if the dimension d of the latent space appears in the minimax rate of convergence. Moreover, the number of observations in the estimation problem considered is n^2 since the full adjacency matrix is known. Nevertheless, the problem suffers from the presence of unobserved latent variables. This all contributes to a nonstandard estimation problem, and finding the optimal rate of convergence is an open problem.

5 Growth Model in RGGs

5.1 Description of the Model

In [40], a new growth model was introduced for RGGs. The so-called Markov Random Geometric Graph (MRGG) already presented in Definition 4 is a Soft-RGG where latent points are sampled with Markovian jumps. Namely, [40] consider n points $X_1, X_2, \dots, X_n$ sampled on the Euclidean sphere $\mathbb{S}^{d-1}$ using a Markovian dynamic. They start by sampling uniformly X_1 on $\mathbb{S}^{d-1}$. Then, for any $i \in \{2, \dots, n\}$, they sample

- A unit vector $Y_i \in \mathbb{S}^{d-1}$ uniformly, orthogonal to X_{i-1}
- A real $r_i \in [-1, 1]$ encoding the distance between X_{i-1} and X_i, see (11). r_i is sampled from a distribution $f_{\mathcal{L}} : [-1, 1] \to [0, 1]$, called the *latitude function*

then X_i is defined by

$$X_i = r_i \times X_{i-1} + \sqrt{1 - r_i^2} \times Y_i \, . \tag{10}$$

This dynamic is illustrated in Fig. 4 and can be understood as follows. Consider that X_{i-1} is the north pole, then choose uniformly a direction (i.e., a longitude) and, in an independent manner, randomly move along the latitudes (the longitude being fixed by the previous step). The geodesic distance γ_i drawn on the latitudes satisfies

$$\gamma_i = \arccos(r_i) \, , \tag{11}$$

where random variable $r_i = \langle X_i, X_{i-1} \rangle$ has density $f_{\mathcal{L}}(r_i)$.

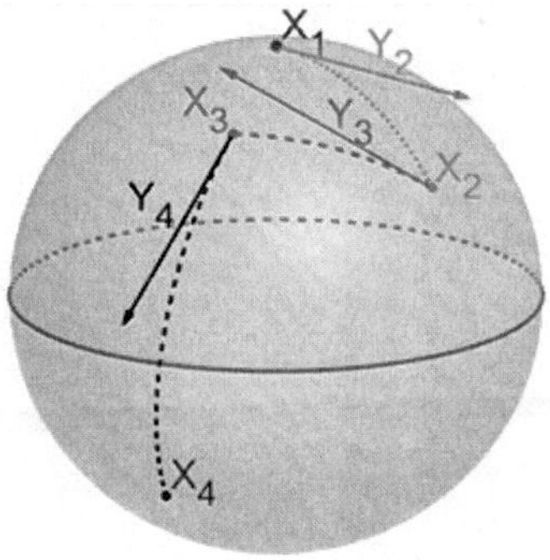

Fig. 4 Visualization of the sampling scheme in $\mathbb{S}^2$

5.2 Spectral Convergences

In this framework and keeping the notations of the previous section, one can show that if $\mathbf{p} \in Z^s_{w_\beta}((-1, 1))$ and if $f_{\mathscr{L}}$ satisfies the condition

$$(\mathscr{H}) \qquad \|f_{\mathscr{L}}\|_\infty := \sup_{t\in[-1,1]} |f_{\mathscr{L}}(t)| < \infty \text{ and } f_{\mathscr{L}} \text{ is bounded away from zero,}$$

then

$$\mathbb{E}\left[\delta_2^2(\lambda(T_n), \lambda^*) \vee \delta_2^2(\lambda^{R_{opt}}(\widehat{T}_n), \lambda^*)\right] = \mathscr{O}\left(\left[\frac{n}{\log^2(n)}\right]^{-\frac{2s}{2s+d-1}}\right), \tag{12}$$

with $\lambda^{R_{opt}}(\hat{T}_n) = (\hat{\lambda}_1, \ldots, \hat{\lambda}_{\tilde{R}_{opt}}, 0, 0, \ldots)$ and $R_{opt} = \lfloor (n/\log^2(n))^{\frac{1}{2s+d-1}} \rfloor$ where $\hat{\lambda}_1, \ldots, \hat{\lambda}_n$ are the eigenvalues of $\widehat{T}_n$ sorted in decreasing order of magnitude. This result is the counterpart of Proposition 2 in this Markovian framework. The proof follows closely the steps of the one of the previous section, but one needs to deal with the dependency of the latent positions. Results from [90] are no longer suited to control the operator norm of (9) since $(X_i)_{i\geq 0}$ is a Markov chain. Nevertheless, this can be achieved by using concentration inequalities for sum of functions of Markov chains and by exploiting the rank one structure of the random matrices $\mathbf{Y}(X_i)\mathbf{Y}(X_i)^\top$ together with a covering set argument. Another difficulty induced by the latent dynamic is the control of a U-statistic of order 2 of the Markov chain $(X_i)_{i\geq 0}$ with a bounded kernel. Non-asymptotic results regarding the tail behavior of U-statistics of a Markov chain have been so far very little touched. In a recent work, [41] proved a concentration inequality for order 2 U-statistics with bounded kernels for uniformly ergodic Markov chain. Theorem 10 gives a simplified version of their main result. Assuming that the condition $(\mathscr{H})$ is fulfilled, the Markov chain $(X_i)_{i\geq 1}$ satisfies the assumptions of Theorem 10 and one can show that (12) holds true.

Theorem 10 ([41, Theorem 2]) *Let us consider a Markov chain $(X_i)_{i\geq 1}$ on some measurable space $(E, \mathscr{E})$ (with E Polish) with transition kernel $P : E \times E \to \mathbb{R}$ and a function $h : E \times E \to \mathbb{R}$. We assume that*

1. *$(X_i)_{i\geq 1}$ is a uniformly ergodic Markov chain with invariant distribution π,*
2. *h is bounded and π-canonical, namely*

$$\forall x \in E, \quad \mathbb{E}_{X\sim\pi}[h(X, x)] = \mathbb{E}_{X\sim\pi}[h(x, X)] = 0,$$

3. *There exist $\delta > 0$ and some probability measure ν on $(E, \mathscr{E})$ such that*

$$\forall x \in E,\ \forall A \in \mathscr{E}, \quad P(x, A) \leq \delta\nu(A).$$

Then there exist constants $\beta, \kappa > 0$ such that for any $u \geq 1$, it holds with probability at least $1 - \beta e^{-u}\log n$,

$$\frac{1}{n(n-1)} \sum_{1\leq i,j\leq n,\ i\neq j} h(X_i, X_j) \leq \kappa \|h\|_\infty \log n \left\{\frac{u}{n} + \left[\frac{u}{n}\right]^2\right\},$$

where κ and β only depend on constants related to the Markov chain $(X_i)_{i\geq 1}$.

Remark Note that Theorem 10 holds for any initial distribution of the Markov chain. In their paper, [41] go beyond the previous Hoeffding tail control by providing a Bernstein-type concentration inequality under the additional assumption that the chain is stationary. For the sake of simplicity, we presented Theorem 10 for a single kernel h, but we point out that their results allow for the dependence of the kernels—say $h_{i,j}$—on the indexes in the sums which bring technical difficulties since standard blocking techniques can no longer be applied. The interest for this concentration result goes beyond the scope of random graphs since U-statistics naturally arise in online learning [29] or testing procedures [47].

5.3 Estimation Procedure

Recalling the notation of the truncated spectrum λ^{*R} (resp. $\lambda^R(\widehat{T}_n)$) of λ^* (resp. $\lambda(\widehat{T}_n)$) from Sect. 4.4, [40] introduce a new procedure (namely, the SCCHEi algorithm) based on a Hierarchical Agglomerative Clustering that returns a partition $\mathscr{C}_{d_0}, \ldots, \mathscr{C}_{d_R}, \Lambda$ of the n eigenvalues of $\widehat{T}_n$ where for any $i \in \{0, \ldots, R\}$, $|\mathscr{C}_{d_i}| = d_i$ (where we recall that d_i is the dimension of the space of Spherical Harmonics of degree i). The authors prove that for any fixed resolution level R, n can be chosen large enough so that the clusters obtained in polynomial time from the SCCHEi algorithm satisfy

$$\delta_2^2(\lambda^{*R}, \lambda^R(\widehat{T}_n)) = \sum_{k=0}^{R} \sum_{\hat{\lambda} \in \mathscr{C}_{d_k}} (\hat{\lambda} - p_k^*)^2. \tag{13}$$

The final estimate of the envelope function with resolution level R is defined as

$$\widehat{\mathbf{p}} := \sum_{k=0}^{R} \widehat{p}_k G_k^{\beta}, \quad \text{where} \quad \forall k \in \mathbb{N}, \quad \widehat{p}_k = \begin{cases} \frac{1}{d_k} \sum_{\lambda \in \mathscr{C}_{d_k}} \lambda \text{ if } k \in \{0, \dots, R\} \\ 0 \qquad \text{otherwise.} \end{cases} \tag{14}$$

Equation (13) is not a sufficient condition to ensure that the L^2 error between the true envelope function and the plug-in estimator $\widehat{\mathbf{p}}$ (see Eq. (14)) goes to 0 has $n \to +\infty$. This is due to identifiability issues coming from the δ_2 metric. In [40, Theorem 3], the author obtain a theoretical guarantee on the L^2 error between the true envelope function and the plug-in estimate by considering additional assumptions on the eigenvalues $(p_k^*)_{k\geq 0}$. Let us finally mention that the optimal resolution level R_{opt} is unknown in practice. To bypass this issue, the authors propose a model selection procedure based on the slope heuristic (see [7]).

5.4 Nonparametric Link Prediction

We are now interested in solving link prediction tasks. Namely, from the observation of the graph at time n, we want to estimate the probabilities of connection between the upcoming node $n+1$ and the nodes already present in the graph. Recalling the definition of the random variables $(Y_i)_{i\geq 2}$ from Sect. 5.1 and denoting further $\text{proj}_{X_n^\perp}(\cdot)$ the orthogonal projection onto the orthogonal complement of $\text{Span}(\text{X}_\text{n})$, the decomposition

$$\begin{aligned} \langle X_i, X_{n+1}\rangle &= \langle X_i, X_n\rangle\langle X_n, X_{n+1}\rangle + \sqrt{1-\langle X_n, X_{n+1}\rangle^2}\sqrt{1-\langle X_i, X_n\rangle^2} \\ &\quad \times \langle \frac{\text{proj}_{X_n^\perp}(X_i)}{\|\text{proj}_{X_n^\perp}(X_i)\|_2}, Y_{n+1}\rangle, \end{aligned} \tag{15}$$

shows that latent distances $\mathbf{D}_{1:n} = (\langle X_i, X_j\rangle)_{1\leq i,j\leq n} \in [-1,1]^{n\times n}$ are enough for link prediction. Indeed, it can be achieved by estimating the posterior probabilities defined for any $i \in [n]$ by

$$\eta_i(\mathbf{D}_{1:n}) = \mathbb{P}\left(A_{i,n+1} = 1 \mid \mathbf{D}_{1:n}\right)$$

$$\begin{aligned} \eta_i(\mathbf{D}_{1:n}) &= \int_{r,u\in(-1,1)} \mathbf{p}\left(\langle X_i, X_n\rangle r + \sqrt{1-r^2}\sqrt{1-\langle X_i, X_n\rangle^2}u\right) \\ &\quad \times f_{\mathscr{L}}(r) w_{\frac{d-3}{2}}(u) \frac{\Gamma(\frac{d-1}{2})}{\Gamma(\frac{d-2}{2})\sqrt{\pi}} dr du, \end{aligned} \tag{16}$$

where $A_{i,n+1} \in \{0, 1\}$ is one if and only if node $n + 1$ is connected to node i, $w_{\frac{d-3}{2}}(u) := (1 - u^2)^{\frac{d-3}{2} - \frac{1}{2}}$ and where $\Gamma : a \in]0, +\infty[\mapsto \int_0^{+\infty} t^{a-1} e^{-t} dt$. Using an approach similar to [5, 40] proved that one can get a consistent estimator $\widehat{G}$ of the Gram matrix of the latent positions $G = \big(\langle X_i, X_j\rangle\big)_{1 \le i,j \le n}$ in Frobenius norm. Hence, one can use a traditional plug-in estimator for $\eta_i(\mathbf{D}_{1:n})$ by replacing in (16) (i) the envelope function $\mathbf{p}$ by $\widehat{\mathbf{p}}$ from (14), (ii) the pairwise distances by their estimates $\big(\widehat{G}_{i,j}\big)_{1 \le i,j \le n}$ and (iii) the latitude function $f_{\mathcal{L}}$ by a non-parametric kernel density estimator built from the latent distances between consecutive nodes $(\langle X_i, X_{i+1}\rangle)_{i \in [n-1]}$ estimated by $\big(\widehat{G}_{i,i+1}\big)_{i \in [n-1]}$.

Through the example of MRGG, one can easily grasp the interest of growth model for random graphs with a geometric structure. Modeling the time evolution of networks, one can hope to solve tasks such as link prediction or collaborative filtering. An interesting research direction would be to extend the previous work to an anisotropic Markov kernel.

6 Connections with Community-Based Models

We have already described open problems and interesting directions to pursue regarding the questions tackled in the Sects. 3, 4 and 5. In this last section, we want to look at RGGs from a different lens by highlighting a recently born line of research that investigates the connections between RGGs and community-based models. Without aiming at presenting in a comprehensive manner the literature on this question, we rather focus on a few recent works that could inspire the reader to contribute in this emerging field.

A plenty number of random graph models have been so far studied. However, real-world problems never match a particular model and most of the time present several internal structures. To take into account this complexity, a growing number of works have been trying to take the best of several known random graph models. Papadopoulos et al. [78] introduced a growth model where new connections with the upcoming node are drawn taking into account both popularity and similarity of vertices. The motivation is to find a balance between two trends for new connections in social networks, namely, *homophily* and *popularity*. One can also mention [61] who consider a growth model that interpolates between pure preferential attachment (essentially the well-known Barabasi–Albert model) and a purely geometric model (the online nearest-neighbor graph). As pointed out by Barthélemy [12, Section II.B.3.a], *it is clear that community detection in spatial networks is a very interesting problem which might receive a specific answer.*

6.1 Extension of RGGs to Take into Account Community Structure

Galhotra et al. [48] proposed a new random graph model that incorporates community membership in standard RGGs. More precisely, they introduce the geometric block model which is defined as follows. Consider $V = V_1 \sqcup V_2 \sqcup \cdots \sqcup V_k$ a partition of $[n]$ in k clusters, $(X_u)_{u\in[n]}$ independent and identical random vectors uniformly distributed on $\mathbb{S}^{d-1}$ and let $\left(r_{i,j}\right)_{1\le i,j\le k} \in [0,2]^{k\times k}$. The geometric block model is a random graph with vertices V and an edge exists between $v \in V_i$ and $u \in V_j$ if and only if $\|X_u - X_v\| \le r_{i,j}$. Focusing on the case where $r_{i,i} = r_s, \forall i$ and $r_{i,j} = r_d, \ \forall i \neq j$, the authors want to recover the partition V observing only the adjacency matrix of the graph. They proved that in the relatively sparse regime (i.e., when $r_s, r_d = \Omega_n\left(\frac{\log n}{n}\right)$), a simple motif-counting algorithm allows to detect communities in the geometric block model and is near-optimal. The proposed greedy algorithm affects two nodes to the same community if the number of their common neighbors lies in a prescribed range whose bounds depend on r_s and r_d that are assumed to be known. The method is proved to recover the correct partition of the nodes with probability tending to 1 as n goes to $+\infty$.

In [87], the previous work is extended by considering arbitrary connection function. The paper sheds light on interesting differences between the standard SBMs and community models that incorporates some geometric structure. We start by presenting their model before highlighting some interesting results. Their model is the planted partition random connection model (PPCM) that relies on a Poisson point process on $\mathbb{R}^d$ with intensity $\lambda > 0$ $\varphi := \{X_1, X_2, \dots\}$ where it is assumed that the enumeration of the points X_i is such that for all $i, j \in \mathbb{N}$, $i > j \implies \|X_i\|_\infty \ge \|X_j\|_\infty$. Each atom $i \in \mathbb{N}$ is marked with a random variable $Z_i \in \{-1, +1\}$. $\overline{\varphi}$ is the marked Poisson point process. The sequence $\{Z_i\}_{i\in\mathbb{N}}$ is i.i.d. with each element being uniformly distributed in $\{-1, +1\}$. The interpretation of this marked point process is that for any node $i \in \mathbb{N}$, its location label is X_i and its community label is Z_i. Considering two connection functions $f_{in}, f_{out} : \mathbb{R}_+ \to [0, 1]$, they first construct an infinite graph G with vertex set $\mathbb{N}$ and place an edge between any two nodes $i, j \in \mathbb{N}$ with probability $f_{in}(\|X_i - X_j\|)\mathbb{1}_{Z_i=Z_j} + f_{out}(\|X_i - X_j\|)\mathbb{1}_{Z_i\neq Z_j}$. The graph G_n is then the induced subgraph of G consisting of the nodes 1 through N_n where $N_n := \sup\left\{i \ge 0 : X_i \in B_n := [-\frac{n^{1/d}}{2}, \frac{n^{1/d}}{2}]^d\right\}$.

Considering that the graph is observed and that the connections functions f_{in}, f_{out} and the location labels $(X_i)_i$ are known, the authors investigate conditions on the parameters of their model allowing to extract information on the community structure from the observed data.

Weak Recovery Weak recovery is said to be solvable if for every $n \in \mathbb{N}\backslash\{0\}$, there exists some algorithm that—based on the observed data G_n and φ—provides a sequence of $\{-1, +1\}$ valued random variables $\{\tau_i^{(n)}\}_{i=1}^{N_n}$ such that there exists a constant $\gamma > 0$ such that the *overlap* between $\{\tau_i^{(n)}\}_{i=1}^{N_n}$ and $\{Z_i\}_{i=1}^{N_n}$ is asymptotically almost surely larger than γ, namely,

$$\lim_{n\to\infty} \mathbb{P}\left(\frac{\sum_{i=1}^{N_n} \tau_i^{(n)} Z_i}{N_n} \geq \gamma\right) = 1.$$

The authors identify regimes where weak recovery can be solved or not. We summarize their results with Proposition 3.

Proposition 3 ([87, Proposition 1 - Corollary 2 - Theorem 2]) *For every $f_{in}(\cdot)$, $f_{out}(\cdot)$ such that $\{r \in \mathbb{R}^+ : f_{in}(r) \neq f_{out}(r)\}$ has positive Lebesgue measure and any $d \geq 2$, there exists a $\lambda_c \in (0, \infty)$ such that*

- *For any $\lambda < \lambda_c$, weak recovery is not solvable.*
- *For any $\lambda > \lambda_c$, there exists an algorithm (which could possibly take exponential time) to solve weak recovery.*

Moreover, there exists $\tilde{\lambda}_c < \infty$ (possibly larger than λ_c) depending on $f_{in}(\cdot)$, $f_{out}(\cdot)$ and d, such that for all $\lambda > \tilde{\lambda}_c$, weak recovery is solvable in polynomial time.

The intrinsic nature of the problem of weak recovery is completely different in the PPCM model compared to the standard sparse SBM. Sparse SBMs are known to be locally treelike with very few short cycles. Efficient algorithms that solve weak recovery in the sparse SBM (such as message passing algorithm, convex relaxation, or spectral methods) deeply rely on the local treelike structure. On the contrary, PPCMs are locally dense even if they are globally sparse. This is due to the presence of a lot of short loops (such as triangles). As a consequence, the standard tools used for SBMs are not relevant to solve weak recovery in PPCMs. Nevertheless, the local density allows to design a polynomial time algorithm that solves weak recovery for $\lambda > \tilde{\lambda}_c$ (see Proposition 3) by simply considering the neighbors of each node. Proposition 3 lets us open the question of the existence of a gap between information versus computation thresholds. Namely, is it always possible to solve weak recovery in polynomial time when $\lambda > \lambda_c$? In the sparse and symmetric SBM, it is known that there is no information-computation gap for $k = 2$ communities, while for $k \geq 4$ a non-polynomial algorithm is known to cross the Kesten-Stigum threshold which was conjectured by Decelle et al. [34] to be the threshold at which weak recovery can be solved efficiently.

Distinguishability The distinguishability problem asks how well one can solve a hypothesis testing problem that consists in finding if a given graph has been sampled from the PPCM model or from the null, which is given by a plain random connection model with connection function $(f_{in}(\cdot)+f_{out}(\cdot))/2$ without communities but having the same average degree and distribution for spatial locations. Sankararaman and Baccelli [87] prove that for every $\lambda > 0$, $d \in \mathbb{N}$ and connection functions $f_{in}(\cdot)$ and $f_{out}(\cdot)$ satisfying $1 \geq f_{in}(r) \geq f_{out}(r) \geq 0$ for all $r \geq 0$, and $\{r \geq 0 : f_{in}(r) \neq f_{out}(r)\}$ having positive Lebesgue measure, the probability distribution of the null and the alternative of the hypothesis test are mutually singular. As a consequence, there exist some regimes (such as $\lambda < \lambda_c$ and $d \geq 2$) where we can be very sure by observing the data that a partition exists but cannot identify it better than at random. In these cases, it is out of reach to bring together the small partitions of nodes in different regions of the space into one coherent. Such behavior does not exist in the sparse SBM with two communities as proved by Mossel et al. [73] and was conjectured to hold also for $k \geq 3$ communities in [34].

6.2 Robustness of Spectral Methods for Community Detection with Geometric Perturbations

In another line of work, [79] are studying robustness of spectral methods for community detection when connections between nodes are perturbed by some latent random geometric graph. They identify specific regimes in which spectral methods are still efficient to solve community detection problems despite geometric perturbations, and we give an overview of their work in what follows. Let us consider some fixed parameter $\kappa \in [0, 1]$ that drives the balance between strength of the community signal and the noise coming from the geometric perturbations. For the sake of simplicity, they consider a model with two communities where each vertex i in the network is characterized by some vector $X_i \in \mathbb{R}^2$ with distribution $\mathcal{N}(0, I_2)$. They consider $p_1, p_2 \in (0, 1)$ that may depend on the number of nodes n with $p_1 > p_2$ and $\sup_n p_1/p_2 < \infty$. Assuming for technical reason $\kappa + \max\{p_1, p_2\} \leq 1$, the probability of connection between i and j is

$$\mathbb{P}\left\{i \sim j \mid X_i, X_j\right\} = \kappa \exp\left(-\gamma \|X_i - X_j\|^2\right) + \begin{cases} p_1 \text{ if } i \text{ and } j \text{ belong to the same community} \\ p_2 \text{ otherwise.} \end{cases},$$

where the inverse width $\gamma > 0$ may depend on n. We denote by $\sigma \in \{\pm 1/\sqrt{n}\}^n$ the normalized community vector illustrating to which community each vertex belong ($\sigma_i = -1/\sqrt{n}$ if i belongs to the first community and $\sigma_i = 1/\sqrt{n}$ otherwise). The matrix of probabilities of this model is given by $Q := P_0 + P_1$ where

$$P_0 := \begin{bmatrix} p_1 J & p_2 J \\ p_2 J & p_1 J \end{bmatrix} \quad \text{and} \quad P_1 := \kappa P = \kappa \left((1 - \delta_{i,j}) e^{-\gamma \|X_i - X_j\|^2}\right)_{1 \leq i,j \leq n}.$$

The adjacency matrix A of the graph can, thus, be written as $A = P_0 + P_1 + A_c$ where A_c is, conditionnally on the X_i's, a random matrix with independent Bernoulli entries which are centered. Given the graph-adjacency matrix A, the objective is to output a normalized vector $x \in \{\pm 1/\sqrt{n}\}^n$ such that, for some $\varepsilon > 0$,

- Exact recovery: with probability tending to 1, $|\sigma^\top x| = 1$,
- Weak recovery (also called detection): with probability tending to 1, $|\sigma^\top x| > \varepsilon$.

Let us highlight that contrary to the previous section, the latent variables $(X_i)_i$ are not observed. When $\kappa = 0$, we recover the standard SBM: $Q = P_0$ has two nonzero eigenvalues which are $\lambda_1 = n(p_1 + p_2)/2$ with associated normalized eigenvector $v_1 = \frac{1}{\sqrt{n}}(1, 1, \ldots, 1)^\top$ and $\lambda_2 = n(p_1 - p_2)/2$ associated to $v_2 = \sigma = \frac{1}{\sqrt{n}}(1, \ldots, 1, -1, \ldots, -1)^\top$. Spectral methods can, thus, be used to recover communities by computing the second eigenvector of the adjacency matrix A. To prove that spectral methods still work in the presence of geometric perturbations, one needs to identify regimes in which the eigenvalues of A are well separated and the second eigenvector is approximately v_2.

In the regime where $\gamma \gg n/\log n$, the spectral radius $\rho(P_1)$ of P_1 vanishes, and we asymptotically recover a standard SBM. Hence, they focus on the following regime:

$$\gamma \underset{n\to\infty}{\to} \infty \quad \text{and} \quad \frac{1}{\gamma}\frac{n}{\ln n} \underset{n\to\infty}{\to} \infty. \tag{A_1}$$

Under Assumption (A_1), [79, Proposition 2] states that with probability tending to one, $\rho(P_1)$ is of order $\frac{\kappa n}{2\gamma}$. Using [15, Theorem 2.7] to get an asymptotic upper bound on the spectral radius of A_c, basic perturbation arguments would prove that standard techniques for community detection work in the regime where

$$\frac{\kappa n}{2\gamma} \ll \sqrt{\frac{n(p_1 + p_2)}{2}} = \sqrt{\lambda_1}.$$

Indeed, it is now well known that weak recovery in the SBM can be solved efficiently as soon as $\lambda_2 > \sqrt{\lambda_1}$ (for example, using the power iteration algorithm on the non-backtracking matrix from [18]). Hence, the regime of interest corresponds to the case where

$$\exists c, C > 0 \quad \text{s.t.} \quad \lambda_2^{-1}\frac{\kappa n}{2\gamma} \in [c, C], \quad \frac{\lambda_2}{\lambda_1} \in [c, C] \quad \text{and} \quad \lambda_2 \gg \sqrt{\lambda_1}, \tag{A_2}$$

which corresponds to the case where the noise induced by the latent random graph is of the same order of magnitude as the signal. Under (A_2), the problem of weak recovery can be tackled using spectral methods on the matrix $S = P_0 + P_1$: the goal is to reconstruct communities based on the second eigenvector of S. To prove that these methods work, the authors first find conditions ensuring that two eigenvalues of S exit the support of the spectrum of P_1. Then, they provide an asymptotic lower bound for the level of correlation between $v_2 = \sigma$ and the second eigenvector w_2 of S, which leads to Theorem 11.

Theorem 11 ([79, Theorem 10]) *Suppose that Assumptions (A_1) and (A_2) hold and that $\lambda_1 > \lambda_2 + 2\frac{\kappa}{2\gamma}$. Then the correlation $|w_2^\top v_2|$ is uniformly bounded away from 0. Moreover, denoting μ_1 the largest eigenvalue of P_1, if the ratio λ_2/μ_1 goes to infinity, then $|w_2^\top v_2|$ tends to 1, which gives weak (and even exact at the limit) recovery.*

6.3 Recovering Latent Positions

From another viewpoint, one can think RGGs as an extension of stochastic block models where the discrete community structure is replaced by an underlying geometry. With this mindset, it is natural to directly transport concepts and questions from clustered random graphs to RGGs. For instance, the task consisting in estimating the communities in SBMs may correspond to the estimation of latent point neighborhoods in RGGs. More precisely, community detection can be understood in RGGs as the problem of recovering the geometric representation of the nodes (e.g., through the Gram matrix of the latent positions). This question has been tackled by Eldan et al. [44] and Valdivia [4]. Both works consider random graphs sampled from the TIRGG model on the Euclidean sphere $\mathbb{S}^{d-1}$ with some envelope function $\mathbf{p}$ (see Definition 3), leading to a graphon model similar to the one presented in Sect. 4.1. While the result from [4] holds in the dense and relatively sparse regimes, the one from [44] covers the sparse case. Thanks to harmonic properties of $\mathbb{S}^{d-1}$, the graphon eigenspace composed only of linear eigenfunctions (harmonic polynomials of degree one) directly relates to the pairwise distances of the latent positions. This allows [44] and [4] to provide a consistent estimate of the Gram matrix of the latent positions in Frobenius norm using a spectral method. Their results hold under the following two key assumptions.

1. *An eigenvalue gap condition.* They assume that the d eigenvalues of the integral operator $\mathbb{T}_W$—associated with the graphon $W := \mathbf{p}(\langle \cdot, \cdot \rangle)$ (see (4))—corresponding to the Spherical Harmonics of degree one is well separated from the rest of the spectrum.
2. *A regularity condition.* They assume that the envelope function $\mathbf{p}$ belongs to some weighted Sobolev space, meaning that the sequence of eigenvalues of $\mathbb{T}_W$ goes to zero fast enough.

In addition to similar assumptions, [4] and [44] share the same proof structure. First, they need to recover the d eigenvectors from the adjacency matrix corresponding to the space of Spherical Harmonics of degree one. Then, the Davis-Kahan Theorem is used to prove that the estimate of the Gram matrix based on the previously selected eigenvectors is consistent in Frobenius norm. To do so, they require a concentration result ensuring that the adjacency matrix A (or some proxy of it) converges in operator norm toward the matrix of probabilities Θ with entries $\Theta_{i,j} = \mathbf{p}\left(\langle X_i, X_j \rangle\right)$ for $1 \leq i \neq j \leq n$ and zero diagonal entries. Valdivia [4] relies on [10, Corollary 3.12], already discussed in (6), that provides the convergence $\|A - \Theta\| \to 0$ as $n \to \infty$ in the dense and relatively sparse regimes. In the sparse regime, such concentration no longer holds. Indeed, in that case, degrees of some vertices are much higher than the expected degree, say deg. As a consequence, some rows of the adjacency matrix A have Euclidean norms much larger than $\sqrt{\text{deg}}$, which implies that for n large enough, it holds with high probability $\|A - \Theta\| \gg \sqrt{\text{deg}}$. To cope with this issue, [44] do not work directly on the adjacency matrix but rather on a slightly amended version of it—say A'—where one reduces the weights of the edges incident to high degree vertices. In that way, all degrees of the new (weighted) network become bounded, and [66, Theorem 5.1] ensures that A' converges to Θ in spectral norm as n goes to $+\infty$. Hence, in the sparse regime, the adjacency matrix converges toward its expectation *after regularization*. The proof of this random matrix theory tool is based on a famous result in functional analysis known as the Grothendieck-Pietsch factorization.

Let us finally mention that this change of behavior of the extreme eigenvalues of the adjacency matrix according to the maximal mean degree has been studied in details for inhomogeneous Erdös-Rényi graphs in [15] and [14].

6.4 Some Perspectives

The paper [87] makes the strong assumption that the locations' labels $(X_i)_{i\geq 1}$ are known. Hence, it should be considered as an initial work calling for future theoretical and practical investigations. Keeping the same model, it would be of great interest to design algorithms able to deal with unobserved latent variables to allow real-data applications. The first step in this direction was made by Avrachenkov et al. [9] where the authors propose a spectral method to recover hidden clusters in the soft geometric block model where latent positions are not observed. On the theoretical side, [87] describe at the end of their paper several open problems. Their suggestions for future works include (i) the extension of their work to a larger number of communities, (ii) the estimation from the data of the parameters of their model (namely f_{in} and f_{out} that they assumed to be known), and (iii) the existence of a possible gap between information versus computation thresholds; namely, they wonder if there is a regime where community detection is solvable, but without any polynomial (in n) time and space algorithms.

Another possible research direction is the extension of the work from Sect. 6.2 to study the same kind of robustness results for more than 2 communities and especially in the sparse regime where $\frac{1}{\gamma} \sim p_i \sim \frac{1}{n}$. As highlighted by Péché and Perchet [79], the sparse case may bring additional difficulties since *standard spectral techniques in this regime involve the non-backtracking matrix (see [18]), and its concentration properties are quite challenging to establish.* Regarding Sect. 6.3, for some applications, it may be interesting to go beyond the recovery of the pairwise distances by embedding the graph in the latent space while preserving the Gram structure. Such question has been tackled, for example, by Perry et al. [83] but only for the Euclidean sphere in small dimensions.

Acknowledgments The authors are in debt to Tselil Schramm who gave a great talk at the S. S. Wilks Memorial Seminar in Statistics (at Princeton University) providing insightful comments on the problem of geometry detection or more specifically on her paper [69].

Appendix: Outline of the Proofs of Theorems 6 and 7

The proofs of Theorems 6 and 7 (cf. Sect. 3.3) are quite complex, and giving their formal descriptions would require heavy technical considerations. In the following, we provide an overview of the proofs highlighting the nice mathematical tools used by Liu et al. [69] and their innovative combination while putting under the rug some technical aspects.

Step 1. Relate the TV distance of the whole graphs to single vertex neighborhood.

$$\begin{aligned} &2\mathrm{TV}(G(n,p,d),G(n,p))^2 \\ &\quad\le \mathrm{KL}(G(n,p,d)||G(n,p)) \qquad \text{from Pinsker's inequality} \\ &\quad\le n\times \mathbb{E}_{G_{n-1}\sim G(n-1,p,d)}\left[\mathrm{KL}\left(\nu_n(\cdot|G_{n-1}),\mathrm{Bern}(p)^{\otimes(n-1)}\right)\right] \qquad \text{from Lemma 2} \\ &\quad= \mathbb{E}_{G_{n-1}\sim G(n-1,p,d)}\mathbb{E}_{S\sim\nu_n(\cdot|G_{n-1})}\log\left(\frac{\nu_n(S|G_{n-1})}{p^{|S|}(1-p)^{n-1-|S|}}\right), \end{aligned} \tag{17}$$

where $\nu_n(\cdot|G_{n-1})$ denotes the distribution of the neighborhood of vertex n when the graph is sampled from $G(n,p,d)$ conditional on the knowledge of the connections between pairs of nodes in $[n-1]$ given by G_{n-1}. Hence, the main difference with [21] is that the tensorization argument from Lemma 2 is used node-wise (and not edge-wise). We are reduced to understand how a vertex incorporates a given graph of size $n-1$ sampled from the distribution $G(n-1,p,d)$. At a high level, the authors show that if one can prove that for some $\varepsilon>0$, with high probability over $G_{n-1}\sim G(n-1,p,d)$, it holds

$$\begin{aligned} \forall S\subseteq[n-1],\quad \nu_n(S|G_{n-1}) &= \mathbb{P}_{G\sim G(n,p,d)}(N_G(n)=S\mid G_{n-1}) \\ &= (1\pm\varepsilon)p^{|S|}(1-p)^{n-1-|S|}, \end{aligned} \tag{18}$$

where $N_G(n)$ denotes the set of nodes connected to node n in the graph G, then

$$\mathrm{TV}(G(n,p,d),G(n,p)) = o_n(n\varepsilon^2). \tag{19}$$

Step 2. Geometric interpretation of neighbrhood probabilities from Eq. (18).

For $G\sim G(n,p,d)$, if vertex i is associated to a (random) vector X_i, and (i,j) is an edge, we consequently know that $\langle X_i,X_j\rangle\ge t_{p,d}$. On the sphere $\mathbb{S}^{d-1}$, the locus of points where X_j can be conditioned on (i,j) being an edge is a sphere cap centered at X_i with a p fraction of the sphere's surface area, which we denote by $\mathrm{cap}(X_i)$. Similarly, if we know that i and j are not adjacent, the locus of points where X_j can fall is the complement of a sphere cap with measure $1-p$ namely, $\overline{\mathrm{cap}(X_i)}$, which we call an "anti-cap." Let us denote σ is the normalized Lebesgue measure on $\mathbb{S}^{d-1}$ so that $\sigma(\mathbb{S}^{d-1}) = 1$. Equipped with this geometric picture, we can view the probability that vertex n's neighborhood is exactly equal to $S \subseteq [n-1]$ as $\sigma(L_S)$, where $L_S \subseteq \mathbb{S}^{d-1}$ is a random set defined by

$$L_S := \Big(\bigcap_{i\in S} \mathrm{cap}(X_i)\Big) \cap \Big(\bigcap_{j\notin S} \overline{\mathrm{cap}(X_j)}\Big).$$

To show that the TV distance between $G(n, p, d)$ and $G(n, p)$ is small, we need to prove that $\sigma(L_S)$ concentrates around $p^{|S|}(1-p)^{n-1-|S|}$ as suggested by Eqs. (18) and (19).

Step 3. Concentration of measure of intersections of sets in $\mathbb{S}^{d-1}$ with random spherical caps.

An essential contribution of [69] is a novel concentration inequality for the area of the intersection of a random spherical cap with any subset $L \subseteq \mathbb{S}^{d-1}$.

Lemma 3 ([see 69, Corollary 4.10]) ***Set-cap intersection concentration Lemma.*** *Suppose $L \subseteq \mathbb{S}^{d-1}$ and let us denote by σ the uniform probability measure on $\mathbb{S}^{d-1}$. Then with high probability over $z \sim \sigma$, it holds*

$$\Big|\frac{\sigma(L \cap \mathrm{cap}(z))}{p\sigma(L)} - 1\Big| = \mathcal{O}_n\big(\delta_n(L)\big) \text{ and } \Big|\frac{\sigma(L \cap \overline{\mathrm{cap}(z)})}{(1-p)\sigma(L)} - 1\Big| = \mathcal{O}_n\Big(\frac{p}{1-p}\delta_n(L)\Big),$$

where $\delta_n(L) = \sqrt{\frac{\log\frac{1}{p} + \log\frac{1}{\sigma(L)}}{\sqrt{d}}}\mathrm{polylog}(n)$.

Sketch of proof of Lemma 3 We give an overview of the proof of Lemma 3, highlighting its interesting connection with optimal transport. Let us consider some probability distribution ν on $\mathbb{S}^{d-1}$. Let us denote $\mathcal{D}$ the optimal coupling between the measures ν and σ, i.e., $\mathcal{D}$ is a probability measure on $\mathbb{S}^{d-1} \times \mathbb{S}^{d-1}$ with marginals ν and σ such that

$$W_2(\nu, \sigma)^2 = \int \|x - y\|_2^2 d\mathcal{D}(x, y),$$

where $W_2(\nu, \sigma)$ is the 2-Wasserstein distance between the measures σ and ν. Then, for any $z \in \mathbb{S}^{d-1}$, it holds

$$\begin{aligned}\mathbb{P}_{x\sim\nu}(\langle z, x\rangle > t_{p,d}) &= \mathbb{P}_{(x,y)\sim\mathcal{D}}(\langle z, y\rangle > t_{p,d} - \langle z, x - y\rangle)\\ &\le \mathbb{P}_{y\sim\sigma}(\langle z, y\rangle > t_{p,d} - u(p, d))\\ &\quad + \mathbb{P}_{(x,y)\sim\mathcal{D}}(|\langle z, x - y\rangle| > u(p, d)),\end{aligned} \tag{20}$$

for some well-chosen threshold $u(p, d)$ depending on p and d. The first term in the right-hand side of Eq. (20) can be proven to concentrate around p with high probability over $z \sim \sigma$ with standard arguments. The second term in Eq. (20)

quantifies how often a randomly chosen transport vector $x - y$ with $(x, y) \sim \mathcal{D}$ has a large projection in the direction z. One can prove that the optimal transport map $\mathcal{D}$ between $x \sim \nu$ and $y \sim \sigma$ has bounded length with high probability and then translate this into a tail bound for the inner product $\langle z, x - y\rangle$ for a random vector $z \sim \sigma$. As a consequence, one can bound with high probability over $z \sim \sigma$ the fluctuations of $\left|\mathbb{P}_{x\sim\nu}(\langle z, x\rangle > t_{p,d}) - p\right|$ which gives Lemma 3 if we take for ν the uniform measure on the set $L \subseteq \mathbb{S}^{d-1}$.

Applying Lemma 3 inductively and using a martingale argument, the authors prove that intersecting j random caps and $(k - j)$ random anticaps, we get a multiplicative fluctuation for $\sigma(L_S)$ around $p^{|S|}(1 - p)^{n-1-|S|}$ that is of the order of $(1 \pm \sqrt{j}\delta + \sqrt{k - j}\frac{p}{1-p}\delta)$. Going back to Eq. (19), this approach is sufficient to prove that

$$\mathrm{TV}(G(n, p, d), G(n, p)) = o_n\Big(\frac{n^3 p^2}{d}\Big),$$

leading to the first statement of Theorem 7.

Step 4. The sparse case and the use of the cavity method.

To get down to a polylogarithmic threshold in the sparse regime, the authors changed paradigm. Previously, they were bounding the quantity

$$\begin{aligned}\mathbb{P}_{G\sim G(n,p,d)}(N_G(n) = S \mid G_{n-1}) &= \mathbb{E}_{X_1,\dots,X_{n-1}\mid G_{n-1}}\mathbb{E}_{X_n\sim\sigma}\big[\mathbb{1}_{N_G(n)=S}\big]\\ &= \mathbb{E}_{X_1,\dots,X_{n-1}\mid G_{n-1}}\big[\sigma(L_S)\big],\end{aligned} \tag{21}$$

by fixing a specific realization of latent positions $X_1, \dots X_{n-1}$ and then analyzing the probability that the node n connects to some $S \subseteq [n - 1]$. The probability that vertex n is adjacent to all vertices in $S \subseteq [n - 1]$ is exactly equal to the measure of the set-caps intersection, which appears to be tight. At a high level, this is a *worst case approach* to upper bound Eq. (21) in the sense that the bound obtained from this analysis may be due to an unlikely latent configuration conditioned on $X_1, \dots, X_{n-1}$ producing G_{n-1}. To obtain a polylogarithmic threshold in the sparse case, one needs to analyze the concentration of $\sigma(L_S)$ on average over vector embeddings of G_{n-1}. To do so, the authors rely on the so-called *cavity method* borrowed from the field of statistical physics. The cavity method allows to understand the distribution of $(X_i)_{i\in S}$ conditional on forming G_{n-1} for any $S \subseteq [n - 1]$ with size of the order $pn = \Theta(1)$. We provide further details on this approach in the following.

A Simplification Using Tight Concentration for Intersections Involving Anticaps Liu et al. [69] first prove that due to tight concentration for the measure of the intersection of random anticaps with sets of lower bounded measure, one can get high-probability estimates for $\nu_n(S \mid G_{n-1})$ by studying the probability that $S \subseteq N_G(n)$, namely,

$$\mathbb{P}(S \subseteq N_G(n) \mid G_{n-1}) = \mathbb{P}\big(\forall i \in S,\ \langle X_i, X_n\rangle \geq t_{p,d} \mid G_{n-1}\big)$$
$$= \mathop{\mathbb{E}}_{\substack{X_n \sim \sigma \\ (X_i)_{i\in[n-1]} \sim \sigma^{G_{n-1}}}} \prod_{i\in S} \mathbb{1}_{\langle X_i, X_n\rangle \geq t_{p,d}}, \tag{22}$$

where $\sigma^{G_{n-1}} := \left[\sigma^{\otimes(n-1)} \mid G_{n-1}\right]$. If $(X_i)_{i\in S}$ in Eq. (22) was a collection of *independent* random vectors distributed *uniformly* on the sphere, then Eq. (22) would be exactly equal to $p^{|S|}$. In the following, we explain how the authors prove that both of these properties are approximately true.

The Cavity Method To bound the fluctuation of Eq. (22) around $p^{|S|}$, [69] use the cavity method. Let us consider $S \subseteq [n-1]$, G_{n-1} sampled from $G(n-1, p, d)$ and its corresponding latent vectors. Let us denote by $\mathcal{B}_{G_{n-1}}(i, \ell)$ the ball of radius-ℓ around a vertex $i \in [n-1]$ in the graph G_{n-1}. Fixing all vectors except those in $K := \bigcup_{i\in S} \mathcal{B}_{G_{n-1}}(i, \ell - 1)$, the cavity method aims at computing the joint distribution of $(X_i)_{i\in S}$ conditional to $(X_i)_{i\notin K}$ and G_{n-1}. Informally speaking, we "carve out" a cavity of depth ℓ around each vertex $i \in S$, and we fix all latent vectors outside of these cavities as presented with Fig. 5. The choice of the depth ℓ results from the following tradeoff:

- We want to choose the depth ℓ small enough so that the balls $\mathcal{B}_{G_{n-1}}(i, \ell)$ for $i \in S$ are all trees and are pairwise disjoint with high probability.
- We want to choose ℓ as large as possible in order to get a bound on the fluctuations of Eq. (22) around $p^{|S|}$ as small as possible.

To formally analyze the distribution of the unfixed vectors upon resampling them, the authors set up a constraint satisfaction problem instance over a continuous alphabet that encodes the edges of G_{n-1} within the trees around S: each node has a

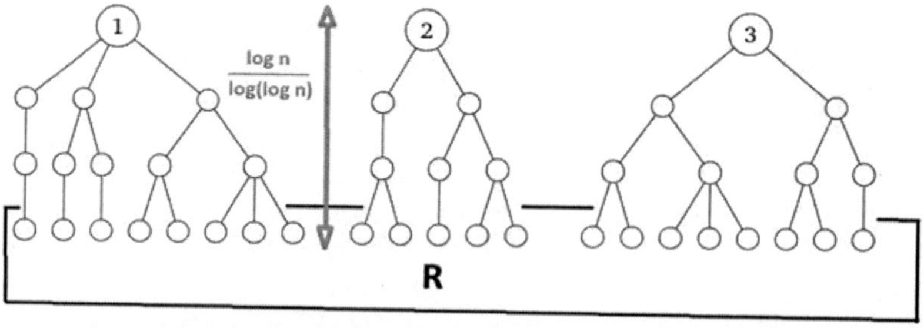

Fig. 5 Illustration of the cavity method to bound the fluctuation of Eq. (22) around $p^{|S|}$ i.e., to bound the deviation of the random variable $\sigma(L_S)$ conditioned on $X_1, \ldots, X_{n-1}$ producing G_{n-1}. With high probability, the neighborhood until depth $\ell = \frac{\log n}{\log \log n}$ of vertices in S are disjoint trees. We fix the latent representation of vertices in the set $R := [n-1] \backslash K$. Using the belief propagation algorithm, one can compute the distribution of $(X_i)_{i\in S} \mid (X_j)_{j\in R}$ where the latent positions $(X_j)_{j\in[n-1]}$ are sampled according to $\sigma^{G_{n-1}}$. This allows to bound the fluctuation of Eq. (22) around $p^{|S|}$

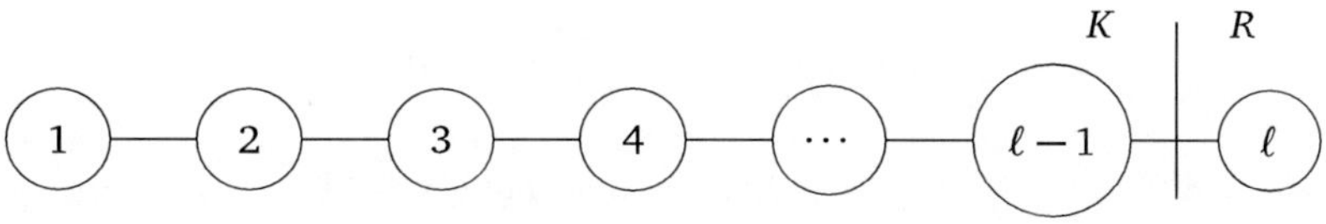

Fig. 6 Simple analysis of the belief propagation algorithm when the neighborhood of vertex $1 \in S$ at depth ℓ is a path

vector-valued variable in $\mathbb{S}^{d-1}$, and the constraints are that nodes joined by an edge must have vectors with inner product at least $t_{p,d}$. The marginal of the latent vectors X_i for $i \in S$ can be obtained using the belief propagation algorithm. Let us recall that belief propagation computes marginal distributions over labels of constraints satisfaction problems when the constraints graph is a tree.

A Simple Analysis of Belief Propagation To ease the reasoning, let us suppose that $1 \in S$ is such that $\mathscr{B}_{G_{n-1}}(1, \ell - 1)$ is a path. Without loss of generality, we consider that the path is given by Fig. 6. Every vector is passing to its parent along the path a convolution of its own measure (corresponding to its "message") with a cap of measure p. Denoting by P, the linear operator is defined so that for any function $h : \mathbb{S}^{d-1} \to \mathbb{R}$,

$$Ph(x) = \frac{1}{p} \int_{\mathrm{cap}(x)} h(y) d\sigma(y),$$

the authors prove that for some $a > 0$, for any probability measure μ on $\mathbb{S}^{d-1}$ with density h with respect to σ,

$$\mathrm{TV}(Ph, \sigma) \leq \mathscr{O}_n\left(\frac{\log^a n}{\sqrt{d}}\right) \mathrm{TV}(\mu, \sigma), \tag{23}$$

which is a contraction result. Since at every step of the belief propagation algorithm, a vertex sends to its parent the image by the operator P of its own measure, we deduce from Eq. (23) that the parent receives a measure which is getting closer to the uniform distribution by a multiplicative factor equal to $\frac{1}{\sqrt{d}}$. The proof of Eq. (23) relies on the set-cap intersection concentration result (see Lemma 3). To get an intuition of this connection, let us consider that h is the density of the uniform probability measure μ on some set $L \subseteq \mathbb{S}^{d-1}$, then

$$Ph(x) = \frac{1}{p} \mathbb{P}_{Y \sim \mu}(Y \in \mathrm{cap}(x)) = \frac{1}{p} \frac{\sigma(L \cap \mathrm{cap}(x))}{\sigma(L)},$$

and we can conclude using Lemma 3 that ensures that with high probability over $x \sim \sigma$, $\sigma(L \cap \mathrm{cap}(x)) = (1 \pm \mathscr{O}_n(\frac{\log^a n}{\sqrt{d}})) p\sigma(L)$. Applying Eq. (23) $\ell = \frac{\log n}{\log \log n}$ times for d being some power of $\log n$, one can show that

$$\mathrm{TV}(P^{\ell}\mu,\sigma)=\mathscr{O}_n\Big[\Big(\frac{\log^a n}{\sqrt{d}}\Big)^{\ell}\Big]=o_n\Big(\frac{1}{\sqrt{n}}\Big).$$

With this approach, one can prove that the distribution of $(X_i)_{i\in S}$ is approximately $\sigma^{\otimes|S|}$. This allows to bound the fluctuations of Eq. (22) around $p^{|S|}$ which leads to Theorem 6 using Eqs. (18) and (19).

As a concluding remark, we mention that [69] demonstrate a coupling of $G_- \sim G(n, p - o_n(p))$, $G \sim G(n, p, d)$, and $G_+ \sim G(n, p + o_n(p))$ that satisfies $G_- \subseteq G \subseteq G_+$ with high probability. This sandwich-type result holds for a proper choice of the latent dimension and allows to transfer known properties of Erdös-Renyi random graphs to RGGs in the studied regime. For example, the authors use this coupling result to upper bound the probability that the depth-ℓ neighborhood of some $i \in [n]$ forms a tree under $G(n, p, d)$ in the sparse regime with $d = \mathrm{polylog}(n)$.

References

1. E. Abbe, Community detection and Stochastic Block models. Found. Trends Commun. Inf. Theory **14**(1–2), 1–162 (2018)
2. R. Aguilar-Sánchez, J.A. Méndez-Bermúdez, F.A. Rodrigues, J.M. Sigarreta, Topological versus spectral properties of random geometric graphs. Phys. Rev. E **102**, 042306 (2020)
3. A. Allen-Perkins, Random spherical graphs. Phys. Rev. E **98**(3), 032310 (2018)
4. E. Araya Valdivia, Random geometric graphs on euclidean balls (2020). arXiv e-prints, pages arXiv–2010
5. E. Araya Valdivia, Y. De Castro, Latent distance estimation for random geometric graphs, in ed. by H. Wallach, H. Larochelle, A. Beygelzimer, F. d'Alché-Buc, E. Fox, R. Garnett, *Advances in Neural Information Processing Systems*, vol. 32 (Curran Associates, Red Hook, 2019), pp. 8724–8734
6. M.A. Arcones, E. Gine, Limit theorems for U-processes. Ann. Probab. **21**(3), 1494–1542 (1993)
7. S. Arlot, Minimal penalties and the slope heuristics: a survey. J. de la Société Française de Statistique **160**(3), 1–106 (2019)
8. K. Avrachenkov, A. Bobu, Cliques in high-dimensional random geometric graphs, in ed. by H. Cherifi, S. Gaito, J.F. Mendes, E. Moro, L.M. Rocha, *Complex Networks and Their Applications VIII* (Springer International Publishing, Cham, 2020), pp. 591–600
9. K. Avrachenkov, A. Bobu, M. Dreveton, Higher-order spectral clustering for geometric graphs. J. Fourier Analy. Appl. **27**(2), 22 (2021)
10. A.S. Bandeira, R. van Handel, Sharp nonasymptotic bounds on the norm of random matrices with independent entries. Ann. Probab. **44**(4), 2479–2506 (2016)
11. A.-L. Barabási, Scale-free networks: a decade and beyond. Science **325**(5939), 412–413 (2009)
12. M. Barthélemy, Spatial networks. Phys. Repor. **499**(1–3), 1–101 (2011)
13. G. Ben Arous, R. Gheissari, A. Jagannath, Algorithmic thresholds for tensor PCA. Ann. Probab. **48**, 2052–2087 (2020)
14. F. Benaych-Georges, C. Bordenave, A. Knowles, et al. Largest eigenvalues of sparse inhomogeneous Erdős–Rényi graphs. Ann. Probab. **47**(3), 1653–1676 (2019)
15. F. Benaych-Georges, C. Bordenave, A. Knowles, et al. Spectral radii of sparse random matrices, in *Annales de l'Institut Henri Poincaré, Probabilités et Statistiques*, vol. 56 (Institut Henri Poincaré, Paris, 2020), pp. 2141–2161

16. P. Blackwell, M. Edmondson-Jones, J. Jordan, *Spectra of Adjacency Matrices of Random Geometric Graphs*. University of Sheffield. Department of Probability and Statistics (2007)
17. B. Bollobás, *Random Graphs*. Cambridge Studies in Advanced Mathematics, 2nd edn. (Cambridge University Press, Cambridge, 2001)
18. C. Bordenave, M. Lelarge, L. Massoulié, Non-backtracking spectrum of random graphs: Community detection and non-regular Ramanujan graphs, in *2015 IEEE 56th Annual Symposium on Foundations of Computer Science* (IEEE, Piscataway, 2015), pp. 1347–1357
19. M. Brennan, G. Bresler, Optimal average-case reductions to sparse PCA: From weak assumptions to strong hardness (2019). arXiv preprint arXiv:1902.07380
20. M. Brennan, G. Bresler, Reducibility and statistical-computational gaps from secret leakage, in *Conference on Learning Theory*, PMLR (2020), pp. 648–847
21. M. Brennan, G. Bresler, D. Nagaraj, Phase transitions for detecting latent geometry in random graphs. Probab. Theory Related Fields **178**, 1215–1289 (2020)
22. M. Brennan, G. Bresler, B. Huang. De finetti-style results for Wishart matrices: Combinatorial structure and phase transitions (2021) https://arxiv.org/abs/2103.14011
23. G. Bresler, D. Nagaraj, Optimal single sample tests for structured versus unstructured network data, in ed. by S. Bubeck, V. Perchet, P. Rigollet, *Proceedings of the 31st Conference On Learning Theory. Proceedings of Machine Learning Research*. PMLR, vol. 75 (2018), pp. 1657–1690
24. H. Breu, D.G. Kirkpatrick, Unit disk graph recognition is NP-hard. Comput. Geom. **9**(1), 3–24 (1998). Special Issue on Geometric Representations of Graphs
25. S. Bubeck, S. Ganguly, Entropic CLT and phase transition in high-dimensional Wishart matrices. CoRR, abs/1509.03258 (2015)
26. S. Bubeck, J. Ding, R. Eldan, M.Z. Racz, Testing for high-dimensional geometry in random graphs. Random Struct. Algor. **49**, 503–532 (2016)
27. A. Channarond, Random graph models: an overview of modeling approaches. J. de la Société Française de Statistique **156**(3), 56–94 (2015)
28. S. Chatterjee, Matrix estimation by universal singular value thresholding. Ann. Statist. **43**(1), 177–214 (2015)
29. S. Clémençon, G. Lugosi, N. Vayatis, Ranking and empirical minimization of U-statistics. Ann. Statist. **36**(2), 844- 874 (2008)
30. F. Dai, Y. Xu, *Approximation Theory and Harmonic Analysis on Spheres and Balls*, vol. 23 (Springer, Berlin, 2013)
31. J. Dall, M. Christensen, Random geometric graphs. Phys. Rev. E **66**, 016121 (2002)
32. Y. De Castro, C. Lacour, T.M.P. Ngoc, Adaptive estimation of nonparametric geometric graphs. Math. Statist. Learn. **2**, 217–274 (2020)
33. V. De la Pena, E. Giné, *Decoupling: From Dependence to Independence* (Springer Science & Business Media, Berlin, 2012)
34. A. Decelle, F. Krzakala, C. Moore, L. Zdeborová, Asymptotic analysis of the Stochastic Block model for modular networks and its algorithmic applications. Phys. Rev. E **84**, 066106 (2011)
35. C.P. Dettmann, O. Georgiou, Random geometric graphs with general connection functions. Phys. Rev. E **93**(3), 032313 (2016)
36. C.P. Dettmann, O. Georgiou, G. Knight, Spectral statistics of random geometric graphs. EPL (Europhys. Lett.) **118**(1), 18003 (2017)
37. L. Devroye, A. György, G. Lugosi, F. Udina, High-dimensional random geometric graphs and their clique number. Electron. J. Probab. **16**, 2481–2508 (2011)
38. P. Diaconis, S. Janson, Graph limits and exchangeable random graphs (2007). arXiv preprint arXiv:0712.2749
39. Q. Duchemin, Reliable time prediction in the Markov stochastic block model. Working paper or preprint. ESAIM: PS **27**, 80–135 (2023). https://doi.org/10.1051/ps/2022019
40. Q. Duchemin, Y. De Castro, Markov random geometric graph (MRGG): a growth model for temporal dynamic networks. Electron. J. Statist. **16**(1), 671–699 (2022)

41. Q. Duchemin, Y. De Castro, C. Lacour, Concentration inequality for U-statistics of order two for uniformly ergodic Markov chains. Bernoulli, **29**(2), 929–956 (Bernoulli Society for Mathematical Statistics and Probability, 2023)
42. R. Eldan, An efficiency upper bound for inverse covariance estimation. Israel J. Math. **207**(1), 1–9 (2015)
43. R. Eldan, D. Mikulincer, Information and dimensionality of anisotropic random geometric graphs, in *Geometric Aspects of Functional Analysis* (Springer, Berlin, 2020), pp. 273–324
44. R. Eldan, D. Mikulincer, H. Pieters, Community detection and percolation of information in a geometric setting (2020). arXiv preprint arXiv:2006.15574
45. V. Erba, S. Ariosto, M. Gherardi, P. Rotondo, Random Geometric Graphs in high dimension (2020). arXiv preprint arXiv:2002.12272
46. E. Estrada, M. Sheerin, Consensus dynamics on random rectangular graphs. Physica D Nonlinear Phenomena **323–324**, 20–26 (2016). Nonlinear Dynamics on Interconnected Networks
47. M. Fromont, B. Laurent, Adaptive goodness-of-fit tests in a density model. Ann. Statist. **34**(2), 680–720 (2006)
48. S. Galhotra, A. Mazumdar, S. Pal, B. Saha, The Geometric Block Model (2017). arXiv preprint arXiv:1709.05510
49. C. Gao, J. Lafferty, Testing network structure using relations between small subgraph probabilities (2017). arXiv preprint arXiv:1704.06742
50. D. Ghoshdastidar, M. Gutzeit, A. Carpentier, U. Von Luxburg, et al., Two-sample hypothesis testing for inhomogeneous random graphs. Ann. Statist. **48**(4), 2208–2229 (2020)
51. E.N. Gilbert, Random plane networks. J. Soc. Ind. Appl. Math. **9**(4), 533–543 (1961)
52. E. Giné, R. Latala, J. Zinn, Exponential and moment inequalities for U-statistics, in *High Dimensional Probability II* (Birkhäuser, Boston, 2000), pp. 13–38
53. A. Goel, S. Rai, B. Krishnamachari, Monotone properties of random geometric graphs have sharp thresholds. Ann. Appl. Probab. **15**(4), 2535–2552 (2005)
54. J. Grygierek, C. Thäle, Gaussian fluctuations for edge counts in high-dimensional random geometric graphs. Statist. Probab. Lett. **158**, 108674 (2020)
55. M. Haenggi, J.G. Andrews, F. Baccelli, O. Dousse, M. Franceschetti, Stochastic geometry and random graphs for the analysis and design of wireless networks. IEEE J. Sel. Areas Commun. **27**(7), 1029–1046 (2009)
56. D. Higham, M. Rasajski, N. Przulj, Fitting a geometric graph to a protein-protein interaction network. Bioinf. **24**, 1093–9 (2008)
57. C. Houdré, P. Reynaud-Bouret, Exponential inequalities for U-statistics of order two with constants, in *Stochastic Inequalities and Applications*. Progress in Probability, vol. 56 (Birkhäuser, Basel, 2002)
58. Y. Issartel, C. Giraud, N. Verzelen, Localization in 1D Non-parametric Latent Space Models from Pairwise Affinities (2021). https://arxiv.org/abs/2108.03098
59. J. Jin, Z.T. Ke, S. Luo, Optimal adaptivity of signed-polygon statistics for network testing. Ann. Stat. **49**(6), 3408–3433 (Institute of Mathematical Statistics, 2021)
60. E. Joly, G. Lugosi, Robust estimation of U-statistics. Stochastic Process. Appl. **126**, 3760–3773 (2016). In Memoriam: Evarist Giné
61. J. Jordan, A.R. Wade, Phase transitions for random geometric preferential attachment graphs. Adv. Appl. Probab. **47**(2), 565–588 (2015)
62. O. Klopp, N. Verzelen, Optimal graphon estimation in cut distance. Probab. Theory Related Fields **174**(3), 1033–1090 (2019)
63. V. Koltchinskii, E. Giné, Random matrix approximation of spectra of integral operators. Bernoulli **6**, 113–167 (2000)
64. A. Kontorovich, M. Raginsky, *Concentration of Measure Without Independence: A Unified Approach via the Martingale Method* (Springer, New York, 2017), pp. 183–210
65. D. Krioukov, F. Papadopoulos, M. Kitsak, A. Vahdat, M. Boguná, Hyperbolic geometry of complex networks. Phys. Rev. E **82**(3), 036106 (2010)
66. C.M. Le, E. Levina, R. Vershynin, *Concentration of random graphs and application to community detection*, in *Proceedings of the International Congress of Mathematicians: Rio de Janeiro 2018* (World Scientific, Singapore, 2018)

67. S. Liu, M.Z. Racz, Phase transition in noisy high-dimensional Random Geometric Graphs (2021)
68. S. Liu, M.Z. Racz, A Probabilistic View of Latent Space Graphs and Phase Transitions. (2021: in press)
69. S. Liu, S. Mohanty, T. Schramm, E. Yang, Testing thresholds for high-dimensional sparse random geometric graphs (2021). ArXiv, abs/2111.11316
70. L. Lovász, *Large Networks and Graph Limits*, vol. 60 (American Mathematical Society, Providence, 2012)
71. S. Lunagómez, S. Mukherjee, R.L. Wolpert, E.M. Airoldi, Geometric representations of random hypergraphs. J. Amer. Statist. Assoc. **112**(517), 363–383 (2017)
72. G. Mao, B. Anderson, Connectivity of large wireless networks under a general connection model. IEEE Trans. Inf. Theory **59**, 1761–1772 (2012)
73. E. Mossel, J. Neeman, A. Sly, Reconstruction and estimation in the planted partition model. Probab. Theory Related Fields **162**, 431–461 (2014)
74. P.-L. Méliot, Asymptotic representation theory and the spectrum of a Random geometric graph on a compact Lie group. Electron. J. Probab. **24**, 85 (2019)
75. T. Müller, P. Pralat, The acquaintance time of (percolated) random geometric graphs. Euro. J. Combin. **48**, 198–214 (2015). Selected Papers of EuroComb'13
76. A. Nyberg, T. Gross, K.E. Bassler, Mesoscopic structures and the Laplacian spectra of random geometric graphs. J. Compl. Netw. **3**(4), 543–551 (2015)
77. M. Ostilli, G. Bianconi, Statistical mechanics of random geometric graphs: geometry-induced first-order phase transition. Phys. Rev. E **91**(4), 042136 (2015)
78. F. Papadopoulos, M. Kitsak, M. Serrano, M. Boguna, D. Krioukov, Popularity versus similarity in growing networks. Nature **489**(7417), 537–540 (2012)
79. S. Péché, V. Perchet, Robustness of community detection to random geometric perturbations. Adv. Neural Inf. Process. Syst. **33**, 17827–17837 (2020)
80. M. Penrose, et al., *Random Geometric Graphs*, vol. 5 (Oxford University Press, Oxford, 2003)
81. M.D. Penrose, Connectivity of soft random geometric graphs. Ann. Appl. Probab. **26**(2), 986–1028 (2016)
82. M. Pereda, E. Estrada, Visualization and machine learning analysis of complex networks in hyperspherical space. Pattern Recognit. **86**, 320–331 (2019)
83. S. Perry, M.S. Yin, K. Gray, S. Kobourov, Drawing graphs on the sphere, in *Proceedings of the International Conference on Advanced Visual Interfaces* (2020), pp. 1–9
84. V. Preciado, A. Jadbabaie, Spectral analysis of virus spreading in random geometric networks, in *Proceedings of the 48h IEEE Conference on Decision and Control (CDC) Held Jointly with 2009 28th Chinese Control Conference* (2009), pp. 4802–4807
85. M. Racz, S. Bubeck, Basic models and questions in statistical network analysis. Statist. Surv. **11**, 1–47 (2016)
86. S. Rai, The spectrum of a random geometric graph is concentrated. J. Theoret. Probab. **20**, 119–132 (2004)
87. A. Sankararaman, F. Baccelli, Community detection on Euclidean random graphs, in *2017 55th Annual Allerton Conference on Communication, Control, and Computing (Allerton)* (2017), pp. 510–517
88. A.L. Smith, D.M. Asta, C.A. Calder, The geometry of continuous latent space models for network data. Statist. Sci. **34**(3), 428–453 (2019)
89. K. Solovey, O. Salzman, D. Halperin, New perspective on sampling-based motion planning via random geometric graphs. Int. J. Robot. Res. **37**(10), 1117–1133 (2018)
90. J.A. Tropp, An introduction to matrix concentration inequalities. Found. Trends Mach. Learn. **8**(1–2), 1–230 (2015)
91. M. Walters, *Random Geometric Graphs*. London Mathematical Society Lecture Note Series (Cambridge University Press, Cambridge, 2011), pp. 365–402
92. G. Wang, Z. Lin, On the performance of multi-message algebraic gossip algorithms in dynamic random geometric graphs. IEEE Commun. Lett. **22**, 1–1

93. D.J. Watts, S.H. Strogatz, Collective dynamics of 'small-world' networks. Nature **393**(6684), 440–442 (1998)
94. P.J. Wolfe, S.C. Olhede, Nonparametric graphon estimation (2013). arXiv e-prints, page arXiv:1309.5936
95. Z. Xie, T. Rogers, Scale-invariant geometric random graphs. Phys. Rev. E **93**(3), 032310 (2016)
96. Z. Xie, J. Zhu, D. Kong, J. Li, A random geometric graph built on a time-varying Riemannian manifold. Phys. A Statist. Mech. Appl. **436**, 492–498 (2015)
97. Z. Xie, Z. Ouyang, Q. Liu, J. Li, A geometric graph model for citation networks of exponentially growing scientific papers. Phys. A Statist. Mech. Appl. **456**, 167–175 (2016)
98. J. Xu, Rates of convergence of spectral methods for graphon estimation, in *International Conference on Machine Learning*, PMLR (2018), pp. 5433–5442
99. K. Zuev, M. Boguna, G. Bianconi, D. Krioukov, Emergence of soft communities from geometric preferential attachment. Sci. Rep. **5**, 9421 (2015)

Functional Estimation in Log-Concave Location Families

Vladimir Koltchinskii and Martin Wahl

2020 Mathematics Subject Classification: 62H12; 62G20; 62H25; 60B20

1 Introduction

Let P be a log-concave probability distribution in $\mathbb{R}^d$ with density e^{-V}, $V : \mathbb{R}^d \mapsto \mathbb{R}$ being a convex function, and let P_θ, $\theta \in \mathbb{R}^d$ be a location family generated by P : $P_\theta(dx) = p_\theta(x)dx = e^{-V(x-\theta)}dx$, $\theta \in \mathbb{R}^d$. In other words, a random variable $X \sim P_\theta$ could be represented as $X = \theta + \xi$, where $\theta \in \mathbb{R}^d$ is a location parameter and $\xi \sim P$ is a random noise with log-concave distribution. Without loss of generality, one can assume that $\mathbb{E}\xi = 0$ (otherwise, one can replace function $V(\cdot)$ by $V(\cdot + \mathbb{E}\xi)$). We will also assume that function V is known and θ is an unknown parameter of the model to be estimated based on i.i.d. observations $X_1, \ldots, X_n$ of X. We will refer to this statistical model as a *log-concave location family*. Our main goal is to study the estimation of $f(\theta)$ for a given smooth functional $f : \mathbb{R}^d \mapsto \mathbb{R}$. A natural estimator of location parameter is the maximum likelihood estimator (MLE) defined as

$$\hat{\theta} := \operatorname*{argmax}_{\theta \in \mathbb{R}^d} \prod_{j=1}^n p_\theta(X_j) = \operatorname*{argmin}_{\theta \in \mathbb{R}^d} \frac{1}{n} \sum_{j=1}^n V(X_j - \theta).$$

Note that by Lemma 2.2.1 in [7] for a log-concave density e^{-V}, $V : \mathbb{R}^d \mapsto \mathbb{R}$ there exist constants $A, B > 0$ such that $e^{-V(x)} \leq Ae^{-B\|x\|}$ for all $x \in \mathbb{R}^d$, implying that $V(x) \to \infty$ as $\|x\| \to \infty$. It is easy to conclude from this fact that MLE does

V. Koltchinskii (✉)
School of Mathematics, Georgia Institute of Technology, Atlanta, GA, USA
e-mail: vlad@math.gatech.edu

M. Wahl
Faculty of Mathematics, Bielefeld University, Bielefeld, Germany
e-mail: martin.wahl@math.uni-bielefeld.de

R. Adamczak et al. (eds.), *High Dimensional Probability IX*, Progress in Probability 80, https://doi.org/10.1007/978-3-031-26979-0_15

exist. Moreover, it is unique if V is strictly convex (this condition is assumed in what follows). In addition, MLE $\hat{\theta}$ is an equivariant estimator with respect to the translation group in $\mathbb{R}^d$:

$$\hat{\theta}(X_1 + u, \ldots, X_n + u) = \hat{\theta}(X_1, \ldots, X_n) + u, u \in \mathbb{R}^d.$$

Also note that

$$\mathbb{E}_\theta V(X - \theta') - \mathbb{E}_\theta V(X - \theta) = K(P_\theta \| P_{\theta'}),$$

where $K(P_\theta \| P_{\theta'})$ is the Kullback-Leibler divergence between P_θ and $P_{\theta'}$, implying that θ is the unique minimal point of $\theta' \mapsto \mathbb{E}_\theta V(X - \theta')$. The uniqueness follows from the identifiability of parameter θ : if θ were not identifiable, we would have $V(x) = V(x + h)$, $x \in \mathbb{R}^d$ for some $h \neq 0$, which would contradict the assumption that $V(x) \to \infty$ as $\|x\| \to \infty$.

Moreover, for differentiable V, the score function of location family is $\frac{\partial}{\partial\theta} \log p_\theta(X) = -V'(X - \theta)$, and, under some regularity, $\mathbb{E}_\theta V'(X - \theta) = 0$. In addition, the Fisher information matrix of such a log-concave location family is well defined, does not depend on θ, and is given by

$$\begin{aligned}\mathcal{I} &= \mathbb{E}_\theta \frac{\partial}{\partial\theta} \log p_\theta(X) \otimes \frac{\partial}{\partial\theta} \log p_\theta(X) = \mathbb{E}_\theta V'(X - \theta) \otimes V'(X - \theta) \\ &= \mathbb{E} V'(\xi) \otimes V'(\xi) = \int_{\mathbb{R}^d} V'(x) \otimes V'(x) e^{-V(x)} dx\end{aligned}$$

(provided that the integral in the right-hand side exists). Under further regularity, for twice differentiable V, we also have (via integration by parts)

$$\mathcal{I} = \mathbb{E} V''(\xi) = \int_{\mathbb{R}^d} V''(x) e^{-V(x)} dx.$$

Finally, if Fisher information $\mathcal{I}$ is non-singular, then, for a fixed d and $n \to \infty$, MLE $\hat{\theta}$ is an asymptotically normal estimator of θ with limit covariance $\mathcal{I}^{-1}$:

$$\sqrt{n}(\hat{\theta} - \theta) \xrightarrow{d} N(0; \mathcal{I}^{-1}) \text{ as } n \to \infty.$$

Assumption 1 below suffices for all the above properties to hold.

It seems natural to estimate $f(\theta)$ by the plug-in estimator $f(\hat{\theta})$, where $\hat{\theta}$ is the MLE. Such an approach yields asymptotically efficient estimators for regular statistical models in the case of fixed dimension d and $n \to \infty$. In particular, for our location family, we have by a standard application of the delta method that

$$\sqrt{n}(f(\hat{\theta}) - f(\theta)) \xrightarrow{d} N(0; \sigma_f^2(\theta)),$$

where $\sigma_f^2(\theta) := \langle \mathcal{I}^{-1} f'(\theta), f'(\theta)\rangle$. However, it is well known that plug-in estimators are suboptimal in high-dimensional problems mainly due to their large bias, and, often, nontrivial bias reduction methods are needed to achieve an optimal error rate. This has been one of the difficulties in the problem of estimation of functionals of parameters of high-dimensional and infinite-dimensional models for a number of years [4, 5, 11, 12, 21, 22, 25, 26].

One approach to this problem is based on replacing f by another function g for which the bias of estimator $g(\hat{\theta})$ is small. To find such a function g, one has to solve approximately the "bias equation" $\mathbb{E}_\theta g(\hat{\theta}) = f(\theta), \theta \in \mathbb{R}^d$. This equation can be written as $\mathcal{T} g = f$, where

$$(\mathcal{T} g)(\theta) := \mathbb{E}_\theta g(\hat{\theta}) = \int_{\mathbb{R}^d} g(u) P(\theta; du), \theta \in \mathbb{R}^d$$

and $P(\theta; A), \theta \in \mathbb{R}^d, A \subset \mathbb{R}^d$ is a Markov kernel on $\mathbb{R}^d$ (or, more generally, on the parameter space Θ of statistical model), providing the distribution of estimator $\hat{\theta}$. Denoting $\mathcal{B} := \mathcal{T} - \mathcal{I}$, where $\mathcal{I}$ is the identity operator in the space of bounded functions on $\mathbb{R}^d$ (not to be confused with the Fisher information also denoted by $\mathcal{I}$), and assuming that $\hat{\theta}$ is close to θ and, as a consequence, operator $\mathcal{B}$ is "small," one can view $\mathcal{T} = \mathcal{I} + \mathcal{B}$ as a small perturbation of identity. In such cases, one can try to solve the equation $\mathcal{T} g = f$ in terms of Neumann series $g = (\mathcal{I} - \mathcal{B} + \mathcal{B}^2 - \ldots) f$. In what follows, we denote by

$$f_k(\theta) := \sum_{j=0}^{k} (-1)^j (\mathcal{B}^j f)(\theta), \theta \in \mathbb{R}^d$$

the partial sum of this series, and we will use $f_k(\hat{\theta})$ (for a suitable choice of k depending on smoothness of functional f) as an estimator of $f(\theta)$. It is easy to see that its bias is

$$\begin{aligned}\mathbb{E}_\theta f_k(\hat{\theta}) - f(\theta) &= (\mathcal{B} f_k)(\theta) + f_k(\theta) - f(\theta) \\ &= (-1)^k (\mathcal{B}^{k+1} f)(\theta), \theta \in \mathbb{R}^d.\end{aligned}$$

If $\mathcal{B}$ is "small" and k is sufficiently large, one can hope to achieve a bias reduction through estimator $f_k(\hat{\theta})$. Another way to explain this approach is in terms of iterative bias reduction: since the bias of plug-in estimator $f(\hat{\theta})$ is equal to $(\mathcal{B} f)(\theta)$, one can estimate the bias by $(\mathcal{B} f)(\hat{\theta})$, and the first order bias reduction yields the estimator $f_1(\hat{\theta}) = f(\hat{\theta}) - (\mathcal{B} f)(\hat{\theta})$. Its bias is equal to $-(\mathcal{B}^2 f)(\theta)$, and the second order bias reduction yields the estimator $f_2(\hat{\theta}) = f(\hat{\theta}) - (\mathcal{B} f)(\hat{\theta}) + (\mathcal{B}^2 f)(\hat{\theta})$, etc. This is close to the idea of iterative bootstrap bias reduction [9, 10, 13].

Let $\{\hat{\theta}^{(k)} : k \geq 0\}$ be the Markov chain with $\hat{\theta}^{(0)} = \theta$ and with transition probability kernel $P(\theta; A), \theta \in \mathbb{R}^d, A \subset \mathbb{R}^d$. This chain can be viewed as an output of iterative application of parametric bootstrap to estimator $\hat{\theta}$ in the model

$X_1, \dots, X_n$ i.i.d. $\sim P_\theta, \theta \in \mathbb{R}^d$: at the first iteration, the data is sampled from the distribution with parameter $\hat{\theta}^{(0)} = \theta$, and estimator $\hat{\theta}^{(1)} = \hat{\theta}$ is computed; at the second iteration, the data is sampled from the distribution $P_{\hat{\theta}}$ (conditionally on the value of $\hat{\theta}$) and bootstrap estimator $\hat{\theta}^{(2)}$ is computed and so on. We will call $\{\hat{\theta}^{(k)} : k \geq 0\}$ the bootstrap chain of estimator $\hat{\theta}$. Clearly, $(\mathcal{T}^k f)(\theta) = \mathbb{E}_\theta f(\hat{\theta}^{(k)})$ and, by Newton's binomial formula, we also have

$$(\mathcal{B}^k f)(\theta) = ((\mathcal{T} - \mathcal{I})^k f)(\theta) = \sum_{j=0}^{k} (-1)^{k-j} \binom{k}{j} (\mathcal{T}^j f)(\theta)$$

$$= \mathbb{E}_\theta \sum_{j=0}^{k} (-1)^{k-j} \binom{k}{j} f(\hat{\theta}^{(j)}). \tag{1.1}$$

This means that $(\mathcal{B}^k f)(\theta)$ is the expectation of the k-th order difference of function f along the bootstrap chain $\{\hat{\theta}^{(j)} : j \geq 0\}$. In the case when $\|\hat{\theta} - \theta\| \lesssim \sqrt{\frac{d}{n}}$ with a high probability, the same bound also holds for the increments $\hat{\theta}^{(j+1)} - \hat{\theta}^{(j)}$ (conditionally on $\hat{\theta}^{(j)}$). If functional f is k times differentiable and d is small comparing with n, one could therefore expect that $(\mathcal{B}^k f)(\theta) \lesssim \left(\frac{d}{n}\right)^{k/2}$ (based on the analogy with the behavior of k-th order differences of k times differentiable functions in the real line). The justification of this heuristic for general parametric models could be rather involved (see [14–16, 18]), but it will be shown below that it is much simpler in the case of equivariant estimators $\hat{\theta}$ (such as the MLE) (see also [17, 19]).

Inserting representation (1.1) into the definition of f_k and using a simple combinatorial identity, we obtain the following useful representation of function $f_k(\theta)$:

$$f_k(\theta) = \mathbb{E}_\theta \sum_{j=0}^{k} (-1)^j \binom{k+1}{j+1} f(\hat{\theta}^{(j)}). \tag{1.2}$$

The following notations will be used throughout the paper (and some of them have been already used). For two variables $A, B \geq 0$, $A \lesssim B$ means that there exists an absolute constant $C > 0$ such that $A \leq CB$. The notation $A \gtrsim B$ means that $B \lesssim A$ and $A \asymp B$ means that $A \lesssim B$ and $B \lesssim A$. If the constants in the relationships $\lesssim, \gtrsim, \asymp$ depend on some parameter(s), say, on γ, this parameter will be used as a subscript of the relationship, say, $A \lesssim_\gamma B$. Given two square matrices A and B, $A \preceq B$ means that $B - A$ is positive semi-definite and $A \succeq B$ means that $B \preceq A$. The norm notation $\|\cdot\|$ (without further subscripts or superscripts) will be used by default in certain spaces. For instance, it will always denote the canonical Euclidean norm of $\mathbb{R}^d$, the operator norm of matrices (linear transformations) and the operator norm of multilinear forms. In some other cases, in particular for functional spaces L_∞, C^s, etc., the corresponding subscripts will be used.

2 Main Results

Recall that, for a convex nondecreasing function $\psi : \mathbb{R}_+ \mapsto \mathbb{R}_+$ with $\psi(0) = 0$, the Orlicz ψ-norm of a r.v. η is defined as

$$\|\eta\|_\psi := \inf\Big\{c \geq 0 : \mathbb{E}\psi\Big(\frac{|\eta|}{c}\Big) \leq 1\Big\}.$$

The Banach space of all r.v. on a probability space $(\Omega, \Sigma, \mathbb{P})$ with finite ψ-norm is denoted by $L_\psi(\mathbb{P})$, and to emphasize the dependence on the underlying probability measure $\mathbb{P}$, we also write $\|\cdot\|_\psi = \|\cdot\|_{L_\psi(\mathbb{P})}$. If $\psi(u) = u^p, u \geq 0, p \geq 1$, then the ψ-norm coincides with the L_p-norm. Another important choice is $\psi_\alpha(u) = e^{u^\alpha} - 1, u \geq 0, \alpha \geq 1$. In particular, for $\alpha = 1$, L_{ψ_1} is the space of sub-exponential r.v. and, for $\alpha = 2$, L_{ψ_2} is the space of sub-gaussian r.v. It is also well known that the ψ_α-norm is equivalent to the following norm defined in terms of moments (or the L_p-norms):

$$\|\eta\|_{\psi_\alpha} \asymp \sup_{p\geq 1} p^{-1/\alpha}\mathbb{E}^{1/p}|\eta|^p, \alpha \geq 1. \tag{2.1}$$

Note that the right-hand side defines a norm for $0 < \alpha < 1$, too, whereas the left-hand side is not a norm in this case since function ψ_α is not convex for $0 < \alpha < 1$. Relationship (2.1) still holds for $0 < \alpha < 1$, but with constants depending on α as α approaches 0. With a slight abuse of notations, we will define $\|\eta\|_{\psi_\alpha}$ by the right-hand side of (2.1) for all $\alpha > 0$.

We will use the following definition of Hölder C^s-norms of functions $f : \mathbb{R}^d \mapsto \mathbb{R}$. For $j \geq 0$, $f^{(j)}$ denotes the j-th Fréchet derivative of f. For $x \in \mathbb{R}^d$, $f^{(j)}(x)$ is a j-linear form on $\mathbb{R}^d$ and the space of such forms will be equipped with the operator norm. Clearly, $f^{(0)} = f$ and $f^{(1)} = f'$ coincides with the gradient ∇f. If f is l times differentiable and $s = l + \rho$, $\rho \in (0, 1]$, define

$$\|f\|_{C^s} := \max_{0\leq j\leq l} \sup_{x\in\mathbb{R}^d} \|f^{(j)}(x)\| \vee \sup_{x,y\in\mathbb{R}^d, x\neq y} \frac{\|f^{(l)}(x) - f^{(l)}(y)\|}{\|x-y\|^\rho}.$$

We will also frequently use L_∞ and Lipschitz norms of functions and their derivatives. For instance, $\|f^{(j)}\|_{L_\infty} = \sup_{x\in\mathbb{R}^d} \|f^{(j)}(x)\|$ and $\|f^{(j)}\|_{\mathrm{Lip}} = \sup_{x,x'\in\mathbb{R}^d, x\neq x'} \frac{\|f^{(j)}(x)-f^{(j)}(x')\|}{\|x-x'\|}$.

In what follows, we will use some facts related to isoperimetry and concentration properties of log-concave measures. Given a Borel probability measure μ on $\mathbb{R}^d$, let

$$\mu^+(A) := \liminf_{\varepsilon\to 0} \frac{\mu(A_\varepsilon) - \mu(A)}{\varepsilon}, \quad A \in \mathcal{B}(\mathbb{R}^d),$$

where A_ε denotes the ε-neighborhood of A and $\mathcal{B}(\mathbb{R}^d)$ is the Borel σ-algebra in $\mathbb{R}^d$. The so-called Cheeger isoperimetric constant of μ is defined as

$$I_C(\mu) := \inf_{A\in\mathcal{B}(\mathbb{R}^d)} \frac{\mu^+(A)}{\mu(A) \wedge (1 - \mu(A))}.$$

According to the well-known Kannan-Lovàsz-Simonovits (KLS) conjecture, for a log-concave probability measure $\mu(dx) = e^{-V(x)}dx$ on $\mathbb{R}^d$ with covariance operator Σ, $I_C(\mu) \gtrsim \|\Sigma\|^{-1/2}$ with a dimension-free constant. This conjecture remains open, but the following deep recent result by Chen [8] provides a lower bound on $I_C(\mu)$ that is almost dimension-free.

Theorem 2.1 *There exists a constant $b > 0$ such that, for all $d \geq 3$ and for all log-concave distributions $\mu(dx) = e^{-V(x)}dx$ in $\mathbb{R}^d$ with covariance Σ,*

$$I_C(\mu) \geq \|\Sigma\|^{-1/2} d^{-b(\frac{\log\log d}{\log d})^{1/2}}.$$

Isoperimetric constants $I_C(\mu)$ are known to be closely related to important functional inequalities, in particular to Poincaré inequality and its generalizations (see, e.g., [6, 24]). It is said that Poincaré inequality holds for a r.v. ξ in $\mathbb{R}^d$ iff, for some constant $C > 0$ and for all locally Lipschitz functions $g : \mathbb{R}^d \mapsto \mathbb{R}$ (which, by Rademacher theorem, are differentiable almost everywhere),

$$\mathrm{Var}(g(\xi)) \leq C\mathbb{E}\|\nabla g(\xi)\|^2.$$

The smallest value $c(\xi)$ of constant C in the above inequality is called the Poincaré constant of ξ (clearly, it depends only on the distribution of ξ). The following property of Poincaré constant will be frequently used: if r.v. $\xi = (\xi_1, \ldots, \xi_n)$ has independent components (with ξ_j being a r.v. in $\mathbb{R}^{d_j}$), then $c(\xi) = \max_{1\leq j\leq n} c(\xi_j)$ (see [20], Corollary 5.7).

If now $\xi \sim \mu$ in $\mathbb{R}^d$, then the following Cheeger's inequality holds (see, e.g., [24], Theorem 1.1):

$$c(\xi) \leq \frac{4}{I_C^2(\mu)}.$$

Moreover, the following L_p-version of Poincaré inequality holds for all $p \geq 1$ and for all locally Lipschitz functions $g : \mathbb{R}^d \mapsto \mathbb{R}$ (see [6], Theorem 3.1)

$$\|g(\xi) - \mathbb{E}g(\xi)\|_{L_p} \lesssim \frac{p}{I_C(\mu)} \|\|\nabla g(\xi)\|\|_{L_p}. \tag{2.2}$$

Remark 2.1 Note that, if $\xi \sim \mu$ and $\mu(dx) = e^{-V(x)}dx$ is log-concave, then (see [24], Theorem 1.5)

$$c(\xi) \asymp \frac{1}{I_C^2(\mu)}$$

and, by Theorem 2.1, we have

$$c(\xi) \leq \|\Sigma\| \, d^{2b(\frac{\log\log d}{\log d})^{1/2}}. \tag{2.3}$$

In what follows, we denote the Poincaré constant $c(\xi)$ of r.v. $\xi \sim \mu$ with log-concave distribution $\mu(dx) = e^{-V(x)}dx$ by $c(V)$. Bound (2.3) implies that $c(V) \lesssim_\epsilon \|\Sigma\| d^\epsilon$ for all $\epsilon > 0$. Also, with this notation, we can rewrite the L_p-version (2.2) of Poincaré inequality as follows:

$$\|g(\xi) - \mathbb{E}g(\xi)\|_{L_p} \lesssim \sqrt{c(V)}p \; \|\|\nabla g(\xi)\|\|_{L_p}. \tag{2.4}$$

This concentration bound will be our main tool in Sect. 4. It will be convenient for our purposes to express it in terms of local Lipschitz constants of g defined as follows:

$$(Lg)(x) := \inf_{U \ni x} \sup_{x', x'' \in U, x' \neq x''} \frac{|g(x') - g(x'')|}{\|x' - x''\|}, \; x \in \mathbb{R}^d$$

where the infimum is taken over all the balls U centered at x. Similar definition could be also used for vector valued functions g. Clearly, $\|\nabla g(x)\| \leq (Lg)(x), x \in \mathbb{R}^d$ and (2.4) implies that

$$\|g(\xi) - \mathbb{E}g(\xi)\|_{L_p} \lesssim \sqrt{c(V)}p \; \|(Lg)(\xi)\|_{L_p}. \tag{2.5}$$

The following assumptions on V will be used throughout the paper.

Assumption 1 *Suppose that*

(i) V is strictly convex and twice continuously differentiable such that, for some constants $M, L > 0$, $\|V''\|_{L_\infty} \leq M$ and $\|V''\|_{\mathrm{Lip}} \leq L$.
(ii) For some constant $m > 0$, $\mathcal{I} \succeq mI_d$.

Under Assumption 1, we have $\mathcal{I} = \mathbb{E}V''(\xi) \preceq MI_d$ and thus $m \leq M$.

Remark 2.2 Obviously, Assumption 1 holds in the Gaussian case, when $V(x) = c_1 + c_2\|x\|^2, x \in \mathbb{R}^d$. In this case, $V''(x) = mI_d, x \in \mathbb{R}^d$ for $m = 2c_2 > 0$, which is much stronger than Assumption 1, (ii). Assumption 1 also holds, for instance, for $V(x) = \varphi(\|x\|^2), x \in \mathbb{R}^d$, where φ is a C^∞ function in $\mathbb{R}$ such that φ'' is supported in $[0, 1]$, $\varphi''(t) \geq 0, t \in \mathbb{R}$ and $\varphi'(0) = 0$. Of course, in this case, the condition $V''(x) \succeq mI_d$ does not hold uniformly in x for any positive m, but Assumption 1,

(ii) holds. If $V(x) = c_1 + \|x\|^{2p}, x \in \mathbb{R}^d$ for some $p \geq 1/2$, it is easy to check that Assumption 1 holds only for $p = 1$.

We are now ready to state our main result.

Theorem 2.2 *Suppose Assumption 1 holds and $d \leq \gamma n$, where*

$$\gamma := c\Big(\frac{m}{M} \wedge \frac{m^2}{L\sqrt{M}}\Big)^2$$

with a small enough constant $c > 0$. Let $f \in C^s$ for some $s = k + 1 + \rho$, $k \geq 0$, $\rho \in (0, 1]$. Then

$$\sup_{\theta \in \mathbb{R}^d} \Big\| f_k(\hat\theta) - f(\theta) - n^{-1} \sum_{j=1}^n \langle V'(\xi_j), \mathcal{I}^{-1} f'(\theta)\rangle \Big\|_{L_{\psi_{2/3}}(\mathbb{P}_\theta)}$$

$$\lesssim_{L,M,m,s} \|f\|_{C^s}\Big[\sqrt{\frac{c(V)}{n}}\Big(\frac{d}{n}\Big)^{\rho/2} + \Big(\sqrt{\frac{d}{n}}\Big)^s\Big].$$

Remark 2.3 Note that, for $k = 0$, $f_k(\hat\theta)$ coincides with the plug-in estimator $f(\hat\theta)$. In particular, it means that, for $s = 2$,

$$\sup_{\theta \in \mathbb{R}^d} \Big\| f(\hat\theta) - f(\theta) - n^{-1} \sum_{j=1}^n \langle V'(\xi_j), \mathcal{I}^{-1} f'(\theta)\rangle \Big\|_{L_{\psi_{2/3}}(\mathbb{P}_\theta)}$$

$$\lesssim_{L,M,m,s} \|f\|_{C^2}\Big[\sqrt{\frac{c(V)}{n}}\Big(\frac{d}{n}\Big)^{\rho/2} + \frac{d}{n}\Big].$$

It will be clear from the proofs that the term $\frac{d}{n}$ in the above bound controls the bias of the plug-in estimator $f(\hat\theta)$. It is easy to construct examples in which this bound on the bias of $f(\hat\theta)$ is optimal. For instance, consider the case of normal model $X_1, \ldots, X_n$ i.i.d. $\sim N(\theta, I_d), \theta \in \mathbb{R}^d$, for which MLE is just $\bar X$. Let $f(\theta) = \|\theta\|^2\varphi(\|\theta\|^2), \theta \in \mathbb{R}^d$, where $\varphi : \mathbb{R} \mapsto [0, 1]$ is a C^∞-function supported in $[-2, 2]$ with $\varphi(u) = 1, u \in [0, 1]$. Then, $\|f\|_{C^s} \lesssim_s 1$ for all $s > 0$ and it is also easy to check that the bias of the plug-in estimator $f(\bar X)$ of functional $f(\theta)$ is $\asymp \frac{d}{n}$ for all θ in a neighborhood of 0. Therefore, the term $\frac{d}{n}$ in the above bound for the plug-in estimator (as well as in the risk bound (2.7) of Proposition 2.1 below) could not be improved to become $(\frac{d}{n})^{s/2}$ for functionals f of smoothness $s > 2$ and the bias reduction is needed to achieve the optimal error rates.

Theorem 2.2 shows that $f_k(\hat\theta) - f(\theta)$ can be approximated by a normalized sum of i.i.d. mean zero r.v. $n^{-1}\sum_{j=1}^n \langle V'(\xi_j), \mathcal{I}^{-1} f'(\theta)\rangle$. Moreover, the error of this approximation is of the order $o(n^{-1/2})$ provided that $d \lesssim n^\alpha$ for some $\alpha \in (0, 1)$ satisfying $s > \frac{1}{1-\alpha}$ and that $\|\Sigma\|$ is bounded by a constant. This follows from the fact that $c(V) \lesssim_\epsilon d^\epsilon \|\Sigma\|$ for an arbitrarily small $\epsilon > 0$ (see Remark 2.1). In addition, by Lemma 3.1 below, r.v. $\langle V'(\xi), \mathcal{I}^{-1} f'(\theta)\rangle$ is subgaussian with

$$\|\langle V'(\xi), \mathcal{I}^{-1} f'(\theta)\rangle\|_{\psi_2} \lesssim \sqrt{M}\|\mathcal{I}^{-1} f'(\theta)\| \lesssim \frac{\sqrt{M}}{m}\|f'(\theta)\|. \tag{2.6}$$

As a result, we can obtain the following simple but important corollaries of Theorem 2.2. Recall that $\sigma_f^2(\theta) = \langle \mathcal{I}^{-1} f'(\theta), f'(\theta)\rangle$.

Corollary 2.1 *Under the conditions of Theorem 2.2,*

$$\sup_{\theta\in\mathbb{R}^d}\left|\|f_k(\hat\theta) - f(\theta)\|_{L_2(\mathbb{P}_\theta)} - \frac{\sigma_f(\theta)}{\sqrt{n}}\right| \lesssim_{L,M,m,s} \|f\|_{C^s}\Big[\sqrt{\frac{c(V)}{n}}\Big(\frac{d}{n}\Big)^{\rho/2} + \Big(\sqrt{\frac{d}{n}}\Big)^s\Big].$$

This corollary immediately follows from the bound of Theorem 2.2 and the fact that the L_2-norm is dominated by the $\psi_{2/3}$-norm. It implies the second claim of the following proposition.

Proposition 2.1 *Let $f \in C^s$ for some $s > 0$.*

1. *For $s \in (0, 1]$,*

$$\sup_{\|f\|_{C^s}\le 1}\sup_{\theta\in\mathbb{R}^d}\|f(\hat\theta) - f(\theta)\|_{L_2(\mathbb{P}_\theta)} \lesssim_{L,M,m,s} \Big(\sqrt{\frac{d}{n}}\Big)^s \wedge 1.$$

2. *For $s = k + 1 + \rho$ for some $k \ge 0$ and $\rho \in (0, 1]$, suppose Assumption 1 holds and also $\|\Sigma\| \lesssim 1$. Then*

$$\sup_{\|f\|_{C^s}\le 1}\sup_{\theta\in\mathbb{R}^d}\|f_k(\hat\theta) - f(\theta)\|_{L_2(\mathbb{P}_\theta)} \lesssim_{L,M,m,s} \Big(\frac{1}{\sqrt{n}} \vee \Big(\sqrt{\frac{d}{n}}\Big)^s\Big) \wedge 1. \tag{2.7}$$

Combining bound (2.7) with the following result shows some form of minimax optimality of estimator $f_k(\hat\theta)$.

Proposition 2.2 *Suppose Assumption 1 holds. Then, for all $s > 0$,*

$$\sup_{\|f\|_{C^s}\le 1}\inf_{\hat T_n}\sup_{\|\theta\|\le 1}\|\hat T_n - f(\theta)\|_{L_2(\mathbb{P}_\theta)} \gtrsim_{m,M} \Big(\frac{1}{\sqrt{n}} \vee \Big(\sqrt{\frac{d}{n}}\Big)^s\Big) \wedge 1,$$

where the infimum is taken over all estimators $\hat T_n = \hat T_n(X_1, \ldots, X_n)$.

The proof of this result is similar to the proof of Theorem 2.2 in [18] in the Gaussian case. Some further comments will be provided in Sect. 6.

Corollary 2.1 also implies that, for all $\theta \in \mathbb{R}^d$,

$$\|f_k(\hat{\theta}) - f(\theta)\|_{L_2(\mathbb{P}_\theta)} \le \frac{\sigma_f(\theta)}{\sqrt{n}} + C\|f\|_{C^s}\Big[\sqrt{\frac{c(V)}{n}}\Big(\frac{d}{n}\Big)^{\rho/2} + \Big(\sqrt{\frac{d}{n}}\Big)^s\Big],$$

where C is a constant depending on M, L, m, s. If $d \lesssim n^{\alpha}$ for some $\alpha \in (0,1)$ and $s > \frac{1}{1-\alpha}$, it easily follows that, for all $B > 0$,

$$\limsup_{n\to\infty} \sup_{(f,\theta): \frac{\|f\|_{C^s}}{\sigma_f(\theta)} \le B} \frac{\sqrt{n}\|f_k(\hat{\theta}) - f(\theta)\|_{L_2(\mathbb{P}_\theta)}}{\sigma_f(\theta)} \le 1. \tag{2.8}$$

The following minimax lower bound will be proven in Sect. 6.

Proposition 2.3 *Suppose Assumption 1 holds, and let $f \in C^s$ for some $s = 1 + \rho$, $\rho \in (0, 1]$. Then, for all $c > 0$ and all $\theta_0 \in \mathbb{R}^d$,*

$$\inf_{\hat{T}_n} \sup_{\|\theta-\theta_0\| \le \frac{c}{\sqrt{n}}} \frac{\sqrt{n}\|\hat{T}_n - f(\theta)\|_{L_2(\mathbb{P}_\theta)}}{\sigma_f(\theta)} \ge 1 - \frac{3\pi}{\sqrt{8mc}} - \frac{2}{\sqrt{m}} \frac{\|f\|_{C^s}}{\sigma_f(\theta_0)} \Big(\frac{c}{\sqrt{n}}\Big)^{\rho},$$

where the infimum is taken over all estimators $\hat{T}_n = \hat{T}_n(X_1, \dots, X_n)$.

The bound of Proposition 2.3 easily implies that, for all $B > 0$,

$$\lim_{c\to\infty} \liminf_{n\to\infty} \inf_{(f,\theta_0): \frac{\|f\|_{C^s}}{\sigma_f(\theta_0)} \le B} \inf_{\hat{T}_n} \sup_{\|\theta-\theta_0\| \le \frac{c}{\sqrt{n}}} \frac{\sqrt{n}\|\hat{T}_n - f(\theta)\|_{L_2(\mathbb{P}_\theta)}}{\sigma_f(\theta)} \ge 1.$$

Along with (2.8), it shows local asymptotic minimaxity of estimator $f_k(\hat{\theta})$.

The next corollaries will be based on the results by Rio [27] on convergence rates in CLT in Wasserstein type distances. For r.v. η_1, η_2 and a convex nondecreasing function $\psi : \mathbb{R}_+ \mapsto \mathbb{R}_+$ with $\psi(0) = 0$, define the Wasserstein ψ-distance between η_1 and η_2 as

$$W_\psi(\eta_1, \eta_2) := W_{L_\psi(\mathbb{P})}(\eta_1, \eta_2) := \inf\Big\{\|\eta_1' - \eta_2'\|_\psi : \eta_1' \overset{d}{=} \eta_1, \eta_2' \overset{d}{=} \eta_2\Big\}.$$

For $\psi(u) = u^p, u \ge 0, p \ge 1$, we will use the notation $W_p = W_{L_p(\mathbb{P})}$ instead of W_ψ. For $\psi = \psi_\alpha, \alpha > 0$, we will modify the above definition using a version of ψ-norm defined in terms of the moments. To emphasize the dependence of the underlying probability measure $\mathbb{P}$ involved in the definitions of these distances on the parameter θ of our statistical model, we will write $W_{\psi,\mathbb{P}_\theta} = W_{L_\psi(\mathbb{P}_\theta)}$, $W_{p,\mathbb{P}_\theta} = W_{L_p(\mathbb{P}_\theta)}$, etc.

Let $\eta_1, \dots, \eta_n$ be i.i.d. copies of a mean zero r.v. η with $\mathbb{E}\eta^2 = 1$. It was proven in [27] (see Theorem 4.1 and Equation (4.3)) that for all $r \in (1, 2]$,

$$W_r\Big(\frac{\eta_1+\cdots+\eta_n}{\sqrt{n}}, Z\Big) \lesssim \frac{\mathbb{E}^{1/r}\eta^{r+2}}{\sqrt{n}},$$

where $Z \sim N(0,1)$. Applying this bound to $\eta := \frac{\langle V'(\xi), \mathcal{I}^{-1}f'(\theta)\rangle}{\sigma_f(\theta)}$ yields

$$W_2\Big(\frac{1}{\sqrt{n}}\sum_{j=1}^n \langle V'(\xi_j), \mathcal{I}^{-1}f'(\theta)\rangle, \sigma_f(\theta)Z\Big) \lesssim \frac{\mathbb{E}^{1/2}\langle V'(\xi), \mathcal{I}^{-1}f'(\theta)\rangle^4}{\sigma_f^2(\theta)} n^{-1/2}.$$

Thus, Theorem 2.2 implies the following corollary.

Corollary 2.2 *Under the conditions of Theorem 2.2, for all $\theta \in \mathbb{R}^d$*

$$W_{2,\mathbb{P}_\theta}\Big(\sqrt{n}(f_k(\hat\theta) - f(\theta)), \sigma_f(\theta)Z\Big)$$
$$\le C_1 \frac{\mathbb{E}^{1/2}\langle V'(\xi), \mathcal{I}^{-1}f'(\theta)\rangle^4}{\sigma_f^2(\theta)} n^{-1/2} + C_2\|f\|_{C^s}\Big[\sqrt{c(V)}\Big(\frac{d}{n}\Big)^{\rho/2} + \sqrt{n}\Big(\sqrt{\frac{d}{n}}\Big)^s\Big],$$

where $C_1 > 0$ is an absolute constant and $C_2 > 0$ is a constant that could depend on M, L, m, s.

Using (2.6), it is easy to check that, under Assumption 1,

$$\mathbb{E}^{1/2}\langle V'(\xi), \mathcal{I}^{-1}f'(\theta)\rangle^4 \lesssim \frac{M}{m^2}\|f'(\theta)\|^2$$

and, in addition, $\sigma_f^2(\theta) \ge M^{-1}\|f'(\theta)\|^2$. This yields

$$\frac{\mathbb{E}^{1/2}\langle V'(\xi), \mathcal{I}^{-1}f'(\theta)\rangle^4}{\sigma_f^2(\theta)} \lesssim \frac{M^2}{m^2}.$$

Therefore, if $d \le n^\alpha$ for some $\alpha \in (0,1)$ and $s > \frac{1}{1-\alpha}$, then

$$\sup_{\|f\|_{C^s}\le 1}\ \sup_{\theta\in\mathbb{R}^d} W_{2,\mathbb{P}_\theta}\Big(\sqrt{n}(f_k(\hat\theta) - f(\theta)), \sigma_f(\theta)Z\Big) \to 0 \text{ as } n \to \infty,$$

implying asymptotic normality of estimator $f_k(\hat\theta)$ of $f(\theta)$ with rate $\sqrt{n}$ and limit variance $\sigma_f^2(\theta)$. It is also easy to show that, under the same conditions on d and s, we have, for all $B > 0$,

$$\sup_{(f,\theta):\frac{\|f\|_{C^s}}{\sigma_f(\theta)}\le B} W_{2,\mathbb{P}_\theta}\Big(\frac{\sqrt{n}(f_k(\hat\theta) - f(\theta))}{\sigma_f(\theta)}, Z\Big) \to 0 \text{ as } n \to \infty,$$

which implies

$$\sup_{(f,\theta):\frac{\|f\|_{C^s}}{\sigma_f(\theta)}\leq B}\sup_{x\in\mathbb{R}}\left|\mathbb{P}_\theta\left\{\frac{\sqrt{n}(f_k(\hat{\theta})-f(\theta))}{\sigma_f(\theta)}\leq x\right\}-\mathbb{P}\{Z\leq x\}\right|\to 0 \text{ as } n\to\infty.$$

It was also proven in [27], Theorem 2.1, that, for i.i.d. copies $\eta_1,\ldots,\eta_n$ of mean zero r.v. η with $\mathbb{E}\eta^2=1$ and $\|\eta\|_{\psi_1}<\infty$ and for some constant $C(\|\eta\|_{\psi_1})<\infty$,

$$W_{\psi_1}\Bigl(\frac{\eta_1+\cdots+\eta_n}{\sqrt{n}},Z\Bigr)\lesssim\frac{C(\|\eta\|_{\psi_1})}{\sqrt{n}}.$$

We will again apply this to $\eta:=\frac{\langle V'(\xi),\mathcal{I}^{-1}f'(\theta)\rangle}{\sigma_f(\theta)}$. In this case, by Lemma 3.1, we have

$$\|\langle V'(\xi),\mathcal{I}^{-1}f'(\theta)\rangle\|_{\psi_1}\lesssim\frac{\sqrt{M}}{m}\|f'(\theta)\|.$$

Also, $\sigma_f(\theta)\geq\frac{\|f'(\theta)\|}{\sqrt{M}}$, implying $\|\eta\|_{\psi_1}\lesssim\frac{M}{m}$. As a result, we get

$$W_{\psi_1}\Bigl(\frac{1}{n}\sum_{j=1}^n\langle V'(\xi_j),\mathcal{I}^{-1}f'(\theta)\rangle,\frac{\sigma_f(\theta)Z}{\sqrt{n}}\Bigr)\lesssim_{M,m}\frac{\sigma_f(\theta)}{n}.$$

Combining this with the bound of Theorem 2.2 yields the following extension of Corollary 2.1.

Corollary 2.3 *Under the conditions of Theorem 2.2, for all convex nondecreasing functions* $\psi:\mathbb{R}_+\to\mathbb{R}_+$ *with* $\psi(0)=0$, *satisfying the condition* $\psi(u)\leq\psi_{2/3}(cu)$, $u\geq 0$ *for some constant* $c>0$,

$$\sup_{\theta\in\mathbb{R}^d}\Bigl|\|f_k(\hat{\theta})-f(\theta)\|_{L_\psi(\mathbb{P}_\theta)}-\frac{\sigma_f(\theta)}{\sqrt{n}}\|Z\|_\psi\Bigr|$$
$$\lesssim_{L,M,m,s}\|f\|_{C^s}\Bigl[\sqrt{\frac{c(V)}{n}}\Bigl(\frac{d}{n}\Bigr)^{\rho/2}+\Bigl(\sqrt{\frac{d}{n}}\Bigr)^s\Bigr].$$

Remark 2.4 Similar results were obtained in [17] in the case of Gaussian shift models, in [19] in the case of more general Poincaré random shift models, and in [14, 15, 18] in the case of Gaussian models with unknown covariance and unknown mean and covariance (the analysis becomes much more involved in the case when the functional depends on unknown covariance). In [16], the proposed higher order bias reduction method was studied in the case of general models with a high-dimensional parameter θ for which there exists an estimator $\hat{\theta}$ admitting high-dimensional normal approximation.

Remark 2.5 If $\mathbb{E}\xi = 0$, one can also use $\bar{X} = \frac{X_1+\cdots+X_n}{n}$ as an estimator of θ and construct the corresponding functions $\bar{f}_k$ based on this estimator. In this case, a bound similar to (2.7) holds for estimator $\bar{f}_k(\bar{X})$, so, it is also minimax optimal. This follows from Theorem 2 [19] along with the bound on Poincaré constant $c(V)$ (see Remark 2.1). Normal approximation of estimator $\bar{f}_k(\bar{X})$ similar to Corollary 2.2 also holds (see [19]). However, the limit variance of estimator $\bar{f}_k(\bar{X})$ is not equal to $\sigma_f^2(\theta)$, but it is rather equal to $\langle \Sigma_\xi f'(\theta), f'(\theta)\rangle$. Since X is an unbiased estimator of θ (note that $\mathbb{E}\xi = 0$), it follows from the Cramér-Rao bound that $\Sigma_X = \Sigma_\xi \succeq \mathcal{I}^{-1}$. This fact implies that the limit variance of estimator $\bar{f}_k(\bar{X})$ is suboptimal:

$$\langle \Sigma_\xi f'(\theta), f'(\theta)\rangle \geq \langle \mathcal{I}^{-1} f'(\theta), f'(\theta)\rangle$$

and this estimator is not asymptotically efficient. This was the main motivation for the development of estimators $f_k(\hat{\theta})$ based on the MLE in the current paper. We conjecture that asymptotic efficiency also holds when the MLE is replaced by Pitman's estimator. Since MLE is defined implicitly as a solution of an optimization problem, there is an additional layer of difficulties in the analysis of the problem comparing with the case of $\bar{X}$. Similar problems in the case of log-concave location-scale families seem to be much more challenging.

Remark 2.6 The proof of Theorem 2.2 could be easily modified and, in fact, significantly simplified to obtain the following result under somewhat different assumptions than Assumption 1 (they are stronger in the sense that the eigenvalues of the Hessian $V''(x)$ are assumed to be bounded away from zero uniformly in x).

Theorem 2.3 *Suppose V is twice continuously differentiable and, for some $M, m > 0$, $\|V''\|_{L_\infty} \leq M$ and $V''(x) \succeq mI_d, x \in \mathbb{R}^d$. Let $f \in C^s$ for some $s = k+1+\rho$, $k \geq 0$, $\rho \in (0,1]$. Then, for all $d \lesssim n$,*

$$\sup_{\theta\in\mathbb{R}^d} \Big\| f_k(\hat{\theta}) - f(\theta) - n^{-1}\sum_{j=1}^n \langle V'(\xi_j), \mathcal{I}^{-1} f'(\theta)\rangle \Big\|_{L_{\psi_1}(\mathbb{P}_\theta)}$$

$$\lesssim_{M,m,s} \|f\|_{C^s}\Big[\frac{1}{\sqrt{n}}\Big(\frac{d}{n}\Big)^{\rho/2} + \Big(\sqrt{\frac{d}{n}}\Big)^s\Big].$$

This result implies that, under the conditions of Theorem 2.3, the bound of Corollary 2.3 holds for all convex nondecreasing functions $\psi : \mathbb{R}_+ \mapsto \mathbb{R}_+$ with $\psi(0) = 0$, satisfying the condition $\psi(u) \leq \psi_1(cu)$, $u \geq 0$ for some constant $c > 0$.

3 Error Bounds for the MLE

Our main goal in this section is to obtain upper bounds on the error $\|\hat{\theta} - \theta\|$ of MLE $\hat{\theta}$. Namely, the following result will be proven.

Theorem 3.1 *Suppose Assumption 1 holds and let* $t \geq 1$. *If* $d \vee t \leq \gamma n$ *for*

$$\gamma := c\Big(\frac{m}{M} \wedge \frac{m^2}{L\sqrt{M}}\Big)^2 \tag{3.1}$$

with a small enough constant $c > 0$, *then, with probability at least* $1 - e^{-t}$,

$$\|\hat{\theta} - \theta\| \lesssim \frac{\sqrt{M}}{m}\Big(\sqrt{\frac{d}{n}} \vee \sqrt{\frac{t}{n}}\Big).$$

Several simple facts will be used in the proof.

For a differentiable function $g : \mathbb{R}^d \mapsto \mathbb{R}$, define the remainder of its first order Taylor expansion

$$S_g(x; h) := g(x+h) - g(x) - \langle g'(x), h\rangle, x, h \in \mathbb{R}^d.$$

The next proposition is straightforward.

Proposition 3.1 *Let* g *be twice differentiable. Then, for all* $x, y, h, h' \in \mathbb{R}^d$, *the following properties hold:*

(i) $|S_g(x; h)| \leq \frac{1}{2}\|g''\|_{L_\infty}\|h\|^2$.
(ii) $|S_g(x; h) - \frac{1}{2}\langle g''(x)h, h\rangle| \leq \frac{1}{6}\|g''\|_{\mathrm{Lip}}\|h\|^3$.
(iii) $|S_g(x; h) - S_g(x; h')| \leq \frac{1}{2}\|g''\|_{L_\infty}\|h - h'\|^2 + \|g''\|_{L_\infty}\|h\|\|h - h'\|$.
(iv) $|S_g(x; h) - S_g(y; h)| \leq \frac{1}{4}\|g''\|_{\mathrm{Lip}}\|h\|^2\|x - y\|$.

If $g \in C^s$ *for* $s = 1 + \rho$, $\rho \in (0, 1]$, *then*

(v) $|S_g(x; h) - S_g(x; h')| \lesssim \|g\|_{C^s}(\|h\|^\rho \vee \|h'\|^\rho)\|h - h'\|$.

Let $\xi_1, \ldots, \xi_n$ be i.i.d. copies of ξ (that is, $\xi_j := X_j - \theta$). Define the following convex functions:

$$g(h) := \mathbb{E}V(\xi + h),$$

$$g_n(h) := n^{-1}\sum_{j=1}^{n} V(\xi_j + h), h \in \mathbb{R}^d. \tag{3.2}$$

Note that $\mathbb{E}g_n(h) = g(h)$ and $g''(0) = \mathcal{I}$.

We will need simple probabilistic bounds for r.v. $g_n'(0) = n^{-1}\sum_{j=1}^{n} V'(\xi_j)$ and $g_n''(0) = n^{-1}\sum_{j=1}^{n} V''(\xi_j)$. We start with the following lemma.

Lemma 3.1 *For all* $u \in \mathbb{R}^d$, $\langle V'(\xi), u\rangle$ *is a subgaussian r.v. with*

$$\|\langle V'(\xi), u\rangle\|_{\psi_2} \lesssim \sqrt{M}\|u\|.$$

Proof For all $k \geq 1$, we have

$$\mathbb{E}\langle V'(\xi), u\rangle^{2k} = \int_{\mathbb{R}^d} \langle V'(x), u\rangle^{2k} e^{-V(x)} dx.$$

By Lemma 2.2.1 in [7], there are constants $A, B > 0$ such that $e^{-V(x)} \leq Ae^{-B\|x\|}$ for all $x \in \mathbb{R}^d$. Moreover, by Assumption 1, V' is M-Lipschitz which implies that $\|V'(x)\| \leq \|V'(0)\| + M\|x\|$ for all $x \in \mathbb{R}^d$. Combining these two facts, the above integral is finite for all $k \geq 1$ and we obtain that

$$\begin{aligned}\int_{\mathbb{R}^d} \langle V'(x), u\rangle^{2k} e^{-V(x)} dx &= \int_{\mathbb{R}^d} \langle V'(x), u\rangle^{2k-1} \langle V'(x), u\rangle e^{-V(x)} dx \\ &= \int_{\mathbb{R}^d} (2k-1)\langle V'(x), u\rangle^{2k-2} \langle V''(x)u, u\rangle e^{-V(x)} dx,\end{aligned}$$

where we used integration by parts in the last equality. Therefore,

$$\begin{aligned}\mathbb{E}\langle V'(\xi), u\rangle^{2k} &\leq (2k-1)M\|u\|^2 \int_{\mathbb{R}^d} \langle V'(x), u\rangle^{2k-2} e^{-V(x)} dx \\ &= (2k-1)M\|u\|^2 \mathbb{E}\langle V'(\xi), u\rangle^{2(k-1)}.\end{aligned}$$

It follows by induction that

$$\mathbb{E}\langle V'(\xi), u\rangle^{2k} \leq (2k-1)!!M^k \|u\|^{2k}.$$

It is easy to conclude that, for all $p \geq 1$,

$$\|\langle V'(\xi), u\rangle\|_{L_p} \lesssim \sqrt{p}\sqrt{M}\|u\|,$$

implying the claim. □

An immediate consequence is the following corollary.

Corollary 3.1 *For all* $t \geq 1$, *with probability at least* $1 - e^{-t}$

$$\|g_n'(0)\| \lesssim \sqrt{M}\Big(\sqrt{\frac{d}{n}} \vee \sqrt{\frac{t}{n}}\Big).$$

Proof Let S^{d-1} be the unit sphere in $\mathbb{R}^d$ and let $A \subset S^{d-1}$ be a 1/2-net with $\mathrm{card}(A) \leq 5^d$. Then

$$\|g_n'(0)\| = \sup_{u \in S^{d-1}} \langle g_n'(0), u\rangle \leq 2 \max_{u \in A} \langle g_n'(0), u\rangle.$$

By Lemma 3.1, we have for all $u \in A$,

$$\|\langle g_n'(0), u\rangle\|_{\psi_2} = \Big\|n^{-1}\sum_{j=1}^n \langle V'(\xi_j), u\rangle\Big\|_{\psi_2} \lesssim \frac{\sqrt{M}}{\sqrt{n}},$$

implying that with probability at least $1 - e^{-t}$

$$|\langle g_n'(0), u\rangle| \lesssim \sqrt{M}\sqrt{\frac{t}{n}}.$$

It remains to use the union bound and to replace t by $t + d\log(5)$. □

Proposition 3.2 *For all $t \geq 1$, with probability at least $1 - e^{-t}$*

$$\|g_n''(0) - \mathcal{I}\| \lesssim M\Big(\sqrt{\frac{d}{n}} \vee \sqrt{\frac{t}{n}}\Big).$$

Moreover,

$$\Big\|\|g_n''(0) - \mathcal{I}\|\Big\|_{\psi_2} \lesssim M\sqrt{\frac{d}{n}}.$$

Proof Similarly to the proof of Corollary 3.1, one can use the fact that $\|\langle V''(\xi)u, v\rangle\|_{\psi_2} \lesssim \|\langle V''(\xi)u, v\rangle\|_{L_\infty} \lesssim M, u, v \in S^{d-1}$ and discretization of the unit sphere to prove the first bound.

Moreover, using that $t, d \geq 1$, the first bound implies that

$$\mathbb{P}\Big\{\|g_n''(0) - \mathcal{I}\| \geq C_1 M\sqrt{\frac{d}{n}}\sqrt{t}\Big\} \leq e^{-t}, t \geq 1,$$

which is equivalent to the second bound. □

We now turn to the proof of Theorem 3.1.

Proof Note that the minimum of convex function g from (3.2) is attained at 0 and also

$$\hat{h} := \operatorname*{argmin}_{h\in\mathbb{R}^d} g_n(h) = \theta - \hat{\theta}, \tag{3.3}$$

so, to prove Theorem 3.1, it will be enough to bound $\|\hat{h}\|$. We will use the following elementary lemma.

Lemma 3.2 *Let $q : \mathbb{R}^d \mapsto \mathbb{R}$ be a convex function attaining its minimum at $\bar{x} \in \mathbb{R}^d$. For all $x_0 \in \mathbb{R}^d$ and $\delta > 0$, the condition $\|\bar{x} - x_0\| \geq \delta$ implies that $\inf_{\|x-x_0\|=\delta} q(x) - q(x_0) \leq 0$.*

Proof Indeed, assume that $\|\bar{x} - x_0\| \geq \delta$. Clearly, $q(\bar{x}) - q(x_0) \leq 0$. Let $x^* = \lambda\bar{x} + (1-\lambda)x_0$ with $\lambda := \frac{\delta}{\|\bar{x}-x_0\|}$. Then, $\|x^* - x_0\| = \delta$ and, by convexity of q, $q(x^*) \leq \lambda q(\bar{x}) + (1-\lambda)q(x_0)$, implying that $q(x^*) - q(x_0) \leq \lambda(q(\bar{x}) - q(x_0)) \leq 0$. □

If $\|\hat{h}\| \geq \delta$, then, by Lemma 3.2,

$$\inf_{\|h\|=\delta} g_n(h) - g_n(0) \leq 0. \tag{3.4}$$

Note that

$$\begin{aligned} g_n(h) - g_n(0) &= \langle g_n'(0), h\rangle + S_{g_n}(0; h) \\ &= \langle g_n'(0), h\rangle + S_{g_n}(0; h) - \frac{1}{2}\langle g_n''(0)h, h\rangle + \frac{1}{2}\langle g_n''(0)h, h\rangle. \end{aligned} \tag{3.5}$$

For $\|h\| = \delta$, we have, by Assumption 1, (ii),

$$\langle g_n''(0)h, h\rangle = \langle g_n''(0)h, h\rangle - \langle \mathcal{I}h, h\rangle + \langle \mathcal{I}h, h\rangle \geq m\delta^2 - \delta^2\|g_n''(0) - \mathcal{I}\|$$

and, by Proposition 3.1, (ii),

$$S_{g_n}(0; h) - \frac{1}{2}\langle g_n''(0)h, h\rangle \geq -\frac{L}{2}\delta^3.$$

Inserting these inequalities into (3.5) and using (3.4), we can conclude that if $\|\hat{h}\| \geq \delta$, then

$$\|g_n'(0)\|\delta + \frac{\delta^2}{2}\|g_n''(0) - \mathcal{I}\| \geq \frac{m}{2}\delta^2 - \frac{L}{2}\delta^3. \tag{3.6}$$

To complete the proof, assume that the bound of Corollary 3.1 holds with constant $C_1 \geq 1$ and the bound of Proposition 3.2 holds with constant $C_2 \geq 1$. If constant c in the definition of γ is small enough, then the condition $d \vee t \leq \gamma n$ implies that

$$C\Big(\sqrt{\frac{d}{n}} \vee \sqrt{\frac{t}{n}}\Big) \leq \frac{m}{M} \wedge \frac{m^2}{L\sqrt{M}}$$

with $C := (16C_1) \vee (4C_2)$. Moreover, let

$$\delta := 4C_1\frac{\sqrt{M}}{m}\Big(\sqrt{\frac{d}{n}} \vee \sqrt{\frac{t}{n}}\Big)$$

Then, $\delta \leq \frac{m}{4L}$ and, on the event

$$E := \Big\{\|g_n''(0) - \mathcal{I}\| \leq C_2 M\Big(\sqrt{\frac{d}{n}} \vee \sqrt{\frac{t}{n}}\Big)\Big\},$$

bound (3.6) implies that $\delta \leq \frac{4}{m}\|g_n'(0)\|$. Note also that, by Proposition 3.2, $\mathbb{P}(E^c) \leq e^{-t}$. By Corollary 3.1, the event $\{\delta \leq \frac{4}{m}\|g_n'(0)\|\}$ occurs with probability at most e^{-t}. Recall that bound (3.6) follows from $\|\hat{\theta}-\theta\| = \|\hat{h}\| \geq \delta$. Thus, with probability at least $1 - 2e^{-t}$, $\|\hat{\theta} - \theta\| \leq \delta$. It remains to adjust the constants in order to replace the probability bound $1 - 2e^{-t}$ with $1 - e^{-t}$.

The following fact will be also useful.

Corollary 3.2 *Suppose Assumption 1 holds and that $d \leq \gamma n$, where $\gamma = c\big(\frac{m}{M} \wedge \frac{m^2}{L\sqrt{M}}\big)^2$ with a small enough constant $c > 0$. Then*

$$\Big\| \|\hat{\theta} - \theta\| \wedge \frac{m}{12L}\Big\|_{\psi_2} \lesssim \frac{\sqrt{M}}{m}\sqrt{\frac{d}{n}} + \frac{m}{L\sqrt{\gamma}}\frac{1}{\sqrt{n}}.$$

Proof First, for $d \leq \gamma n$, Theorem 3.1 can be formulated as

$$\mathbb{P}\Big\{\|\hat{\theta} - \theta\| > C_1\Big(\frac{\sqrt{M}}{m}\sqrt{\frac{d}{n}} \vee \frac{\sqrt{M}}{m}\sqrt{\frac{t}{n}}\Big)\Big\} \leq e^{-t}, t \in [1, \gamma n].$$

This implies that

$$\mathbb{P}\Big\{\|\hat{\theta} - \theta\| > \Big(C_1\frac{\sqrt{M}}{m}\sqrt{\frac{d}{n}} \vee \frac{m}{12L\sqrt{\gamma}}\frac{1}{\sqrt{n}}\Big)\sqrt{t}\Big\} \leq e^{-t}, t \in [1, \gamma n],$$

using that $t \geq 1$ and $\sqrt{\gamma} \leq \frac{1}{12C_1}\frac{m^2}{L\sqrt{M}}$ for c sufficiently small. It follows that

$$\mathbb{P}\Big\{\|\hat{\theta} - \theta\| \wedge \frac{m}{12L} > \Big(C_1\frac{\sqrt{M}}{m}\sqrt{\frac{d}{n}} \vee \frac{m}{12L\sqrt{\gamma}}\frac{1}{\sqrt{n}}\Big)\sqrt{t}\Big\} \leq e^{-t}, t \geq 1,$$

which is equivalent to the claim. □

4 Concentration Bounds

In this section, we prove concentration inequalities for $f(\hat{\theta})$, where f is a smooth function on $\mathbb{R}^d$. Namely, we will prove the following result.

Theorem 4.1 *Let $f \in C^s$ for some $s = 1 + \rho$, $\rho \in (0, 1]$. Suppose that $d \leq \gamma n$, where $\gamma := c\big(\frac{m}{M} \wedge \frac{m^2}{L\sqrt{M}}\big)^2$ with a small enough $c > 0$. Then*

$$\sup_{\theta \in \mathbb{R}^d} \Big\| f(\hat{\theta}) - \mathbb{E}_\theta f(\hat{\theta}) - n^{-1} \sum_{j=1}^n \langle V'(\xi_j), \mathcal{I}^{-1} f'(\theta) \rangle \Big\|_{L_{\psi_{2/3}}(\mathbb{P}_\theta)}$$
$$\lesssim_{M,L,m} \sqrt{c(V)} \|f\|_{C^s} \frac{1}{\sqrt{n}} \Big(\frac{d}{n}\Big)^{\rho/2}.$$

To derive concentration bounds for $f(\hat{\theta})$, we need to bound local Lipschitz constants of estimator $\hat{\theta}(X_1, \ldots, X_n)$ as a function of its variables. A good place to start is to show the continuity of this function. The following fact is, probably, well known. We give its proof for completeness.

Proposition 4.1 *Suppose that V is strictly convex. Then, MLE $\hat{\theta}(x_1, \ldots, x_n)$ exists and is unique for all $(x_1, \ldots, x_n) \in \mathbb{R}^d \times \cdots \times \mathbb{R}^d$ and the function*

$$\mathbb{R}^d \times \cdots \times \mathbb{R}^d \ni (x_1, \ldots, x_n) \mapsto \hat{\theta}(x_1, \ldots, x_n) \in \mathbb{R}^d$$

is continuous.

Proof Let $(x_1, \ldots, x_n) \in \mathbb{R}^d \times \cdots \times \mathbb{R}^d$, $(x_1^{(k)}, \ldots, x_n^{(k)}) \in \mathbb{R}^d \times \cdots \times \mathbb{R}^d$, $k \geq 1$ and $(x_1^{(k)}, \ldots, x_n^{(k)}) \to (x_1, \ldots, x_n)$ as $k \to \infty$. Define

$$p(\theta) := n^{-1} \sum_{j=1}^n V(x_j - \theta),\ p_k(\theta) := n^{-1} \sum_{j=1}^n V(x_j^{(k)} - \theta), \theta \in \mathbb{R}^d, k \geq 1.$$

By continuity of V, $p_k(\theta) \to p(\theta)$ as $k \to \infty$ for all $\theta \in \mathbb{R}^d$. Since p_k and p are convex, this implies the uniform convergence on all compact subsets of $\mathbb{R}^d$.

If $\|\hat{\theta}(x_1^{(k)}, \ldots, x_n^{(k)}) - \hat{\theta}(x_1, \ldots, x_n)\| \geq \delta$, then, by Lemma 3.2,

$$\inf_{\|\theta - \hat{\theta}(x_1, \ldots, x_n)\| = \delta} p_k(\theta) - p_k(\hat{\theta}(x_1, \ldots, x_n)) \leq 0$$

By the uniform convergence of p_k to p on compact sets,

$$\inf_{\|\theta - \hat{\theta}(x_1, \ldots, x_n)\| = \delta} p_k(\theta) - p_k(\hat{\theta}(x_1, \ldots, x_n)) \to \inf_{\|\theta - \hat{\theta}(x_1, \ldots, x_n)\| = \delta} p(\theta) - p(\hat{\theta}(x_1, \ldots, x_n))$$

as $k \to \infty$. Due to strict convexity, the minimum $\hat{\theta}(x_1, \ldots, x_n)$ of $p(\theta)$ exists and is unique (see the argument in the introduction), and

$$\inf_{\|\theta-\hat{\theta}(x_1,\ldots,x_n)\|=\delta} p(\theta) - p(\hat{\theta}(x_1,\ldots,x_n)) > 0,$$

implying that $\|\hat{\theta}(x_1^{(k)},\ldots,x_n^{(k)}) - \hat{\theta}(x_1,\ldots,x_n)\| < \delta$ for all large enough k and thus $\hat{\theta}(x_1^{(k)},\ldots,x_n^{(k)}) \to \hat{\theta}(x_1,\ldots,x_n)$ as $k \to \infty$. □

Note that the continuity of $\hat{\theta}$ also follows from the implicit function theorem in the case when V is twice differentiable with V'' being positive definite throughout $\mathbb{R}^d$.

We will now study Lipschitz continuity properties of $\hat{\theta}$ as a function of the data $X_1,\ldots,X_n$ needed to prove concentration inequalities.

Proposition 4.2 *Let*

$$A_1 := \Big\{(x_1,\ldots,x_n) \in \mathbb{R}^d \times \cdots \times \mathbb{R}^d : \|\hat{\theta}(x_1,\ldots,x_n) - \theta\| \le \frac{m}{12L}\Big\}$$

and

$$A_2 := \Big\{(x_1,\ldots,x_n) \in \mathbb{R}^d \times \cdots \times \mathbb{R}^d : \Big\|n^{-1}\sum_{j=1}^n V''(x_j-\theta) - \mathcal{I}\Big\| \le \frac{m}{4}\Big\}$$

and let $A := A_1 \cap A_2$. Then the function $A \ni (x_1,\ldots,x_n) \mapsto \hat{\theta}(x_1,\ldots,x_n)$ is Lipschitz with constant $\frac{4M}{m\sqrt{n}}$: for all $(x_1,\ldots,x_n), (\tilde{x}_1,\ldots,\tilde{x}_n) \in A$,

$$\|\hat{\theta}(x_1,\ldots,x_n) - \hat{\theta}(\tilde{x}_1,\ldots,\tilde{x}_n)\| \le \frac{4M}{m\sqrt{n}}\Big(\sum_{j=1}^n \|x_j - \tilde{x}_j\|^2\Big)^{1/2}.$$

Proof Due to equivariance, we have

$$\hat{\theta}(x_1,\ldots,x_n) - \hat{\theta}(\tilde{x}_1,\ldots,\tilde{x}_n) = \hat{\theta}(\xi_1,\ldots,\xi_n) - \hat{\theta}(\tilde{\xi}_1,\ldots,\tilde{\xi}_n)$$

with $\xi_j = x_j - \theta$ and $\tilde{\xi}_j = \tilde{x}_j - \theta$. Hence, if we abbreviate $\hat{h} = \theta - \hat{\theta}(\xi_1,\ldots,\xi_n)$ and $\tilde{h} = \theta - \hat{\theta}(\tilde{\xi}_1,\ldots,\tilde{\xi}_n)$, then we have $g_n'(\hat{h}) = 0$ with g_n from (3.2) and $\tilde{g}_n'(\tilde{h}) = 0$ with

$$\tilde{g}_n(h) := n^{-1}\sum_{j=1}^n V(\tilde{\xi}_j + h),\, h \in \mathbb{R}^d.$$

Recall that $g'(0) = 0$ and $g''(0) = \mathcal{I}$. By the first order Taylor expansion for function g', $g'(h) = \mathcal{I}h + r(h)$, where

$$r(h) := \int_0^1 (g''(\lambda h) - g''(0))d\lambda\, h$$

is the remainder. Therefore,

$$g'(\hat{h}) = \mathcal{I}\hat{h} + r(\hat{h}),$$

implying that

$$\begin{aligned}\hat{h} &= \mathcal{I}^{-1}(g'(\hat{h}) - g_n'(\hat{h})) - \mathcal{I}^{-1}r(\hat{h}) \\ &= \mathcal{I}^{-1}(g'(0) - g_n'(0)) + \mathcal{I}^{-1}q_n(\hat{h}) - \mathcal{I}^{-1}r(\hat{h}) \\ &= -\mathcal{I}^{-1}g_n'(0) + \mathcal{I}^{-1}q_n(\hat{h}) - \mathcal{I}^{-1}r(\hat{h}),\end{aligned} \tag{4.1}$$

where

$$q_n(h) := (g_n - g)'(h) - (g_n - g)'(0) = \int_0^1 (g_n'' - g'')(\lambda h)d\lambda\, h.$$

Similarly, we have $\tilde{h} = -\mathcal{I}^{-1}\tilde{g}_n'(0) + \mathcal{I}^{-1}\tilde{q}_n(\tilde{h}) - \mathcal{I}^{-1}r(\tilde{h})$ with $\tilde{q}_n(h) := (\tilde{g}_n - g)'(h) - (\tilde{g}_n - g)'(0)$. Using these representations we now bound the difference between $\hat{h}$ and $\tilde{h}$.

First note that

$$\begin{aligned}\|g_n'(0) - \tilde{g}_n'(0)\| &\le n^{-1}\sum_{j=1}^n \|V''(\xi_j) - V''(\tilde{\xi}_j)\| \\ &\le Mn^{-1}\sum_{j=1}^n \|\xi_j - \tilde{\xi}_j\| \le \frac{M}{\sqrt{n}}\Big(\sum_{j=1}^n \|x_j - \tilde{x}_j\|^2\Big)^{1/2}.\end{aligned} \tag{4.2}$$

Also,

$$\begin{aligned}q_n(\hat{h}) - \tilde{q}_n(\tilde{h}) &= ((g_n'' - g'')(0))(\hat{h} - \tilde{h}) \\ &\quad + \int_0^1 ((g_n'' - g'')(\lambda\hat{h}) - (g_n'' - g'')(0))d\lambda\, (\hat{h} - \tilde{h}) \\ &\quad + \int_0^1 [(g_n'' - g'')(\lambda\hat{h}) - (g_n'' - g'')(\lambda\tilde{h})]d\lambda\, \tilde{h} \\ &\quad + \int_0^1 (g_n'' - \tilde{g}_n'')(\lambda\tilde{h})d\lambda\, \tilde{h}.\end{aligned}$$

Since, by Assumption 1, V'' is Lipschitz with constant L, the function $h \mapsto g''(h) = \mathbb{E}V''(\xi + h)$ satisfies the Lipschitz condition with the same constant L and the function $h \mapsto (g_n'' - g'')(h)$ is Lipschitz with constant at most $2L$. In addition,

$$\|(g_n'' - \tilde{g}_n'')(\lambda\tilde{h})\| \leq n^{-1}\sum_{j=1}^{n}\|V''(\xi_j + \lambda\tilde{h}) - V''(\tilde{\xi}_j + \lambda\tilde{h})\|$$

$$\leq \frac{L}{n}\sum_{j=1}^{n}\|\xi_j - \tilde{\xi}_j\| \leq \frac{L}{\sqrt{n}}\Big(\sum_{j=1}^{n}\|x_j - \tilde{x}_j\|^2\Big)^{1/2}.$$

Therefore, we easily get

$$\|q_n(\hat{h}) - \tilde{q}_n(\tilde{h})\| \leq \|g_n''(0) - g''(0)\|\|\hat{h} - \tilde{h}\| + L(\|\hat{h}\| + \|\tilde{h}\|)\|\hat{h} - \tilde{h}\|$$

$$+ \frac{L}{\sqrt{n}}\|\tilde{h}\|\Big(\sum_{j=1}^{n}\|x_j - \tilde{x}_j\|^2\Big)^{1/2}. \tag{4.3}$$

Similarly, note that

$$r(\hat{h}) - r(\tilde{h}) = \int_0^1 (g''(\lambda\hat{h}) - g''(\lambda\tilde{h}))d\lambda\ \hat{h} + \int_0^1 (g''(\lambda\tilde{h}) - g''(0))d\lambda\ (\hat{h} - \tilde{h})$$

which implies the following bound:

$$\|r(\hat{h}) - r(\tilde{h})\| \leq \frac{L}{2}(\|\hat{h}\| + \|\tilde{h}\|)\|\hat{h} - \tilde{h}\|. \tag{4.4}$$

It follows from (4.2), (4.3), and (4.4) that

$$\|\hat{h} - \tilde{h}\| \leq \frac{1}{m}\bigg(\Big(\frac{M}{\sqrt{n}} + \frac{L}{\sqrt{n}}\|\tilde{h}\|\Big)\Big(\sum_{j=1}^{n}\|x_j - \tilde{x}_j\|^2\Big)^{1/2}$$

$$+ \|g_n''(0) - g''(0)\|\|\hat{h} - \tilde{h}\| + \frac{3}{2}L(\|\hat{h}\| + \|\tilde{h}\|)\|\hat{h} - \tilde{h}\|\bigg).$$

If $\|g_n''(0) - g''(0)\| \leq \frac{m}{4}$ and $\|\hat{h}\| \vee \|\tilde{h}\| \leq \frac{m}{12L}$, we easily conclude that

$$\|\hat{h} - \tilde{h}\| \leq \frac{4M}{m\sqrt{n}}\Big(\sum_{j=1}^{n}\|x_j - \tilde{x}_j\|^2\Big)^{1/2},$$

which completes the proof. □

Since the Lipschitz condition holds for $\hat{\theta}$ only on set A, it will be convenient for our purposes to replace $\hat{\theta}$ with its "smoothed truncated" version $\check{\theta}$ that is Lipschitz in the whole space. To this end, let $\phi : \mathbb{R} \mapsto [0, 1]$ be defined as follows: $\phi(s) = 1, s \leq 1$, $\phi(s) = 0, s \geq 2$ and $\phi(s) = 2 - s, s \in (1, 2)$. Clearly, ϕ is Lipschitz with constant 1. By Theorem 3.1, $\|\hat{\theta} - \theta\| \leq \frac{m}{24L}$ with probability at least $1 - e^{-\gamma n}$, where $\gamma := c\big(\frac{m}{M} \wedge \frac{m^2}{L\sqrt{M}}\big)^2$ with a small enough constant $c > 0$ and it is assumed that $d \leq \gamma n$. Similarly, it follows from Proposition 3.2 that $\|g_n''(0) - g''(0)\| \leq \frac{m}{8}$ with probability at least $1 - e^{-\beta n}$, where $\beta = c(\frac{m}{M})^2$ for a small enough constant $c > 0$ (and under the assumption that $d \leq \beta n$. Clearly, we can assume that $\beta \geq \gamma$, so, both properties hold with probability at least $1 - 2e^{-\gamma n}$ provided that $d \leq \gamma n$. Define

$$\varphi(x_1, \ldots, x_n) := \phi\Big(\frac{24L}{m}\|\hat{\theta}(x_1, \ldots, x_n) - \theta\|\Big),$$

$$\psi(x_1, \ldots, x_n) := \phi\Big(\frac{8}{m}\|g_n''(0)(x_1 - \theta, \ldots, x_n - \theta) - g''(0)\|\Big).$$

and let

$$\check{\theta} := (1 - \varphi\psi)\theta + \varphi\psi\hat{\theta}.$$

Note that $\check{\theta} - \theta = (\hat{\theta} - \theta)\varphi\psi$ and $\check{\theta} = \hat{\theta}$ on the event $\{\varphi = \psi = 1\}$ of probability at least $1 - 2e^{-\gamma n}$.

Proposition 4.3 *If $d \leq \gamma n$, then*

$$\|\mathbb{E}_\theta\check{\theta} - \theta\| \lesssim_{M,L,m} \frac{d}{n} \quad \textit{and} \quad \mathbb{E}_\theta^{1/2}\|\check{\theta} - \theta\|^2 \lesssim_{M,L,m} \sqrt{\frac{d}{n}}.$$

Proof By representation (4.1) and the fact that $g'(0) = 0$,

$$\begin{aligned}\check{\theta} - \theta &= (\hat{\theta} - \theta)\varphi\psi = -\hat{h}\varphi\psi \\ &= \mathcal{I}^{-1}g_n'(0) - \mathcal{I}^{-1}g_n'(0)(1 - \varphi\psi) - \mathcal{I}^{-1}q_n(\hat{h})\varphi\psi + \mathcal{I}^{-1}r(\hat{h})\varphi\psi. \qquad (4.5)\end{aligned}$$

Using Corollary 3.1, we get

$$\begin{aligned}\|\|\mathcal{I}^{-1}g_n'(0)(1 - \varphi\psi)\|\|_{L_2} &\leq \|\|\mathcal{I}^{-1}g_n'(0)\|\|_{L_4}\|I(\varphi\psi \neq 1)\|_{L_4} \\ &\lesssim \frac{1}{m}\|\|g_n'(0)\|\|_{L_4}e^{-\gamma n/4} \lesssim \frac{\sqrt{M}}{m}\sqrt{\frac{d}{n}}e^{-\gamma n/4}.\end{aligned}$$

Note also that

$$\|q_n(\hat{h})\| = \left\| \int_0^1 (g_n'' - g'')(\lambda \hat{h}) d\lambda\, \hat{h} \right\| \le \|g_n''(0) - g''(0)\| \|\hat{h}\| + L\|\hat{h}\|^2,$$

so we also have

$$\begin{aligned}\mathbb{E}^{1/2}\|\mathcal{I}^{-1} q_n(\hat{h})\varphi\psi\|^2 &\le \frac{1}{m}\mathbb{E}^{1/2}\|g_n''(0) - g''(0)\|^2 \Big(\|\hat{h}\| \wedge \frac{m}{12L}\Big)^2 \\ &\quad + \frac{L}{m}\mathbb{E}^{1/2}\Big(\|\hat{h}\| \wedge \frac{m}{12L}\Big)^4 \\ &\le \frac{1}{m}\mathbb{E}^{1/4}\|g_n''(0) - g''(0)\|^4 \mathbb{E}^{1/4}\Big(\|\hat{h}\| \wedge \frac{m}{12L}\Big)^4 \\ &\quad + \frac{L}{m}\mathbb{E}^{1/2}\Big(\|\hat{h}\| \wedge \frac{m}{12L}\Big)^4.\end{aligned}$$

Using the second bound of Proposition 3.2 and the bound of Corollary 3.2, we get

$$\mathbb{E}^{1/2}\|\mathcal{I}^{-1} q_n(\hat{h})\varphi\psi\|^2 \lesssim_{M,L,m} \frac{d}{n}.$$

Similarly, we can show that

$$\mathbb{E}^{1/2}\|\mathcal{I}^{-1} r(\hat{h})\varphi\psi\|^2 \lesssim_{M,L,m} \frac{d}{n},$$

using the fact that

$$\|\mathcal{I}^{-1} r(\hat{h})\varphi\psi\| \le \frac{L}{2m}\Big(\|\hat{h}\| \wedge \frac{m}{12L}\Big)^2$$

and the bound of Corollary 3.2.

The above bounds and representation (4.5) imply that

$$\|\mathbb{E}_\theta \check{\theta} - \theta\| \lesssim_{M,L,m} \frac{d}{n}.$$

Using also Corollary 3.1, we get

$$\mathbb{E}_\theta^{1/2}\|\check{\theta} - \theta\|^2 \lesssim_{M,L,m} \sqrt{\frac{d}{n}}.$$

□

Proposition 4.4 *The function* $(x_1, \ldots, x_n) \mapsto \check{\theta}(x_1, \ldots, x_n)$ *is Lipschitz with constant* $\lesssim \frac{M}{m\sqrt{n}}$ *: for all* $(x_1, \ldots, x_n), (\tilde{x}_1, \ldots, \tilde{x}_n) \in \mathbb{R}^d \times \cdots \times \mathbb{R}^d$,

$$\|\breve{\theta}(x_1,\dots,x_n)-\breve{\theta}(\tilde{x}_1,\dots,\tilde{x}_n)\| \lesssim \frac{M}{m\sqrt{n}}\Bigl(\sum_{j=1}^n \|x_j-\tilde{x}_j\|^2\Bigr)^{1/2}.$$

Proof By Proposition 4.2, on the set A, $\hat{\theta}$ is Lipschitz with constant $\frac{4M}{m\sqrt{n}}$. This implies that function φ is also Lipschitz on the same set with constant $\frac{24L}{m}\frac{4M}{m\sqrt{n}}$. Note also that

$$\|g_n''(0)-\tilde{g}_n''(0)\| \le \frac{L}{\sqrt{n}}\Bigl(\sum_{j=1}^n \|x_j-\tilde{x}_j\|^2\Bigr)^{1/2},$$

implying that ψ is a Lipschitz function (on the whole space) with constant $\frac{8}{m}\frac{L}{\sqrt{n}}$. Using also the fact that φ and ψ are both bounded by 1 and $\|\hat{\theta}-\theta\| \le \frac{m}{12L}$ on the set $\{\varphi \ne 0\}$, it is easy to conclude that $\breve{\theta}$ is Lipschitz on A with constant

$$\lesssim \frac{4M}{m\sqrt{n}} + \frac{m}{12L}\frac{24L}{m}\frac{4M}{m\sqrt{n}} + \frac{m}{12L}\frac{8}{m}\frac{L}{\sqrt{n}} \lesssim \frac{M}{m\sqrt{n}}.$$

It remains to consider the case when $\varphi(x_1,\dots,x_n)\in A$ and $\varphi(\tilde{x}_1,\dots,\tilde{x}_n)\in A^c$ (the case when both points are in A^c is trivial). In this case, define $x_j^{\lambda} = \lambda x_j + (1-\lambda)\tilde{x}_j$, $\lambda\in[0,1]$, $j=1,\dots,n$. Note that A is a closed set (by continuity of both $\hat{\theta}(x_1,\dots,x_n)$ and $n^{-1}\sum_{j=1}^n V''(x_j-\theta)$). If $\bar{\lambda}$ denotes the supremum of those λ for which $(x_1^{\lambda},\dots,x_n^{\lambda})\in A$, then $(x_1^{\bar{\lambda}},\dots,x_n^{\bar{\lambda}})\in \partial A$, $(\varphi\psi)(x_1^{\bar{\lambda}},\dots,x_n^{\bar{\lambda}})=0$ and $\breve{\theta}(x_1^{\bar{\lambda}},\dots,x_n^{\bar{\lambda}}) = 0 = \breve{\theta}(\tilde{x}_1,\dots,\tilde{x}_1)$. Therefore,

$$\|\breve{\theta}(x_1,\dots,x_n)-\breve{\theta}(\tilde{x}_1,\dots,\tilde{x}_n)\| = \|\breve{\theta}(x_1,\dots,x_n)-\breve{\theta}(x_1^{\bar{\lambda}},\dots,x_n^{\bar{\lambda}})\|$$
$$\lesssim \frac{M}{m\sqrt{n}}\Bigl(\sum_{j=1}^n \|x_j-x_j^{\bar{\lambda}}\|^2\Bigr)^{1/2} \lesssim \frac{M}{m\sqrt{n}}\Bigl(\sum_{j=1}^n \|x_j-\tilde{x}_j\|^2\Bigr)^{1/2},$$

where we use the fact that point $(x_1^{\bar{\lambda}},\dots,x_n^{\bar{\lambda}})$ is in the line segment between $(x_1,\dots,x_n)$ and $(\tilde{x}_1,\dots,\tilde{x}_n)$.

The Lipschitz condition for $\breve{\theta}(x_1,\dots,x_n)$ now follows. □

We will now consider concentration properties of linear forms $\langle\breve{\theta}-\theta, w\rangle$, $w\in\mathbb{R}^d$. The following result will be proven.

Theorem 4.2 *Suppose $d \le \gamma n$, where $\gamma := c\bigl(\frac{m}{M}\wedge \frac{m^2}{L\sqrt{M}}\bigr)^2$ with a small enough $c>0$. Then*

$$\sup_{\theta\in\mathbb{R}^d}\Big\|\langle\check{\theta}-\theta,w\rangle-\mathbb{E}\langle\check{\theta}-\theta,w\rangle-n^{-1}\sum_{j=1}^n\langle V'(\xi_j),\mathcal{I}^{-1}w\rangle\Big\|_{L_{\psi_{2/3}}(\mathbb{P}_\theta)}$$

$$\lesssim\sqrt{c(V)}\Big(\frac{M^2}{m^2}+\frac{M^{3/2}L}{m^3}\Big)\frac{1}{\sqrt{n}}\sqrt{\frac{d}{n}}\|w\|.$$

Remark 4.1 Some concentration bounds for linear forms of MLE could be found in [23].

Proof Using representation (4.1) and the fact that $g'(0)=0$, we get

$$\begin{aligned}\langle\check{\theta}-\theta,w\rangle&=\langle\hat{\theta}-\theta,w\rangle\varphi\psi=-\langle\hat{h},w\rangle\varphi\psi\\&=\langle g_n'(0),u\rangle-\langle g_n'(0),u\rangle(1-\varphi\psi)-\langle q_n(\hat{h}),u\rangle\varphi\psi+\langle r(\hat{h}),u\rangle\varphi\psi,\end{aligned}\tag{4.6}$$

where $u=\mathcal{I}^{-1}w$. Since

$$\langle g_n'(0),u\rangle=n^{-1}\sum_{j=1}^n\langle V'(\xi_j),\mathcal{I}^{-1}w\rangle$$

has zero mean, it is enough to study the concentration of three other terms in the right-hand side of (4.6). The first of these terms is $\langle g_n'(0),u\rangle(1-\varphi\psi)$ and we have

$$\|\langle g_n'(0),u\rangle(1-\varphi\psi)\|_{\psi_1}\le\|\langle g_n'(0),u\rangle\|_{\psi_2}\|1-\varphi\psi\|_{\psi_2}$$

$$\le\|\langle g_n'(0),u\rangle\|_{\psi_2}\|I(\varphi\psi\neq1)\|_{\psi_2}\lesssim\frac{\sqrt{M}}{\sqrt{n}}\frac{1}{\sqrt{\gamma n}}\|u\|\lesssim\sqrt{\frac{M}{\gamma}\frac{1}{n}}\|u\|,$$

where we used Lemma 3.1 and the fact that $\mathbb{P}\{\varphi\psi\neq1\}\le2e^{-\gamma n}$ with γ from the statement of Theorem 4.2. Clearly, we also have

$$\|\langle g_n'(0),u\rangle(1-\varphi\psi)-\mathbb{E}\langle g_n'(0),u\rangle(1-\varphi\psi)\|_{\psi_1}\lesssim\sqrt{\frac{M}{\gamma}\frac{1}{n}}\|u\|.\tag{4.7}$$

For two other terms in the right-hand side of (4.6), we will provide bounds on their local Lipschitz constants. It follows from bound (4.3) and the bound of Proposition 4.2 that, for all $(x_1,\dots,x_n),(\tilde{x}_1,\dots,\tilde{x}_n)\in A$,

$$\|q_n(\hat{h})-\tilde{q}_n(\tilde{h})\|$$

$$\lesssim\Big(\frac{M}{m\sqrt{n}}\|g_n''(0)-g''(0)\|+\frac{ML}{m\sqrt{n}}(\|\hat{h}\|+\|\tilde{h}\|)\Big)\Big(\sum_{j=1}^n\|x_j-\tilde{x}_j\|^2\Big)^{1/2}.$$

Recall that function φ is Lipschitz on A with constant $\frac{24L}{m}\frac{4M}{m\sqrt{n}}$ and function ψ is Lipschitz on the whole space with constant $\frac{8}{m}\frac{L}{\sqrt{n}}$. Note also that

$$\|q_n(\hat{h})\| = \left\| \int_0^1 (g_n'' - g'')(\lambda \hat{h}) d\lambda \, \hat{h} \right\| \leq \|g_n''(0) - g''(0)\| \|\hat{h}\| + L\|\hat{h}\|^2.$$

Since, on set A, $\|\hat{h}\| \leq \frac{m}{12L}$, we get

$$\|q_n(\hat{h})\| \leq \frac{m}{12L}\|g_n''(0) - g''(0)\| + \frac{m}{12}\|\hat{h}\|.$$

Denoting $\varphi := \varphi(x_1, \ldots, x_n), \tilde{\varphi} := \varphi(\tilde{x}_1, \ldots, \tilde{x}_n), \psi := \psi(x_1, \ldots, x_n), \tilde{\psi} := \psi(\tilde{x}_1, \ldots, \tilde{x}_n)$, it easily follows from the facts mentioned above that, for all $(x_1, \ldots, x_n), (\tilde{x}_1, \ldots, \tilde{x}_n) \in A$,

$$\begin{aligned}&\|q_n(\hat{h})\varphi\psi - \tilde{q}_n(\tilde{h})\tilde{\varphi}\tilde{\psi}\| \\ &\lesssim \left(\frac{M}{m\sqrt{n}}\|g_n''(0) - g''(0)\| + \frac{ML}{m\sqrt{n}}(\|\hat{h}\| + \|\tilde{h}\|)\right)\left(\sum_{j=1}^n \|x_j - \tilde{x}_j\|^2\right)^{1/2}.\end{aligned}$$

This implies the following bound on the local Lipschitz constant of $q_n(\hat{h})\varphi\psi$ on set A :

$$\begin{aligned}&L(q_n(\hat{h})\varphi\psi)(x_1, \ldots, x_n) \\ &\lesssim \left(\frac{M}{m\sqrt{n}}\left(\|g_n''(0) - g''(0)\| \wedge \frac{m}{4}\right) + \frac{ML}{m\sqrt{n}}\left(\|\hat{h}\| \wedge \frac{m}{12L}\right)\right).\end{aligned} \tag{4.8}$$

The same bound trivially holds on the open set A^c (where $q_n(\hat{h})\varphi\psi)(x_1, \ldots, x_n) = 0$) and, by the argument already used at the end of the proof of Proposition 4.4, it is easy to conclude that bound (4.8) holds on the whole space.

Using bound (4.4) and the bound of Proposition 4.2, we get, for all $(x_1, \ldots, x_n), (\tilde{x}_1, \ldots, \tilde{x}_n) \in A$,

$$\|r(\hat{h}) - r(\tilde{h})\| \leq \frac{2ML}{m\sqrt{n}}(\|\hat{h}\| + \|\tilde{h}\|)\left(\sum_{j=1}^n \|x_j - \tilde{x}_j\|^2\right)^{1/2}$$

and we also have $\|r(\hat{h})\| \leq \frac{L}{2}\|\hat{h}\|^2$. As a result, we get the following condition for the function $r(\hat{h})\varphi\psi$ on set A :

$$\|r(\hat{h})\varphi\psi - r(\tilde{h})\tilde{\varphi}\tilde{\psi}\| \lesssim \frac{2ML}{m\sqrt{n}}(\|\hat{h}\| + \|\tilde{h}\|)\left(\sum_{j=1}^n \|x_j - \tilde{x}_j\|^2\right)^{1/2}.$$

This implies a bound on the local Lipschitz constant of $r(\hat{h})\varphi\psi$ on set A that, by the arguments already used, could be extended to the bound that holds on the whole space:

$$L(r(\hat{h})\varphi\psi)(x_1,\ldots,x_n) \lesssim \frac{ML}{m\sqrt{n}}\Big(\|\hat{h}\| \wedge \frac{m}{12L}\Big).$$

Denoting

$$\zeta(x_1,\ldots,x_n) := (-\langle q_n(\hat{h}), u\rangle\varphi\psi + \langle r(\hat{h}), u\rangle\varphi\psi)(x_1,\ldots,x_n),$$

we can conclude that

$$(L\zeta)(x_1,\ldots,x_n) \lesssim \|u\|\bigg(\frac{M}{m\sqrt{n}}\Big(\|g_n''(0)-g''(0)\| \wedge \frac{m}{4}\Big) + \frac{ML}{m\sqrt{n}}\Big(\|\hat{h}\| \wedge \frac{m}{12L}\Big)\bigg). \tag{4.9}$$

By the second bound of Proposition 3.2,

$$\Big\|\|g_n''(0)-g''(0)\|\Big\|_{\psi_2} \lesssim M\sqrt{\frac{d}{n}}.$$

By the bound of Corollary 3.2,

$$\Big\|\|\hat{\theta}-\theta\| \wedge \frac{m}{12L}\Big\|_{\psi_2} \lesssim \frac{\sqrt{M}}{m}\sqrt{\frac{d}{n}} + \frac{m}{L\sqrt{\gamma}}\frac{1}{\sqrt{n}}. \tag{4.10}$$

Substituting the above bounds in (4.9), we conclude that

$$\begin{aligned}\Big\|(L\zeta)(X_1,\ldots,X_n)\Big\|_{\psi_2} &\lesssim \Big(\frac{M^2}{m\sqrt{n}}\sqrt{\frac{d}{n}} + \frac{M^{3/2}L}{m^2\sqrt{n}}\sqrt{\frac{d}{n}} + \frac{M}{\sqrt{\gamma}}\frac{1}{n}\Big)\|u\| \\ &\lesssim \Big(\frac{M^2}{m\sqrt{n}}\sqrt{\frac{d}{n}} + \frac{M^{3/2}L}{m^2\sqrt{n}}\sqrt{\frac{d}{n}}\Big)\|u\|,\end{aligned}$$

where we also used the fact that the last term is dominated by the other terms.

We are now ready to use concentration inequalities for functions of log-concave r.v. to control $\zeta(X_1,\ldots,X_n) - \mathbb{E}\zeta(X_1,\ldots,X_n)$. For all $p \geq 1$, we have

$$\begin{aligned}&\Big\|\zeta(X_1,\ldots,X_n) - \mathbb{E}\zeta(X_1,\ldots,X_n)\Big\|_{L_p} \\ &\lesssim \sqrt{c(V)p}\Big\|(L\zeta)(X_1,\ldots,X_n)\Big\|_{L_p} \lesssim \sqrt{c(V)}p^{3/2}\Big\|(L\zeta)(X_1,\ldots,X_n)\Big\|_{\psi_2}.\end{aligned}$$

It follows that

$$\Big\|\zeta(X_1,\dots,X_n)-\mathbb{E}\zeta(X_1,\dots,X_n)\Big\|_{\psi_{2/3}} \lesssim \sqrt{c(V)}\Big\|(L\zeta)(X_1,\dots,X_n)\Big\|_{\psi_2}$$
$$\lesssim \sqrt{c(V)}\Big(\frac{M^2}{m\sqrt{n}}\sqrt{\frac{d}{n}}+\frac{M^{3/2}L}{m^2\sqrt{n}}\sqrt{\frac{d}{n}}\Big)\|u\|.$$

Recalling representation (4.6) and bound (4.7), we get

$$\Big\|\langle\check\theta-\theta,w\rangle-\mathbb{E}\langle\check\theta-\theta,w\rangle-\langle\mathcal{I}^{-1}g_n'(0),w\rangle\Big\|_{\psi_{2/3}}$$
$$\lesssim \sqrt{c(V)}\Big(\frac{M^2}{m\sqrt{n}}\sqrt{\frac{d}{n}}+\frac{M^{3/2}L}{m^2\sqrt{n}}\sqrt{\frac{d}{n}}\Big)\|u\|+\sqrt{\frac{M}{\gamma}\frac{1}{n}}\|u\|.$$

Since $c(V)\geq\|\Sigma\|\geq\|\mathcal{I}^{-1}\|$ and $M\geq\|\mathcal{I}\|$, we have $c(V)M\geq 1$. Recalling the definition of γ and also that $\|u\|=\|\mathcal{I}^{-1}w\|\leq\frac{1}{m}\|w\|$, it is easy to complete the proof. □

We are ready to prove Theorem 4.1.

Proof Note that

$$f(\check\theta)-f(\theta)=\langle f'(\theta),\check\theta-\theta\rangle+S_f(\theta;\check\theta-\theta).$$

Therefore,

$$f(\check\theta)-\mathbb{E}_\theta f(\check\theta)$$
$$=\langle f'(\theta),\check\theta-\theta\rangle-\mathbb{E}_\theta\langle f'(\theta),\check\theta-\theta\rangle+S_f(\theta;\check\theta-\theta)-\mathbb{E}_\theta S_f(\theta;\check\theta-\theta). \tag{4.11}$$

By the bound of Theorem 4.2,

$$\Big\|\langle f'(\theta),\check\theta-\theta\rangle-\mathbb{E}\langle f'(\theta),\check\theta-\theta\rangle-n^{-1}\sum_{j=1}^n\langle V'(\xi_j),\mathcal{I}^{-1}f'(\theta)\rangle\Big\|_{\psi_{2/3}}$$
$$\leq\sqrt{c(V)}\Big(\frac{M^2}{m^2}+\frac{M^{3/2}L}{m^3}\Big)\frac{1}{\sqrt{n}}\sqrt{\frac{d}{n}}\|f'(\theta)\|. \tag{4.12}$$

Thus, it remains to control $S_f(\theta;\check\theta-\theta)-\mathbb{E}_\theta S_f(\theta;\check\theta-\theta)$. By Proposition 3.1 (v), for function $f\in C^s$, $s=1+\rho$, $\rho\in(0,1]$, we have

$$|S_f(\theta;h)-S_f(\theta;h')|\lesssim\|f\|_{C^s}(\|h\|^\rho\vee\|h'\|^\rho)\|h-h'\|,\theta,h,h'\in\mathbb{R}^d.$$

Combining this with the bound of Proposition 4.4, we easily get

$$\begin{aligned}
&\Big|S_f(\theta;\check\theta(x_1,\dots,x_n)-\theta)-S_f(\theta;\check\theta(\tilde x_1,\dots,\tilde x_n)-\theta)\Big| \\
&\lesssim \|f\|_{C^s}\frac{M}{m\sqrt{n}}(\|\check\theta(x_1,\dots,x_n)-\theta\|^\rho \vee \|\check\theta(\tilde x_1,\dots,\tilde x_n)-\theta\|^\rho) \\
&\times\Big(\sum_{j=1}^n \|x_j-\tilde x_j\|^2\Big)^{1/2},
\end{aligned}$$

which implies the following bound on the local Lipschitz function of function $S_f(\theta;\check\theta-\theta)$:

$$(LS_f(\theta;\check\theta-\theta))(x_1,\dots,x_n)\lesssim \|f\|_{C^s}\frac{M}{m\sqrt{n}}\|\check\theta(x_1,\dots,x_n)-\theta\|^\rho.$$

Using concentration bounds for log-concave r.v., we get

$$\begin{aligned}
&\Big\|S_f(\theta;\check\theta-\theta)-\mathbb{E}_\theta S_f(\theta;\check\theta-\theta)\Big\|_{L_p} \\
&\lesssim \sqrt{c(V)p}\Big\|(LS_f(\theta;\check\theta-\theta))(X_1,\dots,X_n)\Big\|_{L_p} \\
&\lesssim \sqrt{c(V)p}\|f\|_{C^s}\frac{M}{m\sqrt{n}}\Big\|\|\check\theta-\theta\|^\rho\Big\|_{L_p} \\
&\lesssim \sqrt{c(V)p}\|f\|_{C^s}\frac{M}{m\sqrt{n}}\Big\|\|\check\theta-\theta\|\Big\|_{L_p}^\rho \\
&\lesssim \sqrt{c(V)p}\|f\|_{C^s}\frac{M}{m\sqrt{n}}\Big\|\|\hat\theta-\theta\|\wedge\frac{m}{12L}\Big\|_{L_p}^\rho \\
&\lesssim \sqrt{c(V)}p^{1+\rho/2}\|f\|_{C^s}\frac{M}{m\sqrt{n}}\Big\|\|\hat\theta-\theta\|\wedge\frac{m}{12L}\Big\|_{\psi_2}^\rho,
\end{aligned}$$

which, using bound (4.10), implies that

$$\begin{aligned}
&\Big\|S_f(\theta;\check\theta-\theta)-\mathbb{E}_\theta S_f(\theta;\check\theta-\theta)\Big\|_{\psi_{2/(2+\rho)}} \\
&\lesssim \sqrt{c(V)}\|f\|_{C^s}\frac{M}{m\sqrt{n}}\Big\|\|\hat\theta-\theta\|\wedge\frac{m}{12L}\Big\|_{\psi_2}^\rho \\
&\lesssim \sqrt{c(V)}\|f\|_{C^s}\Big(\frac{M^{1+\rho/2}}{m^{1+\rho}}\frac{1}{\sqrt{n}}\Big(\frac{d}{n}\Big)^{\rho/2}+\frac{M}{L^\rho m^{1-\rho}\gamma^{\rho/2}}\frac{1}{n^{(1+\rho)/2}}\Big).
\end{aligned}$$

Combining this with (4.11) and (4.12), we get

$$\Big\| f(\check{\theta}) - \mathbb{E}_\theta f(\check{\theta}) - n^{-1}\sum_{j=1}^n \langle V'(\xi_j), \mathcal{I}^{-1} f'(\theta)\rangle \Big\|_{\psi_{2/3}}$$

$$\lesssim \sqrt{c(V)}\|f'(\theta)\|\Big(\frac{M^2}{m^2} + \frac{M^{3/2}L}{m^3}\Big)\frac{1}{\sqrt{n}}\sqrt{\frac{d}{n}}$$

$$+ \sqrt{c(V)}\|f\|_{C^s}\Big(\frac{M^{1+\rho/2}}{m^{1+\rho}}\frac{1}{\sqrt{n}}\Big(\frac{d}{n}\Big)^{\rho/2} + \frac{M}{L^\rho m^{1-\rho}\gamma^{\rho/2}}\frac{1}{n^{(1+\rho)/2}}\Big)$$

$$\lesssim_{M,L,m} \sqrt{c(V)}\|f\|_{C^s}\frac{1}{\sqrt{n}}\Big(\frac{d}{n}\Big)^{\rho/2}.$$

It remains to replace in the above bound $\check{\theta}$ by $\hat{\theta}$. To this end, observe that $|f(\hat{\theta}) - f(\check{\theta})| \le 2\|f\|_{L_\infty} I(\check{\theta} \neq \hat{\theta})$. This implies

$$\|f(\hat{\theta}) - f(\check{\theta})\|_{\psi_1} \le 2\|f\|_{L_\infty}\|I(\check{\theta} \neq \hat{\theta})\|_{\psi_1} \lesssim \frac{\|f\|_{L_\infty}}{\gamma n} \lesssim \frac{\|f\|_{C^s}}{\gamma n},$$

which allows us to complete the proof. □

5 Bias Reduction

We turn now to the bias reduction method outlined in Sect. 1. The justification of this method is much simpler in the case of equivariant estimators $\hat{\theta}$ of location parameter. Indeed, in this case

$$\hat{\theta}(X_1, \ldots, X_n) = \theta + \hat{\theta}(\xi_1, \ldots, \xi_n),$$

where $\xi_j = X_j - \theta$, $j = 1, \ldots, n$ are i.i.d. $\sim P$, $P(dx) = e^{-V(x)}dx$. Denote $\vartheta := \hat{\theta}(\xi_1, \ldots, \xi_n)$ and let $\{\vartheta_k\}$ be a sequence of i.i.d. copies of ϑ defined as follows: $\vartheta_k := \hat{\theta}(\xi_1^{(k)}, \ldots, \xi_n^{(k)})$, $\xi_j^{(k)}$, $j = 1, \ldots, n, k \ge 1$ being i.i.d. copies of ξ. Then, the bootstrap chain $\{\hat{\theta}^{(k)} : k \ge 0\}$ has the same distribution as the sequence of r.v. $\{\theta + \sum_{j=1}^k \vartheta_j : k \ge 0\}$. Moreover, let

$$\vartheta(t_1, \ldots, t_k) := \sum_{j=1}^k t_j \vartheta_j, \ (t_1, \ldots, t_k) \in [0, 1]^k.$$

Then, for $(t_1, \ldots, t_k) \in \{0, 1\}$ with $\sum_{i=1}^n t_i = j$, we have $\theta + \vartheta(t_1, \ldots, t_k) \overset{d}{=} \hat{\theta}^{(j)}$. Therefore, we can write

$$
\begin{aligned}
(\mathcal{B}^k f)(\theta) &= \mathbb{E}_\theta \sum_{j=0}^{k} (-1)^{k-j} \binom{k}{j} f(\hat{\theta}^{(j)}) \\
&= \mathbb{E} \sum_{(t_1,\dots,t_k)\in\{0,1\}^k} (-1)^{k-\sum_{j=1}^k t_j} f(\theta + \vartheta(t_1,\dots,t_k)) \\
&= \mathbb{E}\Delta_1 \dots \Delta_k f(\theta + \vartheta(t_1,\dots,t_k)),
\end{aligned}
$$

where

$$
\Delta_j \varphi(t_1,\dots,t_k) := \varphi(t_1,\dots,t_k)_{|t_j=1} - \varphi(t_1,\dots,t_k)_{|t_j=0}.
$$

If function φ is k times continuously differentiable, then by Newton-Leibniz formula

$$
\Delta_1 \dots \Delta_k \varphi(t_1,\dots,t_k) = \int_0^1 \cdots \int_0^1 \frac{\partial^k \varphi(t_1,\dots,t_k)}{\partial t_1 \dots \partial t_k} dt_1 \dots dt_k.
$$

If $f : \mathbb{R}^d \mapsto \mathbb{R}$ is k times continuously differentiable, then

$$
\frac{\partial^k}{\partial t_1 \dots \partial t_k} f(\theta + \vartheta(t_1,\dots,t_k)) = f^{(k)}(\theta + \vartheta(t_1,\dots,t_k))[\vartheta_1,\dots,\vartheta_k]
$$

and we end up with the following integral representation formula:

$$
(\mathcal{B}^k f)(\theta) = \mathbb{E} \int_0^1 \cdots \int_0^1 f^{(k)}(\theta + \vartheta(t_1,\dots,t_k))[\vartheta_1,\dots,\vartheta_k] dt_1 \dots dt_k \tag{5.1}
$$

that plays an important role in the analysis of functions $\mathcal{B}^k f$ and f_k.

It will be convenient to apply this formula not directly to MLE $\hat{\theta}$, but to its smoothed and truncated approximation $\check{\theta}$, defined in the previous section. Note that $\check{\theta}(X_1,\dots,X_n) = \theta + \check{\vartheta}$, where

$$
\begin{aligned}
\check{\vartheta} &= \check{\vartheta}(\xi_1,\dots,\xi_n) \\
&:= \hat{\theta}(\xi_1,\dots,\xi_n)\phi\Big(\frac{24L}{m}\|\hat{\theta}(\xi_1,\dots,\xi_n)\|\Big)\phi\Big(\frac{8}{m}\|g_n''(0)(\xi_1,\dots,\xi_n) - g''(0)\|\Big).
\end{aligned}
$$

Let $\check{\vartheta}_k, k \geq 1$ be i.i.d. copies of $\check{\vartheta}$ defined as follows:

$$
\check{\vartheta}_k := \vartheta_k \phi\Big(\frac{24L}{m}\|\vartheta_k\|\Big)\phi\Big(\frac{8}{m}\|g_n''(0)(\xi_1^{(k)},\dots,\xi_n^{(k)}) - g''(0)\|\Big).
$$

Note that, for all $k \geq 1$, $\vartheta_k = \check{\vartheta}_k$ with probability at least $1 - 2e^{-\gamma n}$.

Let $\check{\theta}^{(k)} := \theta + \sum_{j=1}^{k} \check{\vartheta}_j, k \geq 0$. We will also introduce the operators $(\check{\mathcal{T}} g)(\theta) := \mathbb{E}_\theta g(\check{\theta}), \theta \in \mathbb{R}^d$ and $\check{\mathcal{B}} := \check{\mathcal{T}} - \mathcal{I}$. Let $\check{f}_k := \sum_{j=0}^{k} (-1)^j \check{\mathcal{B}}^j f$. Since $\vartheta_j = \check{\vartheta}_j, j = 1, \ldots, k$ with probability at least $1 - 2ke^{-\gamma n}$, we easily conclude that, if we identify $\hat{\theta}^{(k)}$ with $\theta + \sum_{j=1}^{k} \vartheta_j, k \geq 0$, then the event $E := \{\check{\theta}^{(j)} = \hat{\theta}^{(j)}, j = 1, \ldots, k\}$ occurs with the same probability. This immediately implies the following proposition.

Proposition 5.1 *For all* $k \geq 1$,

$$\|f_k - \check{f}_k\|_{L_\infty} \leq k 2^{k+3} \|f\|_{L_\infty} e^{-\gamma n} \tag{5.2}$$

and

$$\|f_k\|_{L_\infty} \leq 2^{k+1} \|f\|_{L_\infty}, \qquad \|\check{f}_k\|_{L_\infty} \leq 2^{k+1} \|f\|_{L_\infty}. \tag{5.3}$$

Proof For all $\theta \in \mathbb{R}^d$ and all $j = 1, \ldots, k$,

$$\begin{aligned} |\mathbb{E}_\theta f(\hat{\theta}^{(j)}) - \mathbb{E}_\theta f(\check{\theta}^{(j)})| &= |\mathbb{E}_\theta f(\hat{\theta}^{(j)}) I_{E^c} - \mathbb{E}_\theta f(\check{\theta}^{(j)}) I_{E^c}| \\ &\leq 2\|f\|_{L_\infty} \mathbb{P}(E^c) \leq 4\|f\|_{L_\infty} k e^{-\gamma n}. \end{aligned}$$

Therefore, applying (1.2) to f_k and $\check{f}_k$, we arrive at

$$\begin{aligned} |f_k(\theta) - \check{f}_k(\theta)| &\leq \sum_{j=0}^{k} \binom{k+1}{j+1} |\mathbb{E}_\theta f(\hat{\theta}^{(j)}) - \mathbb{E}_\theta f(\check{\theta}^{(j)})| \\ &\leq k 2^{k+3} \|f\|_{L_\infty} e^{-\gamma n}, \end{aligned}$$

which proofs Proposition (5.1). Bounds (5.3) follow by a similar argument. □

Similarly to (5.1), we get

$$\begin{aligned} (\check{\mathcal{B}}^k f)(\theta) &= \mathbb{E} \int_0^1 \ldots \int_0^1 f^{(k)}(\theta + \check{\vartheta}(t_1, \ldots, t_k))[\check{\vartheta}_1, \ldots, \check{\vartheta}_k] dt_1 \ldots dt_k \\ &= \mathbb{E} f^{(k)}(\theta + \check{\vartheta}(\tau_1, \ldots, \tau_k))[\check{\vartheta}_1, \ldots, \check{\vartheta}_k], \end{aligned} \tag{5.4}$$

where $\check{\vartheta}(t_1, \ldots, t_k) := \sum_{j=1}^{k} t_j \check{\vartheta}_j$, $(t_1, \ldots, t_k) \in [0, 1]^k$ and $\tau_1, \ldots, \tau_k$ are i.i.d. r.v. with uniform distribution in $[0, 1]$ (independent of $\{\check{\vartheta}_j\}$).

The next proposition follows from representation (5.4) and differentiation under the expectation sign.

Proposition 5.2 *Let* $f \in C^s$ *for* $s = k + 1 + \rho$, *where* $k \geq 1$ *and* $\rho \in (0, 1]$. *Then, for all* $j = 1, \ldots, k$,

$$\|\check{\mathcal{B}}^j f\|_{C^{1+\rho}} \lesssim \|f\|_{C^s}(\mathbb{E}\|\check{\vartheta}\|)^j.$$

If $\mathbb{E}\|\check{\vartheta}\| \leq 1/2$, *then*

$$\|\check{f}_k\|_{C^{1+\rho}} \lesssim \|f\|_{C^s}.$$

We can also use representation (5.4) and smoothness of function $\check{\mathcal{B}}^k f$ to obtain a bound on the bias of "estimator" $\check{f}_k(\check{\theta})$.

Proposition 5.3 *Let* $f \in C^s$ *for* $s = k + 1 + \rho$, *where* $k \geq 1$ *and* $\rho \in (0, 1]$. *Then, for all* $\theta \in \mathbb{R}^d$,

$$|\mathbb{E}_\theta \check{f}_k(\check{\theta}) - f(\theta)| \lesssim \|f\|_{C^s}(\mathbb{E}\|\check{\vartheta}\|)^k(\|\mathbb{E}\check{\vartheta}\| + \mathbb{E}\|\check{\vartheta}\|^{1+\rho}).$$

Moreover,

$$|\mathbb{E}_\theta \check{f}_k(\check{\theta}) - f(\theta)| \lesssim_{M,L,m} \|f\|_{C^s}\left(\sqrt{\frac{d}{n}}\right)^s.$$

Proof Note that

$$\mathbb{E}_\theta \check{f}_k(\check{\theta}) - f(\theta) = (-1)^k(\check{\mathcal{B}}^{k+1} f)(\theta).$$

We also have

$$\begin{aligned}(\check{\mathcal{B}}^{k+1} f)(\theta) &= \mathbb{E}_\theta(\check{\mathcal{B}}^k f)(\check{\theta}) - (\check{\mathcal{B}}^k f)(\theta)\\ &= \langle(\check{\mathcal{B}}^k f)'(\theta), \mathbb{E}\check{\vartheta}\rangle + \mathbb{E}S_{\check{\mathcal{B}}^k f}(\theta; \check{\vartheta}).\end{aligned}$$

Using bounds of Proposition 5.2 and of Proposition 3.1, we get

$$\begin{aligned}|(\check{\mathcal{B}}^{k+1} f)(\theta)| &\lesssim \|(\check{\mathcal{B}}^k f)'\|\|\mathbb{E}\check{\vartheta}\| + \|\check{\mathcal{B}}^k f\|_{C^{1+\rho}}\mathbb{E}\|\check{\vartheta}\|^{1+\rho}\\ &\lesssim \|f\|_{C^s}(\mathbb{E}\|\check{\vartheta}\|)^k(\|\mathbb{E}\check{\vartheta}\| + \mathbb{E}\|\check{\vartheta}\|^{1+\rho}).\end{aligned}$$

Using also Proposition 4.3, we get

$$|\mathbb{E}_\theta \check{f}_k(\check{\theta}) - f(\theta)| \lesssim_{M,L,m} \|f\|_{C^s}\left(\frac{d}{n}\right)^{k/2}\left(\frac{d}{n} + \left(\frac{d}{n}\right)^{(1+\rho)/2}\right)$$

which allows to complete the proof. □

In view of bound (5.3), the bound of Proposition 5.1 and the fact that $\check{\theta} = \hat{\theta}$ with probability at least $1 - 2e^{-\gamma n}$, we easily conclude that the following proposition holds:

Proposition 5.4 *Let $f \in C^s$ for $s = k + 1 + \rho$, where $k \geq 1$ and $\rho \in (0, 1]$. Then, for all $\theta \in \mathbb{R}^d$,*

$$|\mathbb{E}_\theta \check{f}_k(\hat{\theta}) - f(\theta)| \lesssim_{M,L,m,s} \|f\|_{C^s} \Big(\sqrt{\frac{d}{n}}\Big)^s$$

and

$$|\mathbb{E}_\theta f_k(\hat{\theta}) - f(\theta)| \lesssim_{M,L,m,s} \|f\|_{C^s} \Big(\sqrt{\frac{d}{n}}\Big)^s. \tag{5.5}$$

It is now easy to prove Theorem 2.2.

Proof For all $\theta \in \mathbb{R}^d$,

$$\begin{aligned}
&\Big\| f_k(\hat{\theta}) - f(\theta) - n^{-1} \sum_{j=1}^n \langle V'(\xi_j), \mathcal{I}^{-1} f'(\theta)\rangle \Big\|_{\psi_{2/3}} \\
&\leq \Big\| \check{f}_k(\hat{\theta}) - \mathbb{E}_\theta \check{f}_k(\hat{\theta}) - n^{-1} \sum_{j=1}^n \langle V'(\xi_j), \mathcal{I}^{-1} \check{f}_k'(\theta)\rangle \Big\|_{\psi_{2/3}} \\
&+ \Big\| n^{-1} \sum_{j=1}^n \langle V'(\xi_j), \mathcal{I}^{-1} \check{f}_k'(\theta) - \mathcal{I}^{-1} f'(\theta)\rangle \Big\|_{\psi_{2/3}} \\
&+ \|f_k - \check{f}_k\|_{L_\infty} + |\mathbb{E}_\theta \check{f}_k(\hat{\theta}) - f(\theta)|.
\end{aligned} \tag{5.6}$$

Applying Theorem 4.1 to function $\check{f}_k$ and using the second bound of Proposition 5.2, we get

$$\begin{aligned}
&\Big\| \check{f}_k(\hat{\theta}) - \mathbb{E}_\theta \check{f}_k(\hat{\theta}) - n^{-1} \sum_{j=1}^n \langle V'(\xi_j), \mathcal{I}^{-1} \check{f}_k'(\theta)\rangle \Big\|_{\psi_{2/3}} \\
&\lesssim_{M,L,m} \sqrt{c(V)} \|\check{f}_k\|_{C^{1+\rho}} \frac{1}{\sqrt{n}} \Big(\frac{d}{n}\Big)^{\rho/2} \lesssim_{M,L,m} \sqrt{c(V)} \|f\|_{C^s} \frac{1}{\sqrt{n}} \Big(\frac{d}{n}\Big)^{\rho/2}.
\end{aligned}$$

Moreover, by Lemma 3.1 and the first bound in Proposition 5.2, we have

$$\begin{aligned}
&\Big\| n^{-1} \sum_{j=1}^n \langle V'(\xi_j), \mathcal{I}^{-1} \check{f}_k'(\theta) - \mathcal{I}^{-1} f'(\theta)\rangle \Big\|_{\psi_{2/3}} \\
&\lesssim \frac{1}{m} \frac{\sqrt{M}}{\sqrt{n}} \|\check{f}_k'(\theta) - f'(\theta)\| \leq \frac{1}{m} \frac{\sqrt{M}}{\sqrt{n}} \sum_{j=1}^k \|(\check{\mathcal{B}}^j f)'(\theta)\| \lesssim \|f\|_{C^s} \frac{1}{m} \frac{\sqrt{M}}{\sqrt{n}}.
\end{aligned}$$

Inserting these inequalities into (5.6) and applying Propositions 5.1 and 5.3 to the last two terms in (5.6) allow to complete the proof. □

Next we provide the proof of Proposition 2.1.

Proof The minimum with 1 in both bounds is due to the fact that $\|f_k\|_{L_\infty} \lesssim \|f\|_{L_\infty} \le \|f\|_{C^s}$; so the left-hand side is trivially bounded up to a constant by $\|f\|_{C^s}$.

To prove the first claim, note that, for f with $\|f\|_{C^s} \le 1$,

$$\begin{aligned}\|f(\hat\theta) - f(\theta)\|_{L_2(\mathbb{P}_\theta)} &\le \Big\|\Big(\|\hat\theta - \theta\| \wedge \frac{m}{12L}\Big)^s\Big\|_{L_2(\mathbb{P}_\theta)} \\ &\quad + 2\Big\|I(\|\hat\theta - \theta\| \ge m/(12L))\Big\|_{L_2(\mathbb{P}_\theta)} \\ &\le \Big(\Big\|\|\hat\theta - \theta\| \wedge \frac{m}{12L}\Big\|_{L_2(\mathbb{P}_\theta)}\Big)^s \\ &\quad + 2\mathbb{P}_\theta^{1/2}\Big\{\|\hat\theta - \theta\| \ge m/(12L)\Big\}.\end{aligned}$$

Using the bound of Corollary 3.2, we get

$$\Big(\Big\|\|\hat\theta - \theta\| \wedge \frac{m}{12L}\Big\|_{L_2(\mathbb{P}_\theta)}\Big)^s \lesssim_{M,L,m} \Big(\sqrt{\frac{d}{n}}\Big)^s,$$

and, by the bound of Theorem 3.1, we easily get

$$\mathbb{P}_\theta\Big\{\|\hat\theta - \theta\| \ge m/(12L)\Big\} \le e^{-\gamma n}.$$

The first claim now easily follows.

The proof of the second claim easily follows from Corollary 2.1. We can assume that $d \le \gamma n$ (otherwise, the bound is obvious), and we can drop the term $\sqrt{\frac{c(V)}{n}}\left(\frac{d}{n}\right)^{\rho/2}$ in the bound of Corollary 2.1: it is smaller than $\frac{1}{\sqrt{n}} + \left(\sqrt{\frac{d}{n}}\right)^s$ since $c(V) \lesssim_\epsilon d^\epsilon \|\Sigma\|$ for all $\epsilon > 0$ and $\|\Sigma\| \lesssim 1$. □

We will sketch the proof of Theorem 2.3.

Proof Under the stronger condition $V''(x) \succeq mI_d$, the proof of Theorem 2.2 could be significantly simplified. Recall that, for $\hat h = \theta - \hat\theta$, $g_n'(\hat h) = 0$. This implies that

$$g_n'(0) = g_n'(0) - g_n'(\hat h) = -\int_0^1 g_n''(\lambda \hat h) d\lambda\, \hat h. \tag{5.7}$$

The condition $V''(x) \succeq mI_d$ easily implies that

$$\int_0^1 g_n''(\lambda\hat{h})d\lambda = \int_0^1 n^{-1}\sum_{j=1}^n V''(\xi_j+\lambda\hat{h})d\lambda \succeq mI_d.$$

Therefore, $\left\| \int_0^1 g_n''(\lambda\hat{h})d\lambda\ u\right\| \geq m\|u\|, u\in\mathbb{R}^d$. Combining this with (5.7) yields $\|\hat{h}\|\leq \frac{\|g_n'(0)\|}{m}$. By Corollary 3.1, we get that for all $t\geq 1$, with probability at least $1-e^{-t}$

$$\|\hat{\theta}-\theta\| = \|\hat{h}\| \lesssim \frac{\sqrt{M}}{m}\Big(\sqrt{\frac{d}{n}}\vee\sqrt{\frac{t}{n}}\Big).$$

Unlike the case of Theorem 3.1, the above bound holds in the whole range of $t\geq 1$, and it immediately implies that $\mathbb{E}_\theta^{1/2}\|\hat{\theta}-\theta\|^2 \lesssim \frac{\sqrt{M}}{m}\sqrt{\frac{d}{n}}$, and, moreover, $\|\|\hat{\theta}-\theta\|\|_{L_{\psi_2}(\mathbb{P}_\theta)} \lesssim \frac{\sqrt{M}}{m}\sqrt{\frac{d}{n}}$.

Quite similarly, one can show that, unlike the case of Proposition 4.2, the Lipschitz condition for the function $\mathbb{R}^d\times\cdots\times\mathbb{R}^d \ni (x_1,\dots,x_n)\mapsto\hat{\theta}(x_1,\dots,x_n)\in\mathbb{R}^d$ holds not just on set A, but on the whole space. Indeed, recall that $g_n'(\hat{h})=0$ and $\tilde{g}_n'(\tilde{h})=0$. This implies that

$$\tilde{g}_n'(\tilde{h})-g_n'(\tilde{h}) = g_n'(\hat{h})-g_n'(\tilde{h}) = \int_0^1 g_n''(\tilde{h}+\lambda(\hat{h}-\tilde{h}))d\lambda\ (\hat{h}-\tilde{h}).$$

Since $\int_0^1 g_n''(\tilde{h}+\lambda(\hat{h}-\tilde{h}))d\lambda \succeq mI_d$, we get

$$\|\tilde{g}_n'(\tilde{h})-g_n'(\tilde{h})\| \geq m\|\hat{h}-\tilde{h}\|,$$

which implies

$$\begin{aligned}\|\tilde{\theta}-\hat{\theta}\| = \|\hat{h}-\tilde{h}\| &\leq m^{-1}n^{-1}\sum_{j=1}^n \|V'(\tilde{h}+\tilde{\xi}_j)-V'(\tilde{h}+\xi_j)\|\\ &\leq m^{-1}n^{-1}M\sum_{j=1}^n\|\tilde{\xi}_j-\xi_j\| \leq \frac{M}{m\sqrt{n}}\Big(\sum_{j=1}^n\|\tilde{\xi}_j-\xi_j\|^2\Big)^{1/2}\\ &= \frac{M}{m\sqrt{n}}\Big(\sum_{j=1}^n\|\tilde{x}_j-x_j\|^2\Big)^{1/2},\end{aligned}$$

and the Lipschitz condition holds for the function $\mathbb{R}^d\times\cdots\times\mathbb{R}^d\ni(x_1,\dots,x_n)\mapsto\hat{\theta}(x_1,\dots,x_n)\in\mathbb{R}^d$ with constant $\frac{M}{m\sqrt{n}}$. Due to this fact, there is no need to consider a "smoothed version" $\check{\theta}$ of function $\hat{\theta}$ in the remainder of the proof (as it was done in the proof of Theorem 2.2). All the arguments could be applied directly to $\hat{\theta}$.

Finally, recall that, if $\xi \sim P$, $P(dx) = e^{-V(x)}dx$ with $V''(x) \succeq mI_d, x \in \mathbb{R}^d$, then, for all locally Lipschitz functions $g : \mathbb{R}^d \mapsto \mathbb{R}$, the following logarirthmic Sobolev inequality holds:

$$\mathbb{E}g^2(\xi)\log g^2(\xi) - \mathbb{E}g^2(\xi)\log \mathbb{E}g^2(\xi) \lesssim \frac{1}{m}\mathbb{E}\|\nabla g(\xi)\|^2$$

(see, e.g., [20], Theorem 5.2). It was proven in [2] (see also [1]) that this implies the following moment bound:

$$\|g(\xi) - \mathbb{E}g(\xi)\|_{L_p} \lesssim_m \sqrt{p}\|\|\nabla g(\xi)\|\|_{L_p}, \ p \geq 2.$$

This bound is used to modify the concentration inequalities of Sect. 4, which yields the claim of Theorem 2.3. □

6 Minimax Lower Bounds

In this section, we provide lower bounds for the estimation of the location parameter and functionals thereof that match the upper bounds obtained in the previous sections up to constants.

We start with a comment on the proof of Proposition 2.2.

Proof The proof follows the same line of arguments as in the proof of Theorem 2.2 in [18]. It is based on a construction of a set $\mathcal{F}$ of smooth functionals such that the existence of estimators of $f(\theta)$ for all $f \in \mathcal{F}$ with some error rate would allow one to design an estimator of parameter θ itself with a certain error rate. This rate is then compared with a minimax lower bound $\inf_{\hat{\theta}} \max_{\theta\in\Theta} \mathbb{E}_\theta\|\hat{\theta} - \theta\|^2$ in the parameter estimation problem, where Θ is a maximal ε-net of the unit sphere (for a suitable ε), yielding as a result a minimax lower bound in the functional estimation. Minimax lower bound in the parameter estimation problem can be deduced in a standard way from Theorem 2.5 of [28] using KL divergence (Fano's type argument). In fact, while in the Gaussian location model KL divergence coincides with 1/2 times the squared Euclidean distance, a similar property also holds for our log-concave location models:

$$\begin{aligned} K(P_\theta\|P_{\theta'}) &= \mathbb{E}(V(\xi + \theta - \theta') - V(\xi)) \\ &= \mathbb{E}(V(\xi + \theta - \theta') - V(\xi) - \langle V'(\xi), \theta - \theta'\rangle) \leq M\|\theta - \theta'\|^2/2, \end{aligned}$$

where we used Proposition 3.1 and Assumption 1 in the inequality. □

Our next goal is to provide the proof of the local minimax lower bound of Proposition 2.3. It will be based on Bayes risk lower bounds for the estimation of location parameter as well as functionals thereof that might be of independent interest.

Let $(\mathcal{X}, \mathcal{F}, (\mathbb{P}_\theta)_{\theta\in\Theta})$ be a statistical model, and let G be a topological group acting on the measurable space $(\mathcal{X}, \mathcal{F})$ and Θ. Let Θ be such that $(\mathbb{P}_\theta)_{\theta\in\Theta}$ is G-equivariant, i.e., $\mathbb{P}_{g\theta}(gA) = \mathbb{P}_\theta(A)$ for all $g \in G, \theta \in \Theta$ and $A \in \mathcal{F}$. Suppose that $g \mapsto \mathbb{P}_{g\theta}(A)$ is measurable for every $A \in \mathcal{F}, \theta \in \Theta$.

Recall that, for two probability measures μ, ν on an arbitrary measurable space with μ being absolutely continuous w.r.t. ν, the χ^2-divergence $\chi^2(\mu, \nu)$ is defined as

$$\chi^2(\mu, \nu) := \int \Big(\frac{d\mu}{d\nu} - 1\Big)^2 d\nu = \int \Big(\frac{d\mu}{d\nu}\Big)^2 d\nu - 1.$$

The key ingredient in our proofs is the following version of an equivariant van Trees type inequality established in [30], Proposition 1.

Lemma 6.1 *Let Π be a Borel probability measure on G, let $\psi : \Theta \to \mathbb{R}^m$ be a derived parameter such that $\int_G \|\psi(g\theta)\|^2\Pi(dg) < \infty$ for all $\theta \in \Theta$, and let $\hat{\psi} : \mathcal{X} \to \mathbb{R}^m$ be an estimator of $\psi(\theta)$, based on an observation $X \sim \mathbb{P}_\theta, \theta \in \Theta$. Then, for all $\theta \in \Theta$ and all $h_1, \ldots, h_m \in G$, we have*

$$\int_G \mathbb{E}_{g\theta}\|\hat{\psi}(X) - \psi(g\theta)\|^2\, \Pi(dg)$$
$$\geq \frac{\big(\sum_{j=1}^m \int_G(\psi_j(gh_j^{-1}\theta) - \psi_j(g\theta))\, \Pi(dg)\big)^2}{\sum_{j=1}^m \big(\chi^2(\mathbb{P}_{h_j\theta}, \mathbb{P}_\theta) + \chi^2(\Pi\circ R_{h_j}, \Pi) + \chi^2(\mathbb{P}_{h_j\theta}, \mathbb{P}_\theta)\chi^2(\Pi\circ R_{h_j}, \Pi)\big)},$$

with $\Pi \circ R_{h_j}$ defined by $(\Pi \circ R_{h_j})(B) = \Pi(Bh_j)$ for all Borel sets $B \subset G$.

This lemma will be applied to our log-concave location model with G being the group of all translations of $\mathbb{R}^d$. The Bayes risk lower bound will be formulated for the class of all prior density functions $\pi : \mathbb{R}^d \mapsto \mathbb{R}_+$ with respect to the Lebesgue measure on $\mathbb{R}^d$ satisfying one of the following two conditions:

(P1) $\pi = e^{-W}$ with $W : \mathbb{R}^d \mapsto \mathbb{R}$ being twice differentiable such that $\|W''\|_{L_\infty}$ and $\|W''\|_{\mathrm{Lip}}$ are finite,

(P2) π has bounded support and is twice differentiable such that $\|\pi^{(j)}\|_{L_\infty}$ and $\|\pi^{(j)}\|_{\mathrm{Lip}}$ are finite for $j = 0, 1, 2$ (actually, it suffices to assume it only for $j = 2$ since the support is bounded).

Our first result deals with the estimation of the location parameter itself. We assume that i.i.d. observations $X_1, \ldots, X_n$ are sampled from a distribution belonging to a log-concave location family $e^{-V(x-\theta)}dx, \theta \in \mathbb{R}^d$ with convex function V satisfying Assumption 1.

Theorem 6.1 *Let Π be a probability measure on $\mathbb{R}^d$ with density $\pi : \mathbb{R}^d \mapsto \mathbb{R}_+$ with respect to the Lebesgue measure satisfying either (P1) or (P2). Suppose that π*

has finite Fisher information matrix:

$$\mathcal{J}_\pi = \int_{\mathbb{R}^d} \frac{\pi'(\theta) \otimes \pi'(\theta)}{\pi(\theta)} d\theta.$$

Then, for all $\delta > 0$*, we have*

$$\inf_{\hat{\theta}_n} \int_{\mathbb{R}^d} \mathbb{E}_\theta \|\hat{\theta}_n - \theta\|^2 \, \Pi_\delta(d\theta) \geq \frac{1}{n} \operatorname{tr}\Big(\Big(\mathcal{I} + \frac{1}{\delta^2 n} \mathcal{J}_\pi\Big)^{-1}\Big), \tag{6.1}$$

where Π_δ *is the prior measure with density* $\pi_\delta(\theta) = \delta^{-d}\pi(\delta^{-1}\theta)$, $\theta \in \mathbb{R}^d$ *and where the infimum is taken over all estimators* $\hat{\theta}_n = \hat{\theta}_n(X_1, \dots, X_n)$ *based on* $(X_1, \dots, X_n)$.

Let us discuss two simple implications. First, if we choose $\pi = e^{-V}$ that satisfies (P1) (in view of Assumption 1) and let $\delta \to \infty$, then Theorem 6.1 implies

$$\inf_{\hat{\theta}_n} \sup_{\theta \in \mathbb{R}^d} \mathbb{E}_\theta \|\hat{\theta}_n - \theta\|^2 \geq \frac{1}{n} \operatorname{tr}(\mathcal{I}^{-1}).$$

Secondly, if we choose

$$\pi(\theta) = \prod_{j=1}^{d} \frac{3}{4} \cos^3(\theta_j) I_{[-\frac{\pi}{2}, \frac{\pi}{2}]}(\theta_j),$$

that satisfies (P2), then we have, by an easy computation, $\mathcal{J}_\pi = \frac{9}{2} I_d$. Moreover, for $\delta = \frac{2c}{\pi\sqrt{n}}$, $c > 0$, the prior π_δ has support in $\{\theta \in \mathbb{R}^d : \max_j |\theta_j| \leq \frac{c}{\sqrt{n}}\} \subseteq \{\theta \in \mathbb{R}^d : \|\theta\| \leq c\sqrt{\frac{d}{n}}\}$, and Theorem 6.1 implies

$$\inf_{\hat{\theta}_n} \sup_{\|\theta\| \leq c\sqrt{\frac{d}{n}}} \mathbb{E}_\theta \|\hat{\theta}_n - \theta\|^2 \geq \frac{1}{n} \operatorname{tr}\Big(\Big(\mathcal{I} + \frac{9\pi^2}{8c^2} I_d\Big)^{-1}\Big).$$

The proof of Theorem 6.1 will be based on the following lemma.

Lemma 6.2 *Let* $\pi : \mathbb{R}^d \mapsto \mathbb{R}_+$ *be a probability density function with respect to the Lebesgue measure* λ *satisfying (P2) and* $\int \frac{\|\pi'\|^2}{\pi} d\lambda < \infty$*. Moreover, let* $p = e^{-W}$, $W : \mathbb{R}^d \mapsto \mathbb{R}$*, be a probability density function with respect* λ *satisfying (P1). Suppose that* p *is constant on the support of* π *and that* $\int_{\mathbb{R}^d} \|\theta\|^2 p(\theta) d\theta < \infty$*. For* $\epsilon > 0$*, set*

$$\pi_\epsilon := \frac{q + \epsilon}{1 + \epsilon} p, \qquad q := \frac{\pi}{\int \pi p \, d\lambda}. \tag{6.2}$$

Then, for every $\epsilon > 0$, π_ϵ *is a probability density function with respect to* λ *satisfying (P1) and* $\int \frac{\|\pi_\epsilon'\|^2}{\pi_\epsilon}\, d\lambda < \infty$. *Moreover, as* $\epsilon \to 0$,

$$\mathcal{J}_{\pi_\epsilon} = \int \frac{\pi_\epsilon' \otimes \pi_\epsilon'}{\pi_\epsilon}\, d\lambda \to \int \frac{\pi' \otimes \pi'}{\pi}\, d\lambda = \mathcal{J}_\pi.$$

Proof To see that π_ϵ satisfies (P1), we have to show that $W_\epsilon : \mathbb{R}^d \mapsto \mathbb{R}$ defined by $\pi_\epsilon = e^{-W_\epsilon}$ is twice differentiable with $\|W_\epsilon''\|_{L_\infty}, \|W_\epsilon''\|_{\mathrm{Lip}} < \infty$. Write $q_\epsilon = \frac{q+\epsilon}{1+\epsilon}$ such that $W_\epsilon = -\log q_\epsilon + W$. Hence,

$$W''\epsilon = W'' - \frac{q_\epsilon''}{q_\epsilon} + \frac{q_\epsilon' \otimes q_\epsilon'}{q_\epsilon^2} = W'' - \frac{q''}{q+\epsilon} + \frac{q' \otimes q'}{(q+\epsilon)^2}.$$

Using the fact that $q + \epsilon$ is lower bounded by ϵ and that all involved functions W'', q', q'' and $q + \epsilon$ are bounded and have bounded Lipschitz constant, it follows that π_ϵ satisfies (P1). Moreover, by the assumptions on π and W, we get

$$\pi_\epsilon' = \frac{q' e^{-W} + (q+\epsilon) W' e^{-W}}{1+\epsilon} = \frac{q' e^{-W} + \epsilon W' e^{-W}}{1+\epsilon},$$

and

$$\int \frac{\|\pi_\epsilon'\|^2}{\pi_\epsilon}\, d\lambda \le \frac{2}{1+\epsilon} \int \Big(\frac{\|q'\|^2 e^{-W}}{q+\epsilon} + \|W'\|^2 e^{-W} \epsilon \Big)\, d\lambda < \infty.$$

Moreover, using again that W is constant on the support of π, we get

$$\begin{aligned}
\mathcal{J}_{\pi_\epsilon} &= \frac{1}{1+\epsilon} \int_{\mathbb{R}^d} \Big(\frac{(q' \otimes q') e^{-W}}{q+\epsilon} + \epsilon (W' \otimes W') e^{-W} \Big)\, d\lambda \\
&\to \int_{\mathbb{R}^d} \frac{(q' \otimes q') e^{-W}}{q}\, d\lambda = \int \frac{\pi' \otimes \pi'}{\pi}\, d\lambda = \mathcal{J}_\pi \qquad \text{as } \epsilon \to 0,
\end{aligned}$$

where we applied the dominated convergence theorem in the last step. □

We are now ready to prove Theorem 6.1.

Proof We first consider the case where π satisfies (P1). We assume that the Bayes risk on the left-hand side in (6.1) is finite because otherwise the result is trivial. Since the location model is an example of an equivariant statistical model (with translation group acting on parameter space and sample space), we can apply Lemma 6.1 to $\psi(\theta) = \theta$, yielding that for any $\theta_1, \dots, \theta_d \in \mathbb{R}^d$,

$$\inf_{\hat{\theta}_n} \int_{\mathbb{R}^d} \mathbb{E}_\theta \|\hat{\theta}_n - \theta\|^2 \, \Pi_\delta(d\theta) \tag{6.3}$$

$$\geq \frac{\left(\sum_{j=1}^d \langle e_j, \theta_j\rangle\right)^2}{\sum_{j=1}^d \left(\chi^2(P_{\theta_j}^{\otimes n}, P_0^{\otimes n}) + \chi^2(\Pi_{\delta,\theta_j}, \Pi_\delta) + \chi^2(P_{\theta_j}^{\otimes n}, P_0^{\otimes n})\chi^2(\Pi_{\delta,\theta_j}, \Pi_\delta)\right)},$$

where $e_1, \ldots, e_d$ is the standard basis in $\mathbb{R}^d$ and Π_{δ,θ_j} is the probability measure with density $\delta^{-d}\pi(\delta^{-1}(\theta + \theta_j))$, $\theta \in \mathbb{R}^d$. Let us now apply a limiting argument. First, we have

$$\begin{aligned}\chi^2(P_{\theta_j}, P_0) &= \mathbb{E}e^{2V(\xi)-2V(\xi-\theta_j)} - 1 \\ &\leq e^{L\|\theta_j\|^3}\mathbb{E}e^{2\langle V'(\xi),\theta_j\rangle - \langle V''(\xi)\theta_j,\theta_j\rangle} - 1, \end{aligned} \tag{6.4}$$

where we used Proposition 3.1 in the inequality. If we set $\theta_j = th_j$ with $t > 0$ and $h_j \in \mathbb{R}^d$, and then combine (6.4) with Lemma 3.1 and Assumption 1, we get

$$\limsup_{t\to 0} \frac{1}{t^2}\chi^2(P_{th_j}, P_0) \leq -\mathbb{E}\langle V''(\xi)h_j, h_j\rangle + 2\mathbb{E}\langle V'(\xi), h_j\rangle^2 = \langle h_j, \mathcal{I}h_j\rangle$$

and thus also

$$\limsup_{t\to 0} \frac{1}{t^2}\chi^2(P_{th_j}^{\otimes n}, P_0^{\otimes n}) \leq n\langle h_j, \mathcal{I}h_j\rangle.$$

Moreover, since the prior density π satisfies (P1), Proposition 3.1 and Lemma 3.1 are still applicable, and we also have

$$\limsup_{t\to 0} \frac{1}{t^2}\chi^2(\Pi_{\delta,th_j}, \Pi_\delta) \leq \langle h_j, \delta^{-2}\mathcal{J}_\pi h_j\rangle.$$

Substituting these formulas into (6.3), we get for every $h_1, \ldots, h_d \in \mathbb{R}^d$,

$$\inf_{\hat{\theta}_n} \int_{\mathbb{R}^d} \mathbb{E}_\theta \|\hat{\theta}_n - \theta\|^2 \, \Pi_\delta(d\theta) \geq \frac{\left(\sum_{j=1}^d \langle e_j, h_j\rangle\right)^2}{\sum_{j=1}^d \langle h_j, (n\mathcal{I} + \delta^{-2}\mathcal{J}_\pi)h_j\rangle}.$$

Setting $h_j = (n\mathcal{I} + \delta^{-2}\mathcal{J}_\pi)^{-1}e_j$, $j \leq d$, we arrive at

$$\begin{aligned}\inf_{\hat{\theta}_n} \int_{\mathbb{R}^d} \mathbb{E}_{\delta\theta} \|\hat{\theta}_n - \delta\theta\|^2 \, \pi(\theta)d\theta &= \inf_{\hat{\theta}_n} \int_{\mathbb{R}^d} \mathbb{E}_\theta \|\hat{\theta}_n - \theta\|^2 \, \Pi_\delta(d\theta) \\ &\geq \sum_{j=1}^d \langle e_j, (n\mathcal{I} + \delta^{-2}\mathcal{J}_\pi)^{-1}e_j\rangle = \operatorname{tr}((n\mathcal{I} + \delta^{-2}\mathcal{J}_\pi)^{-1}). \end{aligned} \tag{6.5}$$

It remains to extend (6.5) to all densities π satisfying (P2). To this end, we apply Lemma 6.2 to get π_ϵ, $\epsilon > 0$, from (6.2) satisfying (P1) and $\lim_{\epsilon\to 0} \mathcal{J}_{\pi_\epsilon} = \mathcal{J}_\pi$. Note also that $qp = \pi$. Therefore,

$$\inf_{\hat{\theta}_n} \int_{\mathbb{R}^d} \mathbb{E}_{\delta\theta} \|\hat{\theta}_n - \delta\theta\|^2 \, \pi(\theta) d\theta = \lim_{\epsilon\to 0} \inf_{\hat{\theta}_n} \int_{\mathbb{R}^d} \mathbb{E}_{\delta\theta} \|\hat{\theta}_n - \delta\theta\|^2 \, (\pi(\theta) + \epsilon p(\theta)) d\theta$$

$$= \lim_{\epsilon\to 0} \frac{1}{1+\epsilon} \inf_{\hat{\theta}_n} \int_{\mathbb{R}^d} \mathbb{E}_{\delta\theta} \|\hat{\theta}_n - \delta\theta\|^2 \, (q(\theta)p(\theta) + \epsilon p(\theta)) d\theta$$

$$= \lim_{\epsilon\to 0} \inf_{\hat{\theta}_n} \int_{\mathbb{R}^d} \mathbb{E}_{\delta\theta} \|\hat{\theta}_n - \delta\theta\|^2 \, \pi_\epsilon(\theta) d\theta \geq \lim_{\epsilon\to 0} \operatorname{tr}((n\mathcal{I} + \delta^{-2}\mathcal{J}_{\pi_\epsilon})^{-1})$$

$$= \operatorname{tr}((n\mathcal{I} + \delta^{-2}\mathcal{J}_\pi)^{-1}),$$

where we applied (6.5) to π_ϵ. □

We now turn to the estimation of functionals of the location parameter. For a continuous function $g : \mathbb{R}^d \to \mathbb{R}^d$ and $x_0 \in \mathbb{R}^d$, the local continuity modulus of g at point x_0 is defined by

$$\omega_g(x_0, \delta) = \sup_{\|x - x_0\| \leq \delta} \|g(x) - g(x_0)\|, \ \delta \geq 0.$$

Theorem 6.2 *Let $f : \mathbb{R}^d \mapsto \mathbb{R}$ be a continuously differentiable function, and let $\theta_0 \in \mathbb{R}^d$. Let $\pi : \mathbb{R} \mapsto \mathbb{R}_+$ be a probability density function satisfying (P2) for $d = 1$. Suppose that*

$$\mathcal{J}_\pi = \int_{\mathbb{R}} \frac{(\pi'(s))^2}{\pi(s)} ds < \infty.$$

Then there exists $v \in \mathbb{R}^d$ with $\|v\| = 1$, such that, for every $\delta > 0$,

$$\inf_{\hat{T}_n} \Big(\int_{\mathbb{R}} n\mathbb{E}_{\theta_0 + sv} (\hat{T}_n - f(\theta_0 + sv))^2 \, \Pi_\delta(ds) \Big)^{1/2}$$

$$\geq \|\mathcal{I}^{-1/2} f'(\theta_0)\| - \sqrt{\frac{\mathcal{J}_\pi}{\delta^2 n}} \|\mathcal{I}^{-1} f'(\theta_0)\| - \int_{\mathbb{R}} \omega_{\mathcal{I}^{-1/2} f'}(\theta_0, |s|) \, \Pi_\delta(ds),$$

where Π_δ is the prior distribution with density $\delta^{-1}\pi(\delta^{-1}s)$, $s \in \mathbb{R}$ and where the infimum is taken over all estimators $\hat{T}_n = \hat{T}_n(X_1, \ldots, X_n)$ based on $(X_1, \ldots, X_n)$.

Proof Without loss of generality, we may assume that $\theta_0 = 0$. Our goal is to apply Lemma 6.1 to $\psi = f$ and to the one-dimensional subgroup $G = \mathbb{R}v = \{sv : s \in \mathbb{R}\}$ with direction $v \in \mathbb{R}^d$, $\|v\| = 1$, to be determined later. As in the proof of Theorem 6.1, we first establish a lower bound for densities $\pi = e^{-W}$ satisfying (P1) with $d = 1$ and for the special case that f and f' are bounded on $\mathbb{R}v$. In this case, applying Lemma 6.1, we get for every $t \in \mathbb{R}$,

$$\inf_{\hat{T}_n} \int_{\mathbb{R}} \mathbb{E}_{sv}(\hat{T}_n - f(sv))^2\, \Pi_\delta(ds)$$

$$\geq \frac{\left(\int_{\mathbb{R}}(f((s-t)v) - f(sv))\, \Pi_\delta(ds)\right)^2}{\chi^2(P_{tv}^{\otimes n}, P_0^{\otimes n}) + \chi^2(\Pi_{\delta,t}, \Pi_\delta) + \chi^2(P_{tv}^{\otimes n}, P_0^{\otimes n})\chi^2(\Pi_{\delta,t}, \Pi_\delta)}, \tag{6.6}$$

where $\Pi_{\delta,t}$ is the probability measure with density $\pi_{\delta,t}(s) = \delta^{-1}\pi(\delta^{-1}(s+t))$, $s \in \mathbb{R}$. Now, using that π satisfies (P1), we have

$$\limsup_{t\to 0} \frac{1}{t^2}\chi^2(P_{tv}^{\otimes n}, P_0^{\otimes n}) \leq n\langle v, \mathcal{I}v\rangle,$$

$$\limsup_{t\to 0} \frac{1}{t^2}\chi^2(\Pi_{\delta,t}, \Pi_\delta) \leq \delta^{-2}\mathcal{J}_\pi = \delta^{-2}\int_{\mathbb{R}} W''(s)e^{-W(s)}\, ds,$$

as shown in the proof of Theorem 6.1. Moreover, using that f and f' are bounded on $\mathbb{R}v$, standard results on the differentiation of integrals where the integrand depends on a real parameter (e.g., [3, Corollary 5.9]) yield

$$\lim_{t\to 0}\frac{1}{t}\int_{\mathbb{R}}(f((s-t)v) - f(sv))\, \Pi_\delta(ds) = -\int_{\mathbb{R}}\langle f'(sv), v\rangle\, \Pi_\delta(ds).$$

Substituting these formulas into (6.3) and letting t go to zero, we get

$$\inf_{\hat{T}_n}\int_{\mathbb{R}} \mathbb{E}_{sv}(\hat{T}_n - f(sv))^2\, \Pi_\delta(ds) \geq \frac{\left(\int_{\mathbb{R}}\langle f'(sv), v\rangle\, \Pi_\delta(ds)\right)^2}{n\langle v, \mathcal{I}v\rangle + \delta^{-2}\mathcal{J}_\pi}. \tag{6.7}$$

While this inequality holds for all densities π satisfying (P1), we can apply Lemma 6.2 to extend this inequality to all probability densities satisfying (P2) (see the proof of Theorem 6.1 for the detailed argument). Moreover, for densities π with bounded support, we can also drop the boundedness conditions on f and f'. In fact, the latter can be achieved by applying (6.7) to a functional g with g, g' bounded on $\mathbb{R}v$ and $g = f$ on the support of Π_δ (times v). As a consequence, under the assumptions of Theorem 6.2, we have for every $v \in \mathbb{R}$, $\|v\| = 1$,

$$\inf_{\hat{T}_n}\int_{\mathbb{R}} \mathbb{E}_{sv}(\hat{T}_n - f(sv))^2\, \Pi_\delta(ds)$$

$$\geq \frac{\left(\int_{\mathbb{R}}\langle f'(sv), v\rangle\, \Pi_\delta(ds)\right)^2}{n\langle v, \mathcal{I}v\rangle + \delta^{-2}\mathcal{J}_\pi} = \frac{1}{n}\frac{\left(\int_{\mathbb{R}}\langle f'(sv), v\rangle\, \Pi_\delta(ds)\right)^2}{\langle v, (\mathcal{I} + \frac{\mathcal{J}_\pi}{\delta^2 n}I_d)v\rangle}.$$

Choosing

$$v = \frac{A^{-1} f'(0)}{\|A^{-1} f'(0)\|}, \qquad A = \mathcal{I} + \frac{\mathcal{J}_\pi}{\delta^2 n} I_d,$$

we obtain

$$\inf_{\hat{T}_n} \Big(\int_{\mathbb{R}} n\mathbb{E}_{sv}(\hat{T}_n - f(sv))^2 \, \Pi_\delta(ds) \Big)^{1/2} \geq \frac{|\int_{\mathbb{R}} \langle A^{-1/2} f'(sv), A^{-1/2} f'(0)\rangle \, \Pi_\delta(ds)|}{\|A^{-1/2} f'(0)\|}.$$

Using the inequality

$$\begin{aligned} &|\langle A^{-1/2} f'(sv), A^{-1/2} f'(0)\rangle| \\ &\geq \|A^{-1/2} f'(0)\|^2 - \|A^{-1/2} f'(0)\| \|A^{-1/2} f'(0) - A^{-1/2} f'(sv)\| \\ &\geq \|A^{-1/2} f'(0)\| (\|A^{-1/2} f'(0)\| - \omega_{A^{-1/2} f'}(0, |s|)), \end{aligned}$$

we arrive at

$$\inf_{\hat{T}_n} \Big(\int_{\mathbb{R}} n\mathbb{E}_{sv}(\hat{T}_n - f(sv))^2 \, \Pi_\delta(ds) \Big)^{1/2} \geq \|A^{-1/2} f'(0)\| - \int_{\mathbb{R}} \omega_{A^{-1/2} f'}(0, |s|) \, d\Pi_\delta(s). \tag{6.8}$$

Since $A^{-1} = \mathcal{I}^{-1} - \frac{\mathcal{J}_\pi}{\delta^2 n} A^{-1}\mathcal{I}^{-1} \succeq 0$ and $\mathcal{I}^{-2} \succeq A^{-1}\mathcal{I}^{-1}$, we have

$$\begin{aligned} \|A^{-1/2} f'(0)\| &= \langle A^{-1} f'(0), f'(0)\rangle^{1/2} \\ &= (\langle \mathcal{I}^{-1} f'(0), f'(0)\rangle - \langle \frac{\mathcal{J}_\pi}{\delta^2 n} A^{-1}\mathcal{I}^{-1} f'(0), f'(0)\rangle)^{1/2} \\ &\geq \langle \mathcal{I}^{-1} f'(0), f'(0)\rangle^{1/2} - \langle \frac{\mathcal{J}_\pi}{\delta^2 n} A^{-1}\mathcal{I}^{-1} f'(0), f'(0)\rangle^{1/2} \\ &\geq \|\mathcal{I}^{-1/2} f'(0)\| - \sqrt{\frac{\mathcal{J}_\pi}{\delta^2 n}} \|\mathcal{I}^{-1} f'(0)\|. \end{aligned}$$

Note also that $A^{-1} \preceq \mathcal{I}^{-1}$, implying $\|A^{-1}u\| \leq \|\mathcal{I}^{-1}u\|, u \in \mathbb{R}^d$ and, as a consequence, $\omega_{A^{-1/2} f'}(0, |s|) \leq \omega_{\mathcal{I}^{-1/2} f'}(0, |s|)$. These bounds along with (6.8) imply the claim of the theorem. □

Finally, we prove Proposition 2.3.

Proof Let us choose $\pi(s) = \frac{3}{4}\cos^3(\theta) I_{[-\frac{\pi}{2},\frac{\pi}{2}]}(\theta)$ in which case (P2) holds and $\mathcal{J}_\pi = \frac{9}{2}$. Choosing additionally $\delta = \frac{2c}{\pi\sqrt{n}}$, $c > 0$, Theorem 6.2 yields

$$\inf_{\hat{T}_n} \sup_{\|\theta-\theta_0\|\leq \frac{c}{\sqrt{n}}} (n\mathbb{E}_\theta(\hat{T}_n - f(\theta))^2)^{1/2}$$

$$\geq \|\mathcal{I}^{-1/2} f'(\theta_0)\| - \frac{3\pi}{\sqrt{8c}}\|\mathcal{I}^{-1} f'(\theta_0)\| - \omega_{\mathcal{I}^{-1/2} f'}\Big(\theta_0, \frac{c}{\sqrt{n}}\Big). \tag{6.9}$$

Under Assumption 1, $\|\mathcal{I}^{-1} f'(\theta_0)\| \leq \frac{1}{\sqrt{m}}\|\mathcal{I}^{-1/2} f'(\theta_0)\|$. In addition, for $f \in C^s$, where $s = 1 + \rho$, $\rho \in (0, 1]$,

$$\omega_{\mathcal{I}^{-1/2} f'}\Big(\theta_0, \frac{c}{\sqrt{n}}\Big) \leq \frac{1}{\sqrt{m}}\omega_{f'}\Big(\theta_0, \frac{c}{\sqrt{n}}\Big) \leq \frac{1}{\sqrt{m}}\|f\|_{C^s}\Big(\frac{c}{\sqrt{n}}\Big)^\rho.$$

Recalling that $\|\mathcal{I}^{-1/2} f'(\theta_0)\| = \sigma_f(\theta_0)$, bound (6.9) implies

$$\inf_{\hat{T}_n} \sup_{\|\theta-\theta_0\|\leq \frac{c}{\sqrt{n}}} \frac{\sqrt{n}\|\hat{T}_n - f(\theta)\|_{L_2(\mathbb{P}_\theta)}}{\sigma_f(\theta_0)}$$

$$\geq 1 - \frac{3\pi}{\sqrt{8mc}} - \frac{1}{\sqrt{m}}\frac{\|f\|_{C^s}}{\sigma_f(\theta_0)}\Big(\frac{c}{\sqrt{n}}\Big)^\rho. \tag{6.10}$$

Note that, for all θ satisfying $\|\theta - \theta_0\| \leq \frac{c}{\sqrt{n}}$,

$$|\sigma_f(\theta) - \sigma_f(\theta_0)| = \big|\|\mathcal{I}^{-1/2} f'(\theta)\| - \|\mathcal{I}^{-1/2} f'(\theta_0)\|\big|$$

$$\leq \omega_{\mathcal{I}^{-1/2} f'}\Big(\theta_0, \frac{c}{\sqrt{n}}\Big) \leq \frac{1}{\sqrt{m}}\|f\|_{C^s}\Big(\frac{c}{\sqrt{n}}\Big)^\rho.$$

Therefore,

$$\sup_{\|\theta-\theta_0\|\leq \frac{c}{\sqrt{n}}} \frac{\sqrt{n}\|\hat{T}_n - f(\theta)\|_{L_2(\mathbb{P}_\theta)}}{\sigma_f(\theta_0)}$$

$$\leq \sup_{\|\theta-\theta_0\|\leq \frac{c}{\sqrt{n}}} \frac{\sqrt{n}\|\hat{T}_n - f(\theta)\|_{L_2(\mathbb{P}_\theta)}}{\sigma_f(\theta)} \sup_{\|\theta-\theta_0\|\leq \frac{c}{\sqrt{n}}} \frac{\sigma_f(\theta)}{\sigma_f(\theta_0)}$$

$$\leq \sup_{\|\theta-\theta_0\|\leq \frac{c}{\sqrt{n}}} \frac{\sqrt{n}\|\hat{T}_n - f(\theta)\|_{L_2(\mathbb{P}_\theta)}}{\sigma_f(\theta)}\Big(1 + \sup_{\|\theta-\theta_0\|\leq \frac{c}{\sqrt{n}}} \frac{|\sigma_f(\theta) - \sigma_f(\theta_0)|}{\sigma_f(\theta_0)}\Big)$$

$$\leq \sup_{\|\theta-\theta_0\|\leq \frac{c}{\sqrt{n}}} \frac{\sqrt{n}\|\hat{T}_n - f(\theta)\|_{L_2(\mathbb{P}_\theta)}}{\sigma_f(\theta)} \Big(1 + \frac{1}{\sqrt{m}} \frac{\|f\|_{C^s}}{\sigma_f(\theta_0)} \Big(\frac{c}{\sqrt{n}}\Big)^{\rho}\Big).$$

Using this bound together with (6.10) easily yields

$$\sup_{\|\theta-\theta_0\|\leq \frac{c}{\sqrt{n}}} \frac{\sqrt{n}\|\hat{T}_n - f(\theta)\|_{L_2(\mathbb{P}_\theta)}}{\sigma_f(\theta)} \geq 1 - \frac{3\pi}{\sqrt{8mc}} - \frac{2}{\sqrt{m}} \frac{\|f\|_{C^s}}{\sigma_f(\theta_0)} \Big(\frac{c}{\sqrt{n}}\Big)^{\rho}.$$

□

Acknowledgments Vladimir Koltchinskii was supported in part by NSF grants DMS-1810958 and DMS-2113121. Martin Wahl was supported by the Alexander von Humboldt Foundation.

References

1. R. Adamczak, W. Bednorz, P. Wolff, Moment estimates implied by modified log-Sobolev inequalities. ESAIM Probab. Stat. **21**, 467–494 (2017)
2. S. Aida, D. Stroock, Moment estimates derived from Poincaré and logarithmic Sobolev inequalities. Math. Res. Lett. **1**, 75–86 (1994)
3. R.G. Bartle, *The Elements of Integration and Lebesgue Measure* (Wiley, New York, 1995)
4. P. Bickel, Y. Ritov, Estimating integrated square density derivatives: sharp best order of convergence estimates. Sankhya **50**, 381–393 (1988)
5. L. Birgé, P. Massart, Estimation of integral functionals of a density. Ann. Statist. **23**, 11–29 (1995)
6. S. Bobkov, C. Houdré, Isoperimetric constants for product probability measures. Ann. Probab. **25**(1), 184–205 (1997)
7. S. Brazitikos, A. Giannopoulos, P. Valettas, B.-H. Vritsiou, *Geometry of Isotropic Convex Bodies* (American Mathematical Society, Providence, 2014)
8. Y. Chen, An almost constant lower bound of the isoperimetric coefficient in the KLS conjecture. Geom. Funct. Anal. **31**, 34–61 (2021)
9. P. Hall, *The Bootstrap and Edgeworth Expansion* (Springer, New York, 1992)
10. P. Hall, M.A. Martin, On Bootstrap Resampling and Iteration. Biometrika **75**(4), 661–671 (1988)
11. I.A. Ibragimov, R.Z. Khasminskii, *Statistical Estimation: Asymptotic Theory* (Springer, New York, 1981)
12. I.A. Ibragimov, A.S. Nemirovski, R.Z. Khasminskii, Some problems of nonparametric estimation in Gaussian white noise. Theory Probab. Appl. **31**, 391–406 (1987)
13. J. Jiao, Y. Han, T. Weissman, Bias correction with Jackknife, Bootstrap and Taylor series. IEEE Trans. Inf. Theory **66**(7), 4392–4418 (2020)
14. V. Koltchinskii, Asymptotically efficient estimation of smooth functionals of covariance operators. J. Euro. Math. Soc. **23**(3), 765–843 (2021)
15. V. Koltchinskii, Asymptotic efficiency in high-dimensional covariance estimation. *Proceedings of the International Congress of Mathematicians: Rio de Janeiro 2018*, vol. 3 (2018), pp. 2891–2912
16. V. Koltchinskii, Estimation of smooth functionals in high-dimensional models: bootstrap chains and Gaussian approximation. Ann. Statist. **50**, 2386–2415 (2022). arXiv:2011.03789

17. V. Koltchinskii, M. Zhilova, Efficient estimation of smooth functionals in Gaussian shift models. Ann. Inst. Henri Poincaré Probab. Stat. **57**(1), 351–386 (2021)
18. V. Koltchinskii, M. Zhilova, Estimation of smooth functionals in normal models: bias reduction and asymptotic efficiency. Ann. Statist. **49**(5), 2577–2610 (2021)
19. V. Koltchinskii, M. Zhilova, Estimation of smooth functionals of location parameter in Gaussian and Poincaré random shift models. Sankhya **83**, 569–596 (2021)
20. M. Ledoux, *The Concentration of Measure Phenomenon* (American Mathematical Society, Providence, 2001)
21. B. Levit, On the efficiency of a class of non-parametric estimates. Theory Prob. Appl. **20**(4), 723–740 (1975)
22. B. Levit, Asymptotically efficient estimation of nonlinear functionals. Probl. Peredachi Inf. (Probl. Info. Trans.) **14**(3), 65–72 (1978)
23. Y. Miao, Concentration inequality of maximum likelihood estimator. Appl. Math. Lett. **23**(10), 1305–1309 (2010)
24. E. Milman, On the role of convexity in isoperimetry, spectral gap and concentration. Invent. Math. **177**(1), 1–43 (2009)
25. A. Nemirovski, On necessary conditions for the efficient estimation of functionals of a nonparametric signal which is observed in white noise. Theory Probab. Appl. **35**, 94–103 (1990)
26. A. Nemirovski, *Topics in Non-parametric Statistics*. Ecole d'Ete de Probabilités de Saint-Flour. Lecture Notes in Mathematics, vol. 1738 (Springer, New York, 2000)
27. E. Rio, Upper bounds for minimal distances in the central limit theorem. Ann. Inst. Henri Poincaré Probab. Stat. **45**(3), 802–817 (2009)
28. A.B. Tsybakov, *Introduction to Nonparametric Estimation* (Springer, New York, 2009)
29. R. Vershynin, *High-Dimensional Probability: An Introduction with Applications in Data Science*. (Cambridge University Press, Cambridge, 2018)
30. M. Wahl, Lower bounds for invariant statistical models with applications to principal component analysis. Ann. Inst. Henri Poincaré Probab. Stat. **58**(3), 1565–1589 (2022)